*SCHAUM'S OUTLINE OF*

# THEORY AND PROBLEMS

OF

# GENETICS

## Second Edition

•

BY

**WILLIAM D. STANSFIELD, Ph.D.**

*Department of Biological Sciences*
*California Polytechnic State University*
*at San Luis Obispo*

•

**SCHAUM'S OUTLINE SERIES**
McGRAW-HILL BOOK COMPANY

*New York St. Louis San Francisco Auckland Bogotá Guatemala Hamburg Johannesburg*
*Lisbon London Madrid Mexico Montreal New Delhi Panama Paris*
*San Juan São Paulo Singapore Sydney Tokyo Toronto*

William D. Stansfield has degrees in Agriculture (B.S., 1952), Education (M.A., 1960), and Genetics (M.S., 1962; Ph.D., 1963; University of California at Davis). From 1953 to 1957 he served as an officer in the U. S. Navy. His published research is in immunogenetics, twinning, and mouse genetics. From 1957 to 1959 he was an instructor in high school vocational agriculture and in 1963 joined the faculty of California Polytechnic State University where he is now Professor in the Biological Sciences Department. He has written university-level textbooks in evolution and serology/immunology.

Schaum's Outline of Theory and Problems of
GENETICS



3 4 5 6 7 8 9 10 11 12 13 14 15 16 17 18 19 20 SHP SHP 8 7 6 5 4

ISBN 0-07-060845-8

Sponsoring Editor, Elizabeth Zayatz
Editing Supervisor, Marthe Grice
Production Manager, Nick Monti

**Library of Congress Cataloging in Publication Data**

Stansfield, William D., 1930–
Schaum's outline of theory and problems of genetics.
(Schaum's outline series)
Includes index.
1. Genetics—Problems, exercises, etc.
I. Title.
QH440.3.S7 1982 575.1'076 82-15275
ISBN 0-07-060845-8 AACR2

# Preface

Genetics, the science of heredity, is a fundamental discipline in the biological sciences. All living things are the products of both "nature and nurture". The hereditary units (genes) provide the organism with its "nature" (biological potentialities/limitations); the environment provides the "nurture" which interacts with the genes to give the organism its distinctive anatomical, biochemical, physiological and behavioral characteristics.

Johann (Gregor) Mendel laid the foundation of modern genetics with the publication of his pioneering work on peas in 1866, but his work was not appreciated in his lifetime. The science of genetics began in 1900 with the rediscovery of his original paper. Since then, genetics has rapidly proliferated into numerous distinctive subdisciplines. The kind of studies that Mendel performed are now included in the discipline called Transmission Genetics. Other major areas of investigation are now categorized as Cytogenetics, Cytoplasmic (Plastid) Genetics, Quantitative Genetics, Population (Evolutionary) Genetics, Molecular Genetics, Developmental Genetics and Microbial Genetics. Further specializations are indicated by fields labeled *Drosophila* Genetics, Human Genetics, Fungal Genetics, Viral Genetics, Animal and Plant Breeding, Immunogenetics, etc. This revision touches upon problems in all of these fields without necessarily devoting a complete chapter to each of them.

It has been said that a body of knowledge can be called a science to the degree that its principles can be expressed in mathematical terms. If this is true, then genetics reigns as queen among the biological sciences. The mathematics of elementary genetics is not difficult, but solving genetic problems does require a logical mind and some expertise in working with probabilities. The only mathematical background needed for understanding the concepts in this volume is arithmetic and the elements of algebra. At least some exposure to college biology is desired, but even basic biological principles are reviewed to provide a common base of background information.

The first edition of *Genetics* was developed in 1969 as an aid to students in genetics classes dominated by a mathematical, problem-solving mode of teaching. At that time, this was probably the only such aid available. In the intervening years, other study aids became available, microelectronic calculators became relatively inexpensive, the understanding of basic genetic processes deepened profoundly and a new era of genetic engineering emerged. This second edition reflects the rapid growth of genetic knowledge in the last decade. Extensive reorganization and amplification of information has been made in the areas of DNA replication, protein structure and synthesis, regulatory mechanisms of gene activity, mutagenesis, bacterial and viral genetics, human cytogenetics, quantitative genetics and breeding principles. Three new chapters have been added. Chapter 8 incorporates principles from all of the previous chapters and requires the student to devise solutions to problems without any textual clues as to which principles are being tested. Chapter 13 is an introduction to the quantitation of evolutionary forces. The final chapter presents some of the major technological innovations that have advanced our understanding of genetics at the molecular level and the birth of the age of genetic engineering. Breaking with the tradition established in the first edition, many of the new problems in this revision do not have a mathematical solution, but rather require short essay-type answers. While this book is still primarily intended as a supplemental study aid to any standard genetics textbook, the statements of theory and principle are sufficiently complete for it to be used as a text by itself for a one or two semester course, depending on the desired depth of study.

Each chapter begins with a clear statement of pertinent definitions, principles and background information, fully illustrated by examples. This is followed by sets of solved and supplementary problems. Answers to the supplementary problems are given at the end of each chapter. The solved problems illustrate and amplify the theory, bring into sharp focus those fine points without which the student continually feels himself or herself on unsafe ground, and provide the repetition of basic principles so vital to effective learning. The supplementary problems serve as a complete review of the material of each chapter.

I am indebted to the Literary Executor of the late Sir Ronald A. Fisher, to Dr. Frank Yates, and to Oliver & Boyd Ltd., Edinburgh, for permission to publish a portion of Table IV from the 6th edition of *Statistical Tables for Biological, Agricultural, and Medical Research.* I am also indebted to many people around the world (the first edition of this book was translated into five languages) who have offered helpful comments over the years since the first printing. I would appreciate receiving further suggestions about this new edition.

WILLIAM D. STANSFIELD

# CONTENTS

CONTENTS

# Chapter 1

## The Physical Basis of Heredity

### GENETICS

*Genetics* is that branch of biology concerned with heredity and variation. The hereditary units which are transmitted from one generation to the next (inherited) are called *genes.* The genes reside in a long molecule called deoxyribonucleic acid (DNA). The DNA, in conjunction with a protein matrix, forms nucleoprotein and becomes organized into structures with distinctive staining properties called *chromosomes* found in the nucleus of the cell. The behavior of genes is thus paralleled in many ways by the behavior of the chromosomes of which they are a part. A gene contains coded information for the production of proteins. DNA is normally a stable molecule with the capacity for self-replication. On rare occasions a change may occur spontaneously in some part of DNA. This change, called a *mutation,* alters the coded instructions and may result in a defective protein or in the cessation of protein synthesis. The net result of a mutation is often seen as a change in the physical appearance of the individual or a change in some other measurable attribute of the organism called a *character* or *trait.* Through the process of mutation a gene may be changed into two or more alternative forms called *allelomorphs* or *alleles.*

**Example 1.1.** Healthy people have a gene which specifies the normal protein structure of the red blood cell pigment called hemoglobin. Some anemic individuals have an altered form of this gene, i.e. an allele, which makes a defective hemoglobin protein unable to carry the normal amount of oxygen to the body cells.

Each gene occupies a specific position on a chromosome, called the gene *locus* (*loci,* plural). All allelic forms of a gene therefore are found at corresponding positions on genetically similar (*homologous*) chromosomes. The word "locus" is sometimes used interchangeably for "gene". When the science of genetics was in its infancy the gene was thought to behave as a unit particle. These particles were believed to be arranged on the chromosome like beads on a string. This is still a useful concept for beginning students to adopt, but will require considerable modification when we study the chemical basis of heredity in Chapter 14. All the genes on a chromosome are said to be *linked* to one another and belong to the same *linkage group.* Wherever the chromosome goes it carries all of the genes in its linkage group with it. As we shall see later in this chapter, linked genes are not transmitted independently of one another, but genes in different linkage groups (on different chromosomes) are transmitted independently of one another.

### CELLS

The smallest unit of life is the *cell.* All living things are composed of these basic units, from the simple unicellular structures of bacteria and protozoa to the complex structures of trees and humans. Even within an individual all of the cells do not look alike. A muscle cell is obviously different from a nerve cell which in turn is different from a blood cell, etc. Thus there is no such thing as a typical cell type. Fig. 1-1 below is a composite diagram of an animal cell showing subcellular structures called *organelles* which many types of cells share in common. Most organelles are too small to be seen with the light microscope, but their structure can be studied with the electron microscope. These organelles perform distinctive functions which in total produce the characteristics of life associated with the cell (Table 1.1).

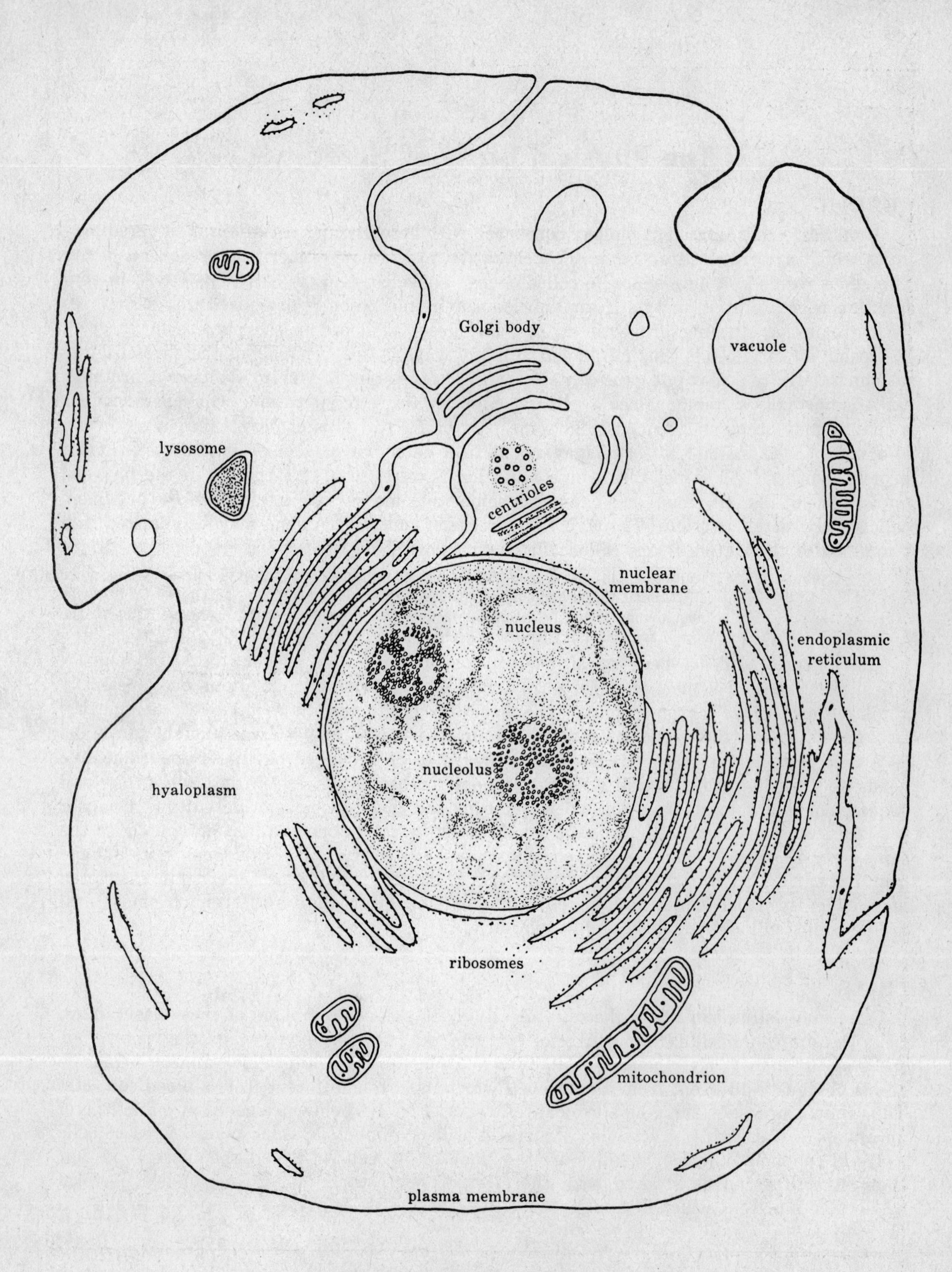


Fig. 1-1. Diagram of an animal cell.

**Table 1.1. Functions of Cellular Organelles**

| *Cell Organelle* | *Functions* |
|---|---|
| **Cell or plasma membrane** | Differentially permeable membrane through which extracellular substances may be selectively sampled and cell products may be liberated. |
| **Cell wall (plants only)** | Thick cellulose wall surrounding the cell membrane giving strength and rigidity to the cell. |
| **Nucleus:** | Regulates growth and reproduction of the cell. |
| Chromosomes | Bearers of hereditary instructions; regulation of cellular processes (seen clearly only during nuclear division). |
| Nucleolus | Synthesizes ribosomal RNA; disappears during cellular replication. |
| Nucleoplasm (nuclear sap) | Contains materials for building DNA and messenger molecules which act as intermediates between nucleus and cytoplasm. |
| Nuclear membrane | Provides selective continuity between nuclear and cytoplasmic materials. |
| **Cytoplasm:** | Contains machinery for carrying out the instructions sent from the nucleus. |
| Endoplasmic reticulum | Greatly expanded surface area for biochemical reactions which normally occur at or across membrane surfaces. |
| Ribosomes | Sites of protein synthesis (shown as black dots lining the endoplasmic reticulum in Fig. 1-1). |
| Centrioles | Form poles for the divisional processes; capable of replication; usually not seen in plant cells. |
| Mitochondria | Energy production (Kreb's cycle, electron transport chain, beta oxidation of fatty acids, etc.). |
| Plastids (plants only) | Structures for storage of starch, pigments, and other cellular products. Photosynthesis occurs in chloroplasts. |
| Golgi body or apparatus | Production of cellular secretions; sometimes called dictyosomes in plants. |
| Lysosome (animals only) | Production of intracellular digestive enzymes which aid in disposal of bacteria and other foreign bodies; may cause cell destruction if ruptured. |
| Vacuoles | Storage depots for excess water, waste products, soluble pigments, etc. |
| Hyaloplasm | Contains enzymes for glycolysis and structural materials such as sugars, amino acids, water, vitamins, nucleotides, etc. (nutrient soup or cell sap). |

## CHROMOSOMES

### 1. Chromosome Number.

In higher organisms, each *somatic* cell (any body cell exclusive of sex cells) contains one set of chromosomes inherited from the *maternal* (female) parent and a comparable set of chromosomes (*homologous* chromosomes or *homologues*) from the *paternal* (male) parent. The number of chromosomes in this dual set is called the *diploid* ($2n$) number. The suffix "-ploid" refers to chromosome "sets". The prefix indicates the degree of ploidy. Sex cells or *gametes*, which contain half the number of chromosome sets found in somatic cells are referred to as *haploid* cells ($n$). A *genome* is a set of chromosomes corresponding to the haploid set of a species. The number of chromosomes in each somatic cell is the same for all members of a given species. For example, human somatic cells contain 46 chromosomes, tobacco has 48, cattle 60, the garden pea 14, the fruit fly 8, etc. The diploid number of a species bears no direct relationship to the species position in the phylogenetic scheme of classification.

### 2. Chromosome Morphology.

The structure of chromosomes becomes most easily visible during certain phases of nuclear division when they are highly coiled. Each chromosome in the genome can usually be distinguished from all others by several criteria, including the relative lengths of the chromosomes, the position of a structure called the *centromere* which divides the chromosome into two arms of varying length, the presence and position of enlarged areas called knobs or chromomeres, the presence of tiny terminal extensions of chromatin material called satellites, etc. A chromosome with a median centromere (*metacentric*) will have arms of approximately equal size. A *submetacentric* or *acrocentric* chromosome has arms of distinctly unequal size. If a chromosome has its centromere at or very near one end of the chromosome it is called *telocentric*. Each chromosome of the genome (with the exception of sex chromosomes) is numbered consecutively according to length, beginning with the longest chromosome first.

### 3. Autosomes vs. Sex Chromosomes.

In the males of some species, including humans, sex is associated with a morphologically dissimilar (*heteromorphic*) pair of chromosomes called *sex chromosomes*. Such a chromosome pair is usually labeled X and Y. Genetic factors on the Y chromosome determine maleness. Females have two morphologically identical X chromosomes. The members of any other homologous pairs of chromosomes (homologues) are morphologically indistinguishable, but usually are visibly different from other pairs (non-homologous chromosomes). All chromosomes exclusive of the sex chromosomes are called *autosomes*. Fig. 1-2 shows the chromosomal complement of the fruit fly *Drosophila melanogaster* ($2n = 8$) with three pairs of autosomes (2, 3, 4) and one pair of sex chromosomes.

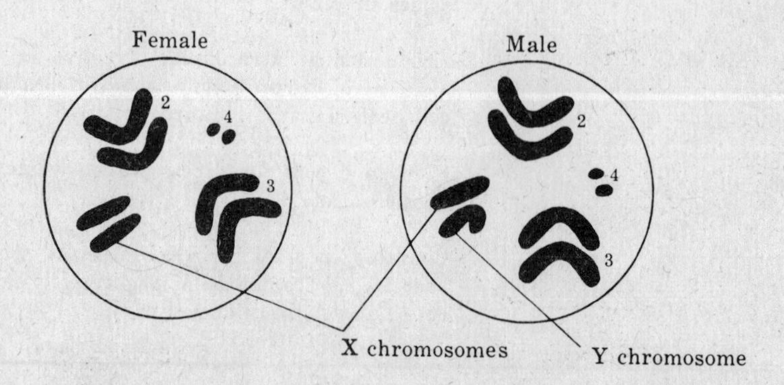


**Fig. 1-2.** Diagram of diploid cells in *Drosophila melanogaster*.

## CELL DIVISION

### 1. Mitosis.

All somatic cells in a multicellular organism are descendants of one original cell, the *zygote,* through a divisional process called *mitosis* (Fig. 1-3). The function of mitosis is first to construct an exact copy of each chromosome and then to distribute, through division of the original (mother) cell, an identical set of chromosomes to each of the two daughter cells. *Interphase* is the period between division cycles. When the cell is ready to begin mitosis each DNA molecule (which is an integral part of the chromosome) *replicates* or makes an exact copy of itself. This copying process produces a chromosome with two identical functional strands called *chromatids,* both attached to a common centromere. At this stage the chromosomes are in a very long attenuated condition and appear only as chromatin granules under the light microscope. A mitotic division has four major stages: prophase,

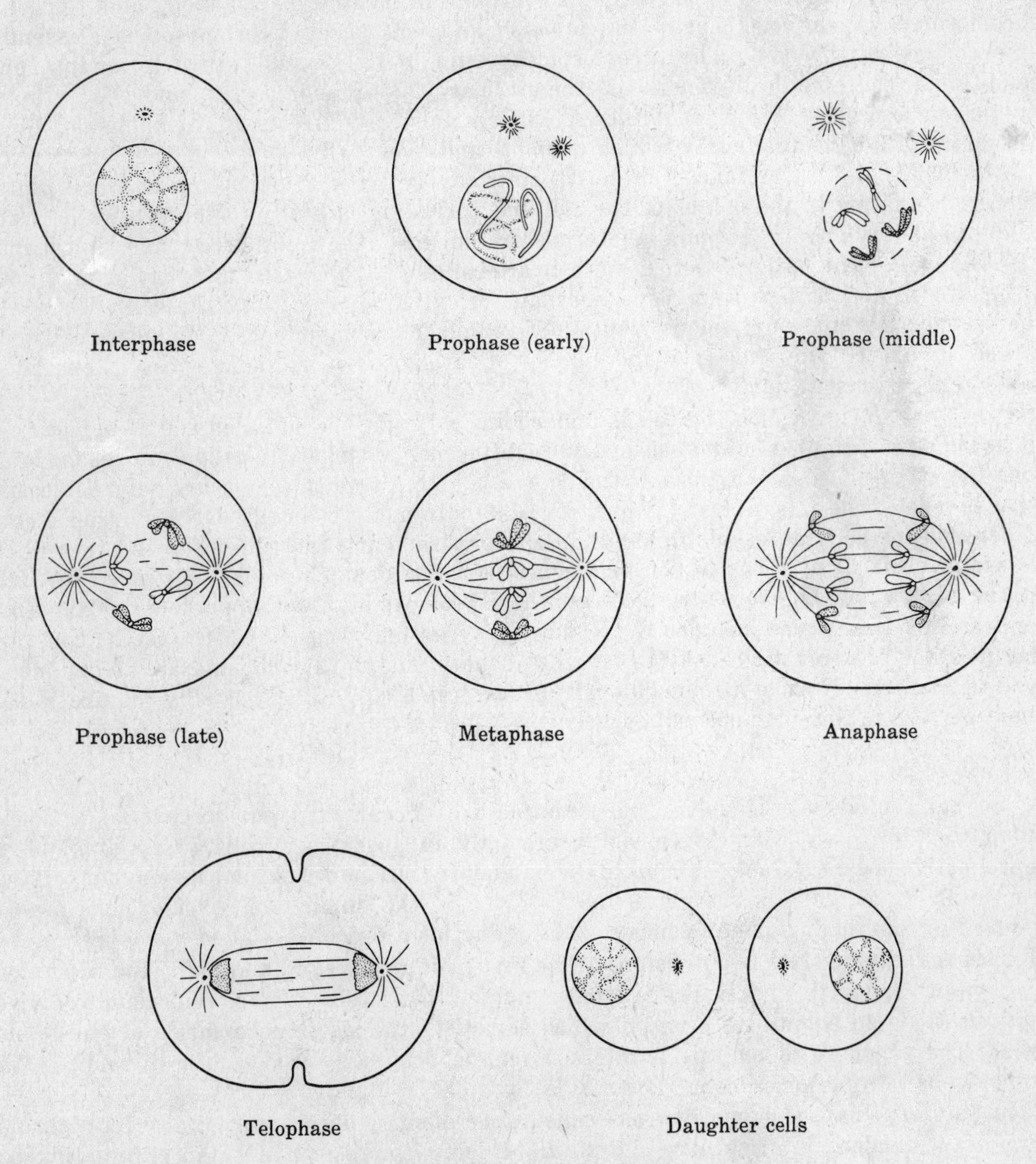


**Fig. 1-3.** Mitosis in animal cells.

metaphase, anaphase and telophase. In *prophase,* the chromosomes become visible in the light microscope presumably due to coiling, shortening and thickening, and adding protein matrix to their mass as the process continues. By late prophase, the two identical or "sister" chromatids may be seen. The centrioles migrate to opposite ends of the cell and there establish mitotic centers (poles) from which spindle fibers organize and extend to the centromeres. The nuclear membrane begins to degenerate and by *metaphase* is completely gone. Now the centromeres move to the center of the cell, a position designated the equatorial plane or metaphase plate, and the spindle fibers, collectively known as the spindle apparatus, become fully formed. *Anaphase* begins when the centromere splits in two, allowing sister chromatids to separate and move towards opposite poles led by their centromeres. At this stage we may arbitrarily designate the separated sisters as new chromosomes. The arms of each chromosome drag behind their centromeres, giving them characteristic shapes depending upon the location of the centromere. Metacentric chromosomes appear V-shaped, submetacentric chromosomes appear J-shaped, and telocentric chromosomes appear rod shaped. In *telophase,* an identical set of chromosomes is assembled at each pole of the cell. The chromosomes begin to uncoil and return to an interphase condition. The spindle degenerates, the nuclear membrane reforms, and the cytoplasm divides in a process called *cytokinesis.* In animals, cytokinesis is accomplished by the formation of a cleavage furrow which deepens and eventually "pinches" the cell in two as shown in Fig. 1-3. Cytokinesis in most plants involves the construction of a *cell plate* of pectin originating in the center of the cell and spreading laterally to the cell wall. Later, cellulose and other strengthening materials are added to the cell plate, converting it into a new cell wall. The two products of mitosis are called *daughter cells* and may or may not be of equal size depending upon where the plane of cytokinesis sections the cell. Thus while there is no assurance of equal distribution of cytoplasmic components to daughter cells, they do contain exactly the same type and number of chromosomes and hence possess exactly the same genetic constitution.

The time during which the cell is undergoing mitosis is designated the *M period.* The times spent in each phase of mitosis are quite different. Prophase usually requires far longer than the other phases; metaphase is the shortest. DNA replication occurs before mitosis in what is termed the *S* (synthesis) *phase* (Fig. 1-4). In nucleated cells, DNA synthesis starts at several positions on each chromosome, thereby reducing the time required to replicate the sister chromatids. The period between M and S is designated the $G_2$ *phase* (post-DNA synthesis). A long $G_1$ *phase* (pre-DNA synthesis) follows mitosis and precedes chromosomal replication. Interphase includes $G_1$, S, and $G_2$. The four phases (M, $G_1$, S, $G_2$) constitute the life cycle of a somatic cell. The lengths of these phases vary considerably from one cell type to another. Normal mammalian cells growing in tissue culture usually require 18 to 24 hours at 37°C to complete the cell cycle.

**2. Meiosis.**

Sexual reproduction involves the manufacture of gametes (*gametogenesis*) and their union (*fertilization*). Gametogenesis occurs only in specialized cells (*germ line*) of the reproductive organs. Gametes contain the haploid ($n$) number of chromosomes, but originate from diploid ($2n$) cells of the germ line. Obviously the number of chromosomes must be reduced by half during gametogenesis. The reductional process is called *meiosis* (Fig. 1-5). Meiosis actually involves two divisions. The first meiotic division (meiosis I) is a reductional division producing two haploid cells from a single diploid cell. The second meiotic division (meiosis II) is an equational division which separates the sister chromatids of the haploid cells. The prophase of meiosis I differs from the prophase of a mitotic division in that homologous chromosomes come to lie side by side in a pairing process called *synapsis.* Each pair of synapsed homologues is called a *bivalent*; since it consists of four chromid strands, a bivalent is also called a *tetrad.* During synapsis non-sister chromatids may break and reunite with one another in a process called *crossing over.* The point of exchange

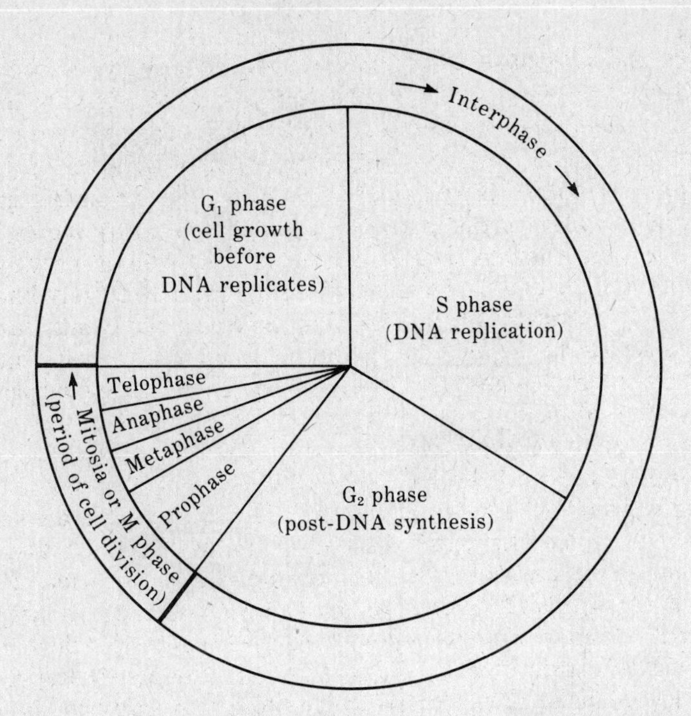


**Fig. 1-4.** Diagram of a typical cell reproductive cycle. (Idea from Jack A. Ward & Howard R. Hetzel, *Biology: Today and Tomorrow*, p. 71, 1980, West Pub.)

appears in the microscope as an overlapping region called a *chiasma* (chiasmata, plural). During metaphase I, the bivalents orient themselves at random on the equatorial plane. At first anaphase the centromeres do not divide, but continue to hold sister chromatids together. The homologues separate and move to opposite poles. That is, whole chromosomes (each consisting of two chromatids) move apart. This in effect is the movement which reduces the chromosome number from the diploid ($2n$) condition to the haploid ($n$) state. Cytokinesis in telophase I divides the diploid mother cell into two haploid daughter cells. This ends the first meiotic division. The brief period between the first and second meiotic divisions is called *interkinesis*. In prophase II, the spindle apparatus reforms. By metaphase II, the centromeres have lined up on the equatorial plane. During anaphase II the centromeres of each chromosome divide, allowing sister chromatids to separate. Cytokinesis in telophase II divides the two cells into four meiotic products.

The distinguishing characteristics of mitosis and meiosis are summarized in Table 1.2 below.

## MENDEL'S LAWS

Gregor Mendel published the results of his genetic studies on the garden pea in 1866 and thereby laid the foundation of modern genetics. In this paper Mendel proposed some basic genetic principles. One of these is known as the *principle of segregation*. He found that from any one parent, only one allelic form of a gene is transmitted through a gamete to the offspring. For example, a plant which had a factor (or gene) for round shaped seed and also an allele for wrinkled shaped seed would transmit only one of these two alleles through a gamete to its offspring. Mendel knew nothing of chromosomes or meiosis, as they had not yet been discovered. We now know that the physical basis for this principle is in first meiotic anaphase where homologous chromosomes segregate or separate from each other. If the gene for round seed is on one chromosome and its allelic form for wrinkled

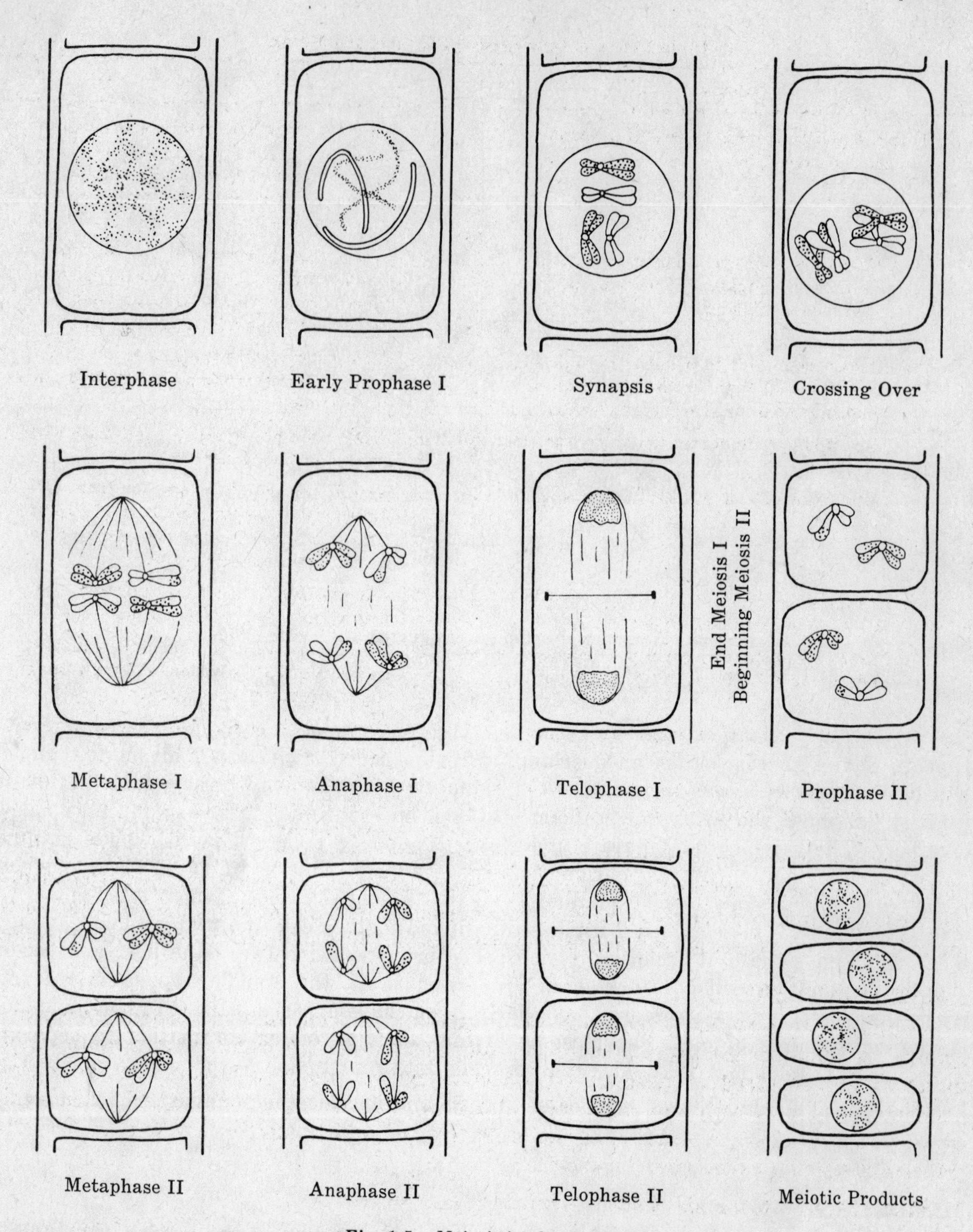


**Fig. 1-5.** Meiosis in plant cells.

seed is on the homologous chromosome, then it becomes clear that alleles normally will not be found in the same gamete.

Mendel's *principle of independent assortment* states that the segregation of one factor pair occurs independently of any other factor pair. We know that this is true only for loci on non-homologous chromosomes. For example, on one homologous pair of chromosomes are the seed shape alleles and on another pair of homologues are the alleles for green and yellow seed color. The segregation of the seed shape alleles occurs independently of the segregation of the seed color alleles because each pair of homologues behaves as an independent

**Table 1.2. Characteristics of Mitosis and Meiosis**

| *Mitosis* | *Meiosis* |
|---|---|
| (1) An equational division which separates sister chromatids. | (1) The first stage is a reductional division which separates homologous chromosomes at first anaphase; sister chromatids separate in an equational division at second anaphase. |
| (2) One division per cycle, i.e. one cytoplasmic division (cytokinesis) per equational chromosomal division. | (2) Two divisions per cycle, i.e. two cytoplasmic divisions, one following reductional chromosomal division and one following equational chromosomal division. |
| (3) Chromosomes fail to synapse; no chiasmata form; genetic exchange between homologous chromosomes does not occur. | (3) Chromosomes synapse and form chiasmata; genetic exchange occurs between homologues. |
| (4) Two products (daughter cells) produced per cycle. | (4) Four cellular products (gametes or spores) produced per cycle. |
| (5) Genetic content of mitotic products are identical. | (5) Genetic content of meiotic products different; centromeres may be replicas of either maternal or paternal centromeres in varying combinations. |
| (6) Chromosome number of daughter cells is the same as that of mother cell. | (6) Chromosome number of meiotic products is half that of the mother cell. |
| (7) Mitotic products are usually capable of undergoing additional mitotic divisions. | (7) Meiotic products cannot undergo another meiotic division although they may undergo mitotic division. |
| (8) Normally occurs in most all somatic cells. | (8) Occurs only in specialized cells of the germ line. |
| (9) Begins at the zygote stage and continues through the life of the organism. | (9) Occurs only after a higher organism has begun to mature; occurs in the zygote of many algae and fungi. |

unit during meiosis. Furthermore, because the orientation of bivalents on the first meiotic metaphase plate is completely at random, four combinations of factors could be found in the meiotic products: (1) round-yellow, (2) wrinkled-green, (3) round-green, (4) wrinkled-yellow.

## GAMETOGENESIS

Usually the immediate end products of meiosis are not fully developed gametes or spores. A period of *maturation* commonly follows meiosis. In plants, one or more mitotic divisions are required to produce reproductive spores, whereas in animals the meiotic products develop directly into gametes through growth and/or differentiation. The entire process of producing mature gametes or spores, of which meiotic division is the most important part, is called *gametogenesis*.

### 1. Animal Gametogenesis (*as represented in mammals*).

Gametogenesis in the male animal is called *spermatogenesis* (Fig. 1-6(*a*)). Mammalian spermatogenesis originates in the germinal epithelium of the seminiferous tubules of the male gonads (testes) from diploid primordial cells. These cells undergo repeated mitotic divisions to form a population of *spermatogonia*. By growth, a spermatogonium may differentiate into a diploid *primary spermatocyte* with the capacity to undergo meiosis. The first meiotic division occurs in these primary spermatocytes, producing haploid *secondary spermatocytes*. From these cells the second meiotic division produces four haploid meiotic products called *spermatids*. Almost the entire amount of cytoplasm then extrudes into a long whiplike tail during maturation and the cell becomes transformed into a mature male gamete called a *sperm cell* or *spermatozoan* (-zoa, plural).

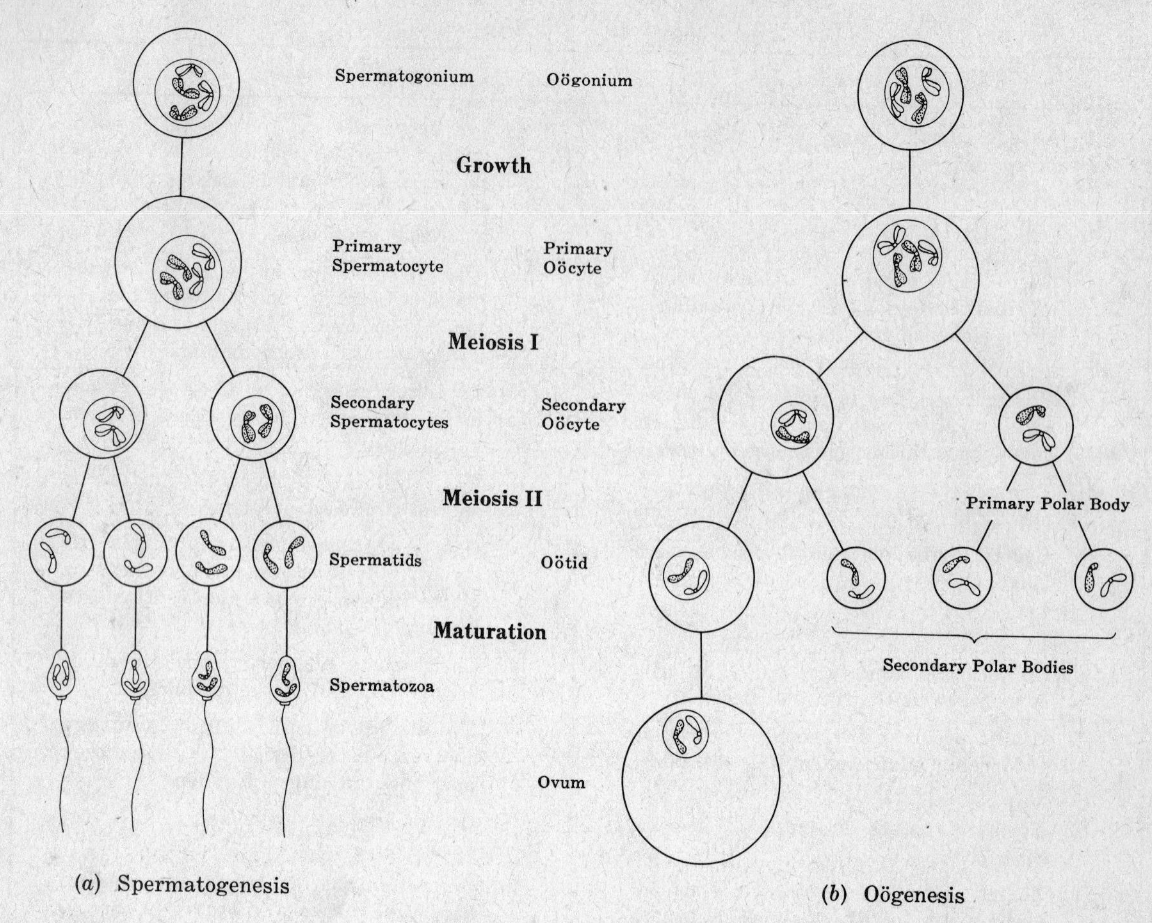


**Fig. 1-6.** Animal gametogenesis.

Gametogenesis in the female animal is called *oögenesis* (Fig. 1-6(*b*)). Mammalian oögenesis originates in the germinal epithelium of the female gonads (ovaries) in diploid primordial cells called *oögonia*. By growth and storage of much cytoplasm or yolk (to be used as food by the early embryo), the oögonium is transformed into a diploid *primary oöcyte* with the capacity to undergo meiosis. The first meiotic division reduces the chromosome number by half and also distributes vastly different amounts of cytoplasm to the two products by a grossly unequal cytokinesis. The larger cell thus produced is called a *secondary oöcyte* and the smaller is a primary *polar body*. In some cases the first polar body may undergo the second meiotic division, producing two secondary polar bodies. All polar bodies degenerate, however, and take no part in fertilization. The second meiotic division of the oöcyte again involves an unequal cytokinesis, producing a large yolky *oötid* and a secondary polar body. By additional growth and differentiation the oötid becomes a mature female gamete called an *ovum* or *egg cell*.

The union of male and female gametes (sperm and egg) is called *fertilization* and reestablishes the diploid number in the resulting cell called a *zygote*. The head of the sperm enters the egg, but the tail piece (the bulk of the cytoplasm of the male gamete) remains outside and degenerates. Subsequent mitotic divisions produce the numerous cells of the embryo which become organized into the tissues and organs of the new individual.

2. **Plant Gametogenesis** (*as represented in angiosperms*).

Gametogenesis in the plant kingdom varies considerably between major groups of plants. The process as described below is that typical of many flowering plants (*angio-*

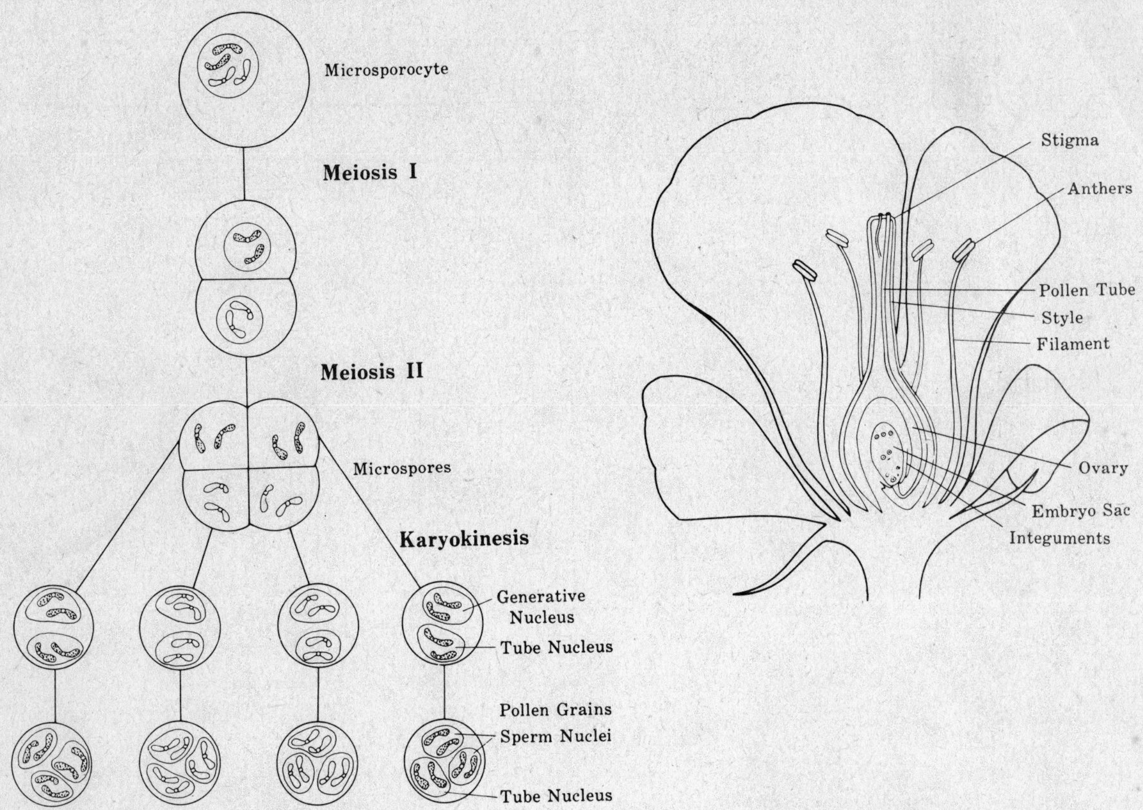


**Fig. 1-7.** Microsporogenesis.

**Fig. 1-8.** Diagram of a flower.

*sperms*). *Microsporogenesis* (Fig. 1-7) is the process of gametogenesis in the male part of the flower (*anther*, Fig. 1-8) resulting in reproductive spores called *pollen grains*. A diploid microspore mother cell (*microsporocyte*) in the anther divides by meiosis, forming at the first division a pair of haploid cells. The second meiotic division produces a cluster of four haploid *microspores*. Following meiosis, each microspore undergoes a mitotic division of the chromosomes without a cytoplasmic division (*karyokinesis*), producing a cell containing two haploid nuclei. Pollen grains are usually shed at this stage. Upon germination of the pollen tube, one of these nuclei (or haploid sets of chromosomes) becomes a *generative nucleus* and divides again by mitosis without cytokinesis to form two *sperm nuclei*. The other nucleus, which does not divide, becomes the *tube nucleus*.

*Megasporogenesis* (Fig. 1-9) is the process of gametogenesis in the female part of the flower (*ovary*, Fig. 1-8) resulting in reproductive cells called *embryo sacs*. A diploid megaspore mother cell (*megasporocyte*) in the ovary divides by meiosis, forming in the first division a pair of haploid cells. The second meiotic division produces a linear group of four haploid *megaspores*. Following meiosis, three of the megaspores degenerate. The remaining megaspore undergoes three mitotic divisions of the chromosomes without intervening cytokineses (*karyokineses*), producing a large cell with eight haploid nuclei (immature embryo sac). The sac is surrounded by maternal tissues of the ovary called *integuments* and by the megasporangium (*nucellus*). At one end of the sac there is an opening in the integuments (*micropyle*) through which the pollen tube will penetrate. Three nuclei of the sac orient themselves near the micropylar end and two of the three (*synergids*) degenerate. The third nucleus develops into an *egg nucleus*. Another group of three nuclei moves to the opposite end of the sac and degenerates (*antipodals*). The two remaining nuclei (*polar nuclei*) unite near the center of the sac, forming a single diploid *fusion nucleus*. The mature embryo sac (*megagametophyte*) is now ready for fertilization.

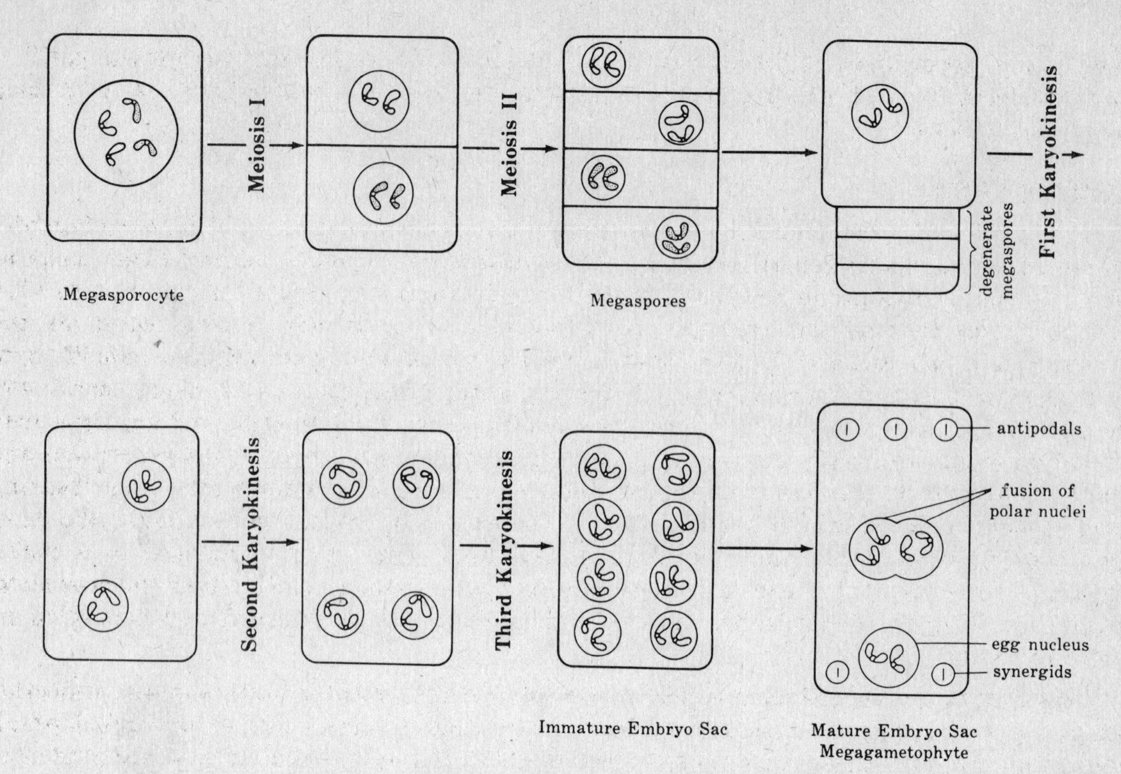


**Fig. 1-9.** Megasporogenesis.

Pollen grains from the anthers are carried by wind or insects to the *stigma*. The pollen grain germinates into a pollen tube which grows down the *style*, presumably under the direction of the tube nucleus. The pollen tube enters the ovary and makes its way through the micropyle of the ovule into the embryo sac (Fig. 1-10). Both sperm nuclei are released into the embryo sac. The pollen tube and the tube nucleus, having served their function, degenerate. One sperm nucleus fuses with the egg nucleus to form a diploid zygote which will then develop into the embryo. The other sperm nucleus unites with the fusion nucleus to form a triploid ($3n$) nucleus which, by subsequent mitotic divisions, forms a starchy nutritive tissue called *endosperm*. The outermost layer of endosperm cells is called *aleurone*. The embryo, surrounded by endosperm tissue, and in some cases such as corn and other grasses where it is also surrounded by a thin outer layer of diploid maternal tissue called *pericarp*, becomes the familiar seed. Since two sperm nuclei are involved, this process is termed *double fertilization*. Upon germination of the seed, the young seedling (the next

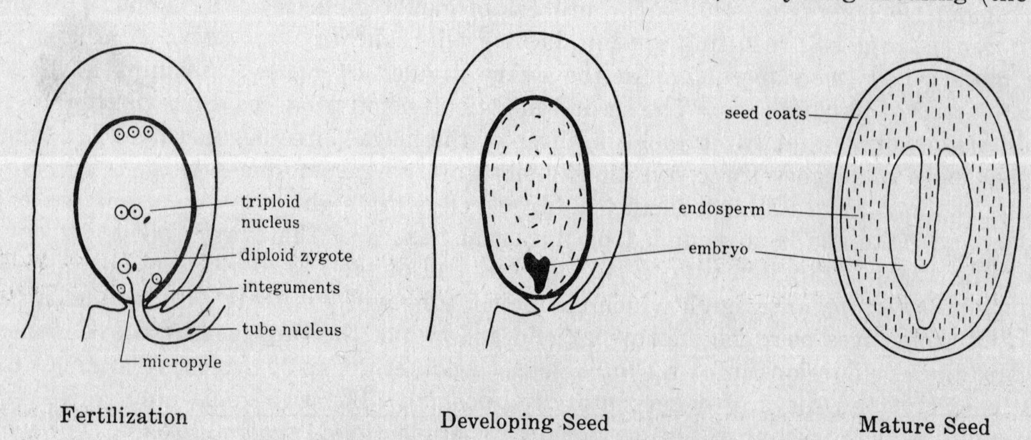


**Fig. 1-10.** Fertilization and development of a seed.

sporophytic generation) utilizes the nutrients stored in the endosperm for growth until it emerges from the soil, at which time it becomes capable of manufacturing its own food by photosynthesis.

## LIFE CYCLES

Life cycles of most plants have two distinctive generations: a haploid *gametophytic* (gamete-bearing plant) generation and a diploid *sporophytic* (spore-bearing plant) generation. Gametophytes produce gametes which unite to form sporophytes which in turn give rise to spores that develop into gametophytes, etc. This process is referred to as the *alternation of generations*. In lower plants, such as mosses and liverworts, the gametophyte is a conspicuous and independently living generation, the sporophyte being small and dependent upon the gametophyte. In higher plants (ferns, gymnosperms and angiosperms), the situation is reversed; the sporophyte is the independent and conspicuous generation and the gametophyte is the less conspicuous and, in the case of gymnosperms (cone bearing plants) and angiosperms (flowering plants), completely dependent generation. We have just seen in angiosperms that the male gametophytic generation is reduced to a pollen tube and three haploid nuclei (*microgametophyte*); the female gametophyte (*megagametophyte*) is a single multinucleated cell called the embryo sac surrounded and nourished by ovarian tissue.

Many simpler organisms such as one-celled animals (protozoa), algae, yeast and other fungi are useful in genetic studies and have interesting life cycles that exhibit considerable variation. Some of these life cycles, as well as those of bacteria and viruses, are presented in later chapters.

# Solved Problems

**1.1.** Consider 3 pairs of homologous chromosomes with centromeres labeled A/a, B/b and C/c where the slash line separates one chromosome from its homologue. How many different kinds of meiotic products can this individual produce?

**Solution:**

For ease in determining all possible combinations, we can use a dichotomous branching system.

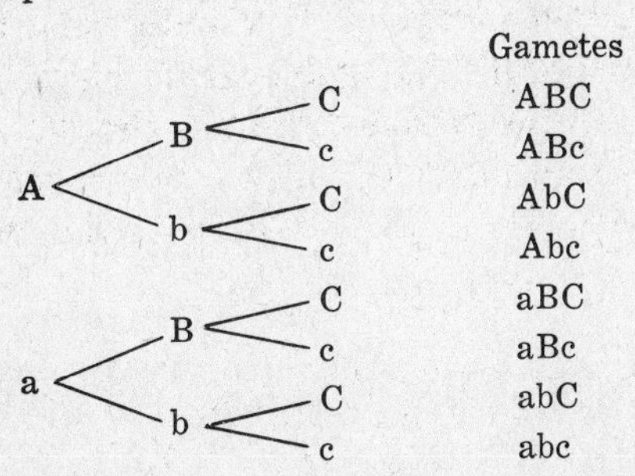


Eight different chromosomal combinations are expected in the gametes.

**1.2.** Develop a general formula which expresses the number of different types of gametic chromosomal combinations which can be formed in an organism with $k$ pairs of chromosomes.

**Solution:**

It is obvious from the solution of the preceding problem that 1 pair of chromosomes gives 2 types of gametes, 2 pairs give 4 types of gametes, 3 pairs give 8 types, etc. The progression 2, 4, 8, . . . can be expressed by the formula $2^k$ where $k$ is the number of chromosome pairs.

**1.3.** The horse (*Equus caballus*) has a diploid complement of 64 chromosomes including 36 acrocentric autosomes; the ass (*Equus asinus*) has 62 chromosomes including 22 acrocentric autosomes. (*a*) Predict the number of chromosomes to be found in the hybrid offspring (mule) produced by mating a male ass (jack) to a female horse (mare). (*b*) Why are mules usually sterile (incapable of producing viable gametes)?

**Solution:**

(*a*) The sperm of the jack carries the haploid number of chromosomes for its species $(62/2 = 31)$; the egg of the mare carries the haploid number for its species $(64/2 = 32)$; the hybrid mule formed by the union of these gametes would have a diploid number of $31 + 32 = 63$.

(*b*) The haploid set of chromosomes of the horse, which includes 18 acrocentric autosomes, is so dissimilar to that of the ass, which includes only 11 acrocentric autosomes, that meiosis in the mule germ line cannot proceed beyond first prophase where synapsis of homologues occurs.

**1.4.** When a plant of chromosomal type *aa* pollinates a plant of type *AA*, what chromosomal type of embryo and endosperm is expected in the resulting seeds?

**Solution:**

The pollen parent produces two sperm nuclei in each pollen grain of type *a*, one combining with the *A* egg nucleus to produce a diploid zygote (embryo) of type *Aa* and the other combining with the maternal fusion nucleus *AA* to produce a triploid endosperm of type *AAa*.

**1.5.** Given the first meiotic metaphase orientation shown on the right, and keeping all products in sequential order as they would be formed from left to right, diagram the embryo sac which develops from the meiotic product at the left and label the chromosomal constitution of all its nuclei.

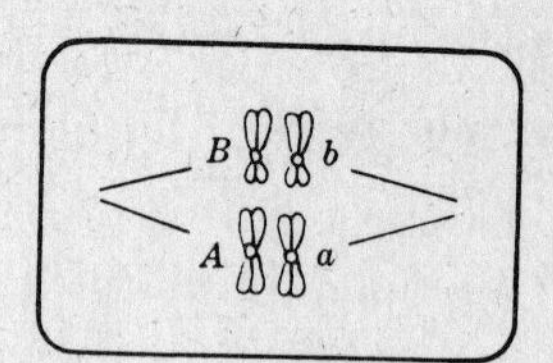


**Solution:**

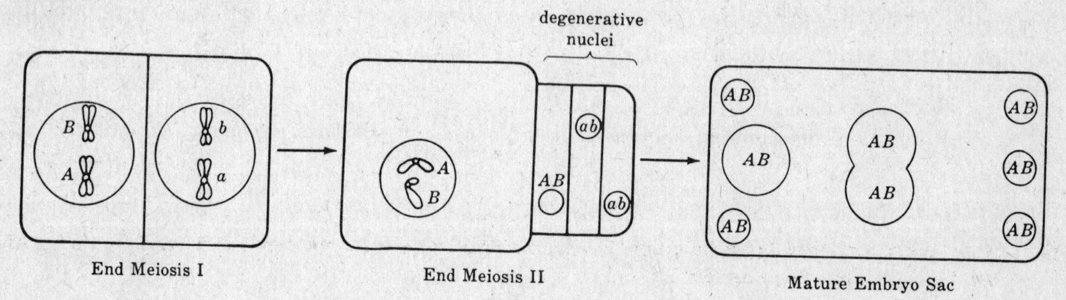


## Supplementary Problems

**1.6.** There are 40 chromosomes in somatic cells of the house mouse. (*a*) How many chromosomes does a mouse receive from its father? (*b*) How many autosomes are present in a mouse gamete? (*c*) How many sex chromosomes are in a mouse ovum? (*d*) How many autosomes are in somatic cells of a female?

**1.7.** Name each stage of mitosis described. (*a*) Chromosomes line up in the equatorial plane. (*b*) Nuclear membrane reforms and cytokinesis occurs. (*c*) Chromosomes become visible, spindle apparatus forms. (*d*) Sister chromatids move to opposite poles of the cell.

**1.8.** Identify the mitotic stage represented in each of the following diagrams of isolated cells from an individual with a genome consisting of one metacentric pair and one acrocentric pair of chromosomes.

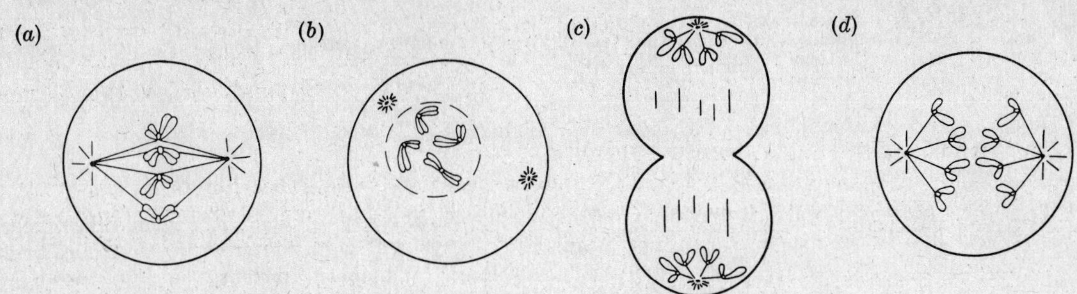


**1.9.** Identify the meiotic stage represented in each of the following diagrams of isolated cells from the germ line of an individual with one pair of acrocentric and one pair of metacentric chromosomes.

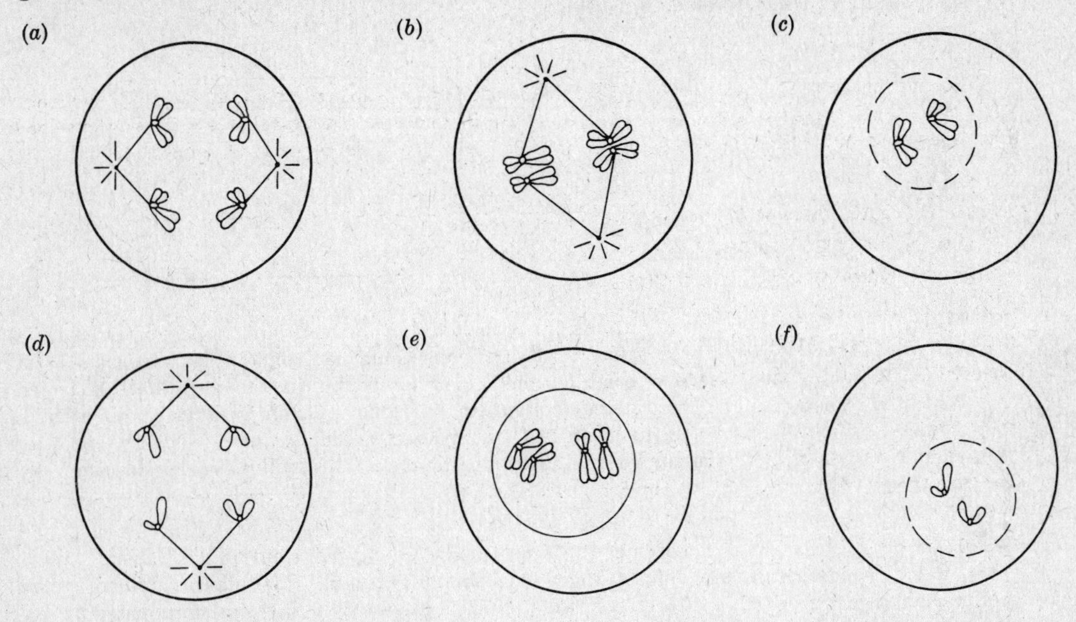


**1.10.** How many different types of gametic chromosomal combinations can be formed in the garden pea ($2n = 14$)? *Hint:* See Problem 1.2.

**1.11.** (*a*) What type of division (equational or reductional) is exemplified by the anaphase chromosomal movements shown below?

(*b*) Does the movement shown at (i) occur in mitosis or meiosis?

(*c*) Does the movement shown at (ii) occur in mitosis or meiosis?

**1.12.** What animal cells correspond to the three megaspores which degenerate following meiosis in plants?

**1.13.** What plant cell corresponds functionally to the primary spermatocyte?

**1.14.** What is the probability of a sperm cell of a man ($n = 23$) containing only replicas of the centromeres which were received from his mother?

**1.15.** How many chromosomes of humans ($2n = 46$) will be found in (*a*) a secondary spermatocyte, (*b*) a spermatid, (*c*) a spermatozoan, (*d*) a spermatogonium, (*e*) a primary spermatocyte?

**1.16.** How many spermatozoa are produced by (*a*) a spermatogonium, (*b*) a secondary spermatocyte, (*c*) a spermatid, (*d*) a primary spermatocyte?

**1.17.** How many human egg cells (ova) are produced by (*a*) an oögonium, (*b*) a primary oöcyte, (*c*) an oötid, (*d*) a polar body?

**1.18.** Corn (*Zea mays*) has a diploid number of 20. How many chromosomes would be expected in (*a*) a meiotic product (microspore or megaspore), (*b*) the cell resulting from the first nuclear division (karyokinesis) of a megaspore, (*c*) a polar nucleus, (*d*) a sperm nucleus, (*e*) a microspore mother cell, (*f*) a leaf cell, (*g*) a mature embryo sac (after degeneration of non-functional nuclei), (*h*) an egg nucleus, (*i*) an endosperm cell, (*j*) a cell of the embryo, (*k*) a cell of the pericarp, (*l*) an aleurone cell?

**1.19.** A pollen grain of corn with nuclei labeled A, B and C fertilized an embryo sac with nuclei labeled D, E, F, G, H, I, J and K as shown below.

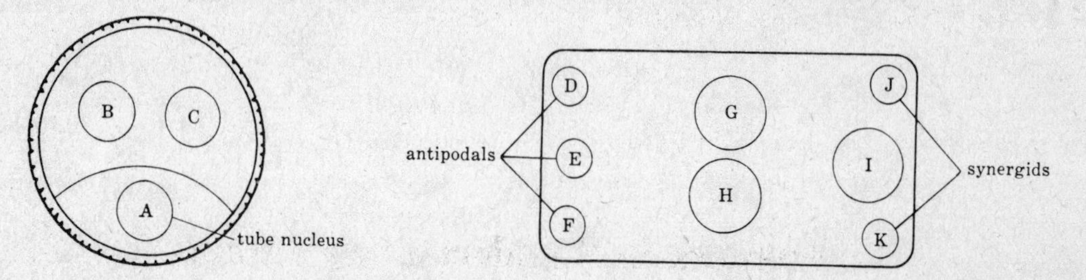


(*a*) Which of the following 5 combinations could be found in the embryo: (1) ABC, (2) BCI, (3) GHC, (4) AI, (5) CI? (*b*) Which of the above 5 combinations could be found in the aleurone layer of the seed? (*c*) Which of the above 5 combinations could be found in the germinating pollen tube? (*d*) Which of the nuclei, if any, in the pollen grain would contain genetically identical sets of chromosomes? (*e*) Which of the nuclei in the embryo sac would be chromosomally and genetically equivalent? (*f*) Which of the nuclei in these two gametophytes will have no descendants in the mature seed?

**1.20.** A certain plant has eight chromosomes in its root cells: a long metacentric pair, a short metacentric pair, a long telocentric pair, and a short telocentric pair. If this plant fertilizes itself (self-pollination), what proportion of the offspring would be expected to have (*a*) four pairs of telocentric chromosomes, (*b*) one telocentric pair and three metacentric pairs of chromosomes, (*c*) two metacentric and two telocentric pairs of chromosomes?

**1.21.** Referring to the preceding problem, what proportion of the meiotic products from such a plant would be expected to contain (*a*) four metacentric pairs of chromosomes, (*b*) two metacentric and two telocentric pairs of chromosomes, (*c*) one metacentric and one telocentric pair of chromosomes, (*d*) two metacentric and two telocentric chromosomes?

**1.22.** How many pollen grains are produced by (*a*) 20 microspore mother cells, (*b*) a cluster of 4 microspores?

**1.23.** How many sperm nuclei are produced by (*a*) a dozen microspore mother cells, (*b*) a generative nucleus, (*c*) 100 tube nuclei?

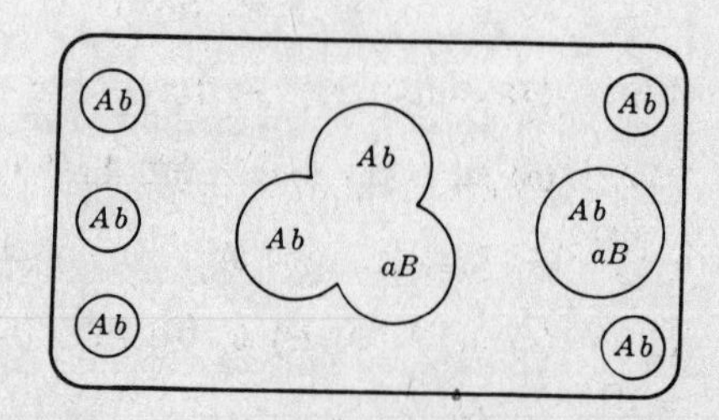


**1.24.** (*a*) Diagram the pollen grain responsible for the doubly fertilized embryo sac shown on the right. (*b*) Diagram the first meiotic metaphase (in an organism with two pairs of homologues labeled *A*, *a* and *B*, *b*) which produced the pollen grain in part (*a*).

For Problems 1.25-1.28, diagram the designated stages of gametogenesis in a diploid organism which has one pair of metacentric and one pair of acrocentric chromosomes. Label each of the chromatids assuming that the locus of gene $A$ is on the metacentric pair (one of which carries the $A$ allele and its homologue carries the $a$ allele) and that the locus of gene $B$ is on the acrocentric chromosome pair (one of which carries the $B$ allele and its homologue carries the $b$ allele).

**1.25.** *Oögenesis*: (*a*) first metaphase; (*b*) first telophase resulting from part (*a*); (*c*) second metaphase resulting from part (*b*); (*d*) second telophase resulting from part (*c*).

**1.26.** *Spermatogenesis*: (*a*) anaphase of a dividing spermatogonium; (*b*) anaphase of a dividing primary spermatocyte; (*c*) anaphase of a secondary spermatocyte derived from part (*b*); (*d*) four sperm cells resulting from part (*b*).

**1.27.** *Microsporogenesis*: (*a*) synapsis in a microsporocyte; (*b*) second meiotic metaphase; (*c*) first meiotic metaphase in the microspore mother cell which produced the cell of part (*b*); (*d*) anaphase of the second nuclear division (karyokinesis) following meiosis in a developing microgametophyte derived from part (*b*).

**1.28.** *Megasporogenesis*: (*a*) second meiotic telophase; (*b*) first meiotic telophase which produced the cell of part (*a*); (*c*) anaphase of the second nuclear division (karyokinesis) in a cell derived from part (*a*); (*d*) mature embryo sac produced from part (*c*).

## Answers to Supplementary Problems

**1.6.** (*a*) 20, (*b*) 19, (*c*) 1, (*d*) 38

**1.7.** (*a*) metaphase, (*b*) telophase, (*c*) prophase, (*d*) anaphase

**1.8.** (*a*) metaphase, (*b*) prophase, (*c*) telophase, (*d*) anaphase

**1.9.** (*a*) 1st anaphase, (*b*) 1st metaphase, (*c*) 2nd prophase or end of 1st telophase, (*d*) 2nd anaphase, (*e*) 1st prophase, (*f*) 2nd telophase (meiotic product)

**1.10.** 128

**1.11.** (*a*) (i) is an equational division, (ii) is a reductional division; (*b*) both; (*c*) meiosis

**1.12.** Polar bodies

**1.13.** Microspore mother cell (microsporocyte); both are diploid cells with the capacity to divide meiotically

**1.14.** $(\frac{1}{2})^{23}$, less than one chance in 8 million

**1.15.** (*a*) 23, (*b*) 23, (*c*) 23, (*d*) 46, (*e*) 46

**1.16.** (*a*) 4, (*b*) 2, (*c*) 1, (*d*) 4

**1.17.** (*a*) 1, (*b*) 1, (*c*) 1, (*d*) 0

**1.18.** (*a*) 10, (*b*) 20, (*c*) 10, (*d*) 10, (*e*) 20, (*f*) 20, (*g*) 30, (*h*) 10, (*i*) 30, (*j*) 20, (*k*) 20, (*l*) 30

**1.19.** (*a*) 5; (*b*) 3; (*c*) 1; (*d*) A, B, C; (*e*) D, E, F, G, H, I, J, K; (*f*) A, D, E, F, J, K

**1.20.** (*a*) 0, (*b*) 0, (*c*) all

**1.21.** (*a*) 0, (*b*) 0, (*c*) 0, (*d*) all

**1.22.** (*a*) 80, (*b*) 4

**1.23.** (*a*) 96, (*b*) 2, (*c*) 0

**1.24.** (*a*)

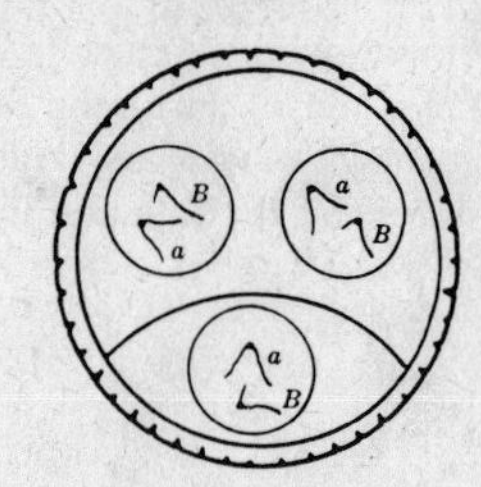

(*b*)

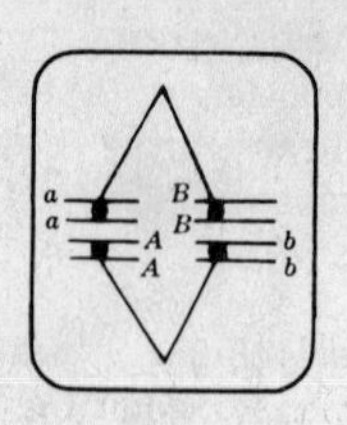

Only one of several possible solutions is shown for each of Problems 1.25-1.28.

**1.25.** (*a*)

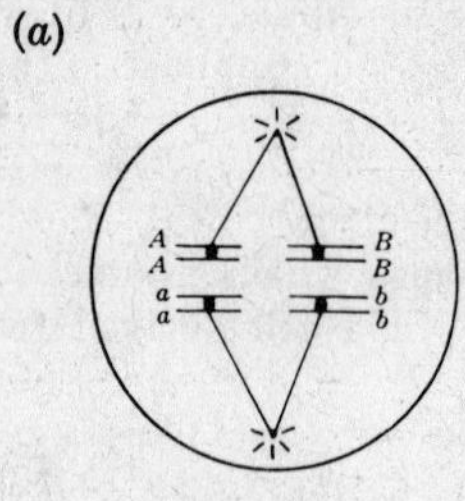

(*b*)

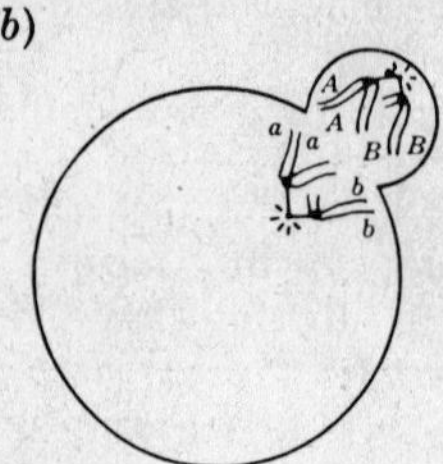

(*c*)

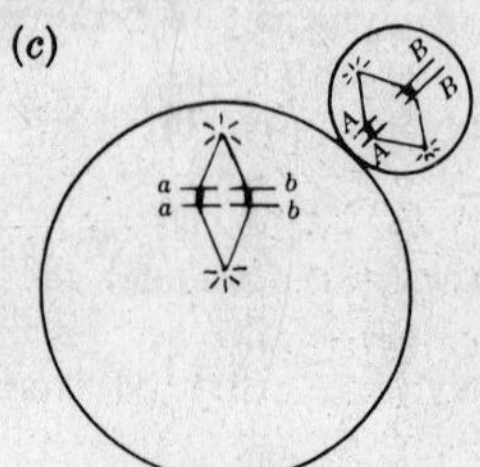

(*d*)

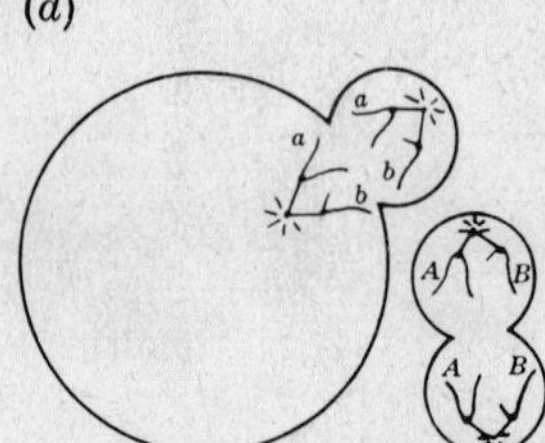

**1.26.** (*a*)

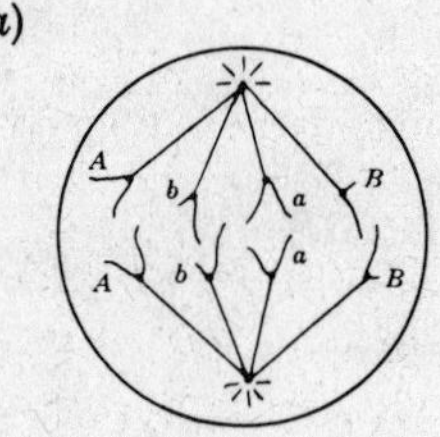

(*b*)

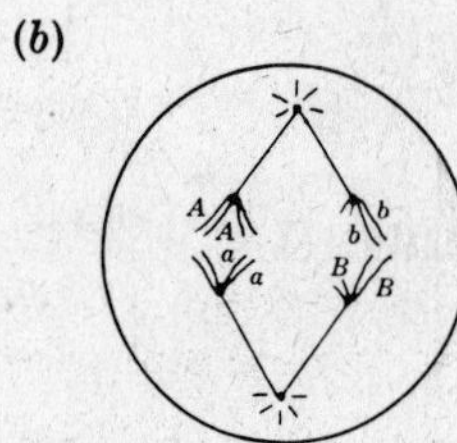

(*c*)

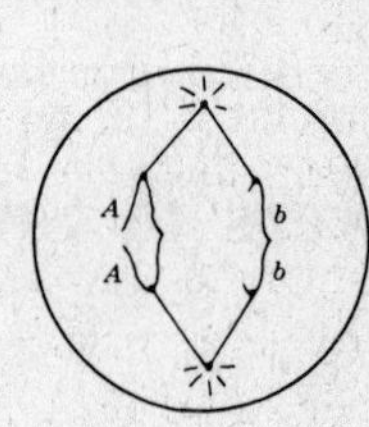

(*d*)

**1.27.** (*a*)

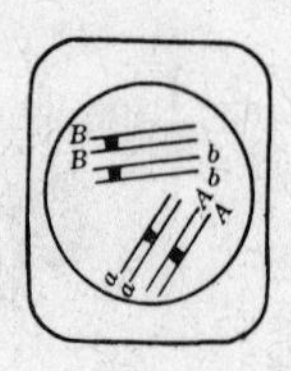

(*b*)

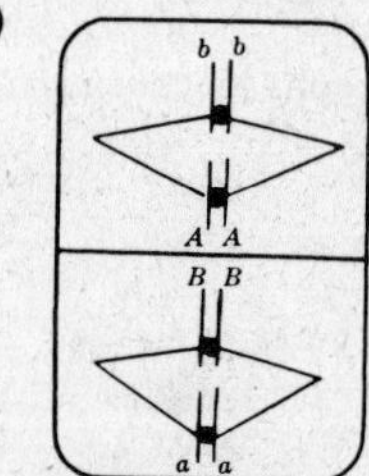

(*c*)

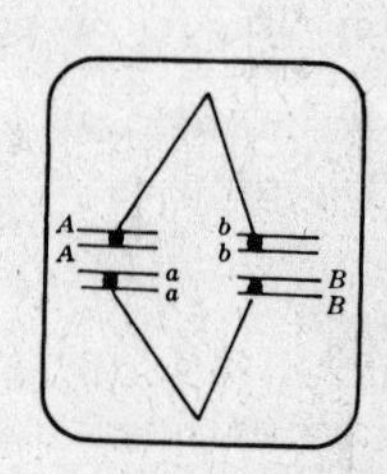

(*d*)

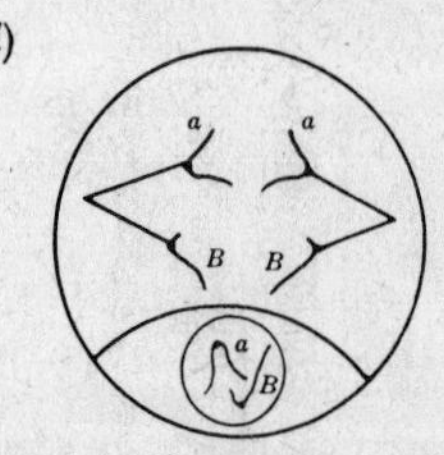

**1.28.** (*a*)

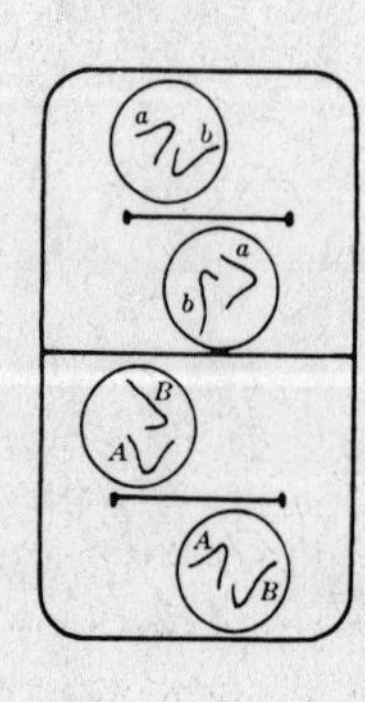

(*b*)

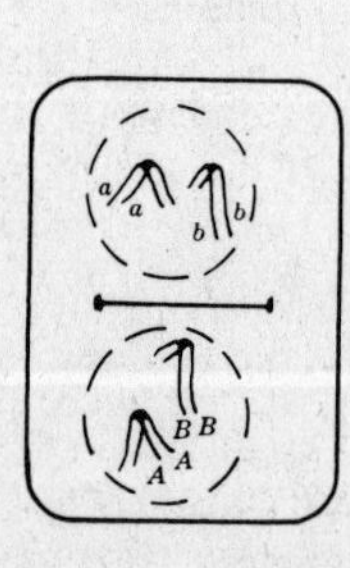

(*c*)

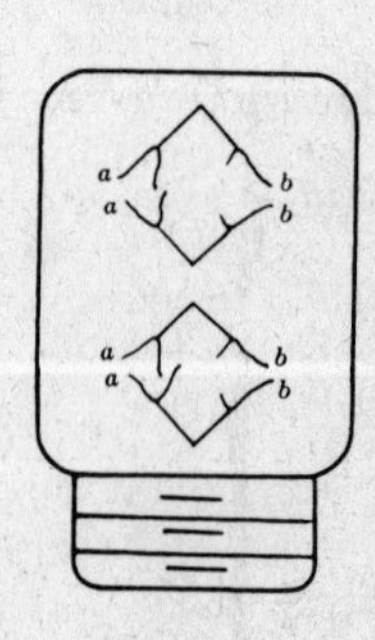

(*d*)

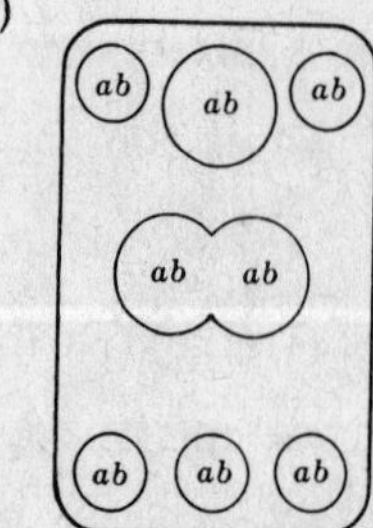

# Chapter 2

## Single Gene Inheritance

### TERMINOLOGY

**1. Phenotype.**

A *phenotype* may be any measurable characteristic or distinctive trait possessed by an organism. The trait may be visible to the eye, such as the color of a flower or the texture of hair, or it may require special tests for its identification, as in the determination of the respiratory quotient or the serological test for blood type. The phenotype is the result of gene products brought to expression in a given environment.

**Example 2.1.** Rabbits of the Himalayan breed in the usual range of environments develop black pigment at the tips of the nose, tail, feet and ears. If raised at very high temperatures, an all white rabbit is produced. The gene for Himalayan color pattern specifies a temperature sensitive enzyme which is inactivated at high temperature, resulting in a loss of pigmentation.

**Example 2.2.** The flowers of hydrangea may be blue if grown in acid soil or pinkish if grown in alkaline soil, due to an interaction of gene products with the hydrogen ion concentration of their environment.

The kinds of traits which we shall encounter in the study of simple Mendelian inheritance will be considered to be relatively unaffected by the normal range of environmental conditions in which the organism is found. It is important, however, to remember that genes establish boundaries within which the environment may modify the phenotype.

**2. Genotype.**

All of the genes possessed by an individual constitute its *genotype*. In this chapter, we shall be concerned only with that portion of the genotype involving alleles at a single locus.

(*a*) ***Homozygous.*** The union of gametes carrying identical alleles produces a *homozygous* genotype. A homozygote produces only one kind of gamete.

**Example 2.3.**

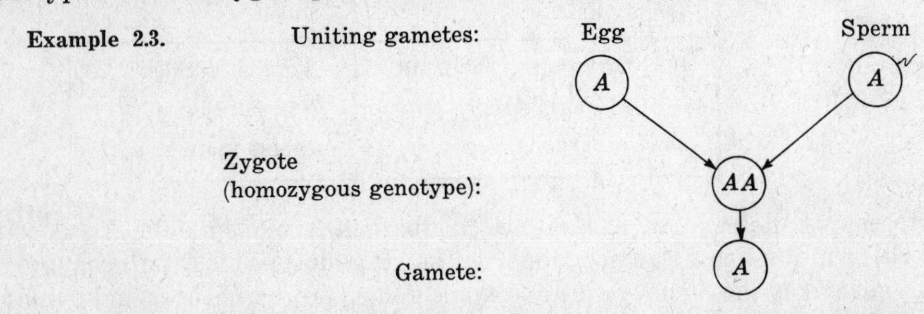


(*b*) ***Pure Line.*** A group of individuals with similar genetic background (breeding) is often referred to as a line or strain or variety or breed. Self-fertilization or mating closely related individuals for many generations (inbreeding) usually produces a population which is homozygous at nearly all loci. Matings between the homozygous individuals of a *pure line* produce only homozygous offspring like the parents. Thus we say that a pure line "breeds true".

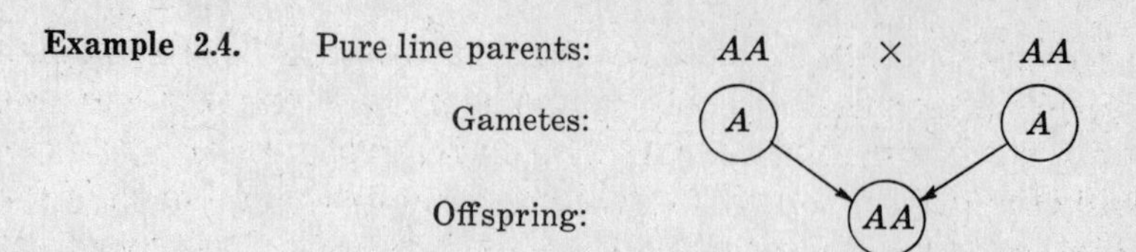


(*c*) ***Heterozygous.*** The union of gametes carrying different alleles produces a *heterozygous* genotype. Different kinds of gametes are produced by a heterozygote.

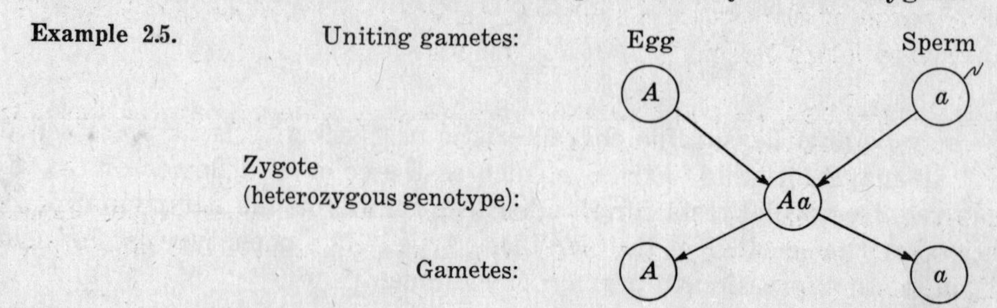


(*d*) ***Hybrid.*** The term *hybrid* as used in the problems of this book is synonymous with the heterozygous condition. Problems in this chapter may involve a single factor hybrid (monohybrid). Problems in the next chapter will consider heterozygosity at two or more loci (polyhybrids).

## ALLELIC RELATIONSHIPS

### 1. Dominant and Recessive Alleles.

Whenever one of a pair of alleles can come to phenotypic expression only in a homozygous genotype, we call that allele a *recessive* factor. The allele which can phenotypically express itself in the heterozygote as well as in the homozygote is called a *dominant* factor. Upper and lower case letters are commonly used to designate dominant and recessive alleles respectively. Usually the genetic symbol corresponds to the first letter in the name of the abnormal (or mutant) trait.

**Example 2.6.** Lack of pigment deposition in the human body is an abnormal recessive trait called "albinism". Using $A$ and $a$ to represent the dominant (normal) allele and the recessive (albino) allele respectively, three genotypes and two phenotypes are possible:

| Genotypes | Phenotypes |
|---|---|
| $AA$ (homozygous dominant) | Normal (pigment) |
| $Aa$ (heterozygote) | Normal (pigment) |
| $aa$ (homozygous recessive) | Albino (no pigment) |

(*a*) ***Carriers.*** Recessive alleles (such as the one for albinism) are often deleterious to those who possess them in duplicate (homozygous recessive genotype). A heterozygote may appear just as normal as the homozygous dominant genotype. A heterozygous individual who possesses a deleterious recessive allele hidden from phenotypic expression by the dominant normal allele is called a *carrier*. Most of the deleterious alleles harbored by a population are found in carrier individuals.

(*b*) ***Wild Type Symbolism.*** A different system for symbolizing dominant and recessive alleles is widely used in numerous organisms from higher plants and animals to the bacteria and viruses. Different genetics texts favor either one or the other system. In the author's opinion, every student should become familiar with both kinds of allelic representation and be able to work genetic problems regardless of the symbolic system used. Throughout the remainder of this book the student will find both systems used extensively. Where one phenotype is obviously of much more common occurrence in the population than its alternative phenotype, the former is usually referred to as *wild type*. The phenotype which is rarely observed is called the *mutant type*. In this system, the symbol + is used to indicate the normal allele for wild type. The base letter for the gene usually is taken from the name of the mutant or abnormal trait. If the mutant gene is recessive the symbol would be a lower case letter(s) corresponding to the initial letter(s) in the name of the trait. Its normal (wild type) dominant allele would have the same lower case letter but with a + as a superscript.

**Example 2.7.** Black body color in *Drosophila* is governed by a recessive gene $b$, and wild type (gray body) by its dominant allele $b^+$.

If the mutant trait is dominant, the base symbol would be an upper case letter without a superscript, and its recessive wild type allele would have the same upper case symbol with a + as a superscript.

**Example 2.8.** Lobe shaped eyes in *Drosophila* are governed by a dominant gene $L$, and wild type (oval eye) by its recessive allele $L^+$.

Remember that the case of the symbol indicates the dominance or recessiveness of the *mutant* allele to which the superscript + for wild type must be referred. After the allelic relationships have been defined, the symbol + by itself may be used for wild type and the letter alone may designate the mutant type.

### 2. Codominant Alleles.

Alleles which lack dominant and recessive relationships may be called intermediate alleles or *codominant* alleles. This means that each allele is capable of some degree of expression when in the heterozygous condition. Hence the heterozygous genotype gives rise to a phenotype distinctly different from either of the homozygous genotypes. Usually the heterozygous phenotype resulting from codominance is intermediate in character between those produced by the homozygous genotypes, hence the erroneous concept of "blending". The phenotype may appear to be a "blend" in heterozygotes but the alleles maintain their individual identities and will segregate from each other in the formation of gametes.

(*a*) ***Symbolism for Codominant Alleles.*** For codominant (or intermediate) alleles, all upper case base symbols with different superscripts should be used. The upper case letters call attention to the fact that each allele can express itself to some degree even when in the presence of its alternative allele (heterozygous).

**Example 2.9.** The alleles governing the M-N blood group system in humans are codominants and may be represented by the symbols $L^M$ and $L^N$, the base letter ($L$) being assigned in honor of its discoverers (Landsteiner and Levine). Two antisera (anti-M and anti-N) are used to distinguish three genotypes and their corresponding phenotypes (blood groups). Agglutination is represented by + and non-agglutination by −.

| Genotype | Reaction with: | | Blood Group |
|---|---|---|---|
| | Anti-M | Anti-N | (Phenotype) |
| $L^ML^M$ | + | − | M |
| $L^ML^N$ | + | + | MN |
| $L^NL^N$ | − | + | N |

### 3. Lethal Alleles.

The phenotypic manifestation of some genes is the death of the individual either in the prenatal or postnatal period prior to maturity. Such factors are called *lethal genes.* A fully dominant lethal allele (i.e. one which kills in both the homozygous and heterozygous conditions) occasionally arises by mutation from a normal allele. Individuals with a dominant lethal die before they can leave progeny. Therefore the mutant dominant lethal is removed from the population in the same generation in which it arose. Recessive lethals kill only when homozygous and may be of two kinds: (1) one which has no obvious phenotypic effect in heterozygotes, and (2) one which exhibits a distinctive phenotype when heterozygous.

**Example 2.10.** By special techniques, a completely recessive lethal ($l$) can sometimes be identified in certain families.

| Genotype | Phenotype |
|---|---|
| $LL$, $Ll$ | normal viability |
| $ll$ | lethal |

**Example 2.11.** The amount of chlorophyll in snapdragons (*Antirrhinum*) is controlled by an incompletely recessive gene which exhibits a lethal effect when homozygous, and a distinctive phenotypic effect when heterozygous.

| Genotype | Phenotype |
|---|---|
| $CC$ | green (normal) |
| $Cc$ | pale green |
| $cc$ | white (lethal) |

### 4. Penetrance and Expressivity.

Differences in environmental conditions or in genetic backgrounds may cause individuals which are genetically identical at a particular locus to exhibit different phenotypes. The percentage of individuals with a particular gene combination which exhibits the corresponding character to any degree represents the penetrance of the trait.

**Example 2.12.** Extra fingers and/or toes (polydactyly) in humans is thought to be produced by a dominant gene ($P$). The normal condition with five digits on each limb is produced by the recessive genotype ($pp$). Some individuals of genotype $Pp$ are not polydactylous, and therefore the gene has a penetrance of less than 100%.

A trait, though penetrant, may be quite variable in its expression. The degree of effect produced by a penetrant genotype is termed *expressivity.*

**Example 2.13.** The polydactylous condition may be penetrant in the left hand (6 fingers) and not in the right (5 fingers), or it may be penetrant in the feet and not in the hands.

A recessive lethal gene which lacks complete penetrance and expressivity will kill less than 100% of the homozygotes before sexual maturity. The terms *semilethal* or *subvital* apply to such genes. The effects which various kinds of lethals have on the reproduction of the next generation form a broad spectrum from complete lethality to sterility in completely viable genotypes. Problems in this book, however, will consider only those lethals which become completely penetrant, usually during the embryonic stage. Genes other than lethals will likewise be assumed completely penetrant.

## 5. Multiple Alleles.

The genetic systems proposed thus far have been limited to a single pair of alleles. The maximum number of alleles at a gene locus which any individual possesses is two, one on each of the homologous chromosomes. But since a gene can be changed to alternative forms by the process of mutation, a large number of alleles is theoretically possible in a population of individuals. Whenever more than two alleles are identified at a gene locus, we have a *multiple allelic series.*

(*a*) ***Symbolism for Multiple Alleles.*** The dominance hierarchy should be defined at the beginning of each problem involving multiple alleles. A capital letter is commonly used to designate the allele which is dominant to all others in the series. The corresponding lower case letter designates the allele which is recessive to all others in the series. Other alleles, intermediate in their degree of dominance between these two extremes, are usually assigned the lower case letter with some suitable superscript.

**Example 2.14.** The color of *Drosophila* eyes is governed by a series of alleles which cause the hue to vary from red or wild type ($w^+$ or $W$) through coral ($w^{co}$), blood ($w^{bl}$), eosin ($w^e$), cherry ($w^{ch}$), apricot ($w^a$), honey ($w^h$), buff ($w^{bf}$), tinged ($w^t$), pearl ($w^p$) and ivory ($w^i$) to white ($w$). Each allele in the system except $w$ can be considered to produce pigment, but successively less is produced by alleles as we proceed down the hierarchy: $w^+ > w^{co} > w^{bl} > w^e > w^{ch} > w^a > w^h > w^{bf} > w^t > w^p > w^i > w$. The wild type allele ($w^+$) is completely dominant and $w$ is completely recessive to all other alleles in the series. *Compounds* are heterozygotes which contain unlike members of an allelic series. The compounds of this series which involve alleles other than $w^+$ tend to be phenotypically intermediate between the eye colors of the parental homozygotes.

**Example 2.15.** A classical example of multiple alleles is found in the ABO blood group system of humans, where the allele $I^A$ for the A antigen is codominant with the allele $I^B$ for the B antigen. Both $I^A$ and $I^B$ are completely dominant to the allele $i$ which fails to specify any detectable antigenic structure. The hierarchy of dominance relationships is symbolized as $(I^A = I^B) > i$. Two antisera (anti-A and anti-B) are required for the detection of four phenotypes.

| Genotypes | Reaction with: | | Phenotype |
|---|---|---|---|
| | Anti-A | Anti-B | (Blood Groups) |
| $I^AI^A$, $I^Ai$ | + | − | A |
| $I^BI^B$, $I^Bi$ | − | + | B |
| $I^AI^B$ | + | + | AB |
| $ii$ | − | − | O |

**Example 2.16.** A slightly different kind of multiple allelic system is encountered in the coat colors of rabbits: $C$ allows full color to be produced (typical gray rabbit); $c^{ch}$, when homozygous, removes yellow pigment from the fur, making a silver-gray color called chinchilla; $c^{ch}$, when heterozygous with alleles lower in the dominance hierarchy, produces light gray fur; $c^h$ produces a white rabbit with black extremities called Himalayan; $c$ fails to produce pigment, resulting in albino. The dominance hierarchy may be symbolized as follows: $C > c^{ch} > c^h > c$.

| Phenotypes | Possible Genotypes |
|---|---|
| full color | $CC$, $Cc^{ch}$, $Cc^h$, $Cc$ |
| chinchilla | $c^{ch}c^{ch}$ |
| light gray | $c^{ch}c^h$, $c^{ch}c$ |
| Himalayan | $c^hc^h$, $c^hc$ |
| albino | $cc$ |

## SINGLE GENE (MONOFACTORIAL) CROSSES

### 1. The Six Basic Types of Matings.

A pair of alleles governs pelage color in the guinea pig; a dominant allele $B$ produces black and its recessive allele $b$ produces white. There are six types of matings possible among the three genotypes. The parental generation is symbolized P and the first filial generation of offspring is symbolized $F_1$.

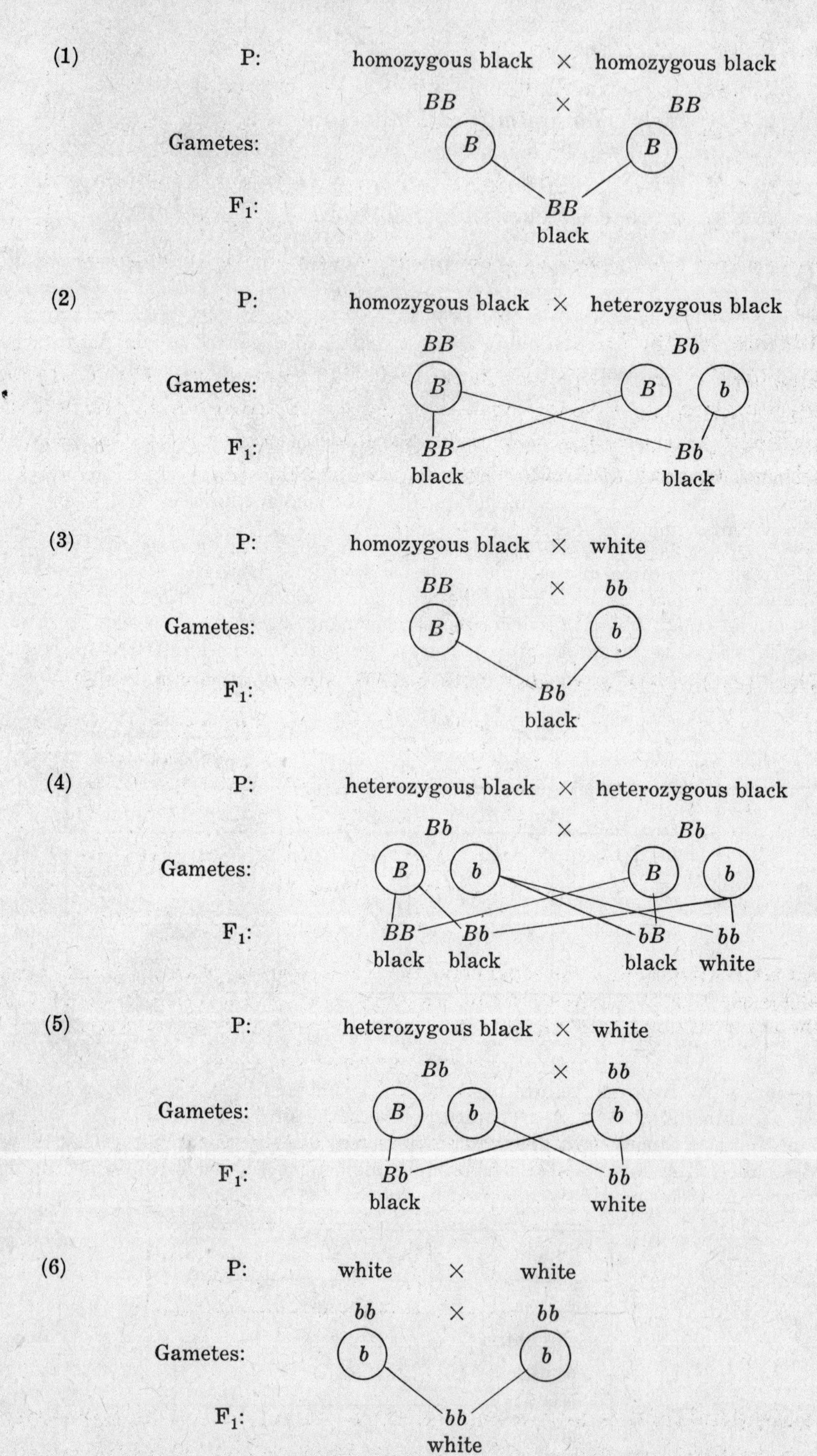

Summary of the six types of matings:

| No. | Matings | Expected $F_1$ Ratios | |
|---|---|---|---|
| | | Genotypes | Phenotypes |
| (1) | $BB \times BB$ | all $BB$ | all black |
| (2) | $BB \times Bb$ | $\frac{1}{2}BB : \frac{1}{2}Bb$ | all black |
| (3) | $BB \times bb$ | all $Bb$ | all black |
| (4) | $Bb \times Bb$ | $\frac{1}{4}BB : \frac{1}{2}Bb : \frac{1}{4}bb$ | $\frac{3}{4}$ black : $\frac{1}{4}$ white |
| (5) | $Bb \times bb$ | $\frac{1}{2}Bb : \frac{1}{2}bb$ | $\frac{1}{2}$ black : $\frac{1}{2}$ white |
| (6) | $bb \times bb$ | all $bb$ | all white |

## 2. Conventional Production of the $F_2$.

Unless otherwise specified in the problem, the second filial generation ($F_2$) is produced by crossing the $F_1$ individuals among themselves randomly. Plants which are normally self-fertilized can be artificially cross pollinated in the parental generation and the resulting $F_1$ progeny are then allowed to pollinate themselves to produce the $F_2$.

**Example 2.17.** P: $BB$ black × $bb$ white

$F_1$: $Bb$ black

**The black $F_1$ males are mated to the black $F_1$ females to produce the $F_2$.**

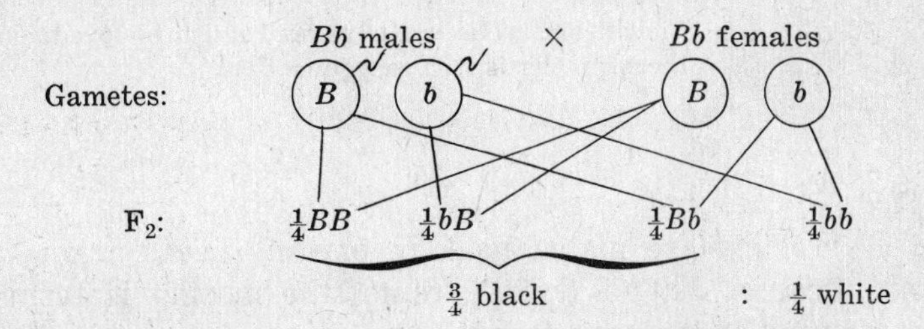


: $\frac{1}{4}$ white

**An alternative method for combining the $F_1$ gametes is to place the female gametes along one side of a "checkerboard" (Punnett square) and the male gametes along the other side and then combine them to form zygotes as shown below.**

| | $B$ | $b$ |
|---|---|---|
| $B$ | $BB$ black | $Bb$ black |
| $b$ | $Bb$ black | $bb$ white |

## 3. Testcross.

Because a homozygous dominant genotype has the same phenotype as the heterozygous genotype, a *testcross* is required to distinguish between them. The testcross parent is always homozygous recessive for all of the genes under consideration. The purpose of a testcross is to discover how many different kinds of gametes are being produced by the individual whose genotype is in question. A homozygous dominant individual will produce only one kind of gamete; a *monohybrid* individual (heterozygous at one locus) produces two kinds of gametes with equal frequency.

**Example 2.18.** Consider the case in which testcrossing a black female produced only black offspring.

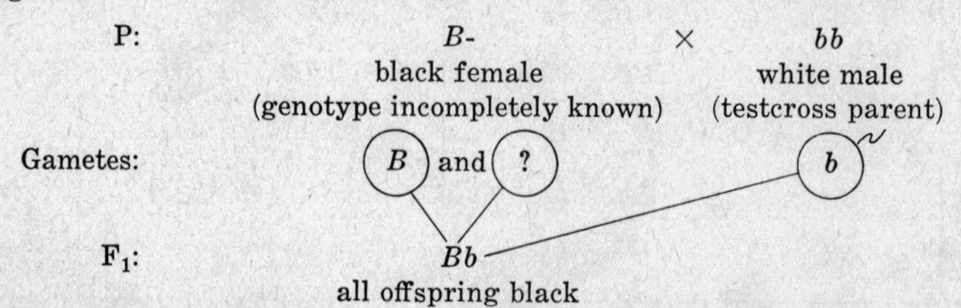


Conclusion: The female parent must be producing only one kind of gamete and therefore she is homozygous dominant $BB$.

**Example 2.19.** Consider the case in which testcrossing a black male produced black and white offspring in approximately equal numbers.

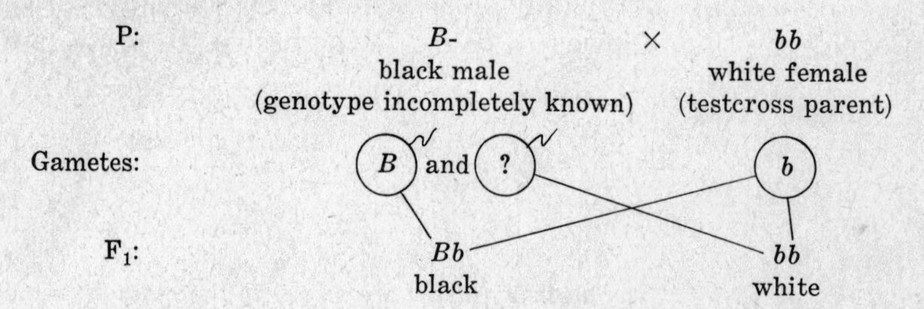


Conclusion: The male parent must be producing two kinds of gametes and therefore he is heterozygous $Bb$.

## 4. Backcross.

If the $F_1$ progeny are mated back to one of their parents (or to individuals with a genotype identical to that of their parents) the mating is termed *backcross*. Sometimes "backcross" is used synonymously with "testcross" in genetic literature, but it will be used exclusively in the former sense in this book.

**Example 2.20.** A homozygous black female guinea pig is crossed to a white male. An $F_1$ son is backcrossed to his mother. Using the symbol ♀ for female and ♂ for male (♀♀ = females, ♂♂ = males), we diagram this backcross as follows:

P: *BB*♀ × *bb*♂
black female   white male

$F_1$: *Bb*♂♂ and ♀♀
black males and females

$F_1$ backcross: *Bb*♂ × *BB*♀
black son   black mother

Backcross progeny: $\left.\begin{array}{l}\frac{1}{2}BB \\ \frac{1}{2}Bb\end{array}\right\}$ all black offspring

## PEDIGREE ANALYSIS

A pedigree is a systematic listing (either as words or as symbols) of the ancestors of a given individual, or it may be the "family tree" for a large number of individuals. It is customary to represent females as circles and males as squares. Matings are shown as horizontal lines between two individuals. The offspring of a mating are connected by a vertical line to the mating line. Different shades or colors added to the symbols can represent various phenotypes. Each generation is listed on a separate row labeled with Roman numerals. Individuals within a generation receive Arabic numerals.

**Example 2.21.** Let solid symbols represent black guinea pigs and open symbols represent white guinea pigs.

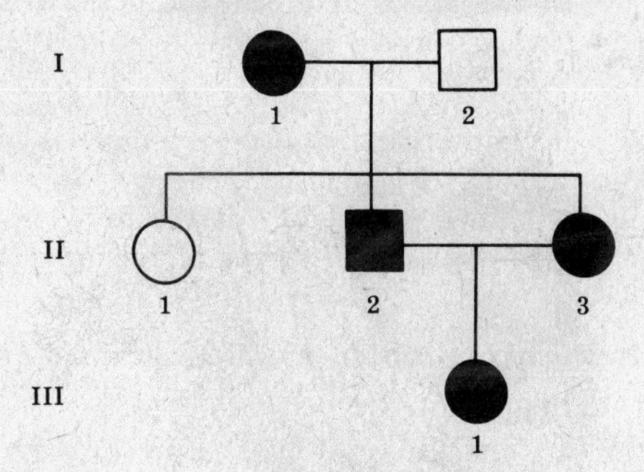


| Individuals | Phenotype | Genotype |
|---|---|---|
| I 1 | black ♀ | *Bb* |
| I 2 | white ♂ | *bb* |
| II 1 | white ♀ | *bb* |
| II 2 | black ♂ | *Bb* |
| II 3 | black ♀ | *Bb* |
| III 1 | black ♀ | *B*-* |

*The dash indicates that the genotype could be either homozygous or heterozygous.

## PROBABILITY THEORY

### 1. Observed vs. Expected Results.

Experimental results seldom conform exactly to the expected ratios. Genetic probabilities derive from the operation of chance events in the meiotic production of gametes and the random union of these gametes in fertilization. Samples from a population of individuals often deviate from the expected ratios, rather widely in very small samples, but usually approaching the expectations more closely with increasing sample size.

**Example 2.22.** Suppose that a testcross of heterozygous black guinea pigs ($Bb \times bb$) produces 5 offspring: 3 black ($Bb$) and 2 white ($bb$). Theoretically we expect half of the total number of offspring to be black and half to be white $= \frac{1}{2}(5) = 2\frac{1}{2}$. Obviously we cannot observe half of an individual, and the results conform as closely to the theoretical expectations as is biologically possible.

**Example 2.23.** Numerous testcrosses of a black guinea pig produced a total of 10 offspring, 8 of which were black and 2 were white. We theoretically expected 5 black and 5 white, but the deviation from the expected numbers which we observed in our small sample of 10 offspring should not be any more surprising than the results of tossing a coin 10 times and observing 8 heads and 2 tails. The fact that at least one white offspring appeared is sufficient to classify the black parent as genetically heterozygous ($Bb$).

**2. Combining Probabilities.**

Two or more events are said to be *independent* if the occurrence or non-occurrence of any one of them does not affect the probabilities of occurrence of any of the others. When two independent events occur with the probabilities $p$ and $q$ respectively, then the probability of their joint occurrence is $pq$. That is, the combined probability is the product of the probabilities of the independent events. If the word "and" is used or implied in the phrasing of a problem solution, a *multiplication* of independent probabilities is usually required.

**Example 2.24.** Theoretically there is an equal opportunity for a tossed coin to land on either heads or tails. Let $p$ = probability of heads $= \frac{1}{2}$, and $q$ = probability of tails $= \frac{1}{2}$. In two tosses of a coin the probability of two heads appearing (i.e. a head on the first toss *and* a head on the second toss) is $p \times p = p^2 = (\frac{1}{2})^2 = \frac{1}{4}$.

**Example 2.25.** In testcrossing a heterozygous black guinea pig ($Bb \times bb$), let the probability of a black ($Bb$) offspring be $p = \frac{1}{2}$ and of a white ($bb$) offspring be $q = \frac{1}{2}$. The combined probability of the first two offspring being white (i.e. the first offspring is white *and* the second offspring is white) $= q \times q = q^2 = (\frac{1}{2})^2 = \frac{1}{4}$.

There is only one way in which two heads may appear in two tosses of a coin, i.e. heads on the first toss and heads on the second toss. The same is true for two tails. There are two ways, however, to obtain one head and one tail in two tosses of a coin. The head may appear on the first toss and the tail on the second *or* the tail may appear on the first toss and the head on the second. *Mutually exclusive events* are those in which the occurrence of any one of them excludes the occurrence of the others. The word "or" is usually required or implied in the phrasing of problem solutions involving mutually exclusive events, signaling that an *addition* of probabilities is to be performed. That is, whenever alternative possibilities exist for the satisfaction of the conditions of a problem, the individual probabilities are combined by addition.

**Example 2.26.** In two tosses of a coin, there are two ways to obtain a head and a tail.

| | First Toss | | Second Toss | | Probability |
|---|---|---|---|---|---|
| First Alternative: | Head ($p$) | (and) | Tail ($q$) | = | $pq$ |
| | | | | | (or) |
| Second Alternative: | Tail ($q$) | (and) | Head ($p$) | = | $qp$ |
| | | Combined probability | | = | $2pq$ |

$p = q = \frac{1}{2}$; hence the combined probability $= 2(\frac{1}{2})(\frac{1}{2}) = \frac{1}{2}$.

**Example 2.27.** In testcrossing heterozygous black guinea pigs ($Bb \times bb$), there are two ways to obtain one black ($Bb$) and one white ($bb$) offspring in a litter of two animals. Let $p$ = probability of black $= \frac{1}{2}$ and $q$ = probability of white $= \frac{1}{2}$.

| | First Offspring | | Second Offspring | | Probability |
|---|---|---|---|---|---|
| First Alternative: | Black ($p$) | (and) | White ($q$) | = | $pq$ |
| | | | | | (or) |
| Second Alternative: | White ($q$) | (and) | Black ($p$) | = | $qp$ |
| | | Combined probability | | = | $2pq$ |

$p = q = \frac{1}{2}$; hence the combined probability $= 2(\frac{1}{2})(\frac{1}{2}) = \frac{1}{2}$.

Many readers will recognize that the application of the above two rules for combining probabilities (independent and mutually exclusive events) is the basis of the binomial distribution, which will be considered in detail in Chapter 7.

# Solved Problems

## DOMINANT AND RECESSIVE ALLELES

**2.1.** Black pelage of guinea pigs is a dominant trait, white is the alternative recessive trait. When a pure black guinea pig is crossed to a white one, what fraction of the black $F_2$ is expected to be heterozygous?

**Solution:**

As shown in Example 2.17, the $F_2$ genotypic ratio is $1BB:2Bb:1bb$. Considering only the black $F_2$, we expect $1BB:2Bb$ or 2 out of every 3 black pigs are expected to be heterozygous; the fraction is $\frac{2}{3}$.

**2.2.** If a black female guinea pig is testcrossed and produces two offspring in each of three litters, all of which are black, what is her probable genotype? With what degree of confidence may her genotype be specified?

**Solution:**

P: $B$- (black female) × $bb$ (white male)

$F_1$: all $Bb$ = all black

The female parent could be homozygous $BB$ or heterozygous $Bb$ and still be phenotypically black, hence the symbol $B$-. If she is heterozygous, each offspring from this testcross has a 50% chance of being black. The probability of 6 offspring being produced, all of which are black, is $(\frac{1}{2})^6 = \frac{1}{64} = 0.0156 = 1.56\%$. In other words, we expect such results to occur by chance less than 2% of the time. Since it is chance that operates in the union of gametes, she might actually be heterozygous and thus far only her $B$ gametes have been the "lucky ones" to unite with the $b$ gametes from the white parent. Since no white offspring have appeared in six of these chance unions we may be approximately 98% confident $(1-0.0156 = 0.9844$ or $98.44\%)$ on the basis of chance, that she is of homozygous genotype ($BB$). It is possible, however, for her very next testcross offspring to be white, in which case we would then become certain that her genotype was heterozygous $Bb$ and not BB.

**2.3.** Heterozygous black guinea pigs ($Bb$) are crossed among themselves. (*a*) What is the probability of the first three offspring being alternately black-white-black or white-black-white? (*b*) What is the probability among 3 offspring of producing 2 black and 1 white in any order?

**Solution:**

(*a*) P: $Bb$ (black) × $Bb$ (black)

$F_1$: $\frac{3}{4}$ black : $\frac{1}{4}$ white

Let $p$ = probability of black = $\frac{3}{4}$, $q$ = probability of white = $\frac{1}{4}$.

Probability of black *and* white *and* black = $p \times q \times p = p^2q$, *or*

Probability of white *and* black *and* white = $q \times p \times q = pq^2$

Combined probability = $p^2q + pq^2 = \frac{3}{16}$

(*b*) Consider the number of ways that 2 black and 1 white offspring could be produced.

| Offspring Order (1st, 2nd, 3rd) | Probability | |
|---|---|---|
| black *and* black *and* white | $= (\frac{3}{4})(\frac{3}{4})(\frac{1}{4}) = \frac{9}{64}$, | *or* |
| black *and* white *and* black | $= (\frac{3}{4})(\frac{1}{4})(\frac{3}{4}) = \frac{9}{64}$, | *or* |
| white *and* black *and* black | $= (\frac{1}{4})(\frac{3}{4})(\frac{3}{4}) = \frac{9}{64}$ | |
| Combined probability | $= \frac{27}{64}$ | |

Once we have ascertained that there are three ways to obtain 2 black and 1 white, the total probability becomes $3(\frac{3}{4})^2(\frac{1}{4}) = \frac{27}{64}$.

**2.4.** A dominant gene $b^+$ is responsible for the wild type body color of *Drosophila*; its recessive allele $b$ produces black body color. A testcross of a wild type female gave 52 black and 58 wild type in the $F_1$. If the wild type $F_1$ females are crossed to their black $F_1$ brothers, what genotypic and phenotypic ratios would be expected in the $F_2$? Diagram the results using the appropriate genetic symbols.

**Solution:**

P: $b^+ -$ ♀ (wild type female) × $bb$ ♂ (black male)

$F_1$: $52bb$ (black) : $58b^+b$ (wild type)

Since the recessive black phenotype appears in the $F_1$ in approximately a 1:1 ratio, we know that the female parent must be heterozygous $b^+b$. Furthermore, we know that the wild type $F_1$ progeny must also be heterozygous. The wild type $F_1$ females are then crossed with their black brothers:

$F_1$ cross: $b^+b$ ♀♀ (wild type females) × $bb$ ♂♂ (black males)

$F_2$: $\frac{1}{2}b^+b$ wild type : $\frac{1}{2}bb$ black

The expected $F_2$ ratio is therefore the same as that observed in the $F_1$, namely 1 wild type : 1 black.

## CODOMINANT ALLELES

**2.5.** Coat colors of the Shorthorn breed of cattle represent a classical example of codominant alleles. Red is governed by the genotype $C^RC^R$, roan (mixture of red and white) by $C^RC^W$, and white by $C^WC^W$. (*a*) When roan Shorthorns are crossed among themselves, what genotypic and phenotypic ratios are expected among their progeny? (*b*) If red Shorthorns are crossed with roans, and the $F_1$ progeny are crossed among themselves to produce the $F_2$, what percentage of the $F_2$ will probably be roan?

**Solution:**

(*a*) P: $C^RC^W$ (roan) × $C^RC^W$ (roan)

$F_1$: $\frac{1}{4}C^RC^R$ red : $\frac{1}{2}C^RC^W$ roan : $\frac{1}{4}C^WC^W$ white

Since each genotype produces a unique phenotype, the phenotypic ratio 1:2:1 corresponds to the same genotypic ratio.

(*b*) P: $C^RC^R$ (red) × $C^RC^W$ (roan)

$F_1$: $\frac{1}{2}C^RC^R$ red : $\frac{1}{2}C^RC^W$ roan

There are three types of matings possible for the production of the $F_2$. Their relative frequencies of occurrence may be calculated by preparing a mating table.

| | $\frac{1}{2}C^RC^R$ ♂ | $\frac{1}{2}C^RC^W$ ♂ |
|---|---|---|
| $\frac{1}{2}C^RC^R$ ♀ | (1) $\frac{1}{4}C^RC^R$ ♀ × $C^RC^R$ ♂ | (2) $\frac{1}{4}C^RC^R$ ♀ × $C^RC^W$ ♂ |
| $\frac{1}{2}C^RC^W$ ♀ | (2) $\frac{1}{4}C^RC^W$ ♀ × $C^RC^R$ ♂ | (3) $\frac{1}{4}C^RC^W$ ♀ × $C^RC^W$ ♂ |

(1) The mating $C^RC^R$ ♀ × $C^RC^R$ ♂ (red × red) produces only red ($C^RC^R$) progeny. But only one-quarter of all matings are of this type. Therefore $\frac{1}{4}$ of all the $F_2$ should be red from this source.

(2) The matings $C^RC^W \times C^RC^R$ (roan female × red male or roan male × red female) are expected to produce $\frac{1}{2}C^RC^R$ (red) and $\frac{1}{2}C^RC^W$ (roan) progeny. Half of all matings are of this kind. Therefore $(\frac{1}{2})(\frac{1}{2}) = \frac{1}{4}$ of all the $F_2$ progeny should be red and $\frac{1}{4}$ should be roan from this source.

(3) The mating $C^RC^W$ ♀ × $C^RC^W$ ♂ (roan × roan) is expected to produce $\frac{1}{4}C^RC^R$ (red), $\frac{1}{2}C^RC^W$ (roan), and $\frac{1}{4}C^WC^W$ (white) progeny. This mating type constitutes $\frac{1}{4}$ of all crosses. Therefore the fraction of all $F_2$ progeny contributed from this source is $(\frac{1}{4})(\frac{1}{4}) = \frac{1}{16}C^RC^R$, $(\frac{1}{4})(\frac{1}{2}) = \frac{1}{8}C^RC^W$, $(\frac{1}{4})(\frac{1}{4}) = \frac{1}{16}C^WC^W$.

The expected $F_2$ contributions from all three types of matings are summarized in the following table.

| Type of Mating | Frequency of Mating | $F_2$ Progeny | | |
|---|---|---|---|---|
| | | Red | Roan | White |
| (1) red × red | $\frac{1}{4}$ | $\frac{1}{4}$ | 0 | 0 |
| (2) red × roan | $\frac{1}{2}$ | $\frac{1}{4}$ | $\frac{1}{4}$ | 0 |
| (3) roan × roan | $\frac{1}{4}$ | $\frac{1}{16}$ | $\frac{1}{8}$ | $\frac{1}{16}$ |
| Totals | | $\frac{9}{16}$ | $\frac{6}{16}$ | $\frac{1}{16}$ |

The fraction of roan progeny in the $F_2$ is $\frac{3}{8}$ or approximately 38%.

## LETHAL ALLELES

**2.6.** The absence of legs in cattle ("amputated") has been attributed to a completely recessive lethal gene. A normal bull is mated with a normal cow and they produce an amputated calf (usually dead at birth). The same parents are mated again.

(*a*) What is the chance of the next calf being amputated?

(*b*) What is the chance of these parents having two calves, both of which are amputated?

(*c*) Bulls carrying the amputated allele (heterozygous) are mated to non-carrier cows. The $F_1$ is allowed to mate at random to produce the $F_2$. What genotypic ratio is expected in the adult $F_2$?

(*d*) Suppose that each $F_1$ female in part (*c*) rears one viable calf, i.e. each of the cows which throws an amputated calf is allowed to remate to a carrier sire until she produces a viable offspring. What genotypic ratio is expected in the adult $F_2$?

**Solution:**

(*a*) If phenotypically normal parents produce an amputated calf, they must both be genetically heterozygous.

P: $Aa$ normal × $Aa$ normal

$F_1$:

| Genotypes | Phenotypes |
|---|---|
| $\frac{1}{4}AA$, $\frac{1}{2}Aa$ | $\frac{3}{4}$ normal |
| $\frac{1}{4}aa$ | = $\frac{1}{4}$ amputated (dies) |

Thus there is a 25% chance of the next offspring being amputated.

(*b*) The chance of the first calf being amputated *and* the second calf also being amputated is the product of the separate probabilities: $(\frac{1}{4})(\frac{1}{4}) = \frac{1}{16}$.

(*c*) The solution to part (*c*) is analogous to that of Problem 2.5(*b*). A summary of the expected $F_2$ follows:

| Type of Mating | $F_2$ Genotypes | | |
|---|---|---|---|
| | $AA$ | $Aa$ | $aa$ |
| $AA \times AA$ | $\frac{4}{16}$ | | |
| $AA \times Aa$ | $\frac{4}{16}$ | $\frac{4}{16}$ | |
| $Aa \times Aa$ | $\frac{1}{16}$ | $\frac{2}{16}$ | $\frac{1}{16}$ |
| Totals | $\frac{9}{16}$ | $\frac{6}{16}$ | $\frac{1}{16}$ |

All $aa$ genotypes die and fail to appear in the adult progeny. Therefore the adult progeny has the genotypic ratio $9AA : 6Aa$ or $3AA : 2Aa$.

(*d*) The results of matings $AA \times AA$ and $AA \times Aa$ remain the same as in part (*c*). The mating of $Aa$ by $Aa$ now is expected to produce $\frac{1}{3}AA$ and $\frac{2}{3}Aa$ adult progeny. Correcting for the frequency of occurrence of this mating, we have $(\frac{1}{4})(\frac{1}{3}) = \frac{1}{12}AA$ and $(\frac{1}{4})(\frac{2}{3}) = \frac{2}{12}Aa$.

Summary of the $F_2$:

| Type of Mating | $F_2$ Genotypes | |
|---|---|---|
| | $AA$ | $Aa$ |
| $AA \times AA$ | $\frac{3}{12}$ | |
| $AA \times Aa$ | $\frac{3}{12}$ | $\frac{3}{12}$ |
| $Aa \times Aa$ | $\frac{1}{12}$ | $\frac{2}{12}$ |
| Totals | $\frac{7}{12}$ | $\frac{5}{12}$ |

The adult $F_2$ genotypic ratio is expected to be $7AA$ to $5Aa$.

**2.7.** A bull, heterozygous for a completely recessive lethal gene, sires 3 calves each out of 32 cows. Twelve of the cows have one or more stillborn calves and therefore must be carriers of this lethal. How many more carrier cows probably exist in this herd undetected?

**Solution:**

The probability that a heterozygous cow will not have a stillborn calf in 3 matings to a heterozygous male is calculated as follows: each calf has a $\frac{3}{4}$ chance of being normal; therefore the probability of 3 calves being normal is $(\frac{3}{4})^3 = \frac{27}{64}$. That is, the probability that we will fail to detect a heterozygous (carrier) cow with 3 calves is $\frac{27}{64}$. The probability that we will detect a carrier cow with 3 calves is $\frac{37}{64}$. Let $x$ = number of heterozygous cows in the herd; then $(\frac{37}{64})x = 12$ or $x = 21$ (to the nearest integer). 21 carrier cows probably exist; since we have detected 12 of them, there are probably 9 carrier cows undetected in this herd.

## MULTIPLE ALLELES

**2.8.** The genetics of rabbit coat colors is given in Example 2.16, page 23. Determine the genotypic and phenotypic ratios expected from mating full colored males of genotype $Cc^{ch}$ to light gray females of genotype $c^{ch}c$.

**Solution:**

P: $Cc^{ch}$ ♂♂ full color × $c^{ch}c$ ♀♀ light gray

$F_1$:

| | $C$ | $c^{ch}$ |
|---|---|---|
| $c^{ch}$ | $Cc^{ch}$ full color | $c^{ch}c^{ch}$ chinchilla |
| $c$ | $Cc$ full color | $c^{ch}c$ light gray |

Thus we have a 1:1:1:1 genotypic ratio, but a phenotypic ratio of 2 full color : 1 chinchilla : 1 light gray.

**2.9.** The coat colors of mice are known to be governed by a multiple allelic series in which the allele $A^y$, when homozygous, is lethal early in embryonic development but produces yellow color when in heterozygous condition with other alleles. Agouti (mousy color) is governed by the $A$ allele, and black by the recessive $a$. The dominance hierarchy is as follows: $A^y > A > a$. What phenotypic and genotypic ratios are expected in the viable $F_1$ from the cross $A^yA \times A^ya$?

**Solution:**

P: $A^yA$ yellow $\times$ $A^ya$ yellow

$F_1$:

| | $(A^y)$ | $(a)$ |
|---|---|---|
| $(A^y)$ | $A^yA^y$ dies | $A^ya$ yellow |
| $(A)$ | $A^yA$ yellow | $Aa$ agouti |

Since $\frac{1}{4}$ of the progeny dies before birth, we should observe two yellow offspring for every one agouti (phenotypic ratio of 2 : 1). However, the genotypic ratio is a 1 : 1 : 1 relationship. That is, $\frac{1}{3}$ of the viable genotypes should be $A^yA$, $\frac{1}{3}A^ya$, and $\frac{1}{3}Aa$.

**2.10.** A man is suing his wife for divorce on the grounds of infidelity. Their first and second child, whom they both claim, are blood groups O and AB respectively. The third child, whom the man disclaims, is blood type B. (*a*) Can this information be used to support the man's case? (*b*) Another test was made in the M-N blood group system. The third child was group M, the man was group N. Can this information be used to support the man's case?

**Solution:**

(*a*) The genetics of the ABO blood group system was presented in Example 2.15, page 23. Because the group O baby has the genotype $ii$, each of the parents must have been carrying the recessive allele. The AB baby indicates that one of his parents had the dominant $I^A$ allele and the other had the codominant allele $I^B$. Any of the four blood groups can appear among the children whose parents are $I^Ai \times I^Bi$. The information given on ABO blood groups is of no use in supporting the man's claim.

(*b*) The genetics of the M-N blood group system was presented in Example 2.9, page 21. The M-N blood groups are governed by a pair of codominant alleles, where groups M and N are produced by homozygous genotypes. A group N father must pass the $L^N$ allele to his offspring; they all would have the N antigen on their red blood cells, and would all be classified serologically as either group MN or N depending upon the genotype of the mother. This man could not be the father of a group M child.

## PEDIGREE ANALYSIS

**2.11.** The black hair of guinea pigs is produced by a dominant gene $B$ and white by its recessive allele $b$. Unless there is evidence to the contrary, assume that II1 and II4 do not carry the recessive allele. Calculate the probability that an offspring of III1 $\times$ III2 will have white hair.

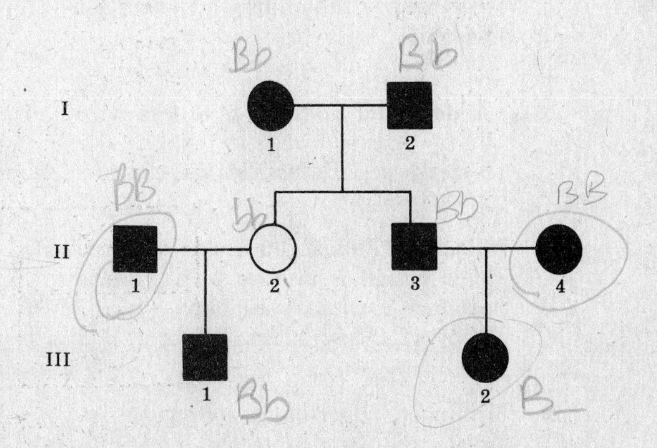


**Solution:**

Both I1 and I2 must be heterozygous ($Bb$) in order to have the white ($bb$) offspring II2. If III1 or III2 had been white, this would constitute evidence that II1 or II4 were heterozygous. In the absence of this evidence the problem tells us to assume that II1 and II4 are homozygous (*BB*). If the offspring of III1 × III2 is to be white, then both III1 and III2 would have to be heterozygous (*Bb*). In this case, II3 would also have to be heterozygous in order to pass the recessive allele on to III2. Under the conditions of the problem, we are certain that III1 is heterozygous because his parents (II1 × II2) are *BB* × *bb*. We notice that II3 is black. The probability that *black* progeny from I1 × I2 are heterozygous is $\frac{2}{3}$. If II3 is heterozygous, the probability that III2 is heterozygous is $\frac{1}{2}$. If III2 is heterozygous, there is a 25% chance that the offspring of III1 × III2 will be white (*bb*). Thus the combined probability that II3 is heterozygous *and* III2 is heterozygous *and* producing a white offspring is the product of the independent probabilities = $(\frac{2}{3})(\frac{1}{2})(\frac{1}{4}) = \frac{2}{24} = \frac{1}{12}$.

# Supplementary Problems

## DOMINANT AND RECESSIVE ALLELES

**2.12.** Several black guinea pigs of the same genotype were mated and produced 29 black and 9 white offspring. What would you predict the genotypes of the parents to be?

**2.13.** If a black female guinea pig is testcrossed and produces at least one white offspring, determine (*a*) the genotype and phenotype of the sire (male parent) which produced the white offspring, (*b*) the genotype of this female.

**2.14.** Heterozygous black (*Bb*) guinea pigs are mated to homozygous recessive (*bb*) whites. Predict the genotypic and phenotypic ratios expected from backcrossing the black $F_1$ progeny to (*a*) the black parent, (*b*) the white parent.

**2.15.** In *Drosophila*, sepia colored eyes are due to a recessive allele *s* and wild type (red eye color) to its dominant allele $s^+$. If sepia eyed females are crossed to pure wild type males, what phenotypic and genotypic ratios are expected if the $F_2$ males are backcrossed to the sepia eyed parental females?

**2.16.** The lack of pigmentation, called albinism, in humans is the result of a recessive allele (*a*) and normal pigmentation is the result of its dominant allele (*A*). Two normal parents have an albino child. Determine the probability that (*a*) the next child is albino, (*b*) the next two children are albinos. (*c*) What is the chance of these parents producing two children, one albino and the other normal?

**2.17.** Short hair is due to a dominant gene *L* in rabbits, and long hair to its recessive allele *l*. A cross between a short haired female and a long haired male produced a litter of 1 long haired and 7 short haired bunnies. (*a*) What are the genotypes of the parents? (*b*) What phenotypic ratio was expected in the offspring generation? (*c*) How many of the eight bunnies were expected to be long haired?

**2.18.** A dominant gene *W* produces wire-haired texture in dogs; its recessive allele *w* produces smooth hair. A group of heterozygous wire-haired individuals are crossed and their $F_1$ progeny are then testcrossed. Determine the expected genotypic and phenotypic ratios among the testcross progeny.

**2.19.** Black wool of sheep is due to a recessive allele *b* and white wool to its dominant allele *B*. A white buck (male) is crossed to a white ewe (female), both animals carrying the allele for black. They produce a white buck lamb which is then backcrossed to the female parent. What is the probability of the backcross offspring being black?

**2.20.** In foxes, silver-black coat color is governed by a recessive allele *b* and red color by its dominant allele *B*. Determine the genotypic and phenotypic ratios expected from the following matings: (*a*) pure red × carrier red, (*b*) carrier red × silver-black, (*c*) pure red × silver-black.

**2.21.** In the Holstein-Friesian breed of dairy cattle, a recessive allele $r$ is known to produce red and white; the dominant allele $R$ is known to produce black and white. If a carrier bull is mated to carrier cows, determine the probability (*a*) of the first offspring being born red and white, (*b*) of the first four offspring born being black and white. (*c*) What is the expected phenotypic ratio among offspring resulting from backcrossing black and white $F_1$ cows to the carrier bull? (*d*) If the carrier bull was mated to homozygous black and white cows, what phenotypic ratio would be expected among the backcross progeny from $F_1$ cows $\times$ carrier bull?

**2.22.** Consider a cross between two heterozygous black guinea pigs ($Bb$). (*a*) In how many ways can three black and two white offspring be produced? (*b*) What is the probability from such a cross of three black and two white offspring appearing in any order?

## CODOMINANT ALLELES

**2.23.** When chickens with splashed white feathers are crossed with black feathered birds, their offspring are all slate blue (Blue Andalusian). When Blue Andalusians are crossed among themselves, they produce splashed white, blue, and black offspring in the ratio of 1 : 2 : 1 respectively. (*a*) How are these feather traits inherited? (*b*) Using any appropriate symbols, indicate the genotypes for each phenotype.

**2.24.** Yellow coat color in guinea pigs is produced by the homozygous genotype $C^YC^Y$, cream color by the heterozygous genotype $C^YC^W$, and white by the homozygous genotype $C^WC^W$. What genotypic and phenotypic ratios are matings between cream colored individuals likely to produce?

**2.25.** The shape of radishes may be long ($S^LS^L$), round ($S^RS^R$), or oval ($S^LS^R$). If long radishes are crossed to oval radishes and the $F_1$ then allowed to cross at random among themselves, what phenotypic ratio is expected in the $F_2$?

**2.26.** The Palomino horse is a hybrid exhibiting a golden color with lighter mane and tail. A pair of codominant alleles ($D^1$ and $D^2$) is known to be involved in the inheritance of these coat colors. Genotypes homozygous for the $D^1$ allele are chestnut colored (reddish), heterozygous genotypes are Palomino colored, and genotypes homozygous for the $D^2$ allele are almost white and called *cremello*. (*a*) From matings between Palominos, determine the expected Palomino : non-Palomino ratio among the offspring. (*b*) What percentage of the non-Palomino offspring in part (*a*) will breed true? (*c*) What kind of mating will produce only Palominos?

## LETHAL ALLELES

**2.27.** Chickens with shortened wings and legs are called creepers. When creepers are mated to normal birds they produce creepers and normals with equal frequency. When creepers are mated to creepers they produce 2 creepers to 1 normal. Crosses between normal birds produce only normal progeny. How can these results be explained?

**2.28.** In the Mexican Hairless breed of dogs, the hairless condition is produced by the heterozygous genotype ($Hh$). Normal dogs are homozygous recessive ($hh$). Puppies homozygous for the $H$ allele are usually born dead with abnormalities of the mouth and absence of external ears. If the average litter size at weaning is 6 in matings between hairless dogs, what would be the average expected *number* of hairless and normal offspring at weaning from matings between hairless and normal dogs?

**2.29.** A pair of codominant alleles is known to govern cotyledon leaf color in soybeans. The homozygous genotype $C^GC^G$ produces dark green, the heterozygous genotype $C^GC^Y$ produces light green, and the other homozygous genotype $C^YC^Y$ produces yellow leaves so deficient in chloroplasts that seedlings do not grow to maturity. If dark green plants are pollinated only by light green plants and the $F_1$ crosses are made at random to produce an $F_2$, what phenotypic and genotypic ratios would be expected in the mature $F_2$ plants?

**2.30.** Thalassemia is a hereditary disease of the blood of humans resulting in anemia. Severe anemia (thalassemia major) is found in homozygotes ($T^MT^M$) and a milder form of anemia (thalassemia minor) is found in heterozygotes ($T^MT^N$). Normal individuals are homozygous $T^NT^N$. If all individuals with thalassemia major die before sexual maturity, (*a*) what proportion of the adult $F_1$ from marriages of thalassemia minors by normals would be expected to be normal, (*b*) what fraction of the adult $F_1$ from marriages of minors by minors would be expected to be anemic?

**2.31.** The Pelger anomaly of rabbits involves abnormal white blood cell nuclear segmentation. Pelgers are heterozygous ($Pp$), normal individuals are homozygous ($PP$). The homozygous recessive genotypes ($pp$) have grossly deformed skeletons and usually die before or soon after birth. If Pelgers are mated together, what phenotypic ratio is expected in the adult $F_2$?

## MULTIPLE ALLELES

**2.32.** A multiple allelic series is known in the Chinese primrose where $A$ (Alexandria type = white eye) $> a^n$ (normal type = yellow eye) $> a$ (Primrose Queen type = large yellow eye). List all of the genotypes possible for each of the phenotypes in this series.

**2.33.** Plumage color in mallard ducks is dependent upon a set of three alleles: $M^R$ for restricted mallard pattern, $M$ for mallard, and $m$ for dusky mallard. The dominance hierarchy is $M^R > M > m$. Determine the genotypic and phenotypic ratios expected in the $F_1$ from the following crosses (*a*) $M^RM^R \times M^RM$, (*b*) $M^RM^R \times M^Rm$, (*c*) $M^RM \times M^Rm$, (*d*) $M^Rm \times Mm$, (*e*) $Mm \times mm$.

**2.34.** A number of self-incompatibility alleles is known in clover such that the growth of a pollen tube down the style of a diploid plant is inhibited when the latter contains the same self-incompatibility allele as that in the pollen tube. Given a series of self-incompatibility alleles $S^1, S^2, S^3, S^4$, what genotypic ratios would be expected in embryos and in endosperms of seeds from the following crosses?

| | Seed Parent | Pollen Parent |
|---|---|---|
| (*a*) | $S^1S^4$ | $S^3S^4$ |
| (*b*) | $S^1S^2$ | $S^1S^2$ |
| (*c*) | $S^1S^3$ | $S^2S^4$ |
| (*d*) | $S^2S^3$ | $S^3S^4$ |

**2.35.** The coat colors of many animals exhibit the "agouti" pattern which is characterized by a yellow band of pigment near the tip of the hair. In rabbits, a multiple allelic series is known where the genotypes $E^DE^D$ and $E^De$ produce only black (non-agouti), but the heterozygous genotype $E^DE$ produces black with a trace of agouti. The genotypes $EE$ or $Ee$ produce full color, and the recessive genotype $ee$ produces reddish-yellow. What phenotypic and genotypic ratios would be expected in the $F_1$ and $F_2$ from the cross (*a*) $E^DE^D \times Ee$, (*b*) $E^De \times ee$?

**2.36.** The inheritance of coat colors of cattle involves a multiple allelic series with a dominance hierarchy as follows: $S > s^h > s^c > s$. The $S$ allele puts a band of white color around the middle of the animal and is referred to as a Dutch belt; the $s^h$ allele produces Hereford-type spotting; solid color is a result of the $s^c$ allele; and Holstein-type spotting is due to the $s$ allele. Homozygous Dutch-belted males are crossed to Holstein-type spotted females. The $F_1$ females are crossed to a Hereford-type spotted male of genotype $s^hs^c$. Predict the genotypic and phenotypic frequencies in the progeny.

**2.37.** The genetics of the ABO human blood groups was presented in Example 2.15, page 23. A man of blood group B is being sued by a woman of blood group A for paternity. The woman's child is blood group O. (*a*) Is this man the father of this child? Explain. (*b*) If this man actually is the father of this child, specify the genotypes of both parents. (*c*) If it was impossible for this group B man to be the father of a type O child, regardless of the mother's genotype, specify his genotype. (*d*) If a man was blood group AB, could he be the father of a group O child?

**2.38.** A multiple allelic series is known to govern the intensity of pigmentation in the mouse such that $D$ = full color, $d$ = dilute color, and $d^l$ is lethal when homozygous. The dominance order is $D > d > d^l$. A full colored mouse carrying the lethal is mated to a dilute colored mouse also carrying the lethal. The $F_1$ is backcrossed to the dilute parent. (*a*) What phenotypic ratio is expected in the viable backcross progeny? (*b*) What percentage of the full colored backcross progeny carry the lethal? (*c*) What fraction of the dilute colored progeny carry the lethal?

## PEDIGREE ANALYSIS

**2.39.** The phenotypic expression of a dominant gene in Ayrshire cattle is a notch in the tips of the ears. In the pedigree below, where solid symbols represent notched individuals, determine the probability of notched progeny being produced from the matings (*a*) III1 × III3, (*b*) III2 × III3, (*c*) III3 × III4, (*d*) III1 × III5, (*e*) III2 × III5.

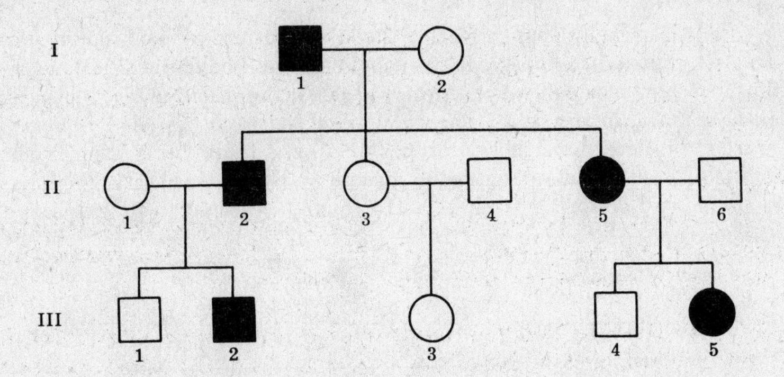


**2.40.** A single recessive gene $r$ is largely responsible for the development of red hair in humans. Dark hair is largely due to its dominant allele $R$. In the family pedigree shown below, unless there is evidence to the contrary, assume that individuals who marry into this family do not carry the $r$ allele. Calculate the probability of red hair appearing in children from the marriages (*a*) III3 × III9, (*b*) III4 × III10, (*c*) IV1 × IV2, (*d*) IV1 × IV3. Solid symbols represent red hair; open symbols represent dark hair.

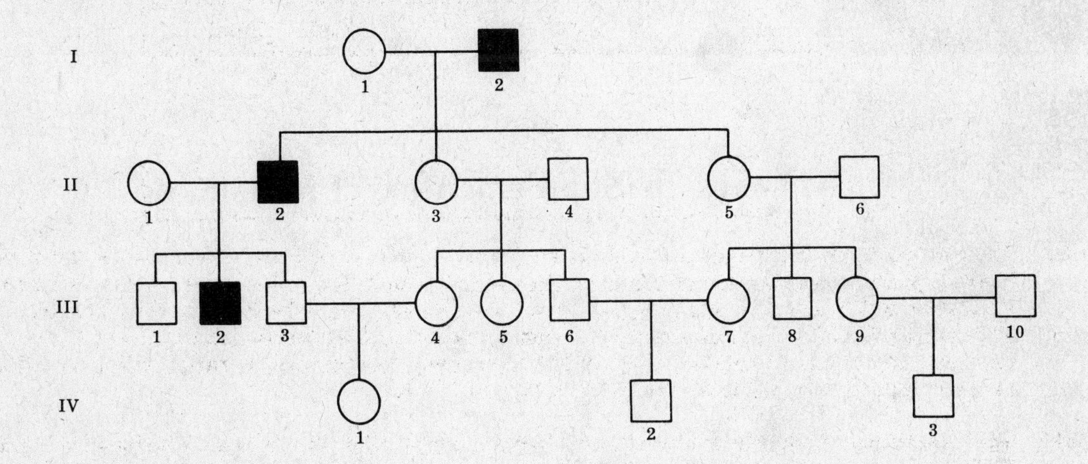


**2.41.** The gene for spotted coat color in rabbits ($S$) is dominant to its allele for solid color ($s$). In the following pedigree assume that those individuals brought into the family from outside do not carry the gene for solid color, unless there is evidence to the contrary. Calculate the probability of solid colored bunnies being produced from the matings (*a*) III1 × III9, (*b*) III1 × III5, (*c*) III3 × III5, (*d*) III4 × III6, (*e*) III6 × III9, (*f*) IV1 × IV2, (*g*) III9 × IV2, (*h*) III5 × IV2, (*i*) III6 × IV1. Solid symbols represent solid colored animals, open symbols represent spotted animals.

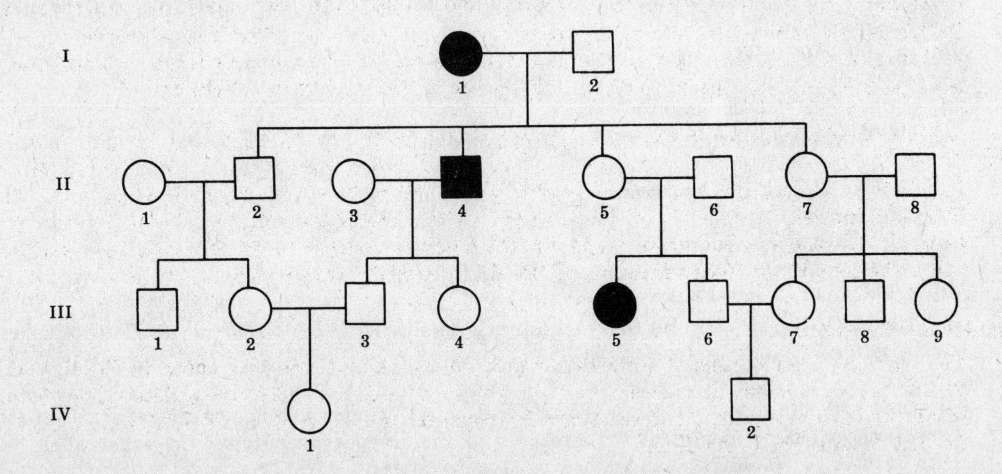

**2.42.** A multiple allelic series in dogs governs the distribution of coat color pigments. The allele $A^s$ produces an even distribution of dark pigment over the body; the allele $a^y$ reduces the intensity of pigmentation and produces sable or tan colored dogs; the allele $a^t$ produces spotted patterns such as tan and black, tan and brown, etc. The dominance hierarchy is $A^s > a^y > a^t$. Given the following family pedigree, (*a*) determine the genotypes of all the individuals insofar as possible, (*b*) calculate the probability of spotted offspring being produced by mating III1 by III2, (*c*) find the fraction of the dark pigmented offspring from I1 × II3 that is expected to be heterozygous.

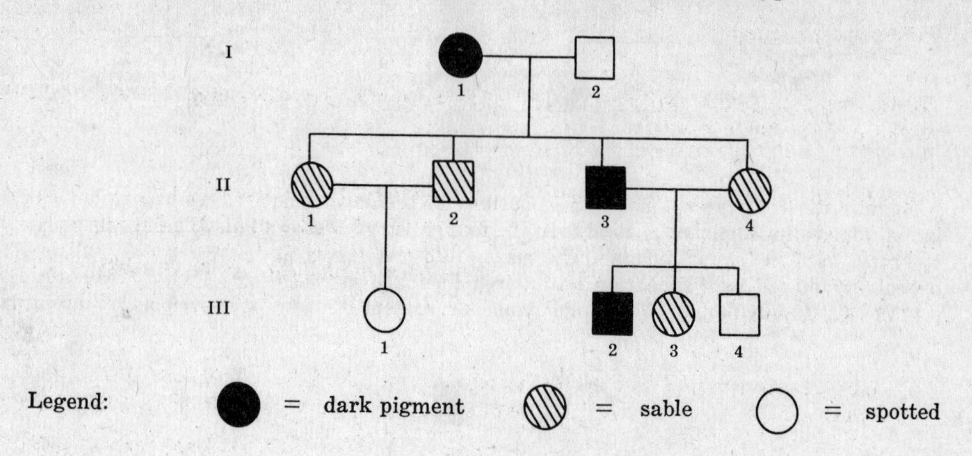


# Answers to Supplementary Problems

**2.12.** $Bb \times Bb$

**2.13.** (*a*) $bb$ = white, (*b*) $Bb$

**2.14.** (*a*) $\frac{1}{4}BB : \frac{1}{2}Bb : \frac{1}{4}bb$, $\frac{3}{4}$ black : $\frac{1}{4}$ white; (*b*) $\frac{1}{2}Bb$ = black : $\frac{1}{2}bb$ = white

**2.15.** $\frac{1}{2}$ wild type $s^+s$ : $\frac{1}{2}$ sepia $ss$

**2.16.** (*a*) $\frac{1}{4}$, (*b*) $\frac{1}{16}$, (*c*) $2(\frac{3}{4} \times \frac{1}{4}) = \frac{3}{8}$

**2.17.** (*a*) $Ll$ female × $ll$ male, (*b*) 1 short : 1 long, (*c*) 4

**2.18.** $\frac{1}{2}Ww$ = wire-haired : $\frac{1}{2}ww$ = smooth

**2.19.** $\frac{1}{6}$

**2.20.** (*a*) $\frac{1}{2}BB : \frac{1}{2}Bb$, all red; (*b*) $\frac{1}{2}Bb$ = red : $\frac{1}{2}bb$ = silver-black; (*c*) all $Bb$ = red

**2.21.** (*a*) $\frac{1}{4}$, (*b*) $\frac{81}{256}$, (*c*) $\frac{5}{6}$ black and white : $\frac{1}{6}$ red and white, (*d*) 7 black and white : 1 red and white

**2.22.** (*a*) 10, (*b*) $10(\frac{3}{4})^3(\frac{1}{4})^2 = \frac{135}{512}$

**2.23.** (*a*) Single pair of codominant alleles
(*b*) $F^SF^S$ = splashed-white : $F^SF^B$ = Blue Andalusian : $F^BF^B$ = black

**2.24.** $\frac{1}{4}C^YC^Y$ = yellow : $\frac{1}{2}C^YC^W$ = cream : $\frac{1}{4}C^WC^W$ = white

**2.25.** $\frac{9}{16}$ long : $\frac{6}{16}$ oval : $\frac{1}{16}$ round

**2.26.** (*a*) 1 Palomino : 1 non-Palomino
(*b*) 100%, $D^1D^1 \times D^1D^1$ = all $D^1D^1$ (chestnut); similarly $D^2D^2 \times D^2D^2$ = all $D^2D^2$ (cremello)
(*c*) $D^1D^1$ (chestnut) × $D^2D^2$ (cremello)

**2.27.** Creepers are heterozygous. Normal birds and lethal zygotes are homozygous for alternative alleles. One of the alleles is dominant with respect to the creeper phenotype; the other allele is dominant with respect to viability.

**2.28.** 4 normal : 4 hairless

**2.29.** $\frac{9}{15}$ dark green $C^GC^G$ : $\frac{6}{15}$ light green $C^GC^Y$

**2.30.** (*a*) $\frac{1}{2}$, (*b*) $\frac{2}{3}$

**2.31.** $\frac{1}{2}$ Pelger : $\frac{1}{2}$ normal

**2.32.** Alexandria type (white eye) = $AA$, $Aa^n$, $Aa$; normal type (yellow eye) = $a^na^n$, $a^na$; Primrose Queen type (large yellow eye) = $aa$

**2.33.** (*a*) $\frac{1}{2}M^RM^R : \frac{1}{2}M^RM$; all restricted

(*b*) $\frac{1}{2}M^RM^R : \frac{1}{2}M^Rm$; all restricted

(*c*) $\frac{1}{4}M^RM^R : \frac{1}{4}M^Rm : \frac{1}{4}M^RM : \frac{1}{4}Mm$; $\frac{3}{4}$ restricted : $\frac{1}{4}$ mallard

(*d*) $\frac{1}{4}M^RM : \frac{1}{4}M^Rm : \frac{1}{4}Mm : \frac{1}{4}mm$; $\frac{1}{2}$ restricted : $\frac{1}{4}$ mallard : $\frac{1}{4}$ dusky

(*e*) $\frac{1}{2}Mm$ = mallard : $\frac{1}{2}mm$ = dusky

**2.34.** (*a*) embryos = $\frac{1}{2}S^1S^3 : \frac{1}{2}S^3S^4$ endosperms = $\frac{1}{2}S^1S^1S^3 : \frac{1}{2}S^4S^4S^3$

(*b*) none

(*c*) embryos = $\frac{1}{4}S^1S^2 : \frac{1}{4}S^1S^4 : \frac{1}{4}S^3S^2 : \frac{1}{4}S^3S^4$; endosperms = $\frac{1}{4}S^1S^1S^2 : \frac{1}{4}S^1S^1S^4 : \frac{1}{4}S^3S^3S^2 : \frac{1}{4}S^3S^3S^4$

(*d*) embryos = $\frac{1}{2}S^2S^4 : \frac{1}{2}S^3S^4$, endosperms = $\frac{1}{2}S^2S^2S^4 : \frac{1}{2}S^3S^3S^4$

**2.35.** (*a*) $F_1$ = $\frac{1}{2}E^DE$ (black with trace of agouti) : $\frac{1}{2}E^De$ (non-agouti black);
$F_2$ = $\frac{1}{4}E^DE^D : \frac{1}{4}E^De : \frac{1}{4}E^DE : \frac{1}{16}EE : \frac{1}{8}Ee : \frac{1}{16}ee$;
$\frac{1}{2}$ non-agouti black : $\frac{1}{4}$ black with trace of agouti : $\frac{3}{16}$ full color : $\frac{1}{16}$ reddish-yellow

(*b*) $F_1$ = $\frac{1}{2}E^De$ (non-agouti black) : $\frac{1}{2}ee$ (reddish-yellow);
$F_2$ = $\frac{1}{16}E^DE^D : \frac{3}{8}E^De : \frac{9}{16}ee$; $\frac{7}{16}$ non-agouti black : $\frac{9}{16}$ reddish-yellow

**2.36.** $\frac{1}{4}Ss^h : \frac{1}{4}Ss^c : \frac{1}{4}s^hs : \frac{1}{4}s^cs$; $\frac{1}{2}$ Dutch-belted : $\frac{1}{4}$ Hereford-type spotting : $\frac{1}{4}$ solid color

**2.37.** (*a*) The man could be the father, but paternity cannot be *proved* by blood type. In certain cases, a man may be *excluded* as a father of a child (see part (*d*)). (*b*) $I^Bi$ man × $I^Ai$ woman, (*c*) $I^BI^B$, (*d*) No

**2.38.** (*a*) $\frac{2}{5}$ full color : $\frac{3}{5}$ dilute, (*b*) 50%, (*c*) $\frac{2}{3}$

**2.39.** (*a*) 0, (*b*) $\frac{1}{2}$, (*c*) 0, (*d*) $\frac{1}{2}$, (*e*) $\frac{3}{4}$

**2.40.** (*a*) $\frac{1}{8}$, (*b*) 0, (*c*) $\frac{35}{576}$, (*d*) $\frac{7}{192}$

**2.41.** (*a*) $\frac{1}{16}$, (*b*) $\frac{1}{4}$, (*c*) $\frac{1}{2}$, (*d*) $\frac{1}{6}$, (*e*) $\frac{1}{12}$, (*f*) $\frac{119}{1728}$, (*g*) $\frac{17}{288}$, (*h*) $\frac{17}{72}$, (*i*) $\frac{7}{72}$

**2.42.** (*a*) I1 = $A^sa^y$, I2 = $a^ta^t$, II1 = $a^ya^t$, II2 = $a^ya^t$, II3 = $A^sa^t$, II4 = $a^ya^t$, III1 = $a^ta^t$, III2 = $A^s$-($A^sa^y$ or $A^sa^t$), III3 = $a^ya^t$, III4 = $a^ta^t$, (*b*) $\frac{1}{4}$, (*c*) $\frac{2}{3}$

# Chapter 3

## Two or More Genes

### INDEPENDENT ASSORTMENT

In this chapter we shall consider simultaneously two or more traits, each specified by a different pair of independently assorting autosomal genes, i.e. genes on different chromosomes other than the sex chromosomes.

**Example 3.1.** In addition to the coat color locus of guinea pigs introduced in Chapter 2 ($B$- = black, $bb$ = white), another locus on a different chromosome (independently assorting) is known to govern length of hair, such that $L$- = short hair and $ll$ = long hair. Any of 4 different genotypes exist for the black, short-haired phenotype: $BBLL$, $BBLl$, $BbLL$, $BbLl$. Two different genotypes produce a black, long-haired pig: $BBll$ or $Bbll$; likewise 2 genotypes for a white, short-haired pig: $bbLL$ or $bbLl$; and only 1 genotype specifies a white, long-haired pig: $bbll$.

A *dihybrid* genotype is heterozygous at two loci. Dihybrids form four genetically different gametes with approximately equal frequencies because of the random orientation of non-homologous chromosome pairs on the first meiotic metaphase plate (Chapter 1).

**Example 3.2.** A dihybrid black, short-haired guinea pig ($BbLl$) produces 4 types of gametes in equal frequencies.

| | | | Gametes | Frequency |
|---|---|---|---|---|
| $B$ | $L$ | = | $BL$ | $\frac{1}{4}$ |
| | $l$ | = | $Bl$ | $\frac{1}{4}$ |
| $b$ | $L$ | = | $bL$ | $\frac{1}{4}$ |
| | $l$ | = | $bl$ | $\frac{1}{4}$ |

A summary of the gametic output for all 9 genotypes involving two pairs of independently assorting factors is shown below.

| Genotypes | Gametes in Relative Frequencies |
|---|---|
| $BBLL$ | all $BL$ |
| $BBLl$ | $\frac{1}{2}BL : \frac{1}{2}Bl$ |
| $BBll$ | all $Bl$ |
| $BbLL$ | $\frac{1}{2}BL : \frac{1}{2}bL$ |
| $BbLl$ | $\frac{1}{4}BL : \frac{1}{4}Bl : \frac{1}{4}bL : \frac{1}{4}bl$ |
| $Bbll$ | $\frac{1}{2}Bl : \frac{1}{2}bl$ |
| $bbLL$ | all $bL$ |
| $bbLl$ | $\frac{1}{2}bL : \frac{1}{2}bl$ |
| $bbll$ | all $bl$ |

A *testcross* is the mating of an incompletely known genotype to a genotype which is homozygous recessive at all of the loci under consideration. The phenotypes of the offspring produced by a testcross reveal the number of different gametes formed by the parental genotype under test. When all of the gametes of an individual are known, the genotype of that individual also becomes known. A monohybrid testcross gives a 1 : 1 phenotypic ratio, indicating that one pair of factors is segregating. A dihybrid testcross gives a 1 : 1 : 1 : 1 ratio, indicating that two pairs of factors are segregating and assorting independently.

**Example 3.3.** Testcrossing a dihybrid yields a 1:1:1:1 genotypic and phenotypic ratio among the progeny.

Parents: *BbLl* × *bbll*
black, short-haired × white, long-haired

$F_1$:
$\frac{1}{4}$*BbLl* black, short-haired
$\frac{1}{4}$*Bbll* black, long-haired
$\frac{1}{4}$*bbLl* white, short-haired
$\frac{1}{4}$*bbll* white, long-haired

## SYSTEMS FOR SOLVING DIHYBRID CROSSES

### 1. Gametic Checkerboard Method.

When two dihybrids are crossed, four kinds of gametes are produced in equal frequencies in both the male and female. A 4×4 gametic checkerboard can be used to show all 16 possible combinations of these gametes. This method is laborious, time-consuming, and offers more opportunities for error than the other methods which follow.

**Example 3.4.** P: *BBLL* × *bbll*
black, short × white, long

$F_1$: *BbLl* = black, short

$F_2$:

| Female Gametes | Male Gametes: *BL* | *Bl* | *bL* | *bl* |
|---|---|---|---|---|
| *BL* | *BBLL* black short | *BBLl* black short | *BbLL* black short | *BbLl* black short |
| *Bl* | *BBLl* black short | *BBll* black long | *BbLl* black short | *Bbll* black long |
| *bL* | *BbLL* black short | *BbLl* black short | *bbLL* white short | *bbLl* white short |
| *bl* | *BbLl* black short | *Bbll* black long | *bbLl* white short | *bbll* white long |

$F_2$ Summary:

| Proportions | Genotypes |
|---|---|
| $\frac{1}{16}$ | *BB LL* |
| $\frac{1}{8}$ | *BB Ll* |
| $\frac{1}{16}$ | *BB ll* |
| $\frac{1}{8}$ | *Bb LL* |
| $\frac{1}{4}$ | *Bb Ll* |
| $\frac{1}{8}$ | *Bb ll* |
| $\frac{1}{16}$ | *bb LL* |
| $\frac{1}{8}$ | *bb Ll* |
| $\frac{1}{16}$ | *bb ll* |

| Proportions | Phenotypes |
|---|---|
| $\frac{9}{16}$ | black, short |
| $\frac{3}{16}$ | black, long |
| $\frac{3}{16}$ | white, short |
| $\frac{1}{16}$ | white, long |

### 2. Genotypic and Phenotypic Checkerboard Methods.

A knowledge of the monohybrid probabilities presented in Chapter 2 may be applied in a simplified genotypic or phenotypic checkerboard.

**Example 3.5.** Genotypic checkerboard.

$F_1$: *BbLl* (black, short) × *BbLl* (black, short)

Considering only the B locus, $Bb \times Bb$ produces $\frac{1}{4}BB$, $\frac{1}{2}Bb$ and $\frac{1}{4}bb$. Likewise for the L locus, $Ll \times Ll$ produces $\frac{1}{4}LL$, $\frac{1}{2}Ll$ and $\frac{1}{4}ll$. Let us place these genotypic probabilities in a checkerboard and combine independent probabilities by multiplication.

$F_2$:

| | $\frac{1}{4}LL$ | $\frac{1}{2}Ll$ | $\frac{1}{4}ll$ |
|---|---|---|---|
| $\frac{1}{4}BB$ | $\frac{1}{16}BBLL$ | $\frac{1}{8}BBLl$ | $\frac{1}{16}BBll$ |
| $\frac{1}{2}Bb$ | $\frac{1}{8}BbLL$ | $\frac{1}{4}BbLl$ | $\frac{1}{8}Bbll$ |
| $\frac{1}{4}bb$ | $\frac{1}{16}bbLL$ | $\frac{1}{8}bbLl$ | $\frac{1}{16}bbll$ |

**Example 3.6.** Phenotypic checkerboard.

$F_1$: *BbLl* (black, short) × *BbLl* (black, short)

Considering the B locus, $Bb \times Bb$ produces $\frac{3}{4}$ black and $\frac{1}{4}$ white. Likewise at the L locus, $Ll \times Ll$ produces $\frac{3}{4}$ short and $\frac{1}{4}$ long. Let us place these independent phenotypic probabilities in a checkerboard and combine them by multiplication.

$F_2$:

| | $\frac{3}{4}$ black | $\frac{1}{4}$ white |
|---|---|---|
| $\frac{3}{4}$ short | $\frac{9}{16}$ black, short | $\frac{3}{16}$ white, short |
| $\frac{1}{4}$ long | $\frac{3}{16}$ black, long | $\frac{1}{16}$ white, long |

## 3. Branching Systems.

This procedure was introduced in Chapter 1 as a means for determining all possible ways in which any number of chromosome pairs could orient themselves on the first meiotic metaphase plate. It can also be used to find all possible genotypic or phenotypic combinations. It will be the method of choice for solving most examples in this and subsequent chapters.

**Example 3.7.** Genotypic trichotomy.

| | | | Ratio | Genotypes |
|---|---|---|---|---|
| $\frac{1}{4}BB$ | $\frac{1}{4}LL$ | = | $\frac{1}{16}$ | *BB LL* |
| | $\frac{1}{2}Ll$ | = | $\frac{1}{8}$ | *BB Ll* |
| | $\frac{1}{4}ll$ | = | $\frac{1}{16}$ | *BB ll* |
| $\frac{1}{2}Bb$ | $\frac{1}{4}LL$ | = | $\frac{1}{8}$ | *Bb LL* |
| | $\frac{1}{2}Ll$ | = | $\frac{1}{4}$ | *Bb Ll* |
| | $\frac{1}{4}ll$ | = | $\frac{1}{8}$ | *Bb ll* |
| $\frac{1}{4}bb$ | $\frac{1}{4}LL$ | = | $\frac{1}{16}$ | *bb LL* |
| | $\frac{1}{2}Ll$ | = | $\frac{1}{8}$ | *bb Ll* |
| | $\frac{1}{4}ll$ | = | $\frac{1}{16}$ | *bb ll* |

**Example 3.8.** Phenotypic dichotomy.

| | | | Ratio | Phenotypes |
|---|---|---|---|---|
| $\frac{3}{4}$ black | $\frac{3}{4}$ short | = | $\frac{9}{16}$ | black, short |
| | $\frac{1}{4}$ long | = | $\frac{3}{16}$ | black, long |
| $\frac{1}{4}$ white | $\frac{3}{4}$ short | = | $\frac{3}{16}$ | white, short |
| | $\frac{1}{4}$ long | = | $\frac{1}{16}$ | white, long |

If only one of the genotypic frequencies or phenotypic frequencies is required, there is no need to be concerned with any other genotypes or phenotypes. A mathematical solution can be readily obtained by combining independent probabilities.

**Example 3.9.** To find the frequency of genotype *BBLl* in the offspring of dihybrid parents, first consider each locus separately: $Bb \times Bb = \frac{1}{4}BB$; $Ll \times Ll = \frac{1}{2}Ll$. Combining these independent probabilities, $\frac{1}{4} \times \frac{1}{2} = \frac{1}{8}BBLl$.

**Example 3.10.** To find the frequency of white, short pigs in the offspring of dihybrid parents, first consider each trait separately: $Bb \times Bb = \frac{1}{4}$ white ($bb$); $Ll \times Ll = \frac{3}{4}$ short ($L$-). Combining these independent probabilities, $\frac{1}{4} \times \frac{3}{4} = \frac{3}{16}$ white, short.

## MODIFIED DIHYBRID RATIOS

The classical phenotypic ratio resulting from the mating of dihybrid genotypes is 9:3:3:1. This ratio appears whenever the alleles at both loci display dominant and recessive relationships. The classical dihybrid ratio may be modified if one or both loci have codominant alleles or lethal alleles. A summary of these modified phenotypic ratios in adult progeny is shown below.

| Allelic Relationships in Dihybrid Parents | | Expected Adult Phenotypic Ratio |
|---|---|---|
| First Locus | Second Locus | |
| dominant-recessive | codominants | 3:6:3:1:2:1 |
| codominants | codominants | 1:2:1:2:4:2:1:2:1 |
| dominant-recessive | lethal* | 3:1:6:2 |
| codominant | lethal* | 1:2:1:2:4:2 |
| lethal* | lethal* | 4:2:2:1 |

*Lethal gene is incompletely recessive (see Example 2.11).

## HIGHER COMBINATIONS

The methods for solving two factor crosses may easily be extended to solve problems involving three or more pairs of independently assorting autosomal factors. Given any number of heterozygous pairs of factors ($n$) in the $F_1$, the following general formulas apply.

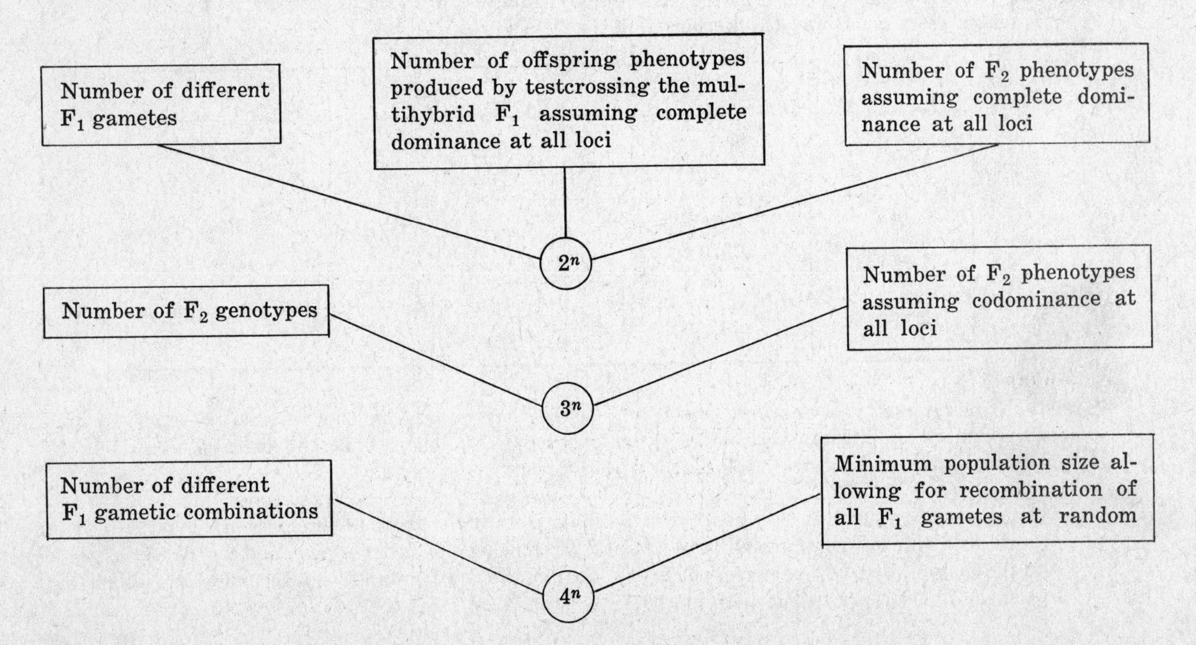

## Solved Problems

**3.1.** Black coat color in Cocker Spaniels is governed by a dominant allele $B$ and red coat color by its recessive allele $b$; solid pattern is governed by the dominant allele of an independently assorting locus $S$, and spotted pattern by its recessive allele $s$. A solid black male is mated to a solid red female and produces a litter of six pups: two solid black, two solid red, one black and white, and one red and white. Determine the genotypes of the parents.

**Solution:**

An unknown portion of a genotype will be indicated by a dash (-).

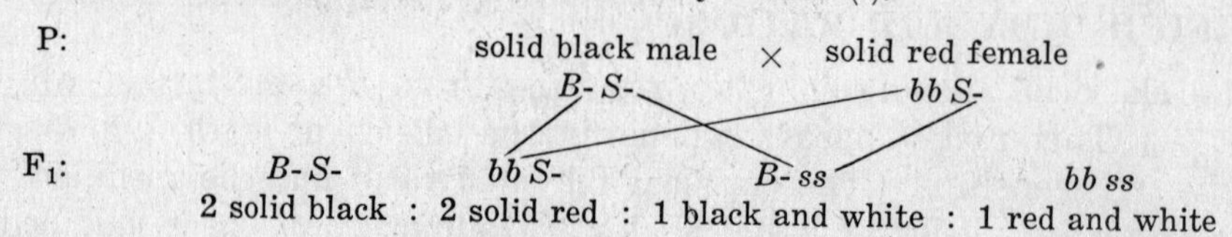


Whenever a homozygous double recessive progeny appears (red and white in this case), each of the parents must have possessed at least one recessive allele at each locus. The black and white pup also indicates that both parents were heterozygous at the S-locus. The solid red pups likewise indicate that the male parent must have been heterozygous at the B-locus. The solid black pups fail to be of any help in determining the genotypes of these parents. Complete genotypes may now be written for both parents and for two of the pups.

P: solid black male × solid red female
*BbSs* *bbSs*

$F_1$: *BbS-* *bbS-* *Bbss* *bbss*
2 solid black : 2 solid red : 1 black and white : 1 red and white

**3.2.** How many different crosses may be made (*a*) from a single pair of factors, (*b*) from two pairs of factors, and (*c*) from any given number $n$ of pairs of factors?

**Solution:**

(*a*) All possible matings of the three genotypes produced by a single pair of factors may be represented in a genotypic checkerboard.

| | *AA* | *Aa* | *aa* |
|---|---|---|---|
| *AA* | *AA* × *AA* (1) | *AA* × *Aa* (2) | *AA* × *aa* (3) |
| *Aa* | *Aa* × *AA* (2) | *Aa* × *Aa* (4) | *Aa* × *aa* (5) |
| *aa* | *aa* × *AA* (3) | *aa* × *Aa* (5) | *aa* × *aa* (6) |

The symmetry of matings above and below the squares on the diagonal becomes obvious. The number of different crosses may be counted as follows: 3 in the first column, 2 in the second, and 1 in the third: $3 + 2 + 1 = 6$ different types of matings.

(*b*) There are $3^2 = 9$ different genotypes possible with two pairs of factors. If a nine by nine checkerboard were constructed, the same symmetry would exist above and below the squares on the diagonal as was shown in part (*a*). Again, we may count the different types of matings as an arithmetic progression from 9 to 1; $9+8+7+6+5+4+3+2+1 = 45$.

(c) The sum of any arithmetic progression of this particular type may be found by the formula $M = \frac{1}{2}(g^2 + g)$ where $M$ = number of different types of matings, and $g$ = number of genotypes possible with $n$ pairs of factors.

**3.3.** In the garden pea, Mendel found that yellow seed color was dominant to green ($Y > y$) and round seed shape was dominant to shrunken ($S > s$). (a) What phenotypic ratio would be expected in the $F_2$ from a cross of a pure yellow, round × green, shrunken? (b) What is the $F_2$ ratio of yellow : green and of round : shrunken?

**Solution:**

(a) P: *YY SS* (yellow, round) × *yy ss* (green, shrunken)

$F_1$: *Yy Ss* (yellow, round)

$F_2$:
$\frac{9}{16}$ *Y-S-* yellow, round
$\frac{3}{16}$ *Y-ss* yellow, shrunken
$\frac{3}{16}$ *yyS-* green, round
$\frac{1}{16}$ *yy ss* green, shrunken

(b) The ratio of yellow : green = ($\frac{9}{16}$ yellow, round + $\frac{3}{16}$ yellow, shrunken) : ($\frac{3}{16}$ green, round + $\frac{1}{16}$ green, shrunken) = 12 : 4 = 3 : 1. The ratio of round : shrunken = ($\frac{9}{16}$ yellow, round + $\frac{3}{16}$ green, round) : ($\frac{3}{16}$ yellow, shrunken + $\frac{1}{16}$ green, shrunken) = 12 : 4 = 3 : 1. Thus at each of the individual loci a 3 : 1 $F_2$ phenotypic ratio is observed, just as would be expected for a monohybrid cross.

**3.4.** Tall tomato plants are produced by the action of a dominant allele *D*, and dwarf plants by its recessive allele *d*. Hairy stems are produced by a dominant gene *H*, and hairless stems by its recessive allele *h*. A dihybrid tall, hairy plant is testcrossed. The $F_1$ progeny were observed to be 118 tall, hairy : 121 dwarf, hairless : 112 tall, hairless : 109 dwarf, hairy. (a) Diagram this cross. (b) What is the ratio of tall : dwarf; of hairy : hairless? (c) Are these two loci assorting independently of one another?

**Solution:**

(a) Parents: *Dd Hh* (tall, hairy) × *dd hh* (dwarf, hairless)

Gametes: (*DH*) (*Dh*) (*dH*) (*dh*) × (*dh*)

$F_1$:

| Genotypes | Number | Phenotypes |
|---|---|---|
| *Dd Hh* | 118 | tall, hairy |
| *Dd hh* | 112 | tall, hairless |
| *dd Hh* | 109 | dwarf, hairy |
| *dd hh* | 121 | dwarf, hairless |

Note that the observed numbers approximate a 1 : 1 : 1 : 1 phenotypic ratio.

(b) The ratio of tall : dwarf = (118 + 112) : (109 + 121) = 230 : 230 or 1 : 1 ratio. The ratio of hairy : hairless = (118 + 109) : (112 + 121) = 227 : 233 or approximately 1 : 1 ratio. Thus the testcross results for each locus individually approximate a 1 : 1 phenotypic ratio.

(c) Whenever the results of a testcross approximate a 1 : 1 : 1 : 1 ratio, it indicates that the two gene loci are assorting independently of each other in the formation of gametes. That is to say, all four types of gametes have an equal opportunity of being produced through the random orientation which non-homologous chromosomes assume on the first meiotic metaphase plate.

**3.5.** A dominant allele $L$ governs short hair in guinea pigs and its recessive allele $l$ governs long hair. Codominant alleles at an independently assorting locus specify hair color, such that $C^YC^Y$ = yellow, $C^YC^W$ = cream, and $C^WC^W$ = white. From matings between dihybrid short, cream pigs ($LlC^YC^W$), predict the phenotypic ratio expected in the progeny.

**Solution:**

$\frac{3}{4}L$- → $\frac{1}{4}C^YC^Y$ = $\frac{3}{16}L\text{-}C^YC^Y$ short, yellow
$\frac{3}{4}L$- → $\frac{1}{2}C^YC^W$ = $\frac{6}{16}L\text{-}C^YC^W$ short, cream
$\frac{3}{4}L$- → $\frac{1}{4}C^WC^W$ = $\frac{3}{16}L\text{-}C^WC^W$ short, white

$\frac{1}{4}ll$ → $\frac{1}{4}C^YC^Y$ = $\frac{1}{16}llC^YC^Y$ long, yellow
$\frac{1}{4}ll$ → $\frac{1}{2}C^YC^W$ = $\frac{2}{16}llC^YC^W$ long, cream
$\frac{1}{4}ll$ → $\frac{1}{4}C^WC^W$ = $\frac{1}{16}llC^WC^W$ long, white

Thus six phenotypes appear in the offspring in the ratio 3:6:3:1:2:1. The dash (-) in the genotypes $L$- indicates that either allele $L$ or $l$ may be present, both combinations resulting in a short-haired phenotype.

**3.6.** Normal leg size, characteristic of the Kerry type of cattle, is produced by the homozygous genotype $DD$. Short legged Dexter type cattle possess the heterozygous genotype $Dd$. The homozygous genotype $dd$ is lethal, producing grossly deformed stillbirths called "bulldog calves". The presence of horns in cattle is governed by the recessive allele of another gene locus $p$, the polled condition (absence of horns) being produced by its dominant allele $P$. In matings between polled Dexter cattle of genotype $DdPp$, what phenotypic ratio is expected in the adult progeny?

**Solution:**

P: $DdPp$ (Dexter, polled) × $DdPp$ (Dexter, polled)

$F_1$:

$\frac{1}{4}DD$ → $\frac{3}{4}P$- = $\frac{3}{16}DD\,P$- polled, Kerry
$\frac{1}{4}DD$ → $\frac{1}{4}pp$ = $\frac{1}{16}DD\,pp$ horned, Kerry

$\frac{1}{2}Dd$ → $\frac{3}{4}P$- = $\frac{6}{16}Dd\,P$- polled, Dexter
$\frac{1}{2}Dd$ → $\frac{1}{4}pp$ = $\frac{2}{16}Dd\,pp$ horned, Dexter

$\frac{1}{4}dd$ → $\frac{3}{4}P$- = $\frac{3}{16}dd\,P$- lethal
$\frac{1}{4}dd$ → $\frac{1}{4}pp$ = $\frac{1}{16}dd\,pp$ lethal

The phenotypic ratio of viable offspring thus becomes: $\frac{3}{12}$ polled, Kerry; $\frac{1}{12}$ horned, Kerry; $\frac{6}{12}$ polled, Dexter; $\frac{2}{12}$ horned, Dexter.

**3.7.** Stem color of tomato plants is known to be under the genetic control of at least one pair of alleles such that $A$- results in the production of anthocyanin pigment (purple stem). The recessive genotype $aa$ lacks this pigment and hence is green. The edge of the tomato leaf may be deeply cut under the influence of a dominant allele $C$. The recessive genotype $cc$ produces smooth edged leaves called "potato leaf". The production of two locules in the tomato fruit is a characteristic of the dominant allele $M$; multiple locules are produced by the recessive genotype $mm$. A cross is made between two pure lines : purple, potato, biloculed × green, cut, multiloculed. What phenotypic ratio is expected in the $F_2$?

**Solution:**

P: $AA\,cc\,MM$ (purple, potato, biloculed) × $aa\,CC\,mm$ (green, cut, multiloculed)

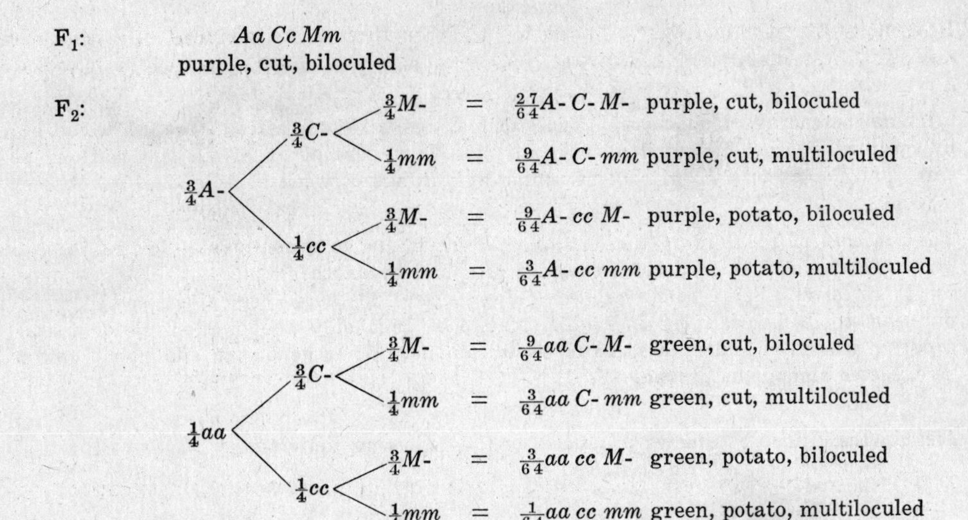


# Supplementary Problems

## DIHYBRID CROSSES WITH DOMINANT AND RECESSIVE ALLELES

**3.8.** The position of the flower on the stem of the garden pea is governed by a pair of alleles. Flowers growing in the axils (upper angle between petiole and stem) are produced by the action of a dominant allele *T*, those growing only at the tip of the stem by its recessive allele *t*. Colored flowers are produced by a dominant gene *C*, and white flowers by its recessive allele *c*. A dihybrid plant with colored flowers in the leaf axils is crossed to a pure strain of the same phenotype. What genotypic and phenotypic ratios are expected in the $F_1$ progeny?

**3.9.** In summer squash, white fruit color is governed by a dominant allele (*W*) and yellow fruit color by the recessive (*w*). A dominant allele at another locus (*S*) produces disc shaped fruit and its recessive allele (*s*) yields sphere-shaped fruit. If a homozygous white disc variety of genotype *WWSS* is crossed with a homozygous yellow sphere variety (*wwss*), the $F_1$ are all white disc dihybrids of genotype *WwSs*. If the $F_1$ is allowed to mate at random, what would be the phenotypic ratio expected in the $F_2$ generation?

**3.10.** In *Drosophila*, ebony body color is produced by a recessive gene *e* and wild type (gray) body color by its dominant allele $e^+$. Vestigial wings are governed by a recessive gene *vg*, and normal wing size (wild type) by its dominant allele $vg^+$. If wild type dihybrid flies are crossed and produce 256 progeny, how many of these progeny flies are expected in each phenotypic class?

**3.11.** Short hair in rabbits is governed by a dominant gene (*L*) and long hair by its recessive allele (*l*). Black hair results from the action of the dominant genotype (*B-*) and brown from the recessive genotype (*bb*). (*a*) In crosses between dihybrid short, black and homozygous short, brown rabbits, what genotypic and phenotypic ratios are expected among their progeny? (*b*) Determine the expected genotypic and phenotypic ratios in progeny from the cross $LlBb \times Llbb$.

**3.12.** The genetic information for the following eight parts is found in Problem 3.11. (*a*) What phenotypic ratio is expected among progeny from crosses of $LlBb \times LlBb$? (*b*) What percentage of the $F_1$ genotypes in part (*a*) breeds true (i.e. what percentage is of homozygous genotypes)? (*c*) What percentage of the $F_1$ genotypes is heterozygous for only one pair of genes? (*d*) What percentage of the $F_1$ genotypes is heterozygous at both loci? (*e*) What percentage of the $F_1$ genotypes could be used for testcross purposes (i.e. homozygous double recessive)? (*f*) What percentage of the $F_1$ progeny could be used for testcross purposes at the B locus (i.e. homozygous recessive *bb*)? (*g*) What percentage of all short haired $F_1$ individuals is expected to be brown? (*h*) What percentage of all black $F_1$ individuals will breed true for both black and short hair?

**3.13.** How many different matings are possible (*a*) when three pairs of factors are considered simultaneously, (*b*) when four pairs of factors are considered simultaneously? *Hint:* See Problem 3.2(*c*).

**3.14.** (*a*) What percentage of all possible types of matings with two pairs of factors would be represented by matings between identical genotypes? (*b*) What percentage of all the matings possible with three pairs of factors would be represented by matings between non-identical genotypes?

**3.15.** The presence of feathers on the legs of chickens is due to a dominant allele ($F$) and clean legs to its recessive allele ($f$). Pea comb shape is produced by another dominant allele ($P$) and single comb by its recessive allele ($p$). In crosses between pure feathered leg, single-combed individuals and pure pea-combed, clean leg individuals, suppose that only the single-combed, feathered leg $F_2$ progeny are saved and allowed to mate at random. What genotypic and phenotypic ratios would be expected among the progeny ($F_3$)?

**3.16.** List all the different gametes produced by the following individuals: (*a*) $AA\,BB\,Cc$, (*b*) $aa\,Bb\,Cc$, (*c*) $Aa\,Bb\,cc\,Dd$, (*d*) $AA\,Bb\,Cc\,dd\,Ee\,Ff$.

**3.17.** The normal cloven-footed condition in swine is produced by the homozygous recessive genotype $mm$. A mule-footed condition is produced by the dominant genotype $M$-. White coat color is governed by the dominant allele of another locus $B$ and black by its recessive allele $b$. A white, mule-footed sow (female) is mated to a black, cloven-footed boar (male) and produces several litters. Among 26 offspring produced by this mating, all were found to be white with mule feet. (*a*) What is the most probable genotype of the sow? (*b*) The next litter produced 8 white, mule-footed offspring and one white cloven-footed pig. Now, what is the most probable genotype of the sow?

**3.18.** A white, mule-footed boar (see Problem 3.17) is crossed to a sow of the same phenotype. Among the $F_1$ offspring there were found 6 white, cloven-footed : 7 black, mule-footed : 15 white, mule-footed : 3 black, cloven-footed pigs. (*a*) If all the black mule-footed $F_1$ offspring from this type of mating were to be testcrossed, what phenotypic ratio would be expected among the testcross progeny? (*b*) If the sow was to be testcrossed, what phenotypic ratio of progeny is expected?

**3.19.** In poultry, a crested head is produced by a dominant gene $C$ and plain head by its recessive allele $c$. Black feather color $R$- is dominant to red $rr$. A homozygous black-feathered, plain-headed bird is crossed to a homozygous red-feathered, crested-headed bird. What phenotypic and genotypic ratios are expected from testcrossing only the $F_2$ black-crested birds? *Hint:* Remember to account for the relative frequencies of the different genotypes in this one phenotypic class.

**3.20.** Bronze turkeys have at least one dominant allele $R$. Red turkeys are homozygous for its recessive allele $rr$. Another dominant gene $H$ produces normal feathers, and the recessive genotype $hh$ produces feathers lacking webbing, a condition termed "hairy". In crosses between homozygous bronze, hairy birds and homozygous red, normal-feathered birds, what proportion of the $F_2$ progeny will be (*a*) genotype $Rrhh$, (*b*) phenotype bronze, hairy, (*c*) genotype $rrHH$, (*d*) phenotype red, normal-feathered, (*e*) genotype $RrHh$, (*f*) phenotype bronze, normal feathered, (*g*) genotype $rrhh$, (*h*) phenotype red, normal-feathered, (*i*) genotype $RRHh$?

## MODIFIED DIHYBRID RATIOS

**3.21.** In peaches, the homozygous genotype $G^oG^o$ produces oval glands at the base of the leaves, the heterozygous genotype $G^oG^a$ produces round glands, and the homozygous genotype $G^aG^a$ results in the absence of glands. At another locus, a dominant gene $S$ produces fuzzy peach skin and its recessive allele $s$ produces smooth (nectarine) skin. A homozygous variety with oval glands and smooth skin is crossed to a homozygous variety with fuzzy skin lacking glands at the base of its leaves. What genotypic and phenotypic proportions are expected in the $F_2$?

**3.22.** In Shorthorn cattle, coat colors are governed by a codominant pair of alleles $C^R$ and $C^W$. The homozygous genotype $C^RC^R$ produces red, the other homozygote produces white and the heterozygote produces roan (a mixture of red and white). The presence of horns is produced by the homozygous recessive genotype $pp$ and the polled condition by its dominant allele $P$. If roan cows heterozygous for the horned gene are mated to a horned, roan bull, what phenotypic ratio is expected in the offspring?

**3.23.** A gene locus with codominant alleles is known to govern feather color in chickens such that the genotype $F^BF^B$ = black, $F^WF^W$ = splashed white and $F^BF^W$ = blue. Another locus with codominant alleles governs feather morphology such that $M^NM^N$ = normal feather shape, $M^NM^F$ = slightly abnormal feathers called "mild frizzle", and $M^FM^F$ = grossly abnormal feathers called "extreme frizzle". If blue, mildly-frizzled birds are crossed among themselves, what phenotypic proportions are expected among their offspring?

**3.24.** In the above problem, if all the blue offspring with normal feathers and all the splashed-white, extremely-frizzled offspring are isolated and allowed to mate at random, what phenotypic ratio would be expected among their progeny?

**3.25.** The shape of radishes may be long ($LL$), round ($L'L'$) or oval ($LL'$). Color may be red ($RR$), white ($R'R'$) or purple ($RR'$). If a long, white strain is crossed with a round, red strain, what phenotypic proportions are expected in the $F_1$ and $F_2$?

**3.26.** Suppose that two strains of radishes are crossed (see above problem) and produce a progeny consisting of 16 long white, 31 oval purple, 16 oval white, 15 long red, 17 oval red and 32 long purple. What would be the phenotypes of the parental strains?

**3.27.** A dominant gene in mice, $K$, produces a kinked tail; recessive genotypes at this locus, $kk$, have normal tails. The homozygous condition of another locus, $AA$, produces a gray color called agouti; the heterozygous condition $A^yA$ produces yellow color; the homozygous genotype $A^yA^y$ is lethal. (*a*) If yellow mice, heterozygous for kinky tail, are crossed together, what phenotypic proportions are expected in their offspring? (*b*) What proportion of the offspring is expected to be of genotype $A^yAKk$? (*c*) If all the yellow offspring were allowed to mate at random, what would be the genotypic and phenotypic ratios among their adult progeny?

**3.28.** An incompletely dominant gene $N$ in the Romney Marsh breed of sheep causes the fleece of homozygotes to be "hairy", i.e. containing fibers which lack the normal amount of crimp. Normal wool is produced by the homozygous genotype $N'N'$. Heterozygotes, $NN'$, can be distinguished at birth by the presence of large, medulated fibers called "halo-hairs" scattered over the body. A gene known as "lethal gray" causes homozygous gray fetuses ($G^lG^l$) to die before 15 weeks in gestation. The heterozygous genotype $G^lG$ produces gray fleece, and the homozygous genotype $GG$ produces black. If heterozygous halo, gray individuals are mated together, (*a*) what would be the phenotypic proportions expected in the live progeny, (*b*) what proportion of the live progeny would carry the lethal gene, (*c*) what proportion of the live progeny with halo-hairs would carry the lethal gene, (*d*) what proportion of all the zygotes would be expected to be of genotype $NN'G^lG^l$?

**3.29.** Infantile amaurotic idiocy (Tay Sachs disease) is a recessive hereditary abnormality causing death within the first few years of life only when homozygous ($ii$). The dominant condition at this locus produces a normal phenotype ($I$-). Abnormally shortened fingers (brachyphalangy) is thought to be due to a genotype heterozygous for a lethal gene ($BB^L$), the homozygote ($BB$) being normal, and the other homozygote ($B^LB^L$) being lethal. What are the phenotypic expectations among teen-age children from parents who are both brachyphalangic and heterozygous for infantile amaurotic idiocy?

**3.30.** In addition to the gene governing infantile amaurotic idiocy in the above problem, the recessive genotype of another locus ($jj$) results in death before age 18 due to a condition called "juvenile amaurotic idiocy". Only individuals of genotype $I$-$J$- will survive to adulthood. (*a*) What proportion of the children from parents of genotype ($IiJj$) would probably not survive to adulthood? (*b*) What proportion of the adult survivors in part (*a*) would not be carriers of either hereditary abnormality?

**3.31.** A genetic condition on chromosome 2 in the fruit fly *Drosophila melanogaster* is lethal when homozygous ($Pm/Pm$), but when heterozygous ($Pm/Pm^+$) produces a purplish eye color called "plum". The other homozygous condition ($Pm^+/Pm^+$) produces wild type eye color. On chromosome 3, a gene called "stubble" produces short, thick bristles when heterozygous ($Sb/Sb^+$), but is lethal when homozygous ($Sb/Sb$). The homozygous condition of its alternative allele ($Sb^+/Sb^+$), produces bristles of normal size (wild type). (*a*) What phenotypic ratio is expected among progeny from crosses between plum, stubble parents? (*b*) If the offspring of part (*a*) are allowed to mate at random to produce an $F_2$, what phenotypic ratio is expected?

**3.32.** Feather color in chickens is governed by a pair of codominant alleles such that $F^BF^B$ produces black, $F^WF^W$ produces splashed white and $F^BF^W$ produces blue. An independently segregating locus governs the length of leg; $CC$ genotypes possess normal leg length, $CC^L$ genotypes produce squatty, shortlegged types called "creepers", but homozygous $C^LC^L$ genotypes are lethal. Determine the kinds of progeny phenotypes and their expected ratios which crosses between dihybrid blue creepers are likely to produce.

**3.33.** Fat mice can be produced by two independently assorting genes. The recessive genotype *ob/ob* produces a fat, sterile mouse called "obese". Its dominant allele *Ob* produces normal growth. The recessive genotype *ad/ad* also produces a fat, sterile mouse called "adipose" and its dominant allele *Ad* produces normal growth. What phenotypic proportions of fat versus normal would be expected among the $F_1$ and $F_2$ from parents of genotype *Ob/ob, Ad/ad*?

**HIGHER COMBINATIONS**

**3.34.** The seeds from Mendel's tall plants were round and yellow, all three characters due to a dominant gene at each of three independently assorting loci. The recessive genotypes *dd, ww* and *gg* produce dwarf plants with wrinkled and green seeds respectively. (*a*) If a pure tall, wrinkled, yellow variety is crossed with a pure dwarf, round, green variety, what phenotypic ratio is expected in the $F_1$ and $F_2$? (*b*) What percentage of the $F_2$ is expected to be of genotype *Dd WW gg*? (*c*) If all the dwarf, round, green individuals in the $F_2$ are isolated and artificially crossed at random, what phenotypic ratio of offspring is expected?

**3.35.** The coat colors of mice are known to be governed by several genes. The presence of a yellow band of pigment near the tip of the hair is called "agouti" pattern and is produced by the dominant allele *A*. The recessive condition at this locus (*aa*) does not have this subapical band and is termed non-agouti. The dominant allele of another locus, *B*, produces black and the recessive genotype *bb* produces brown. The homozygous genotype $c^hc^h$ restricts pigment production to the extremities in a pattern called Himalayan, whereas the genotype *C-* allows pigment to be distributed over the entire body. (*a*) In crosses between pure brown, agouti, Himalayan and pure black mice, what are the phenotypic expectations of the $F_1$ and $F_2$? (*b*) What proportion of the black-agouti, full colored $F_2$ would be expected to be of genotype *AaBBCc*? (*c*) What percentage of all the Himalayans in the $F_2$ would be expected to show brown pigment? (*d*) What percentage of all the agoutis in the $F_2$ would be expected to exhibit black pigment?

**3.36.** In addition to the information given in the problem above, a fourth locus in mice is known to govern the density of pigment deposition. The genotype *D-* produces full color, but the recessive genotype *dd* produces a dilution of pigment. Another allele at this locus, $d^l$, is lethal when homozygous, produces a dilution of pigment in the genotype $dd^l$, and produces full color when in heterozygous condition with the dominant allele $Dd^l$. (*a*) What phenotypic ratio would be expected among the live $F_2$ progeny if the $F_1$ from the cross $aabbCCDd \times AABBccdd^l$ were allowed to mate at random? (*b*) What proportion of the live $F_2$ would be expected to be of genotype $AABbccdd^l$?

**3.37.** In the parental cross $AABBCCDDEE \times aabbccddee$, (*a*) how many different $F_1$ gametes can be formed, (*b*) how many different genotypes are expected in the $F_2$, (*c*) how many squares would be necessary in a gametic checkerboard to accommodate the $F_2$?

**3.38.** A pure strain of Mendel's peas, dominant for all seven of his independently assorting genes, was testcrossed. (*a*) How many different kinds of gametes could each of the parents produce? (*b*) How many different gametes could the $F_1$ produce? (*c*) If the $F_1$ was testcrossed, how many phenotypes would be expected in the offspring and in what proportions? (*d*) How many genotypes would be expected in the $F_2$? (*e*) How many combinations of $F_1$ gametes are theoretically possible (considering, e.g., *AABBCCDDEEFFGG* sperm nucleus $\times$ *aabbccddeeffgg* egg nucleus a different combination than *AABBCCDDEEFFGG* egg nucleus $\times$ *aabbccddeeffgg* sperm nucleus)? (*f*) How many different kinds of matings could theoretically be made among the $F_2$? *Hint:* See solution to Problem 3.2(*c*).

## Answers to Supplementary Problems

**3.8.** $\frac{1}{4}CCTT : \frac{1}{4}CCTt : \frac{1}{4}CcTT : \frac{1}{4}CcTt$; all axial, colored

**3.9.** $\frac{9}{16}$ white, disc : $\frac{3}{16}$ white, sphere : $\frac{3}{16}$ yellow, disc : $\frac{1}{16}$ yellow, sphere

**3.10.** 144 wild type : 48 vestigial : 48 ebony : 16 ebony, vestigial

**3.11.** (*a*) $\frac{1}{4}LLBb : \frac{1}{4}LlBb : \frac{1}{4}LLbb : \frac{1}{4}Llbb$; $\frac{1}{2}$ short, black : $\frac{1}{2}$ short, brown
(*b*) $\frac{1}{8}LLBb : \frac{1}{4}LlBb : \frac{1}{8}LLbb : \frac{1}{4}Llbb : \frac{1}{8}llBb : \frac{1}{8}llbb$;
$\frac{3}{8}$ short, black : $\frac{3}{8}$ short, brown : $\frac{1}{8}$ long, black : $\frac{1}{8}$ long, brown

**3.12.** (*a*) $\frac{9}{16}$ short, black : $\frac{3}{16}$ short, brown : $\frac{3}{16}$ long, black : $\frac{1}{16}$ long, brown, (*b*) 25%, (*c*) 50%, (*d*) 25%, (*e*) 6.25%, (*f*) 25%, (*g*) 25%, (*h*) 8.33%

**3.13.** (*a*) 378, (*b*) 3321 **3.14.** (*a*) 20%, (*b*) 92.86%

**3.15.** $4FFpp : 4Ffpp : 1ffpp$; 8 feathered leg, single comb : 1 clean leg, single comb

**3.16.** (*a*) *ABC*, *ABc*, (*b*) *aBC*, *aBc*, *abC*, *abc*, (*c*) *ABcD*, *ABcd*, *AbcD*, *Abcd*, *aBcD*, *aBcd*, *abcD*, *abcd*, (*d*) *ABCdEF*, *ABCdEf*, *ABCdeF*, *ABCdef*, *ABcdEF*, *ABcdEf*, *ABcdeF*, *ABcdef*, *AbCdEF*, *AbCdEf*, *AbCdeF*, *AbCdef*, *AbcdEF*, *AbcdEf*, *AbcdeF*, *Abcdef*

**3.17.** (*a*) *BBMM*, (*b*) *BBMm*

**3.18.** (*a*) 2 black, mule-foot : 1 black, cloven foot, (*b*) $\frac{1}{4}$ white, mule-foot : $\frac{1}{4}$ white, cloven-foot : $\frac{1}{4}$ black, mule foot : $\frac{1}{4}$ black, cloven-foot

**3.19.** $4RrCc$ = black, crested : $2Rrcc$ = black, plain : $2rrCc$ = red, crested : $1rrcc$ = red, plain

**3.20.** (*a*) $\frac{1}{8}$, (*b*) $\frac{3}{16}$, (*c*) $\frac{1}{16}$, (*d*) $\frac{3}{16}$, (*e*) $\frac{1}{4}$, (*f*) $\frac{9}{16}$, (*g*) $\frac{1}{16}$, (*h*) $\frac{3}{16}$, (*i*) $\frac{1}{8}$

**3.21.** $\frac{1}{16}G^AG^ASS : \frac{2}{16}G^AG^ASs : \frac{1}{16}G^AG^Ass : \frac{2}{16}G^AG^OSS : \frac{4}{16}G^AG^OSs : \frac{2}{16}G^AG^Oss : \frac{1}{16}G^OG^OSS : \frac{2}{16}G^OG^OSs : \frac{1}{16}G^OG^Oss$; $\frac{3}{16}$ fuzzy, glandless : $\frac{1}{16}$ smooth, glandless : $\frac{6}{16}$ round gland, fuzzy : $\frac{2}{16}$ round gland, smooth : $\frac{3}{16}$ oval gland, fuzzy : $\frac{1}{16}$ oval gland, smooth

**3.22.** 1 red, polled : 1 red, horned : 2 roan, polled : 2 roan, horned : 1 white, polled : 1 white, horned

**3.23.** $\frac{1}{16}$ black : $\frac{1}{8}$ black, mildly frizzled : $\frac{1}{16}$ black, extremely frizzled : $\frac{1}{8}$ blue : $\frac{1}{4}$ blue, mildly frizzled : $\frac{1}{8}$ blue, extremely frizzled : $\frac{1}{16}$ splashed-white : $\frac{1}{8}$ splashed-white, mildly frizzled : $\frac{1}{16}$ splashed-white, extremely frizzled

**3.24.** 1 black : 2 blue : 1 splashed-white : 2 blue, mildly frizzled : 2 splashed-white, mildly frizzled : 1 splashed-white, extremely frizzled

**3.25.** $F_1$ is all oval, purple; $F_2$ is $\frac{1}{16}$ long, red : $\frac{1}{8}$ long, purple : $\frac{1}{16}$ long, white : $\frac{1}{8}$ oval, red : $\frac{1}{4}$ oval, purple : $\frac{1}{8}$ oval, white : $\frac{1}{16}$ round, red : $\frac{1}{8}$ round, purple : $\frac{1}{16}$ round, white

**3.26.** long, purple × oval, purple

**3.27.** (*a*) $\frac{1}{2}$ yellow, kinky : $\frac{1}{6}$ yellow : $\frac{1}{4}$ agouti, kinky : $\frac{1}{12}$ agouti, (*b*) $\frac{1}{3}$, (*c*) $\frac{1}{6}A^yAKK : \frac{1}{3}A^yAKk : \frac{1}{6}A^yAkk : \frac{1}{12}AAKK : \frac{1}{6}AAKk : \frac{1}{12}AAkk$; $\frac{1}{2}$ yellow, kinky : $\frac{1}{6}$ yellow : $\frac{1}{4}$ agouti, kinky : $\frac{1}{12}$ agouti

**3.28.** (*a*) $\frac{1}{12}$ black, hairy : $\frac{1}{6}$ black, halo-haired : $\frac{1}{12}$ black : $\frac{1}{6}$ gray, hairy : $\frac{1}{3}$ gray, halo-haired : $\frac{1}{6}$ gray, (*b*) $\frac{2}{3}$, (*c*) $\frac{2}{3}$, (*d*) $\frac{1}{8}$

**3.29.** $\frac{1}{3}$ normal : $\frac{2}{3}$ brachyphalangic **3.30.** (*a*) $\frac{7}{16}$, (*b*) $\frac{1}{9}$

**3.31.** (*a*) $\frac{4}{9}$ plum, stubble : $\frac{2}{9}$ plum : $\frac{2}{9}$ stubble : $\frac{1}{9}$ wild type, (*b*) 1 plum, stubble : 1 plum : 1 stubble : 1 wild type

**3.32.** $\frac{1}{12}$ black : $\frac{1}{6}$ blue : $\frac{1}{12}$ splashed-white : $\frac{1}{6}$ black, creeper : $\frac{1}{3}$ blue, creeper : $\frac{1}{6}$ splashed-white, creeper

**3.33.** $F_1 = \frac{9}{16}$ normal : $\frac{7}{16}$ fat; $F_2 = \frac{64}{81}$ normal : $\frac{17}{81}$ fat

**3.34.** (*a*) $F_1$ is all tall, round, yellow; $F_2$ is 27 tall, round, yellow : 9 tall, round, green : 9 tall, wrinkled, yellow : 9 dwarf, round, yellow : 3 tall, wrinkled, green : 3 dwarf, round, green : 3 dwarf, wrinkled, yellow : 1 dwarf, wrinkled, green, (*b*) 3.12%, (*c*) 8 round : 1 wrinkled

**3.35.** (*a*) $F_1$ is all agouti, black; $F_2$ is 27 agouti, black : 9 agouti, black, Himalayan : 9 agouti, brown : 9 black : 3 agouti, brown, Himalayan : 3 black, Himalayan : 3 brown : 1 brown, Himalayan, (*b*) $\frac{4}{27}$, (*c*) 25%, (*d*) 75%

**3.36.** (*a*) 189 agouti, black : 216 agouti, black, dilute : 63 agouti, black, Himalayan : 72 agouti, black, Himalayan, dilute : 63 agouti, brown : 72 agouti, brown, dilute : 63 black : 72 black, dilute : 21 agouti, brown, Himalayan : 24 agouti, brown, Himalayan, dilute : 21 black, Himalayan : 24 black, Himalayan, dilute : 21 brown : 24 brown, dilute : 7 brown, Himalayan : 8 brown, Himalayan, dilute, (*b*) $\frac{1}{120}$

**3.37.** (*a*) 32, (*b*) 243, (*c*) 1024

**3.38.** (*a*) one each, (*b*) 128, (*c*) 128, each with equal frequency, (*d*) 2187, (*e*) 16,384, (*f*) 2,392,578

# Chapter 4

## Genetic Interaction

### TWO FACTOR INTERACTIONS

The phenotype is a result of gene products brought to expression in a given environment. The environment includes not only external factors such as temperature and the amount or quality of light, but also internal factors such as hormones and enzymes. Genes specify the structure of proteins. All known enzymes are proteins. *Enzymes* perform catalytic functions, causing the splitting or union of various molecules. All of the chemical reactions which occur in the cell constitute the subject of *intermediary metabolism*. These reactions occur as stepwise conversions of one substance into another, each step being mediated by a specific enzyme. All of the steps which transform a precursor substance to its end product constitute a *biosynthetic pathway*.

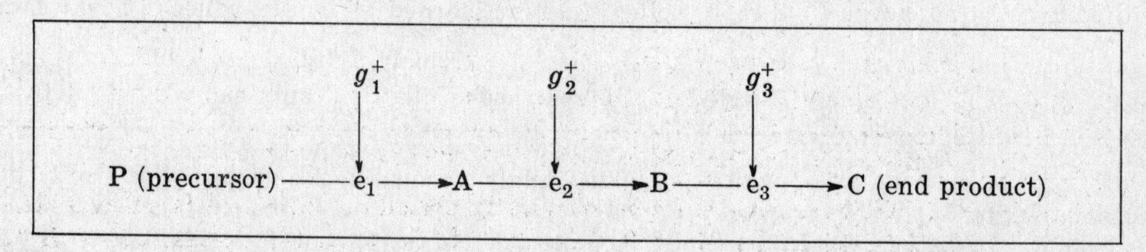


Several genes are usually required to specify the enzymes involved in even the simplest pathways. Each metabolite (A, B, C) is produced by the catalytic action of different enzymes ($e_x$) specified by different wild type genes ($g_x^+$). *Genetic interaction* occurs whenever two or more genes specify enzymes which catalyze steps in a common pathway. If substance C is essential for the production of a normal phenotype, and the recessive mutant alleles $g_1$, $g_2$ and $g_3$ produce defective enzymes, then a mutant (abnormal) phenotype would result from a genotype homozygous recessive at any of the three loci. If $g_3$ is mutant, the conversion of B to C does not occur and substance B tends to accumulate in excessive quantity; if $g_2$ is mutant, substance A will accumulate. Thus mutants are said to produce "metabolic blocks". An organism with a mutation only in $g_2$ could produce a normal phenotype if it was given either substance B or C, but an organism with a mutation in $g_3$ has a specific requirement for C. Thus gene $g_3^+$ becomes dependent upon gene $g_2^+$ for its expression as a normal phenotype. If the genotype is homozygous for the recessive $g_2$ allele, then the pathway ends with substance A. Neither $g_3^+$ nor its recessive allele $g_3$ has any effect on the phenotype. Thus genotype $g_2g_2$ can hide or mask the phenotypic expression of alleles at the $g_3$ locus. Originally a gene or locus which suppressed or masked the action of a gene at another locus was termed *epistatic*. The gene or locus suppressed was *hypostatic*. Later it was found that both loci could be mutually epistatic to one another. Now the term epistasis has come to be synonymous with almost any type of gene interaction. Dominance involves *intra*-allelic gene suppression, or the masking effect which one allele has upon the expression of another allele at the same locus. Epistasis involves *inter*-allelic gene suppression, or the masking effect which one gene locus has upon the expression of another. The classical phenotypic ratio of 9:3:3:1 observed in the progeny of dihybrid parents becomes modified by epistasis into ratios which are various combinations of the 9:3:3:1 groupings.

**Example 4.1.** A particularly illuminating example of gene interaction occurs in white clover. Some strains have a high cyanide content, others have a low cyanide content. Crosses between two strains with low cyanide have produced an $F_1$ with a high concentration of cyanide in their leaves. The $F_2$ shows a ratio of 9 high cyanide : 7 low cyanide. Cyanide is known to be produced from the substrate cyanogenic glucoside by enzymatic catalysis. One strain of clover has the enzyme but not the substrate. The other strain makes substrate but is unable to convert it to cyanide. The pathway may be diagrammed as follows where $G^x$ produces an enzyme and $g^x$ results in a metabolic block.

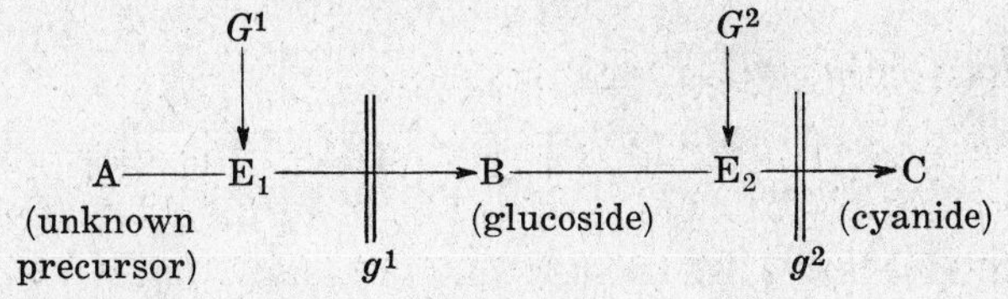

$G^1$ → $E_1$; $G^2$ → $E_2$

A —— $E_1$ —‖→ B —— $E_2$ —‖→ C

(unknown precursor) $g^1$ (glucoside) $g^2$ (cyanide)

Tests on leaf extracts have been made for cyanide content before and after the addition of either glucoside or the enzyme $E_2$.

| $F_2$ Ratio | Genotype | Leaf Extract Alone | Leaf Extract Plus Glucoside | Leaf Extract Plus $E_2$ |
|---|---|---|---|---|
| 9 | $G^1$-$G^2$- | + | + | + |
| 3 | $G^1$-$g^2g^2$ | 0 | 0 | + |
| 3 | $g^1g^1G^2$- | 0 | + | 0 |
| 1 | $g^1g^1g^2g^2$ | 0 | 0 | 0 |

Legend: + = cyanide present, 0 = no cyanide present.

If the leaves are phenotypically classified on the basis of cyanide content of extract alone, a ratio of 9 : 7 results. If the phenotypic classification is based either on extract plus glucoside or on extract plus $E_2$, a ratio of 12 : 4 is produced. If all of these tests form the basis of phenotypic classification, the classical 9 : 3 : 3 : 1 ratio emerges.

## EPISTATIC INTERACTIONS

When epistasis is operative between two gene loci, the number of phenotypes appearing in the offspring from dihybrid parents will be less than four. There are six types of epistatic ratios commonly recognized, three of which have 3 phenotypes and the other three having only 2 phenotypes.

### 1. Dominant Epistasis (12 : 3 : 1).

When the dominant allele at one locus, for example the *A* allele, produces a certain phenotype regardless of the allelic condition of the other locus, then the A-locus is said to be epistatic to the B-locus. Furthermore, since the dominant allele *A* is able to express itself in the presence of either *B* or *b*, this is a case of dominant epistasis. Only when the genotype of the individual is homozygous recessive at the epistatic locus (*aa*) can the alleles of the hypostatic locus (*B* or *b*) be expressed. Thus the genotypes *A-B-* and *A-bb* produce the same phenotype, whereas *aaB-* and *aabb* produce two additional phenotypes. The classical 9 : 3 : 3 : 1 ratio becomes modified into a 12 : 3 : 1 ratio.

### 2. Recessive Epistasis (9 : 3 : 4).

If the recessive genotype at one locus (e.g. *aa*) suppresses the expression of alleles at the B-locus, the A-locus is said to exhibit recessive epistasis over the B-locus. Only if the dominant allele is present at the A-locus can the alleles of the hypostatic B-locus be expressed. The genotypes *A-B-* and *A-bb* produce two additional phenotypes. The 9 : 3 : 3 : 1 ratio becomes a 9 : 3 : 4 ratio.

**3. Duplicate Genes with Cumulative Effect (9 : 6 : 1).**

If the dominant condition (either homozygous or heterozygous) at either locus (but not both) produces the same phenotype, the $F_2$ ratio becomes 9:6:1. For example, where the epistatic genes are involved in producing various amounts of a substance such as pigment, the dominant genotypes of each locus may be considered to produce one unit of pigment independently. Thus genotypes *A-bb* and *aaB-* produce one unit of pigment each and therefore have the same phenotype. The genotype *aabb* produces no pigment, but in the genotype *A-B-* the effect is cumulative and two units of pigment are produced.

**4. Duplicate Dominant Genes (15 : 1).**

The 9:3:3:1 ratio is modified into a 15:1 ratio if the dominant alleles of both loci each produce the same phenotype without cumulative effect.

**5. Duplicate Recessive Genes (9 : 7).**

In the case where identical phenotypes are produced by both homozygous recessive genotypes, the $F_2$ ratio becomes 9:7. The genotypes *aaB-*, *A-bb* and *aabb* produce one phenotype. Both dominant alleles, when present together, complement each other and produce a different phenotype.

**6. Dominant and Recessive Interaction (13 : 3).**

Only two $F_2$ phenotypes result when a dominant genotype at one locus (e.g. *A-*) and the recessive genotype at the other (*bb*) produce the same phenotypic effect. Thus *A-B-*, *A-bb* and *aabb* produce one phenotype and *aaB-* produces another in the ratio 13:3.

**Table 4.1. Summary of Epistatic Ratios**

<table>
<tr><th>Genotypes</th><th>A-B-</th><th>A-bb</th><th>aaB-</th><th>aabb</th></tr>
<tr><td>Classical ratio</td><td>9</td><td>3</td><td>3</td><td>1</td></tr>
<tr><td>Dominant epistasis</td><td colspan="2">12</td><td>3</td><td>1</td></tr>
<tr><td>Recessive epistasis</td><td>9</td><td>3</td><td colspan="2">4</td></tr>
<tr><td>Duplicate genes with cumulative effect</td><td>9</td><td colspan="2">6</td><td>1</td></tr>
<tr><td>Duplicate dominant genes</td><td colspan="3">15</td><td>1</td></tr>
<tr><td>Duplicate recessive genes</td><td>9</td><td colspan="3">7</td></tr>
<tr><td>Dominant and recessive interaction</td><td colspan="2">13</td><td>3</td><td></td></tr>
</table>

## NON-EPISTATIC INTERACTIONS

Genetic interaction may also occur without epistasis if the end products of different pathways each contribute to the same trait.

**Example 4.2.** The dull-red eye color characteristic of wild type flies is a mixture of two kinds of pigments (B and D) each produced from non-pigmented compounds (A and C) by the action of different enzymes ($e_1$ and $e_2$) specified by different wild type genes ($g_1^+$ and $g_2^+$).

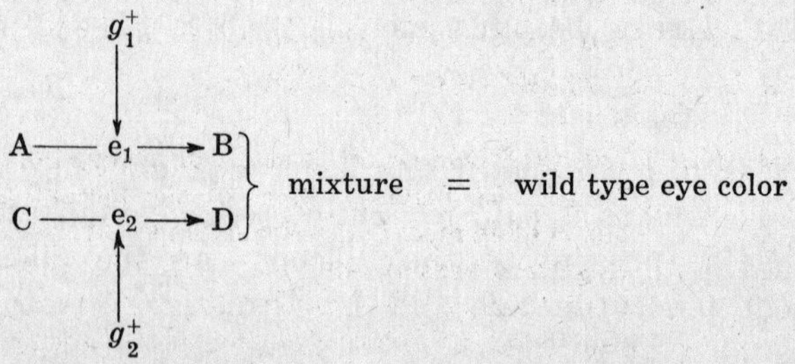

$g_1^+$ ↓

A —— $e_1$ ——→ B
C —— $e_2$ ——→ D
} mixture = wild type eye color

↑ $g_2^+$

The recessive alleles at these two loci ($g_1$ and $g_2$) specify enzymatically inactive proteins. Thus a genotype without either dominant allele would not produce any pigmented compounds and the eye color would be white.

| Phenotypes | Genotypes | End Products |
|---|---|---|
| wild type | $g_1^+/-$, $g_2^+/-$ | B and D |
| color B | $g_1^+/-$, $g_2/g_2$ | B and C |
| color D | $g_1/g_1$, $g_2^+/-$ | D and A |
| white | $g_1/g_1$, $g_2/g_2$ | A and C |

In the above example, the genes for color B and color D are both dominant to white, but when they occur together they produce a novel phenotype (wild type) by interaction. If the two genes are assorting independently, the classical 9:3:3:1 ratio will not be disturbed.

**Example 4.3.** A brown ommochrome pigment is produced in *Drosophila melanogaster* by a dominant gene $st^+$ on chromosome 3. A scarlet pterin pigment is produced by a dominant gene $bw^+$ on chromosome 2. The recessive alleles at these two loci produce no pigment. When pure scarlet flies are mated to pure brown flies, a novel phenotype (wild type) appears in the progeny.

P: brown $st^+/st^+$, $bw/bw$ × scarlet $st/st$, $bw^+/bw^+$

$F_1$: wild type $st^+/st$, $bw^+/bw$

$F_2$:
9 $st^+/-$, $bw^+/-$ wild type
3 $st^+/-$, $bw/bw$ brown
3 $st/st$, $bw^+/-$ scarlet
1 $st/st$, $bw/bw$ white

## INTERACTIONS WITH THREE OR MORE FACTORS

Recall from Chapter 3 that the progeny from trihybrid parents are expected in the phenotypic ratio 27:9:9:9:3:3:3:1. This classical ratio can also be modified whenever two or all three of the loci interact. Interactions involving four or more loci are also possible. Most genes probably depend to some extent upon other genes in the total genotype. The total phenotype depends upon interactions of the total genotype with the environment.

## PLEIOTROPISM

Many and perhaps most of the biochemical pathways in the living organism are interconnected and often interdependent. Products of one reaction chain may be used in several other metabolic schemes. It is not surprising, therefore, that the phenotypic expression of a gene usually involves more than one trait. Sometimes one trait will be clearly evident (major effect) and other, perhaps seemingly unrelated ramifications (secondary effects) will be less evident to the casual observer. In other cases, a number of related changes may be considered together as a *syndrome*. All of the manifold phenotypic expressions of a single gene are spoken of as *pleiotropic* gene effects.

**Example 4.4.** The syndrome called "sickle cell anemia" in humans is due to an abnormal hemoglobin. This is the primary effect of the mutant gene. Subsidiary effects of the abnormal hemoglobin include the sickle-shape of the cells and their tendency to clump together and clog blood vessels in various organs of the body. As a result, heart, kidney, spleen and brain damage are common elements of the syndrome. Defective corpuscles are readily destroyed in the body, causing severe anemia.

# Solved Problems

## TWO FACTOR INTERACTIONS

**4.1.** Coat colors of dogs depend upon the action of at least two genes. At one locus a dominant epistatic inhibitor of coat color pigment (*I*-) prevents the expression of color alleles at another independently assorting locus, producing white coat color. When the recessive condition exists at the inhibitor locus (*ii*), the alleles of the hypostatic locus may be expressed, *iiB*- producing black and *iibb* producing brown. When dihybrid white dogs are mated together, determine (*a*) the phenotypic proportions expected in the progeny, (*b*) the chance of choosing, from among the white progeny, a genotype homozygous at both loci.

**Solution:**

(*a*) P: *IiBb* × *IiBb*
white white

$F_1$: 9/16 *I-B-* } = 12/16 white
3/16 *I-bb* }
3/16 *iiB-* = 3/16 black
1/16 *iibb* = 1/16 brown

(*b*) The genotypic proportions among the white progeny are as follows:

| | | Proportion of Total $F_1$ | Proportion of White $F_1$ |
|---|---|---|---|
| $\frac{1}{4}II$ | $\frac{1}{4}BB$ | 1/16 *IIBB* | 1/12 |
| | $\frac{1}{2}Bb$ | 2/16 *IIBb* | 2/12 |
| | $\frac{1}{4}bb$ | 1/16 *IIbb* | 1/12 |
| $\frac{1}{2}Ii$ | $\frac{1}{4}BB$ | 2/16 *IiBB* | 2/12 |
| | $\frac{1}{2}Bb$ | 4/16 *IiBb* | 4/12 |
| | $\frac{1}{4}bb$ | 2/16 *Iibb* | 2/12 |
| Totals: | | 12/16 | 12/12 |

The only homozygous genotypes at both loci in the above list are $\frac{1}{12}IIBB$ and $\frac{1}{12}IIbb = \frac{2}{12}$ or $\frac{1}{6}$ of all the white progeny. Thus there is one chance in six of choosing a homozygous genotype from among the white progeny.

**4.2.** Two white flowered strains of the sweet pea (*Lathyrus odoratus*) were crossed, producing an $F_1$ with only purple flowers. Random crossing among the $F_1$ produced 96 progeny plants, 53 exhibiting purple flowers and 43 with white flowers. (*a*) What phenotypic ratio is approximated by the $F_2$? (*b*) What type of interaction is involved? (*c*) What were the probable genotypes of the parental strains?

**Solution:**

(*a*) To determine the phenotypic ratio in terms of familiar sixteenths, the following proportion for white flowers may be made: $43/96 = x/16$, from which $x = 7.2$. That is, 7.2 white : 8.8 purple, or approximately a 7 : 9 ratio. We might just as well have arrived at the same conclusion by establishing the proportion for purple flowers: $53/96 = x/16$, from which $x = 8.8$ purple.

(*b*) A 7 : 9 ratio is characteristic of duplicate recessive genes where the recessive genotype at either or both of the loci produces the same phenotype.

(*c*) If *aa* or *bb* or both could produce white flowers, then only the genotype *A-B-* could produce purple. For two white parental strains (pure lines) to be able to produce an all purple $F_1$, they must be homozygous for different dominant-recessive combinations. Thus

P: *aaBB* × *AAbb*
white white

$F_1$: *AaBb*
purple

$F_2$: 9/16 *A-B-* = 9/16 purple
3/16 *A-bb*
3/16 *aaB-* } = 7/16 white
1/16 *aabb*

**4.3.** Red color in wheat kernels is produced by the genotype *R-B-*, white by the double recessive genotype (*rrbb*). The genotypes *R-bb* and *rrB-* produce brown kernels. A homozygous red variety is crossed to a white variety. (*a*) What phenotypic results are expected in the $F_1$ and $F_2$? (*b*) If the brown $F_2$ is artificially crossed at random (wheat is normally self-fertilized), what phenotypic and genotypic proportions are expected in the offspring?

**Solution:**

(*a*) P: *RRBB* × *rrbb*
red white

$F_1$: *RrBb*
red

$F_2$: 9/16 *R-B-* = 9/16 red
3/16 *R-bb*
3/16 *rrB-* } = 6/16 brown
1/16 *rrbb* = 1/16 white

(*b*) The proportion of genotypes represented among the brown $F_2$ must first be determined.

| | Proportion of Total $F_2$ | Proportion of Brown $F_2$ |
|---|---|---|
| $(\frac{1}{4}RR)(\frac{1}{4}bb)$ | 1/16 *RRbb* | 1/6 |
| $(\frac{1}{2}Rr)(\frac{1}{4}bb)$ | 2/16 *Rrbb* | 2/6 |
| $(\frac{1}{4}rr)(\frac{1}{4}BB)$ | 1/16 *rrBB* | 1/6 |
| $(\frac{1}{4}rr)(\frac{1}{2}Bb)$ | 2/16 *rrBb* | 2/6 |
| Totals: | 6/16 | 6/6 |

Next, the relative frequencies of the various matings may be calculated in a checkerboard.

| | 1/6 *RRbb* | 2/6 *Rrbb* | 1/6 *rrBB* | 2/6 *rrBb* |
|---|---|---|---|---|
| 1/6 *RRbb* | 1/36 *RRbb* × *RRbb* (1) | 2/36 *RRbb* × *Rrbb* (2) | 1/36 *RRbb* × *rrBB* (3) | 2/36 *RRbb* × *rrBb* (4) |
| 2/6 *Rrbb* | 2/36 *Rrbb* × *RRbb* (2) | 4/36 *Rrbb* × *Rrbb* (5) | 2/36 *Rrbb* × *rrBB* (6) | 4/36 *Rrbb* × *rrBb* (7) |
| 1/6 *rrBB* | 1/36 *rrBB* × *RRbb* (3) | 2/36 *rrBB* × *Rrbb* (6) | 1/36 *rrBB* × *rrBB* (8) | 2/36 *rrBB* × *rrBb* (9) |
| 2/6 *rrBb* | 2/36 *rrBb* × *RRbb* (4) | 4/36 *rrBb* × *Rrbb* (7) | 2/36 *rrBb* × *rrBB* (9) | 4/36 *rrBb* × *rrBb* (10) |

| Matings | Progeny | Genotypic Proportions (f) | Mating Frequency (m) | m·f |
|---|---|---|---|---|
| (1) *RRbb* × *RRbb* | *RRbb* | 100% | 1/36 | 1/36 |
| (2) *RRbb* × *Rrbb* | *RRbb* | 1/2 | 4/36 | 4/72 |
| | *Rrbb* | 1/2 | | 4/72 |
| (3) *RRbb* × *rrBB* | *RrBb* | 100% | 2/36 | 2/36 |
| (4) *RRbb* × *rrBb* | *RrBb* | 1/2 | 4/36 | 4/72 |
| | *Rrbb* | 1/2 | | 4/72 |
| (5) *Rrbb* × *Rrbb* | *RRbb* | 1/4 | 4/36 | 4/144 |
| | *Rrbb* | 1/2 | | 4/72 |
| | *rrbb* | 1/4 | | 4/144 |
| (6) *Rrbb* × *rrBB* | *RrBb* | 1/2 | 4/36 | 4/72 |
| | *rrBb* | 1/2 | | 4/72 |
| (7) *Rrbb* × *rrBb* | *RrBb* | 1/4 | 8/36 | 8/144 |
| | *rrBb* | 1/4 | | 8/144 |
| | *Rrbb* | 1/4 | | 8/144 |
| | *rrbb* | 1/4 | | 8/144 |
| (8) *rrBB* × *rrBB* | *rrBB* | 100% | 1/36 | 1/36 |
| (9) *rrBb* × *rrBB* | *rrBB* | 1/2 | 4/36 | 4/72 |
| | *rrBb* | 1/2 | | 4/72 |
| (10) *rrBb* × *rrBb* | *rrBB* | 1/4 | 4/36 | 4/144 |
| | *rrBb* | 1/2 | | 4/72 |
| | *rrbb* | 1/4 | | 4/144 |

Summary of progeny genotypes:

1/9 *RRbb*
2/9 *RrBb*
2/9 *Rrbb*
1/9 *rrBB*
2/9 *rrBb*
1/9 *rrbb*

Summary of progeny phenotypes:

2/9 *R-B-* = 2/9 red
1/3 *R-bb*, 1/3 *rrB-* = 2/3 brown
1/9 *rrbb* = 1/9 white

**4.4.** The following pedigree shows the transmission of swine coat colors through three generations:

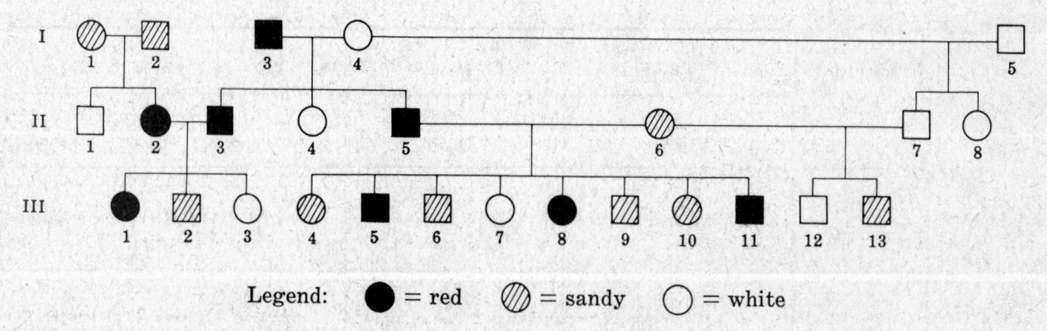


Assume the offspring of II5 × II6 shown in this pedigree occur in the ratio expected from the genotypes represented by their parents. How are these colors most likely inherited?

**Solution:**

Notice first that three phenotypes are expressed in this pedigree. This rules out epistatic combinations producing only two phenotypes such as those expressed in dominant and recessive interaction (13 : 3), duplicate dominant genes (15 : 1), and duplicate recessive genes (9 : 7). Epistatic gene interactions producing three phenotypes are those expressed in dominant epistasis (12 : 3 : 1), recessive epistasis (9 : 3 : 4), and dominant genes with cumulative action (9 : 6 : 1). Let us proceed to solve this problem by making an assumption and then applying it to the pedigree to see if the phenotypes shown there can be explained by our hypothesis.

**Case 1.** Assume dominant epistasis is operative. The genotypes responsible for the three phenotypes may be represented as follows: *A-B-* and *A-bb* = 1st phenotype, *aaB-* = 2nd phenotype, and *aabb* = 3rd phenotype. We must now determine which of the phenotypes represented in this pedigree corresponds to each of the genotypic classes. Obviously the only pure line phenotype is the third one. Offspring of the mating *aabb* × *aabb* would all be phenotypically identical to the parents. The mating I4 × I5 appears to qualify in this respect and we shall tentatively assume that white coat color is represented by the genotype *aabb*. Certain matings between individuals with the dominant epistatic gene (*A*) could produce three phenotypically different types of offspring (e.g. *AaBb* × *AaBb*). Such a mating is observed between II2 and II3. Therefore we might assume red color to be represented by the genotype *A-*. Sandy color must then be represented by genotype *aaB-*. Matings between sandy individuals could produce only sandy (*aaB-*) or white (*aabb*) progeny. However, sandy parents I1 × I2 produce white and red progeny (II1, II2). Therefore the assumption of dominant interaction must be wrong.

**Case 2.** Assume recessive epistasis to be operative. The genotypes responsible for the three phenotypes in this case may be represented as follows: *A-B-* as 1st phenotype, *A-bb* as 2nd phenotype, and *aaB-* and *aabb* as 3rd phenotype. As pointed out in Case 1, matings between individuals of genotype *AaBb* are the only kind among identical phenotypes capable of producing all three phenotypes in the progeny. Thus *A-B-* should represent red (e.g. II2 × II3). The *aa* genotypes breed true, producing only white individuals (I4 × I5). Sandy is produced by genotype *A-bb*. Sandy × sandy (I1 × I2) could not produce the red offspring (II2). Therefore the assumption of recessive interaction must be wrong.

**Case 3.** Assume that duplicate genes with cumulative action are interacting. The genotypes responsible for the three phenotypes in this case may be represented as follows: *A-B-* as 1st phenotype, *A-bb* and *aaB-* as 2nd phenotype, and *aabb* as 3rd phenotype. As explained in the previous two cases, *A-B-* must be red and *aabb* must be white. If we assume that any dominant genotype at either the A-locus or B-locus contributes one unit of pigment to the phenotype, then either the genotype *aaB-* or *A-bb* could be sandy; we further assume that the presence of both dominant genes (*A-B-*) would contribute two units of pigment to produce a red phenotype. Thus the mating II5 (*AaBb*) red × II6 (*aaBb*) sandy would be expected to produce offspring phenotypes in the following proportions:

| | | | |
|---|---|---|---|
| red | *A-B-* | $\frac{1}{2} \cdot \frac{3}{4} = \frac{3}{8}$ | |
| sandy | *A-bb* | $\frac{1}{2} \cdot \frac{1}{4} = \frac{1}{8}$ | $\frac{1}{2}$ |
| | *aaB-* | $\frac{1}{2} \cdot \frac{3}{4} = \frac{3}{8}$ | |
| white | *aabb* | $\frac{1}{2} \cdot \frac{1}{4} = \frac{1}{8}$ | |

The same phenotypic ratio would be expected if II6 was *Aabb*. These expectations correspond to the ones given in the pedigree (III4 through III11) and therefore the hypothesis of dominant genes with cumulative action is consistent with the data.

## INTERACTIONS WITH THREE OR MORE FACTORS

**4.5.** At least three loci are known to govern coat colors in mice. The genotype *C-* will allow pigment to be produced at the other two loci. The recessive genotype *cc* does not allow pigment production, resulting in "albino". The "agouti" pattern depends upon the genotype *A-*, and non-agouti upon the recessive *aa*. The color of the pigment may be black (*B-*) or chocolate (*bb*). Five coat colors may be produced by the action of alleles at these three loci:

| | |
|---|---|
| wild type (agouti, black) | *A-B-C-* |
| black (non-agouti) | *aaB-C-* |
| chocolate (non-agouti) | *aabbC-* |
| cinnamon (agouti, chocolate) | *A-bbC-* |
| albino | *- - - -cc* |

(*a*) What phenotypic frequencies are expected in the $F_2$ from crosses of pure black with albinos of type *AAbbcc*? (*b*) A cinnamon male is mated to a group of albino females of identical genotype and among their progeny were observed 43 wild type, 40 cinnamon, 39 black, 41 chocolate and 168 albino. What are the most probable genotypes of the parents?

**Solution:**

(*a*) P: *aaBBCC* × *AAbbcc*
pure black albino

$F_1$: *AaBbCc*
wild type

$F_2$:

| | |
|---|---|
| 27 *A-B-C-* | wild type |
| 9 *A-B-cc* | albino |
| 9 *A-bbC-* | cinnamon |
| 9 *aaB-C-* | black |
| 3 *A-bbcc* | albino |
| 3 *aaB-cc* | albino |
| 3 *aabbC-* | chocolate |
| 1 *aabbcc* | albino |

Summary of (*a*): 27/64 wild type
16/64 albino
9/64 cinnamon
9/64 black
3/64 chocolate

(*b*) P: *A-bbC-*♂ × ----*cc* ♀♀
cinnamon male albino females

$F_1$:

| | |
|---|---|
| 43 wild type | *A-B-C-* |
| 40 cinnamon | *A-bbC-* |
| 39 black | *aaB-C-* |
| 41 chocolate | *aabbC-* |
| 168 albino | ----*cc* |
| 331 | |

In part (*b*), the cinnamon progeny, *A-bbC-*, indicate *b* in the female parents. The black progeny, *aaB-C-*, indicate *a* in both parents, and *B* in the female parents. The chocolate progeny, *aabbC-*, indicate *a* in both parents, and *b* in the females. The albinos indicate *c* in the male. The genotype of the male is now known to be *AabbCc*. But the genotype of the albino females is known only to be *a-Bbcc*. They could be either *AaBbcc* or *aaBbcc*.

**Case 1.** Assume the females to be *AaBbcc*.

Parents: *AabbCc*♂ × *AaBbcc*♀♀

The expected phenotypic frequencies among the progeny would be:

| | | | | | |
|---|---|---|---|---|---|
| *A-BbCc* | wild type | $\frac{3}{4}\cdot\frac{1}{2}\cdot\frac{1}{2}$ | = $\frac{3}{16}(331)$ | = | approx. 62 |
| *A-bbCc* | cinnamon | $\frac{3}{4}\cdot\frac{1}{2}\cdot\frac{1}{2}$ | = $\frac{3}{16}(331)$ | = | approx. 62 |
| *aaBbCc* | black | $\frac{1}{4}\cdot\frac{1}{2}\cdot\frac{1}{2}$ | = $\frac{1}{16}(331)$ | = | approx. 21 |
| *aabbCc* | chocolate | $\frac{1}{4}\cdot\frac{1}{2}\cdot\frac{1}{2}$ | = $\frac{1}{16}(331)$ | = | approx. 21 |
| ----*cc* | albino | $1\cdot 1\cdot\frac{1}{2}$ | = $\frac{1}{2}(331)$ | = | approx. 166 |

Obviously, the expectations deviate considerably from the observations. Therefore, the females are probably not of genotype *AaBbcc*.

**Case 2.** Assume the females to be of genotype *aaBbcc*.

Parents: *AabbCc*♂ × *aaBbcc*♀♀

The expected phenotypic frequencies among the progeny would be:

| | | | | | |
|---|---|---|---|---|---|
| *AaBbCc* | wild type | $\frac{1}{2}\cdot\frac{1}{2}\cdot\frac{1}{2}$ | = $\frac{1}{8}(331)$ | = | approx. 41 |
| *AabbCc* | cinnamon | $\frac{1}{2}\cdot\frac{1}{2}\cdot\frac{1}{2}$ | = $\frac{1}{8}(331)$ | = | approx. 41 |
| *aaBbCc* | black | $\frac{1}{2}\cdot\frac{1}{2}\cdot\frac{1}{2}$ | = $\frac{1}{8}(331)$ | = | approx. 41 |
| *aabbCc* | chocolate | $\frac{1}{2}\cdot\frac{1}{2}\cdot\frac{1}{2}$ | = $\frac{1}{8}(331)$ | = | approx. 41 |
| ----*cc* | albino | $1\cdot 1\cdot\frac{1}{2}$ | = $\frac{1}{2}(331)$ | = | approx. 166 |

Now the expectations correspond very closely to the observations. Hence the genotype of the parental albino females is probably *aaBbcc*.

**4.6.** Lewis-a blood group substance appears on the human red blood cell when the dominant gene *Le* is present, but is absent if the dominant gene of the "secretor" locus, *Se*, is present. Suppose that from a number of families where both parents are Lewis-a negative of genotype *LeleSese*, we find that most of them have 3 Lewis-a positive : 13 Lewis-a negative children. In a few other families, suppose we find 2 Lewis-a negative : 1 Lewis-a positive. Furthermore, in families where both parents are secretors of genotype *Sese*, we find most of them exhibit a ratio of 3 secretor : 1 non-secretor, but a few of them show 9 secretor : 7 non-secretor. Propose a hypothesis to account for these results.

**Solution:**

If only two loci are interacting, the dominant *Se* gene can suppress the expression of *Le*, resulting in Lewis-a negative blood type. When both parents are dihybrid, we expect a 13 : 3 ratio in the progeny characteristic of dominant and recessive interaction.

| | | | |
|---|---|---|---|
| P: | *LeleSese* | × | *LeleSese* |
| | Lewis-a neg. | | Lewis-a neg. |
| $F_1$: | 9 *Le-Se-*<br>3 *leleSe-*<br>1 *lelesese* | = | 13 Lewis-a negative |
| | 3 *Le-sese* | = | 3 Lewis-a positive |

The 9 : 7 ratio found in some families for the secretor trait indicates that two factors are again interacting. This is the ratio produced by duplicate recessive interaction; i.e. whenever the recessive alleles at either of two loci are present, a non-secretor phenotype results. Let us symbolize the alleles of the second locus by *X* and *x*.

| | | | |
|---|---|---|---|
| P: | *SeseXx* | × | *SeseXx* |
| | secretor | | secretor |
| $F_1$: | 9 *Se-X-* | = | 9 secretors |
| | 3 *Se-xx*<br>3 *seseX-*<br>1 *sesexx* | = | 7 non-secretors; the *xx* genotype suppresses the expression of *Se*. |

If we assume the *x* gene to be relatively rare, then most families will have only the dominant gene *X*, but in a few families both parents will be heterozygous *Xx*. Let us assume that this is the case in those families which produce 2 Lewis-a negative : 1 Lewis-a positive.

| | | | |
|---|---|---|---|
| P: | *LeleSeseXx* | × | *LeleSeseXx* |
| | Lewis-a negative | | Lewis-a negative |

$F_1$:
27 *Le-Se-X-*
*9 *Le-Se-xx*
*9 *Le-seseX-*
9 *leleSe-X-*
*3 *Le-sesexx*
3 *leleSe-xx*
3 *leleseseX-*
1 *lelesesexx*

If *Se* suppresses *Le*, but *xx* suppresses *Se*, then only the genotypes marked with an asterisk (*) will be Lewis-a positive, giving a ratio of 21 Lewis-a positive : 43 Lewis-a negative. This is very close to a 1 : 2 ratio and indeed would appear to be such with limited data.

# Supplementary Problems

### TWO FACTOR INTERACTIONS

**4.7.** When homozygous yellow rats are crossed to homozygous black rats, the $F_1$ is all gray. Mating the $F_1$ among themselves produced an $F_2$ consisting of 10 yellow, 28 gray, 2 cream-colored and 8 black. (*a*) How are these colors inherited? (*b*) Show, using appropriate genetic symbols, the genotypes for each color. (*c*) How many of the 48 $F_2$ rats were expected to be cream-colored? (*d*) How many of the 48 $F_2$ rats were expected to be homozygous?

**4.8.** Four comb shapes in poultry are known to be governed by two gene loci. The genotype *R-P-* produces walnut comb, characteristic of the Malay breed; *R-pp* produces rose comb, characteristic of the Wyandotte breed; *rrP-* produces pea comb, characteristic of the Brahma breed; *rrpp* produces single comb, characteristic of the Leghorn breed. (*a*) If pure Wyandottes are crossed with pure Brahmas, what phenotypic ratios are expected in the $F_1$ and $F_2$? (*b*) A Malay hen was crossed to a Leghorn cock and produced a dozen eggs, 3 of which grew into birds with rose combs and 9 with walnut combs. What is the probable genotype of the hen? (*c*) Determine the proportion of comb types that would be expected in offspring from each of the following crosses: (1) *Rrpp* × *RrPP*, (2) *rrPp* × *RrPp*, (3) *rrPP* × *RRPp*, (4) *RrPp* × *rrpp*, (5) *RrPp* × *RRpp*, (6) *RRpp* × *rrpp*, (7) *RRPP* × *rrpp*, (8) *Rrpp* × *Rrpp*, (9) *rrPp* × *Rrpp*, (10) *rrPp* × *rrpp*.

**4.9.** Listed below are 7 two-factor interaction ratios observed in progeny from various dihybrid parents. Suppose that in each case one of the dihybrid parents is testcrossed (instead of being mated to another dihybrid individual). What phenotypic ratio is expected in the progeny of each testcross? (*a*) 9:6:1, (*b*) 9:3:4, (*c*) 9:7, (*d*) 15:1, (*e*) 12:3:1, (*f*) 9:3:3:1, (*g*) 13:3.

**4.10.** White fruit color in summer squash is governed by a dominant gene (*W*) and colored fruit by its recessive allele (*w*). Yellow fruit is governed by an independently assorting hypostatic gene (*G*) and green by its recessive allele (*g*). When dihybrid plants are crossed, the offspring appear in the ratio 12 white : 3 yellow : 1 green. What fruit color ratios are expected from the crosses (*a*) *Wwgg* × *WwGG*, (*b*) *WwGg* × green, (*c*) *Wwgg* × *wwGg*, (*d*) *WwGg* × *Wwgg*? (*e*) If two plants are crossed producing $\frac{1}{2}$ yellow and $\frac{1}{2}$ green progeny, what are the genotypes and phenotypes of the parents?

**4.11.** Matings between black rats of identical genotype produced offspring as follows: 14 cream-colored, 47 black, and 19 albino. (*a*) What epistatic ratio is approximated by these offspring? (*b*) What type of epistasis is operative? (*c*) What are the genotypes of the parents and the offspring (use your own symbols)?

**4.12.** A dominant gene *S* in *Drosophila* produces a peculiar eye condition called "star". Its recessive allele $S^+$ produces the normal eye of wild type. The expression of *S* can be suppressed by the dominant allele of another locus, *Su-S*. The recessive allele of this locus, $Su\text{-}S^+$, has no effect on $S^+$. (*a*) What type of interaction is operative? (*b*) When a normal-eyed male of genotype *Su-S/Su-S, S/S* is crossed to a homozygous wild type female of genotype $Su\text{-}S^+/Su\text{-}S^+$, $S^+/S^+$, what phenotypic ratio is expected in the $F_1$ and $F_2$? (*c*) What percentage of the wild type $F_2$ is expected to carry the dominant gene for star-eye?

**4.13.** The Black Langshan breed of chickens has feathered shanks. When Langshans are crossed to the Buff Rock breed with unfeathered shanks, all the $F_1$ have feathered shanks. Out of 360 $F_2$ progeny, 24 were found to have non-feathered shanks and 336 had feathered shanks. (*a*) What is the mode of interaction in this trait? (*b*) What proportion of the feathered $F_2$ would be expected to be heterozygous at one locus and homozygous at the other?

**4.14.** On chromosome three of corn there is a dominant gene ($A_1$) which, together with the dominant gene ($A_2$) on chromosome nine, produces colored aleurone. All other genetic combinations produce colorless aleurone. Two pure colorless strains are crossed to produce an all colored $F_1$. (*a*) What were the genotypes of the parental strains and the $F_1$? (*b*) What phenotypic proportions are expected among the $F_2$? (*c*) What genotypic ratio exists among the white $F_2$?

**4.15.** Two pairs of alleles govern the color of onion bulbs. A pure red strain crossed to a pure white strain produces an all red $F_1$. The $F_2$ was found to consist of 47 white, 38 yellow and 109 red bulbs. (*a*) What epistatic ratio is approximated by the data? (*b*) What is the name of this type of gene interaction? (*c*) If another $F_2$ is produced by the same kind of a cross, and 8 bulbs of the $F_2$ are found to be of the double recessive genotype, how many bulbs would be expected in each phenotypic class?

**4.16.** A plant of the genus *Capsella*, commonly called "shepherd's purse", produces a seed capsule the shape of which is controlled by two independently assorting genes. When dihybrid plants were interpollinated, 6% of the progeny were found to possess ovoid-shaped seed capsules. The other 94% of the progeny had triangular-shaped seed capsules. (*a*) What two factor epistatic ratio is approximated by the progeny? (*b*) What type of interaction is operative?

**4.17.** The color of corn aleurone is known to be controlled by several genes; *A*, *C*, and *R* are all necessary for color to be produced. The locus of a dominant inhibitor of aleurone color, I, is very closely linked to that of *C*. Thus any one or more of the genotypes *I-*, *aa-*, *cc-* or *rr-* produces colorless aleurone. (*a*) What would be the colored : colorless ratio among $F_2$ progeny from the cross *AAIICCRR* × *aaiiCCRR*? (*b*) What proportion of the colorless $F_2$ is expected to be homozygous?

**4.18.** A dominant allele *C* must be present in order for any pigment to be developed in mice. The kind of pigment produced depends upon another locus such that *B-* produces black and *bb* produces brown. Individuals of epistatic genotype *cc* are incapable of making pigment and are called "albinos". A homozygous black female is testcrossed to an albino male. (*a*) What phenotypic ratio is expected in the $F_1$ and $F_2$? (*b*) If all the albino $F_2$ mice are allowed to mate at random, what genotypic ratio is expected in the progeny?

**4.19.** Suppose that crossing two homozygous lines of white clover, each with a low content of cyanide, produces only progeny with high levels of cyanide. When these $F_1$ progeny are backcrossed to either parental line, half the progeny has low cyanide content and the other half has high cyanide content. (*a*) What type of interaction may account for these results? (*b*) What phenotypic ratio is expected in the $F_2$? (*c*) If a 12 : 4 ratio is observed among progeny from parents with high cyanide content, what are the parental genotypes? (*d*) If the low cyanide $F_2$, exclusive of the double recessives, are allowed to cross at random among themselves, what proportion of their progeny is expected to contain a high cyanide content?

**4.20.** In cultivated flowers called "stocks", the recessive genotype of one locus (*aa*) prevents the development of pigment in the flower, thus producing a white color. In the presence of the dominant allele *A*, alleles at another locus may be expressed as follows: *C*- = red, *cc* = cream. (*a*) When cream stocks of the genotype *Aacc* are crossed to red stocks of the genotype *AaCc*, what phenotypic and genotypic proportions are expected in the progeny? (*b*) If cream stocks crossed to red stocks produce white progeny, what may be the genotypes of the parents? (*c*) When dihybrid red stocks are crossed together, what phenotypic ratio is expected among the progeny? (*d*) If red stocks crossed to white stocks produce progeny with red, cream and white flowers, what are the genotypes of the parents?

**4.21.** An inhibitor of pigment production in onion bulbs (*I*-) exhibits dominant epistasis over another locus, the genotype *iiR*- producing red bulbs and *iirr* producing yellow bulbs. (*a*) A pure white strain is crossed to a pure red strain and produces an all white $F_1$ and an $F_2$ with $\frac{12}{16}$ white, $\frac{3}{16}$ red and $\frac{1}{16}$ yellow. What were the genotypes of the parents? (*b*) If yellow onions are crossed to a pure white strain of a genotype different from the parental type in part (*a*), what phenotypic ratio is expected in the $F_1$ and $F_2$? (*c*) Among the white $F_2$ of part (*a*), suppose that 32 were found to be of genotype *IiRR*. How many progeny are expected in each of the three $F_2$ phenotypic classes?

**4.22.** For color to be produced in corn kernels, the dominant alleles at three loci must be present (*A, C, R*). Another locus is hypostatic to these three loci; the dominant allele of this locus (*Pr*) yields purple pigment and the recessive allele (*pr*) determines red kernels. (*a*) What phenotypic proportions are expected in the $F_1$ and $F_2$ when a red strain of genotype *AACCRRprpr* is crossed to a colorless strain of genotype *AACCrrPrPr*? (*b*) If the red $F_2$ are crossed among themselves, what genotypic and phenotypic ratios are expected among the progeny?

**4.23.** Crossing certain genetically identical monohybrid white-bulbed onions produces white and colored progeny in the ratio 3 : 1 respectively. Crossing certain genetically identical monohybrid colored-bulbed onions produces colored and white progeny in the ratio 3 : 1 respectively. White is epistatic to color. (*a*) What phenotypic ratio is expected among the $F_1$ and $F_2$ progeny produced by test-crossing a white variety homozygous for the dominant alleles at two independently assorting loci? (*b*) What type of interaction is operative? (*c*) Suppose that the dominant condition (*I*-) at one locus and/or the recessive condition at the other (*cc*) could both produce white onion bulbs. White onions of genotype *I-C-*, in the presence of ammonia fumes, turn yellow. Onions which are white due to the action of *cc* in their genotype fail to change color in the presence of ammonia fumes. A white plant which turns yellow in ammonia fumes is crossed to a white one which does not change color. The progeny occur in the proportions $\frac{7}{8}$ white : $\frac{1}{8}$ colored. What are the genotypes of the parents? (*d*) A white plant which changes color in ammonia fumes is crossed to a colored plant. The progeny occur in the ratio 3 colored : 5 white. Determine the genotypes of the parents.

**4.24.** The color of the flower center in the common yellow daisy may be either purple-centered or yellow-centered. Two genes (*P* and *Y*) are known to interact in this trait. The results of two matings are given below:

(1) P: *PpYY* (purple-centered) × *PpYY* (purple-centered)

$F_1$: 3/4 *P-YY* purple-centered
1/4 *ppYY* yellow-centered

(2) P: *ppYy* (yellow-centered) × *ppYy* (yellow-centered)

$F_1$: 3/4 *ppY-*, 1/4 *ppyy* } all yellow-centered

Determine the phenotypic ratios of progeny from the matings (*a*) *PpYy* × *PpYy*, (*b*) *PpYy* × *ppyy*, (*c*) *PPyy* × *ppYY*.

**4.25.** The aleurone of corn kernels may be either yellow, white or purple. When pollen from a homozygous purple plant is used to fertilize a homozygous white plant, the aleurone of the resulting kernels are all purple. When homozygous yellow plants are crossed to homozygous white plants, only seeds with yellow aleurone are produced. When homozygous purple plants are crossed to homozygous yellow plants, only purple progeny appear. Some crosses between purple plants produce purple, yellow and white progeny. Some crosses between yellow plants produce both yellow and white offspring. Crosses between yellow plants never produce purple progeny. Crosses among plants produced from seeds with white aleurone always produce only white progeny. (*a*) Can these results be explained on the basis of the action of a single gene locus with multiple alleles? (*b*) What is the simplest explanation for the mode of gene action? (*c*) If plants with only dominant alleles at the two loci are crossed to plants grown from white seeds, what phenotypic proportions are expected among their $F_2$ progeny? (*d*) In part (*c*), how many generations of seeds must be planted in order to obtain an $F_2$ progeny phenotypically expressing the aleurone genes derived from the adult parent sporophytes? (*e*) What is the advantage of studying the genetics of seed traits rather than traits of the sporophyte?

**4.26.** Three fruit shapes are recognized in the summer squash (*Cucurbita pepo*): disc-shaped, elongated and sphere-shaped. A pure disc-shaped variety was crossed to a pure elongated variety. The $F_1$ were all disc-shaped. Among 80 $F_2$, there were 30 sphere-shaped, 5 elongated and 45 disc-shaped. (*a*) Reduce the $F_2$ numbers to their lowest ratio. (*b*) What type of interaction is operative? (*c*) If the sphere-shaped $F_2$ cross at random, what phenotypic proportions are expected in the progeny?

**4.27.** When yellow mice are mated together, $\frac{2}{3}$ of the progeny are yellow and $\frac{1}{3}$ are agouti. Another locus is known to govern pigment production. When parents heterozygous at this locus are mated together, $\frac{3}{4}$ of the offspring are colored and $\frac{1}{4}$ are albino. These albino mice cannot express whatever genes they may have at the agouti locus. Yellow mice are crossed with albinos and produce an $F_1$ containing $\frac{1}{2}$ albino, $\frac{1}{3}$ yellow and $\frac{1}{6}$ agouti. (*a*) What are the probable genotypes of the parents (use your own symbols)? (*b*) If the yellow $F_1$ mice are crossed among themselves, what phenotypic ratio is expected among the offspring? (*c*) What proportion of the yellow offspring from part (*b*) would be expected to breed true?

**4.28.** The pedigree on the right illustrates a case of dominant epistasis. (*a*) What symbol represents the genotype $A\text{-}B\text{-}$? (*b*) What symbol represents the genotype $aaB\text{-}$? (*c*) What symbol represents the genotype $aabb$? (*d*) What type of epistasis would be represented if II2 × II3 produced, in addition to □ and ◍, an offspring of type ●? (*e*) What type of interaction would be represented if III5 × III6 produced, in addition to ▨ and ○, an offspring of type ●?

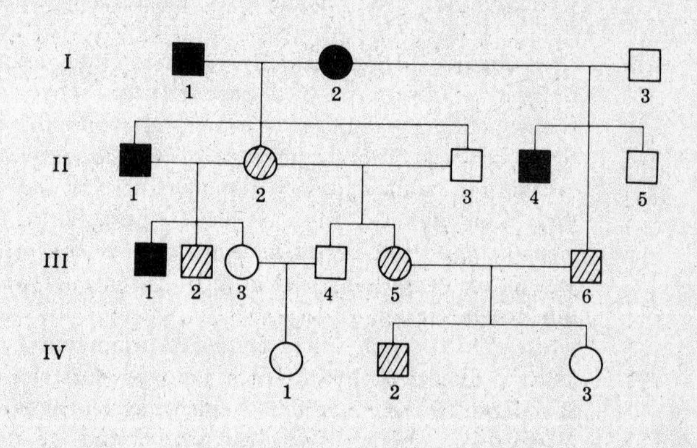


**4.29.** Given the following pedigree showing three generations of mink breeding, where open symbols represent wild type and solid symbols represent platinum, determine (*a*) the mode of inheritance of these coat colors, (*b*) the most probable genotypes of all individuals in the pedigree (use of familiar symbols such as $A,a$ and $B,b$ are suitable), (*c*) what phenotypic proportions are expected in progeny from III1 × III2.

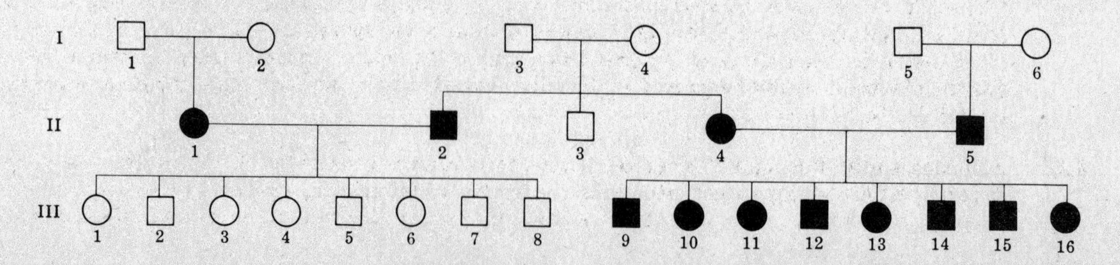

**4.30.** The pedigree in Fig. (*a*) shows the genetic transmission of feather color in chickens. Open symbols represent white feathers, solid symbols represent colored feathers. Under the assumption of dominant and recessive interaction (given *A*- or *bb* or both = white, *aaB*- = color) assign genotypes to each individual in the pedigree. Indicate by (-) whatever genes cannot be determined.

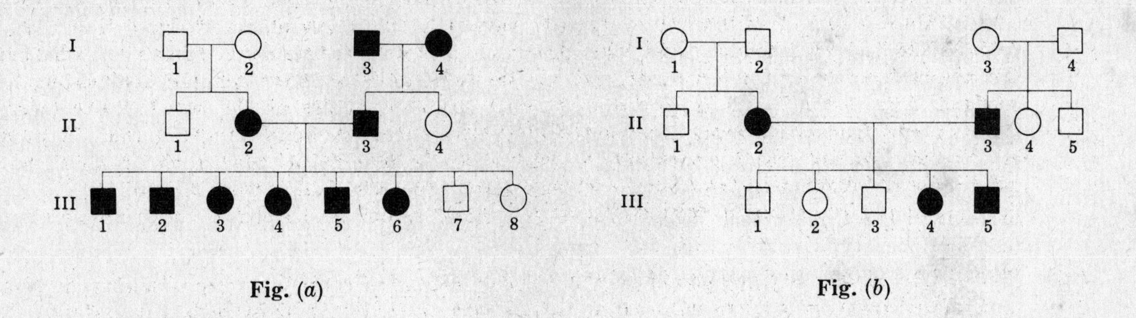


Fig. (*a*) Fig. (*b*)

**4.31.** The pedigree in Fig. (*b*) shows the inheritance of deafness in humans. Open symbols represent normal hearing and solid symbols represent deafness. Under the assumption of duplicate recessive interaction (given *A*-*B*- = normal, *aa* or *bb* or both = deaf) assign genotypes to each individual in the pedigree. Indicate by (-) whatever genes cannot be determined.

## INTERACTIONS WITH THREE OR MORE FACTORS

**4.32.** A wheat variety with colored seeds is crossed to a colorless strain producing an all colored $F_1$. In the $F_2$, $\frac{1}{64}$ of the progeny has colorless seeds. (*a*) How many pairs of genes control seed color? (*b*) What were the genotypes of the parents and the $F_1$ (use your own symbols)?

**4.33.** In mice, spotted coat color is due to a recessive gene *s* and solid coat color to its dominant allele *S*. Colored mice possess a dominant allele *C* whereas albinos are homozygous recessive *cc*. Black is produced by a dominant allele *B* and brown by its recessive allele *b*. The *cc* genotype is epistatic to both the B and S loci. What phenotypic ratio is expected among the progeny of trihybrid parents?

**4.34.** A pure line of corn (*CCRR*) exhibiting colored aleurone is testcrossed to a colorless aleurone strain. Approximately 56% of the $F_2$ has colored aleurone, the other 44% being colorless. A pure line (*AARR*) with colored aleurone, when testcrossed, also produces the same phenotypic ratio in the $F_2$. (*a*) What phenotypic ratio is expected in the $F_2$ when a pure colored line of genotype *AACCRR* is testcrossed? (*b*) What proportion of the colorless $F_2$ is *aaccrr*? (*c*) What genotypic ratio exists among the colored $F_2$?

**4.35.** If a pure white onion strain is crossed to a pure yellow strain, the $F_2$ ratio is 12 white : 3 red : 1 yellow. If another pure white onion is crossed to a pure red onion, the $F_2$ ratio is 9 red : 3 yellow : 4 white. (*a*) What percentage of the white $F_2$ from the second mating would be homozygous for the yellow allele? (*b*) If the white $F_2$ (homozygous for the yellow allele) of part (*a*) is crossed to the pure white parent of the first mating mentioned at the beginning of this problem, determine the $F_1$ and $F_2$ phenotypic expectations.

**4.36.** For any color to be developed in the aleurone layer of corn kernels, the dominant alleles at two loci plus the recessive condition at a third locus (*A*-*R*-*ii*) must be present. Any other genotypes produce colorless aleurone. (*a*) What phenotypic ratio of colored : colorless would be expected in progeny from matings between parental plants of genotype *AaRrIi*? (*b*) What proportion of the colorless progeny in part (*a*) would be expected to be heterozygous at one or more of the three loci? (*c*) What is the probability of picking from among the colored seeds in part (*a*) two which, when grown into adult sporophytes and artificially crossed, would produce some colorless progeny with the triple recessive genotype?

**4.37.** A dominant gene *V* is known in humans which causes certain areas of the skin to become depigmented, a condition called "vitiligo". Albinism is the complete lack of pigment production and is produced by the recessive genotype *aa*. The albino locus is epistatic to the vitiligo locus. Another gene locus, the action of which is independent of the previously mentioned loci, is known to be involved in a mildly anemic condition called "thalassemia". (*a*) When adult progeny from parents both of whom exhibit vitiligo and a mild anemia is examined, the following phenotypic proportions are observed: $\frac{1}{16}$ normal : $\frac{3}{16}$ vitiligo : $\frac{1}{8}$ mildly anemic : $\frac{1}{12}$ albino : $\frac{3}{8}$ vitiligo and mildly anemic : $\frac{1}{6}$ albino and mildly anemic. What is the mode of genetic action of the gene for thalassemia? (*b*) What percentage of the viable albino offspring in part (*a*) would carry the gene for vitiligo? (*c*) What percentage of viable offspring with symptoms of mild anemia also shows vitiligo?

**4.38.** When the White Leghorn breed of chickens is crossed to the White Wyandotte breed, all the $F_1$ birds have white feathers. The $F_2$ birds appear in the ratio 13 white : 3 colored. When the White Leghorn breed is crossed to the White Silkie breed, the $F_1$ is white and the $F_2$ is also 13 white : 3 colored. But when White Wyandottes are crossed to White Silkies, the $F_1$ is all colored and the $F_2$ appears in the ratio 9 colored : 7 white. (*a*) How are feather colors inherited in these breeds (use appropriate symbols in your explanation)? (*b*) Show, by use of your own symbols, the genotypes of each of the three breeds (assume the breed is homozygous for all loci under consideration). (*c*) What phenotypic ratio is expected among progeny from trihybrid parents? (*d*) What proportion of the white offspring of part (*c*) is expected to be dihybrid?

**4.39.** A corn plant which grew from a seed with purple aleurone is self-pollinated. The $F_1$ produces $\frac{9}{16}$ purple aleurone, $\frac{3}{16}$ red, $\frac{3}{16}$ yellow and $\frac{1}{16}$ white. Some of the red $F_1$, when selfed, produce progeny in the ratio 12 red : 3 yellow : 1 white. Some of the purple $F_1$, when selfed, produce progeny in the ratio of 9 purple : 3 red : 4 white. (*a*) How can these results be explained? Using any appropriate symbols, diagram all three of the crosses mentioned above. (*b*) If, instead of being selfed, each of the above three parental types had been testcrossed, what phenotypic results would be expected? (*c*) What phenotypic proportions are observed in progeny from mating trihybrid purple × dihybrid purple (with the third locus homozygous recessive)?

**4.40.** For color to be developed in the aleurone layer of corn, the dominant alleles at four loci must be present ($A_1, A_2, C$ and $R$). If a line pure for colored aleurone is testcrossed to a non-colored line, find (*a*) the phenotypic ratio expected in the $F_1$ and $F_2$, (*b*) the percentage of the colored $F_2$ expected to be genetically like the $F_1$, (*c*) the percentage of the colorless $F_2$ expected to be homozygous for both the $A_1$ and $A_2$ alleles. (*d*) If a colored line is testcrossed and 25% of the offspring are colored, how many loci are heterozygous? (*e*) If the colored line in part (*d*) produces $12\frac{1}{2}$% colored offspring, how many loci are heterozygous? (*f*) If all four loci of the colored line in part (*d*) are heterozygous, what percentage of the offspring are expected to be colorless?

# Answers to Supplementary Problems

**4.7.** (*a*) Two pairs of non-epistatic genes interact to produce these coat colors.
(*b*) *A-B-* (gray), *A-bb* (yellow), *aaB-* (black), *aabb* (cream) (*c*) 3 (*d*) 12

**4.8.** (*a*) $F_1$ : all walnut comb; $F_2$ : $\frac{9}{16}$ walnut : $\frac{3}{16}$ rose : $\frac{3}{16}$ pea : $\frac{1}{16}$ single (*b*) *RRPp*
(*c*)

| | *R-P-* walnut | *R-pp* rose | *rrP-* pea | *rrpp* single |
|---|---|---|---|---|
| (1) | 3/4 | | 1/4 | |
| (2) | 3/8 | 1/8 | 3/8 | 1/8 |
| (3) | all | | | |
| (4) | 1/4 | 1/4 | 1/4 | 1/4 |
| (5) | 1/2 | 1/2 | | |
| (6) | | all | | |
| (7) | all | | | |
| (8) | | 3/4 | | 1/4 |
| (9) | 1/4 | 1/4 | 1/4 | 1/4 |
| (10) | | | 1/2 | 1/2 |

**4.9.** (*a*) 1:2:1 (*b*) 1:1:2 (*c*) 1:3 (*d*) 3:1 (*e*) 2:1:1 (*f*) 1:1:1:1 (*g*) 3:1

**4.10.** (*a*) $\frac{3}{4}$ white : $\frac{1}{4}$ yellow (*b*) and (*c*) $\frac{1}{2}$ white : $\frac{1}{4}$ yellow : $\frac{1}{4}$ green (*d*) $\frac{3}{4}$ white : $\frac{1}{8}$ yellow : $\frac{1}{8}$ green (*e*) yellow (*wwGg*) × green (*wwgg*)

**4.11.** (*a*) 9:3:4 (*b*) recessive epistasis (*c*) P: *BbCc* (black); $F_1$: *B-C-* (black), *bbC-* (cream), *B-cc* and *bbcc* (albino)

**4.12.** (*a*) dominant and recessive interaction (*b*) $F_1$: all wild type; $F_2$: 13 wild type : 3 star (*c*) 69%

**4.13.** (*a*) duplicate dominant genes with only the double recessive genotype producing non-feathered shanks (*b*) $\frac{8}{15}$

**4.14.** (*a*) P: $A_1A_1a_2a_2 \times a_1a_1A_2A_2$; $F_1$: $A_1a_1A_2a_2$
(*b*) $\frac{9}{16}$ colored : $\frac{7}{16}$ colorless (*c*) $\frac{1}{7}A_1A_1a_2a_2 : \frac{2}{7}A_1a_1a_2a_2 : \frac{1}{7}a_1a_1A_2A_2 : \frac{2}{7}a_1a_1A_2a_2 : \frac{1}{7}a_1a_1a_2a_2$

**4.15.** (*a*) 9:3:4 (*b*) recessive epistasis (*c*) 32 white : 24 yellow : 72 red

**4.16.** (*a*) 15 triangular : 1 ovoid (*b*) duplicate dominant interaction

**4.17.** (*a*) 13 colorless : 3 colored (*b*) $\frac{3}{13}$

**4.18.** (*a*) $F_1$: all black; $F_2$: $\frac{9}{16}$ black : $\frac{3}{16}$ brown : $\frac{4}{16}$ albino (*b*) $\frac{1}{4}BBcc : \frac{1}{2}Bbcc : \frac{1}{4}bbcc$

**4.19.** (*a*) duplicate recessive interaction (*b*) 9 high cyanide : 7 low cyanide (*c*) *A-Bb* × *AABb* (*d*) $\frac{2}{9}$

**4.20.** (*a*) $\frac{3}{8}$ red : $\frac{3}{8}$ cream : $\frac{1}{4}$ white; $\frac{1}{8}AACc : \frac{1}{8}AAcc : \frac{1}{4}AaCc : \frac{1}{4}Aacc : \frac{1}{8}aaCc : \frac{1}{8}aacc$
(*b*) *Aacc* × *AaCc* (or) *Aacc* × *AaCC*
(*c*) 9 red : 3 cream : 4 white
(*d*) *AaCc* × *aaCc* (or) *AaCc* × *aacc*

**4.21.** (*a*) *IIrr* × *iiRR* (*b*) $F_1$: all white; $F_2$: $\frac{12}{16}$ white : $\frac{3}{16}$ red : $\frac{1}{16}$ yellow (*c*) 16 yellow : 48 red : 192 white

**4.22.** (*a*) $F_1$: all purple; $F_2$: $\frac{9}{16}$ purple : $\frac{3}{16}$ red : $\frac{4}{16}$ white
(*b*) $\frac{4}{9}$ *AACCprprRR* : $\frac{4}{9}$ *AACCprprRr* : $\frac{1}{9}$ *AACCprprrr*; 8 red : 1 white

**4.23.** (*a*) $F_1$: all white; $F_2$: 13 white : 3 colored (*b*) dominant and recessive interaction (*c*) *IiCc* × *Iicc* (*d*) *IiCc* × *iiCc*

**4.24.** (*a*) $\frac{9}{16}$ purple centered : $\frac{7}{16}$ yellow centered (*b*) $\frac{1}{4}$ purple centered : $\frac{3}{4}$ yellow centered (*c*) all purple centered

**4.25.** (*a*) no (*b*) dominant epistasis where *Y-R-* or *yyR-* produce purple, *Y-rr* = yellow, and *yyrr* = white (*c*) $\frac{3}{4}$ purple : $\frac{3}{16}$ yellow : $\frac{1}{16}$ white (*d*) one (*e*) The appearance of seed traits requires one less generation of rearing than that for tissues found in the sporophyte.

**4.26.** (*a*) 9 disc : 6 sphere : 1 elongated (*b*) duplicate genes with cumulative effect (*c*) $\frac{2}{3}$ sphere : $\frac{2}{9}$ disc : $\frac{1}{9}$ elongate

**4.27.** (*a*) $A^yACc \times A^yAcc$ (where $A^y$ allele is lethal when homozygous) (*b*) 2 yellow : 1 agouti : 1 albino (*c*) none

**4.28.** (*a*) solid symbol (*b*) diagonal lines (*c*) open symbol (*d*) recessive epistasis (*e*) duplicate genes with cumulative effect

**4.29.** (*a*) duplicate recessive interaction (*b*) *A-B-* (wild type), *aa--* or *--bb* or *aabb* (platinum); *AaB-* (I1, 2), *A-Bb* (I3, 4), *A-B-* (I5, 6, II3), *aaBB* (II1), *AAbb* (II2), *AaBb* (III1 thru 8), either *aa* or *bb* or both (II4, 5, III9 thru 16) (*c*) 9 wild type : 7 platinum

**4.30.** The following set of genotypes is only one of several possible solutions: *aaB-* (III1, 2, 3, 4, 5, 6), *A-* or *bb* or both (II1), *a---* (I1, 2), *aaBb* (I3, 4, II2, 3), *aabb* (II4, III7, 8).

**4.31.** The following set of genotypes is only one of several possible solutions: *AaB-* (I1, 2), *A-Bb* (I3, 4), *A-B-* (II1, 4, 5), *aaB-* (II2), *Aabb* (II3), *AaBb* (III1, 2, 3), *aa* or *bb* or both (III4, 5).

**4.32.** (*a*) 3 (*b*) P: $AABBCC \times aabbcc$; $F_1$: *AaBbCc*

**4.33.** 27 solid black : 9 spotted black : 9 solid brown : 3 spotted brown : 16 albino

**4.34.** (*a*) 27 colored : 37 colorless (*b*) $\frac{1}{37}$ (*c*) $\frac{1}{27}AACCRR$ : $\frac{2}{27}AACCRr$ : $\frac{2}{27}AACcRR$ : $\frac{4}{27}AACcRr$ : $\frac{2}{27}AaCCRR$ : $\frac{4}{27}AaCCRr$ : $\frac{4}{27}AaCcRR$ : $\frac{8}{27}AaCcRr$

**4.35.** (*a*) 25% (*b*) $F_1$: all white; $F_2$: 52 white : 9 red : 3 yellow

**4.36.** (*a*) 9 colored : 55 colorless (*b*) $\frac{48}{55}$ (*c*) $\frac{16}{81}$

**4.37.** (*a*) The gene for thalassemia is dominant to its normal allele, causing mild anemia when heterozygous, but is lethal when homozygous (*b*) 75% (*c*) 56.25%

**4.38.** (*a*) 3 loci involved; one possesses a dominant inhibitor of color (*I-*) and the other two possess different recessive inhibitors of color (*cc* and *oo*). Only the genotype *iiC-O-* produces colored birds; all other genotypes produce white feathers. (*b*) White Leghorn (*CCOOII*), White Wyandotte (*ccOOii*), White Silkie (*CCooii*) (*c*) 55 white : 9 colored (*d*) $\frac{20}{55}$

**4.39.** (*a*) Three pairs of factors contribute to seed color: $R$ = color, $r$ = colorless; $Pr$ = purple, $pr$ = red; $Y$ = yellow, $y$ = white. The R-locus exhibits recessive epistasis over the Pr-locus (hence the 9 : 3 : 4 ratio), i.e. the dominant gene $R$ is necessary for any color to be produced by the Pr-alleles. The R-locus exhibits dominant epistasis over the Y-locus (hence the 12 : 3 : 1 ratio), i.e. segregation of alleles at the Y-locus can only be expressed in colorless (*rr*) seeds. First cross: $RrPrprYy \times RrPrprYy = 9:3:3:1$. Second cross: $RrprprYy \times RrprprYy = 12:3:1$. Third cross: $RrPrpryy \times RrPrpryy = 9:3:4$. (*b*) First cross: 1 purple : 1 red : 1 yellow : 1 white; Second cross: 2 red : 1 yellow : 1 white. Third cross: 2 white : 1 red : 1 purple. (*c*) $\frac{9}{16}$ purple : $\frac{3}{16}$ red : $\frac{1}{8}$ yellow : $\frac{1}{8}$ white

**4.40.** (*a*) $F_1$: all colored; $F_2$: $\frac{81}{256}$ colored : $\frac{175}{256}$ non-colored (*b*) 19.75% (*c*) 4% (*d*) 2 (*e*) 3 (*f*) 93.75%

# Chapter 5

## The Genetics of Sex

### THE IMPORTANCE OF SEX

We are probably too accustomed to thinking of sex in terms of the males and females of our own or domestic species. Plants also have sexes; at least we know that there are male and female portions of a flower. All organisms, however, do not possess only two sexes. Some of the lowest forms of plant and animal life may have several sexes. For example, in one variety of the ciliated protozoan *Paramecium bursaria* there are eight sexes or "mating types" all morphologically identical. Each mating type is physiologically incapable of conjugating with its own type, but may exchange genetic material with any of the seven other types within the same variety. In most higher organisms, the number of sexes has been reduced to just two. These sexes may reside in different individuals or within the same individual. An animal possessing both male and female reproductive organs is usually referred to as a *hermaphrodite*. In plants where *staminate* (male) and *pistillate* (female) flowers occur on the same plant, the term of preference is *monoecious*. Moreover, most of our flowering plants have both male and female parts within the same flower (perfect flower). Relatively few angiosperms are *dioecious*, i.e., having the male and female elements in different individuals. Among the common cultivated crops known to be dioecious are asparagus, date palm, hemp, hops, and spinach.

Whether or not there are two or more sexes, or whether or not these sexes reside in the same or different individuals is relatively unimportant. The importance of sex itself is that it is a mechanism which provides for the great amount of genetic variability characterizing most natural populations. The evolutionary process of natural selection depends upon this genetic variability to supply the raw material from which the better adapted types usually survive to reproduce their kind. Many subsidiary mechanisms have evolved to ensure cross fertilization in most species as a means for generating new genetic combinations in each generation.

### SEX DETERMINING MECHANISMS

Most mechanisms for the determination of sex are under genetic control and may be classified into one of the following categories.

**1. Sex Chromosome Mechanisms.**

(*a*) ***Heterogametic Males.*** In humans, and apparently in all other mammals, the presence of the Y chromosome may determine a tendency to maleness. Normal males are chromosomally XY and females are XX. This produces a 1:1 sex ratio in each generation. Since the male produces two kinds of gametes as far as the sex chromosomes are concerned, he is said to be the *heterogametic* sex. The female, producing only one kind of gamete, is the *homogametic* sex. This mode of sex determination is commonly referred to as the XY method.

**Example 5.1.** XY Method of Sex Determination.

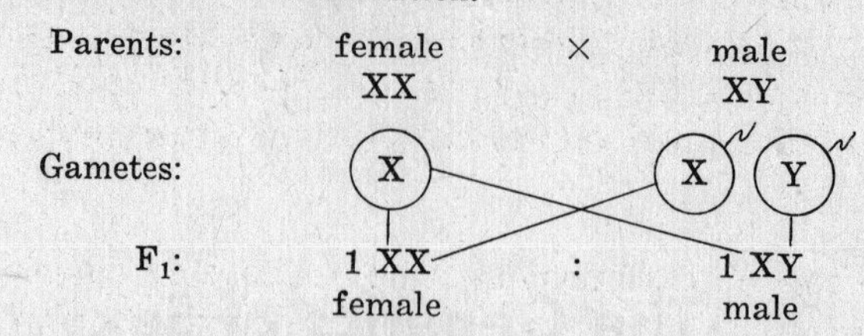


In some insects, especially those of the orders Hemiptera (true bugs) and Orthoptera (grasshoppers and roaches), males are also heterogametic, but produce either X-bearing sperm or gametes without a sex chromosome. In males of these species, the X chromosome has no homologous pairing partner because there is no Y chromosome present. Thus males exhibit an odd number in their chromosome complement. The one-X and two-X condition determines maleness and femaleness respectively. If the single X chromosome of the male is always included in one of the two types of gametes formed, then a 1:1 sex ratio will be produced in the progeny. This mode of sex determination is commonly referred to as the XO method where the O symbolizes the lack of a chromosome analogous to the Y of the XY system.

**Example 5.2.** XO Method of Sex Determination.

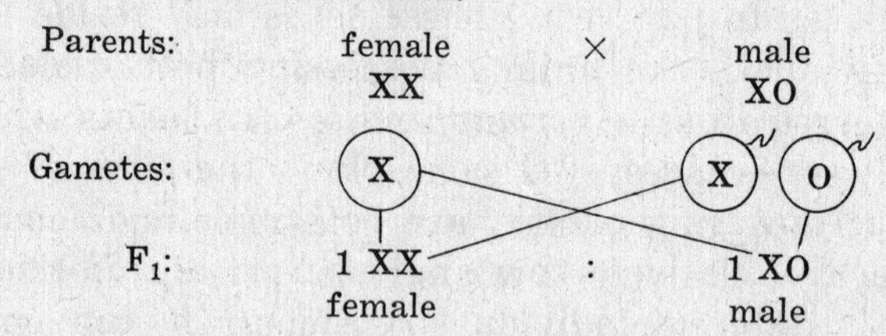


(*b*) ***Heterogametic Females.*** This method of sex determination is found in a comparatively large group of animals including the butterflies, moths, caddis flies, silkworms, and in some birds and fishes. The one-X and two-X condition in these species determines femaleness and maleness respectively. The females of some species (e.g. domestic chickens) have a chromosome similar to that of the Y in humans. In these cases, the chromosomes are sometimes labeled Z and W instead of X and Y respectively in order to call attention to the fact that the female (ZW) is the heterogametic sex and the male (ZZ) is the homogametic sex. The females of other species have no homologue to the single sex chromosome as in the case of the XO mechanism discussed previously. To point out this difference, the symbols ZZ and ZO may be used to designate males and females respectively. A 1:1 sex ratio is expected in either case.

**Example 5.3.** ZO Method of Sex Determination.

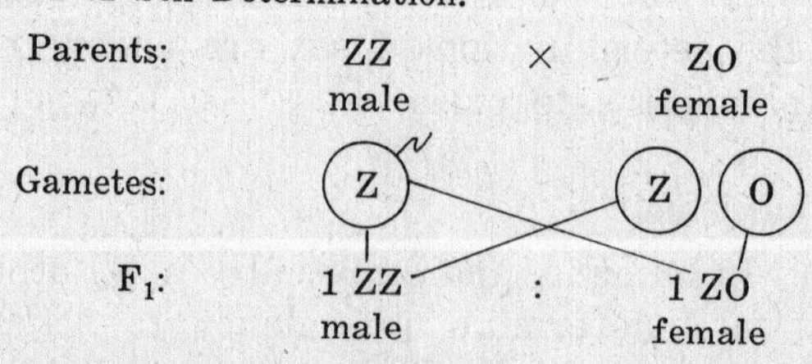


**Example 5.4.** ZW Method of Sex Determination.

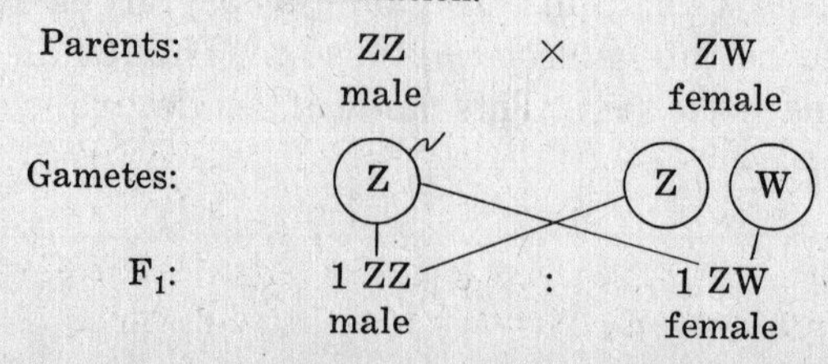

The W chromosome of the chicken is not a strong female determining element. Recent studies indicate that sex determination in chickens, and probably birds in general, is similar to that of *Drosophila*, i.e. dependent upon the ratio between the Z chromosomes and the number of autosomal sets of chromosomes (see next section on genic balance).

**2. Genic Balance.**

The presence of the Y chromosome in *Drosophila*, though it is essential for male fertility, apparently has nothing to do with the determination of sex. Instead, the factors for maleness residing in all of the autosomes are "weighed" against the factors for femaleness residing on the X chromosome(s). If each haploid set of autosomes carries factors with a male-determining value equal to one, then each X chromosome carries factors with a female-determining value of $1\frac{1}{2}$. Let A represent a haploid set of autosomes. In a normal male (AAXY), the male : female determinants are in the ratio $2:1\frac{1}{2}$ and therefore the balance is in favor of maleness. A normal female (AAXX) has a male : female ratio of $2:3$ and therefore the balance is in favor of femaleness. Several abnormal combinations of chromosomes have confirmed this hypothesis. For example, an individual with three sets of autosomes and two X chromosomes has a ratio of $3:3$ which makes its genetic sex neutral, and indeed phenotypically it appears as a sterile intersex.

**3. Haplodiploidy.**

Male bees are known to develop parthenogenetically from unfertilized eggs (arrhenotoky) and are therefore haploid. Females (both workers and queens) originate from fertilized (diploid) eggs. Sex chromosomes are not involved in this mechanism of sex determination which is characteristic of the insect order Hymenoptera including the ants, bees, wasps, etc. The quantity and quality of food available to the diploid larva determines whether that female will become a sterile worker or a fertile queen. Thus environment here determines sterility or fertility but does not alter the genetically determined sex. The sex ratio of the offspring is under the control of the queen. Most of the eggs laid in the hive will be fertilized and develop into worker females. Those eggs which the queen chooses not to fertilize (from her store of sperm in the seminal receptacle) will develop into fertile haploid males. Queen bees usually mate only once during their lifetime.

**4. Single Gene Effects.**

(*a*) ***Complementary Sex Factors.*** At least two members of the insect order Hymenoptera are known to produce males by homozygosity at a single gene locus as well as by haploidy. This has been confirmed in the tiny parasitic wasp *Bracon hebetor* (often called *Habrobracon juglandis*), and more recently in bees also. At least nine sex alleles are known at this locus in *Bracon* and may be represented by $s^a, s^b, s^c, \ldots, s^i$. All females must be heterozygotes such as $s^as^b$, $s^as^c$, $s^ds^f$, etc. If an individual is homozygous for any of these alleles such as $s^as^a$, $s^cs^c$, etc., it develops into a diploid male (usually sterile). Haploid males, of course, would carry only one of the alleles at this locus, e.g. $s^a$, $s^c$, $s^g$, etc.

**Example 5.5.**

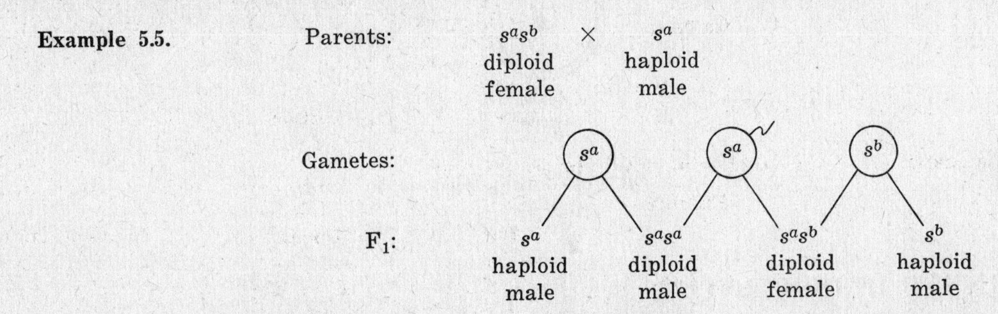


Among the diploid progeny we expect 1 $s^as^a$ male : 1 $s^as^b$ female. Among the haploid progeny we expect 1 $s^a$ male : 1 $s^b$ male.

(*b*) ***The "Transformer" Gene of Drosophila.*** A recessive gene (*tra*) on chromosome 3 of *Drosophila,* when homozygous, transforms a diploid female into a sterile male. The X/X, *tra*/*tra* individuals resemble normal males in external and internal morphology with the exception that the testes are much reduced in size. The gene is without effect in normal males. The presence of this gene can considerably alter the sex ratio (see Problem 5.1). The significance of these kinds of genes resides in the fact that a mechanism of sex determination based on numerous genes throughout the genome can apparently be nullified by a single gene substitution.

(*c*) ***"Mating Type" in Microorganisms.*** In microorganisms such as the alga *Chlamydomonas* and the fungi *Neurospora* and yeast, sex is under control of a single gene. Haploid individuals possessing the same allele of this "mating type" locus usually cannot fuse with each other to form a zygote, but haploid cells of opposite (complementary) allelic constitution at this locus may fuse. Asexual reproduction in the single celled motile alga *Chlamydomonas reinhardi* usually involves two mitotic divisions within the old cell wall (Fig. 5-1). Rupture of the sporangium releases the new generation of haploid *zoospores.* If nutritional requirements are satisfied, asexual reproduction may go on indefinitely. In unfavorable conditions where nitrogen balance is upset, daughter cells may be changed to gametes. Genetically there are two mating types, plus (+) and minus (−), which are morphologically indistinguishable and therefore called *isogametes.* Fusion of gametes unites two entire cells into a diploid nonmotile zygote which is relatively resistant to unfavorable growth conditions. With the return of conditions favoring growth, the zygote experiences meiosis and forms four motile haploid daughter cells (zoospores), two of plus and two of minus mating type.

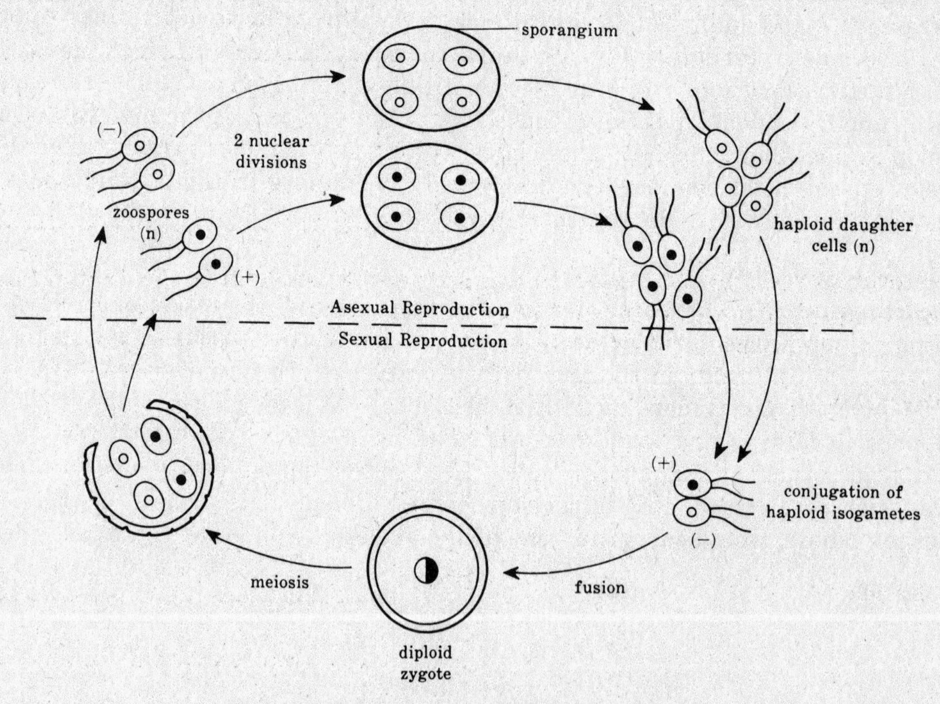


**Fig. 5-1.** Life cycle of *Chlamydomonas reinhardi.*

## SEX-LINKED INHERITANCE

Any gene located on the X chromosome (mammals, *Drosophila,* and others) or on the analogous Z chromosome (in birds and other species with the ZO or ZW mechanism of sex determination) is said to be *sex-linked*. The first sex-linked gene found in *Drosophila* was the recessive white eye mutation. Reciprocal crosses involving autosomal traits yield comparable results. This is not true with sex-linked traits as shown below. When white eyed females are crossed with wild type (red eyed) males, all the male offspring have white eyes like their mother and all the female offspring have red eyes like their father.

**Example 5.6.**

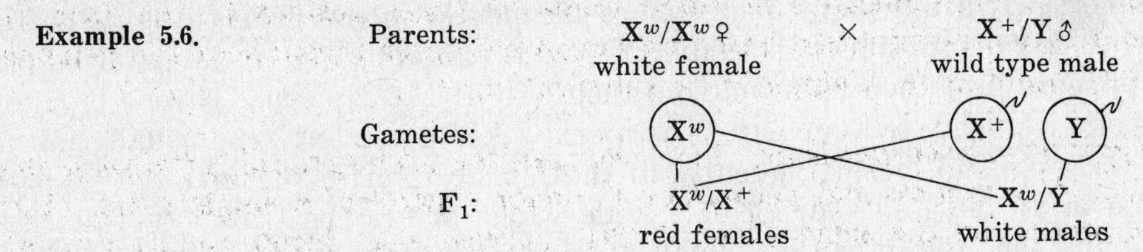


This criss-cross method of inheritance is characteristic of sex-linked genes. This peculiar type of inheritance is due to the fact that the Y chromosome carries no alleles homologous to those at the white locus on the X chromosome. In fact, in most organisms with the Y-type chromosome, the Y is virtually devoid of known genes. Thus males carry only one allele for sex-linked traits. This one-allelic condition is termed *hemizygous* in contrast to the homozygous or heterozygous possibilities in the female. If the $F_1$ of Example 5.6 mate among themselves to produce an $F_2$, a 1 red : 1 white phenotypic ratio is expected in both the males and females.

**Example 5.7.** $F_1$: $X^+/X^w$ red female × $X^w/Y$ white male

$F_2$:

| | $X^+$ | $X^w$ |
|---|---|---|
| $X^w$ ♂ | $X^+/X^w$ red female | $X^w/X^w$ white female |
| Y ♂ | $X^+/Y$ red male | $X^w/Y$ white male |

The reciprocal cross, where the sex-linked mutation appears in the male parent, results in the disappearance of the trait in the $F_1$ and its reappearance only in the males of the $F_2$. This type of skip-generation inheritance also characterizes sex-linked genes.

**Example 5.8.**

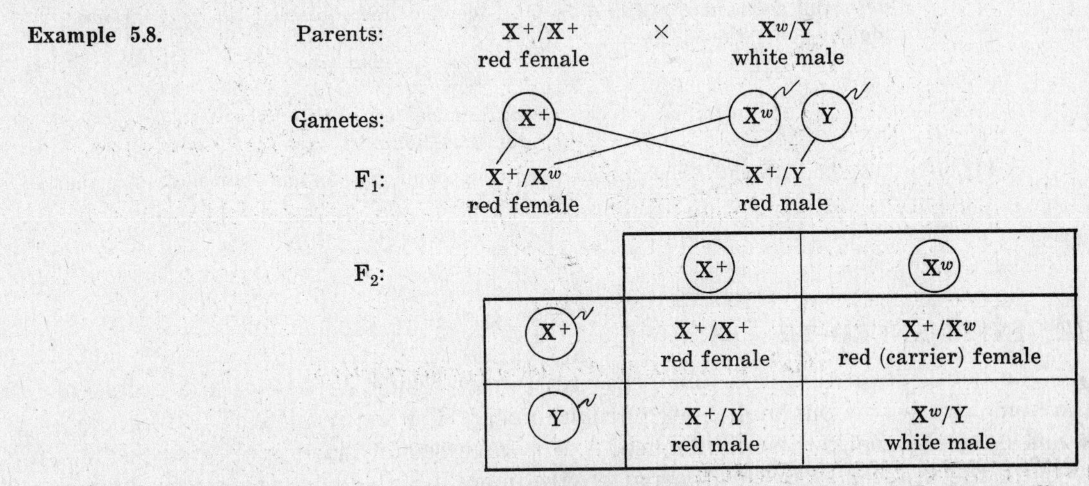


Thus a 3 red : 1 white phenotypic ratio is expected in the total $F_2$ disregarding sex, but only the males show the mutant trait. The phenotypic ratio among the $F_2$ males is 1 red : 1 white. All $F_2$ females are phenotypically wild type.

Whenever working with problems involving sex-linkage in this book, be sure to list the ratios for males and females separately unless specifically directed by the problem to do otherwise.

In normal diploid organisms with sex determining mechanisms like that of humans or *Drosophila*, a trait governed by a sex-linked recessive gene usually manifests itself in the following manner: (1) it is usually found more frequently in the male than in the female of the species, (2) it fails to appear in females unless it also appeared in the paternal parent, (3) it seldom appears in both father and son, then only if the maternal parent is heterozygous. On the other hand, a trait governed by a sex-linked dominant gene usually manifests itself by: (1) being found more frequently in the female than in the male of the species, (2) being found in all female offspring of a male which shows the trait, (3) failing to be transmitted to any son from a mother which did not exhibit the trait herself.

## VARIATIONS OF SEX-LINKAGE

The sex chromosomes (X and Y) often are of unequal size, shape and/or staining qualities. The fact that they pair during meiosis is indication that they contain at least some homologous segments. Genes on the homologous segments are said to be *incompletely sex-linked* or *partially sex-linked* and may recombine by crossing over in both sexes just as do the gene loci on homologous autosomes. Special crosses are required to demonstrate the presence of such genes on the X chromosome and few examples are known. Genes on the non-homologous segment of the X chromosome are said to be *completely sex-linked* and exhibit the peculiar mode of inheritance described in the preceding sections. In humans, a few genes are known to reside in the non-homologous portion of the Y chromosome. In such cases, the trait would be expresssed only in males and would always be transmitted from father to son. Such completely Y-linked genes are called *holandric genes* (Fig. 5-2).

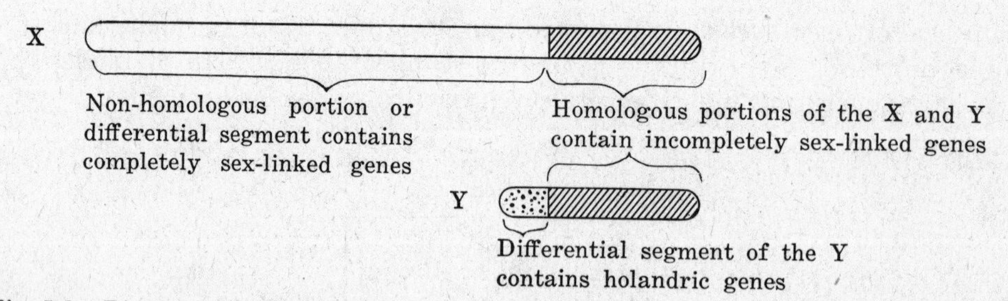


**Fig. 5-2.** Diagram of X and Y chromosomes showing homologous and non-homologous regions.

## SEX-INFLUENCED TRAITS

The genes governing sex-influenced traits may reside on any of the autosomes or on the homologous portions of the sex chromosomes. The expression of dominance or recessiveness by the alleles of sex-influenced loci is reversed in males and females due, in large part, to the difference in the internal environment provided by the sex hormones. Thus examples of sex-influenced traits are most readily found in the higher animals with well-developed endocrine systems.

**Example 5.9.** The gene for pattern baldness in humans exhibits dominance in men, but acts recessively in women.

| Genotypes | Phenotypes | |
|---|---|---|
| | Men | Women |
| $b'b'$ | bald | bald |
| $b'b$ | bald | non-bald |
| $bb$ | non-bald | non-bald |

## SEX-LIMITED TRAITS

Some autosomal genes may only come to expression in one of the sexes either because of differences in the internal hormonal environment or because of anatomical dissimilarities. For example, we know that bulls have many genes for milk production which they may transmit to their daughters, but they or their sons are unable to express this trait. The production of milk is therefore limited to variable expression in only the female sex. When the penetrance of a gene in one sex is zero, the trait will be *sex-limited.*

**Example 5.10.** Chickens have a recessive gene for cock-feathering which is penetrant only in the male environment.

| Genotype | Phenotypes | |
|---|---|---|
| | Males | Females |
| $HH$ | hen-feathering | hen-feathering |
| $Hh$ | hen-feathering | hen-feathering |
| $hh$ | cock-feathering | hen-feathering |

## SEX REVERSAL

Female chickens (ZW) that have laid eggs have been known to undergo not only a reversal of the secondary sexual characteristics such as development of cock-feathering, spurs, and crowing, but also the development of testes and even the production of sperm cells (primary sexual characteristics). This may occur when, for example, disease destroys the ovarian tissue, and in the absence of the female sex hormones the rudimentary testicular tissue present in the center of the ovary is allowed to proliferate. In solving problems involving sex reversals, it must be remembered that the functional male derived through sex reversal will still remain genetically female (ZW).

## SEXUAL PHENOMENA IN PLANTS

Most flowering plants are monoecious and therefore do not have sex chromosomes. Indeed, the ability of mitotically produced cells with exactly the same genetic endowment to produce tissues with different sexual functions in a perfect flower speaks clearly for the bipotentiality of such plant cells. Well-known examples of dioecism usually are under the genetic control of a single gene locus. However, at least one well-documented case of chromosomal sexuality is known in plants, i.e. in the genus *Melandrium* (a member of the pink family). Here the Y chromosome determines a tendency to maleness just as it does in humans. Pistillate plants are XX and staminate plants are XY.

The ability of gametes produced by the same individual to unite and produce viable and fertile offspring is common among many families of flowering plants. Self-fertilization is also known to occur in a few of the lower animal groups. The perfect flowers of some monoecious plants fail to open (*cleistogamy*) until after the pollen has matured and accomplished self-fertilization. Self-fertilization is obligatory in barley, beans, oats, peas, soybeans, tobacco, tomato, wheat and many other crops. In some species, self-fertilization as well as cross-fertilization may occur to varying degrees. For example, cotton and sorghum commonly experience more than 10% cross-fertilization. Still other monoecious species have developed genetic mechanisms which prevent self-fertilization or the development of zygotes produced by the union of identical gametes, making cross-fertilization obligatory. Self-incompatibility in monoecious species can become as efficient in enforcing cross-fertilization as would be exhibited under a dioecious mechanism of sex determination.

# Solved Problems

## SEX DETERMINING MECHANISMS

**5.1.** An autosomal recessive gene (*tra*), when homozygous, transforms a *Drosophila* female (X/X) into a phenotypic male. All such "transformed" males are sterile. The gene is without effect in males (X/Y). A cross is made between a female heterozygous at the *tra*-locus and a male homozygous recessive at the same locus. What is the expected sex ratio in the $F_1$ and $F_2$?

**Solution:**

We will use a slash mark (/) to separate alleles or homologous chromosomes, and a comma (,) to separate one gene locus from another.

Parents: X/X, +/*tra* normal female × X/Y, *tra*/*tra* normal male

Gametes: (X +) (X *tra*) (X *tra*) (Y *tra*)

$F_1$:

| | X + | X *tra* |
|---|---|---|
| X *tra* | X/X, +/*tra* normal females | X/X, *tra*/*tra* "transformed" males |
| Y *tra* | X/Y, +/*tra* normal males | X/Y, *tra*/*tra* normal males |

The $F_1$ phenotypic proportions thus appear as $\frac{3}{4}$ males : $\frac{1}{4}$ females.

$F_2$: The "transformed" $F_1$ males are sterile and hence do not contribute gametes to the $F_2$. Two kinds of matings must be considered. First mating $= \frac{1}{2}$ of all possible matings:

X/X, +/*tra* × X/Y, +/*tra*
females males

Offspring:

| | X + | X *tra* |
|---|---|---|
| X + | X/X, +/+ female | X/X, +/*tra* female |
| X *tra* | X/X, +/*tra* female | X/X, *tra*/*tra* "transformed" male |
| Y + | X/Y, +/+ male | X/Y, +/*tra* male |
| Y *tra* | X/Y, +/*tra* male | X/Y, *tra*/*tra* male |

Thus $F_2$ offspring from this mating type appear in the proportions $\frac{3}{8}$ female : $\frac{5}{8}$ male. But this type of mating constitutes only half of all possible matings. Therefore the contribution to the total $F_2$ from this mating is $\frac{1}{2} \cdot \frac{3}{8} = \frac{3}{16}$ female : $\frac{1}{2} \cdot \frac{5}{8} = \frac{5}{16}$ male. Second mating $= \frac{1}{2}$ of all possible matings:

X/X, +/*tra* × X/Y, *tra*/*tra*
females males

This is the same as the original parental mating and hence we expect $\frac{3}{4}$ males : $\frac{1}{4}$ females. Correcting these proportions for the frequency of this mating, we have $\frac{1}{2} \cdot \frac{3}{4} = \frac{3}{8}$ males : $\frac{1}{2} \cdot \frac{1}{4} = \frac{1}{8}$ females. Summary of the $F_2$ from both matings: males $= \frac{5}{16} + \frac{3}{8} = \frac{11}{16}$; females $= \frac{3}{16} + \frac{1}{8} = \frac{5}{16}$.

## SEX-LINKED INHERITANCE

**5.2.** There is a dominant sex-linked gene *B* which places white bars on an adult black chicken as in the Barred Plymouth Rock breed. Newly hatched chicks, which will become barred later in life, exhibit a white spot on the top of the head. (*a*) Diagram the cross through the $F_2$ between a homozygous barred male and a non-barred female. (*b*) Diagram the reciprocal cross through the $F_2$ between a homozygous non-barred male and a barred female. (*c*) Will both of the above crosses be useful in sexing $F_1$ chicks at hatching?

**Solution:**

(*a*) Parents: $Z^B/Z^B$ × $Z^b$/W
barred male non-barred female

Gametes: (*B*) ♂ (*b*) (W)

$F_1$: *B*/*b* *B*/W
barred male barred female

$F_2$:

| | (*B*) | (W) |
|---|---|---|
| (*B*) ♂ | *B*/*B* barred male | *B*/W barred female |
| (*b*) ♂ | *B*/*b* barred male | *b*/W non-barred female |

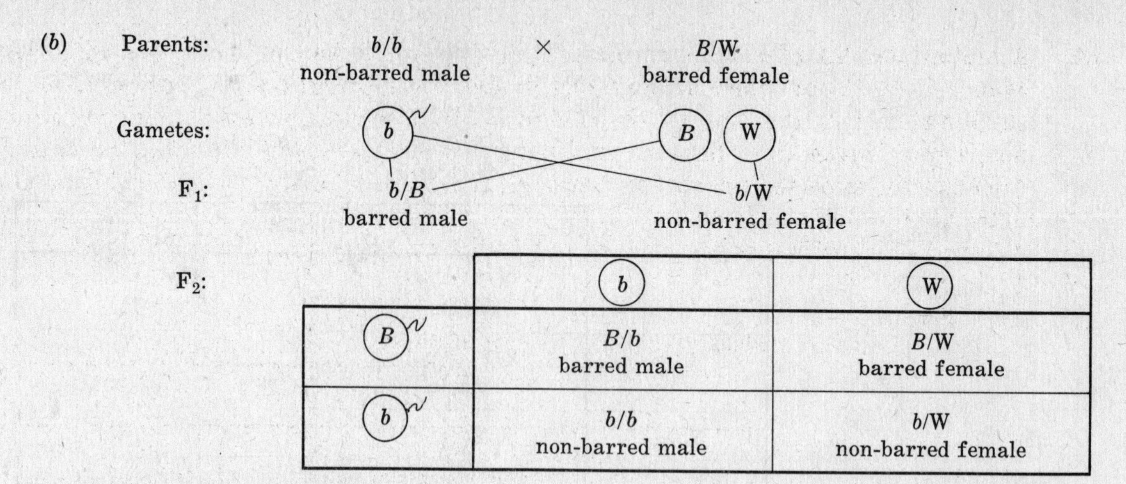


(c) No. Only the cross shown in (b) would be diagnostic in sexing $F_1$ chicks at birth through the use of this genetic marker. Only male chicks will have a light spot on their heads.

**5.3.** A recessive sex-linked gene ($h$) prolongs the blood-clotting time, resulting in what is commonly called "bleeder's disease" (hemophilia). From the information in the pedigree, answer the following questions. (*a*) If II2 marries a normal man, what is the chance of her first child being a hemophilic boy? (*b*) Suppose her first child is actually hemophilic. What is the probability that her second child will be a hemophilic boy? (*c*) If II3 marries a hemophilic man, what is the probability that her first child will be normal? (*d*) If the mother of I1 was phenotypically normal, what phenotype was her father? (*e*) If the mother of I1 was hemophilic, what phenotype was her father?

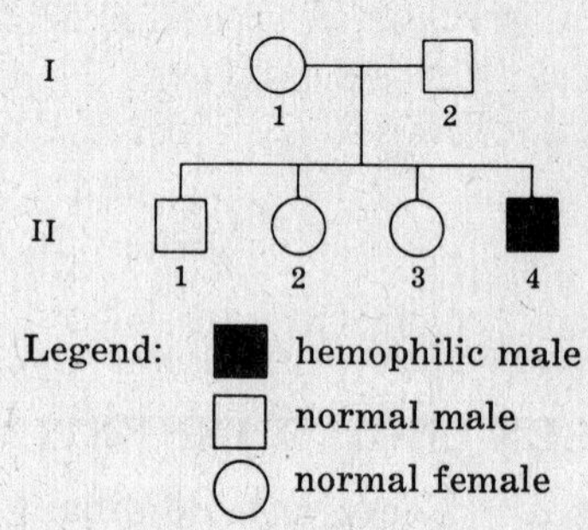


**Solution:**

(*a*) Since II4 is a hemophilic male ($h$Y), the hemophilic allele is on an X chromosome which he received from his mother (I1). But I1 is phenotypically normal and therefore must be heterozygous or a carrier of hemophilia of genotype $Hh$. I2 and II1 are both normal males ($H$Y). Therefore the chance of II2 being a carrier female ($Hh$) is $\frac{1}{2}$. When a carrier woman marries a normal man ($H$Y), 25% of their children are expected to be hemophilic boys ($h$Y). The combined probability that she is a carrier *and* will produce a hemophilic boy is $\frac{1}{2} \cdot \frac{1}{4} = \frac{1}{8}$.

(*b*) Because her first child was hemophilic, she *must* be a carrier. One-quarter of the children from carrier mothers ($Hh$) × normal fathers ($H$Y) are expected to be hemophilic boys ($h$Y).

(*c*) II3 (like II2) has a 50% chance of being a carrier of hemophilia ($Hh$). If she marries a hemophilic man ($h$Y), $\frac{1}{2}$ of their children (both boys and girls) are expected to be hemophilic. The combined chance of II3 being a carrier and producing a hemophilic child is $\frac{1}{2} \cdot \frac{1}{2} = \frac{1}{4}$. Therefore the probability that her first child is normal is the complementary fraction, $\frac{3}{4}$.

(*d*) It is impossible to deduce the phenotype of the father of I1 from the information given because the father could be either normal or hemophilic and still produce a daughter (I1) which is heterozygous normal ($Hh$), depending upon the genotype of the normal mother:

| | | | | | |
|---|---|---|---|---|---|
| (1) | $HH$<br>normal mother | × | $h$Y<br>hemophilic father | = | $Hh$(I1)<br>carrier daughter |
| (2) | $Hh$<br>carrier mother | × | $H$Y<br>normal father | = | $Hh$(I1)<br>carrier daughter |

(*e*) In order for a hemophilic mother ($hh$) to produce a normal daughter ($Hh$), her father must possess the dominant normal allele ($H$Y) and therefore would have normal blood clotting time.

**5.4.** A mutant sex-linked condition called "notch" ($N$) is lethal in *Drosophila* when hemizygous in males or when homozygous in females. Heterozygous females ($Nn$) have small notches in the tips of their wings. Homozygous recessive females ($nn$) or hemizygous males ($n$Y) have normal wings (wild type). (*a*) Other than within sex, calculate the expected phenotypic ratios in the viable $F_1$ and $F_2$ when wild type males are mated to notched females. (*b*) What is the ratio of males : females in the viable $F_1$ and $F_2$? (*c*) What is the ratio of notched : wild type in the viable $F_1$ and $F_2$?

**Solution:**

(*a*) Parents: $n$Y (wild type males) × $Nn$ (notched females)

$F_1$:

| | ($N$) | ($n$) |
|---|---|---|
| ($n$)♂ | $Nn$ notched females | $nn$ wild type females |
| (Y)♂ | $N$Y lethal | $n$Y wild type males |

The ratio of $F_1$ viable phenotypes is 1 wild type female : 1 notched female : 1 wild type male.

$F_2$: There are two kinds of matings to be considered: (1) $Nn \times n$Y, (2) $nn \times n$Y. The first mating gives results identical to those of the $F_1$. The second mating produces equal numbers of wild type males and females. The two kinds of matings are expected to occur with equal frequency and therefore the contribution which each phenotype makes to the total $F_2$ must be halved.

| 1st Mating | 2nd Mating |
|---|---|
| $\frac{1}{4}$ notched females × $\frac{1}{2}$ = $\frac{1}{8}$ | $\frac{1}{2}$ wild females × $\frac{1}{2}$ = $\frac{1}{4}$ = $\frac{2}{8}$ |
| $\frac{1}{4}$ wild females × $\frac{1}{2}$ = $\frac{1}{8}$ | $\frac{1}{2}$ wild males × $\frac{1}{2}$ = $\frac{1}{4}$ = $\frac{2}{8}$ |
| $\frac{1}{4}$ wild males × $\frac{1}{2}$ = $\frac{1}{8}$ | |
| $\frac{1}{4}$ lethal males × $\frac{1}{2}$ = $\frac{1}{8}$ | |

The ratio of viable $F_2$ phenotypes thus becomes 3 wild type females : 3 wild type males : 1 notched female.

(*b*) By inspection of the data in part (*a*), it is clear that the sex ratio in the viable $F_1$ is 2 females : 1 male, and in the $F_2$ it is 4 females : 3 males.

(*c*) Similarly, the ratio for wing phenotypes in the viable $F_1$ is 2 wild : 1 notched, and in the $F_2$ it is 6 wild : 1 notched.

**5.5.** A recessive sex-linked gene in *Drosophila* ($v$) produces vermilion eye color when homozygous in females or when hemizygous in males. An autosomal recessive on chromosome 2 ($bw$) produces brown eye color. Individuals which are homozygous recessive at both the brown and the vermilion loci have white eyes. (*a*) Determine the phenotypic expectations among the $F_1$ and $F_2$ when homozygous wild type females are crossed to males which are white eyed due to the interaction of the brown and vermilion loci. (*b*) Another recessive gene called "scarlet" ($st$) on chromosome 3 also produces white eye when homozygous in combination with homozygous recessive brown ($st/st, bw/bw$). Providing at least one $bw^+$ allele is present, homozygous scarlet, homozygous or hemizygous vermilion, or both produce a nearly identical eye color which we shall call "orange-red". What phenotypic ratio is expected among the $F_1$ and $F_2$ progeny from testcrossing wild type females homozygous at all three loci?

**Solution:**

(*a*) P: $bw^+/bw^+, v^+/v^+$ (wild type females) × $bw/bw, v/$Y (white eyed males)

$F_1$: 1/2 $bw^+/bw, v^+/v$ = wild type females

1/2 $bw^+/bw, v^+/$Y = wild type males

$F_2$:

| | $bw^+ \, v^+$ | $bw \, v^+$ | $bw^+$ Y | $bw$ Y |
|---|---|---|---|---|
| $bw^+ \, v^+$ | $bw^+/bw^+, v^+/v^+$ wild type | $bw^+/bw, v^+/v^+$ wild type | $bw^+/bw^+, v^+/Y$ wild type | $bw^+/bw, v^+/Y$ wild type |
| $bw^+ \, v$ | $bw^+/bw^+, v^+/v$ wild type | $bw^+/bw, v^+/v$ wild type | $bw^+/bw^+, v/Y$ vermilion | $bw^+/bw, v/Y$ vermilion |
| $bw \, v^+$ | $bw^+/bw, v^+/v^+$ wild type | $bw/bw, v^+/v^+$ brown | $bw^+/bw, v^+/Y$ wild type | $bw/bw, v^+/Y$ brown |
| $bw \, v$ | $bw^+/bw, v^+/v$ wild type | $bw/bw, v^+/v$ brown | $bw^+/bw, v/Y$ vermilion | $bw/bw, v/Y$ white |

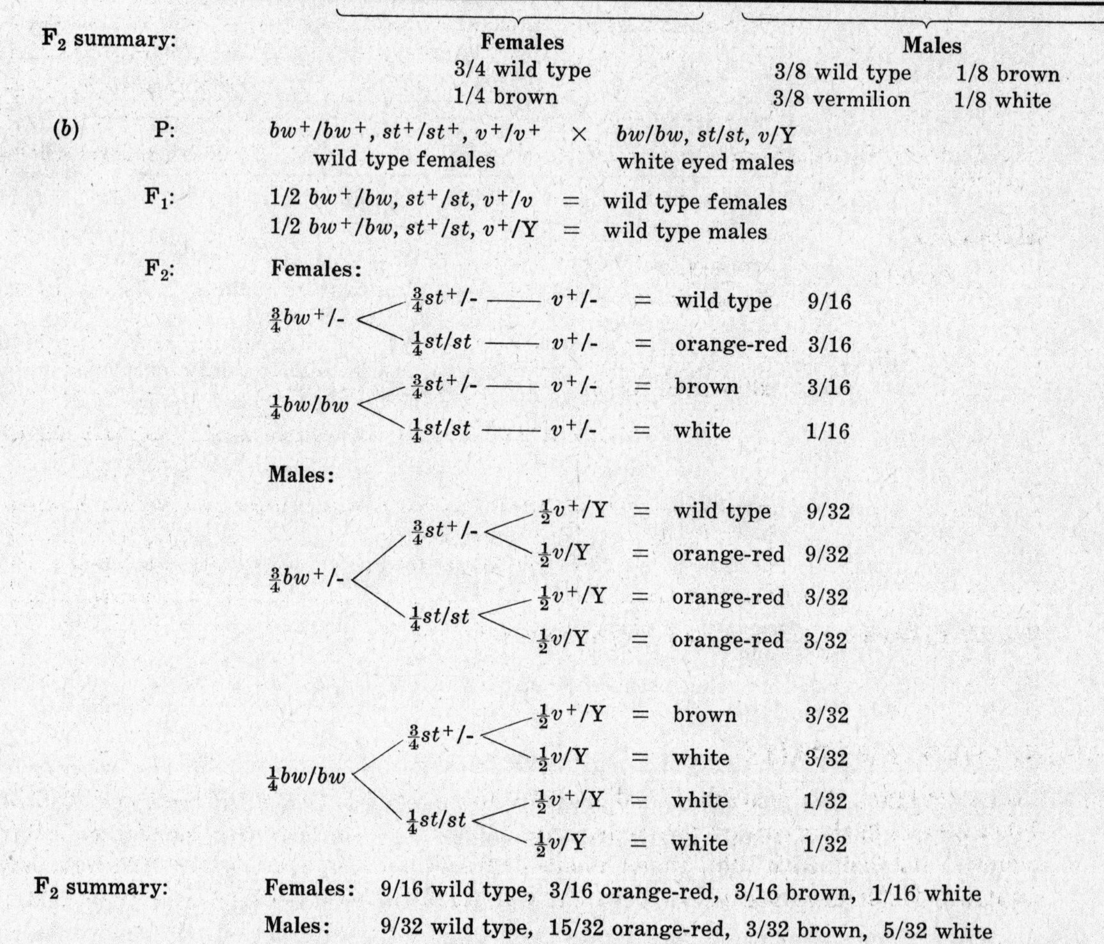

**$F_2$ summary:**

**Females**: 3/4 wild type, 1/4 brown

**Males**: 3/8 wild type, 3/8 vermilion, 1/8 brown, 1/8 white

(*b*) P: $bw^+/bw^+, st^+/st^+, v^+/v^+$ wild type females $\times$ $bw/bw, st/st, v/Y$ white eyed males

$F_1$: 1/2 $bw^+/bw, st^+/st, v^+/v$ = wild type females

1/2 $bw^+/bw, st^+/st, v^+/Y$ = wild type males

$F_2$: **Females:**

$\frac{3}{4}bw^+/-$ — $\frac{3}{4}st^+/-$ —— $v^+/-$ = wild type 9/16

$\frac{3}{4}bw^+/-$ — $\frac{1}{4}st/st$ —— $v^+/-$ = orange-red 3/16

$\frac{1}{4}bw/bw$ — $\frac{3}{4}st^+/-$ —— $v^+/-$ = brown 3/16

$\frac{1}{4}bw/bw$ — $\frac{1}{4}st/st$ —— $v^+/-$ = white 1/16

**Males:**

$\frac{3}{4}bw^+/-$ — $\frac{3}{4}st^+/-$ — $\frac{1}{2}v^+/Y$ = wild type 9/32

$\frac{3}{4}bw^+/-$ — $\frac{3}{4}st^+/-$ — $\frac{1}{2}v/Y$ = orange-red 9/32

$\frac{3}{4}bw^+/-$ — $\frac{1}{4}st/st$ — $\frac{1}{2}v^+/Y$ = orange-red 3/32

$\frac{3}{4}bw^+/-$ — $\frac{1}{4}st/st$ — $\frac{1}{2}v/Y$ = orange-red 3/32

$\frac{1}{4}bw/bw$ — $\frac{3}{4}st^+/-$ — $\frac{1}{2}v^+/Y$ = brown 3/32

$\frac{1}{4}bw/bw$ — $\frac{3}{4}st^+/-$ — $\frac{1}{2}v/Y$ = white 3/32

$\frac{1}{4}bw/bw$ — $\frac{1}{4}st/st$ — $\frac{1}{2}v^+/Y$ = white 1/32

$\frac{1}{4}bw/bw$ — $\frac{1}{4}st/st$ — $\frac{1}{2}v/Y$ = white 1/32

**$F_2$ summary:**

**Females:** 9/16 wild type, 3/16 orange-red, 3/16 brown, 1/16 white

**Males:** 9/32 wild type, 15/32 orange-red, 3/32 brown, 5/32 white

## VARIATIONS OF SEX LINKAGE

**5.6.** The recessive incompletely sex-linked gene called "bobbed" $bb$ causes the bristles of *Drosophila* to be shorter and of smaller diameter than the normal bristles produced by its dominant wild type allele $bb^+$. Determine the phenotypic expectations of the $F_1$ and $F_2$ when bobbed females are crossed with each of the two possible heterozygous males.

**Solution:**

Recall that an incompletely sex-linked gene has an allele on the homologous portion of the Y chromosome in a male. The wild type allele in heterozygous males may be on either the X or the Y, thus making possible two types of crosses.

**First Cross:**

Parents: $X^{bb}X^{bb}$ × $X^{bb+}Y^{bb}$
bobbed females — wild type males

$F_1$: $X^{bb}X^{bb+}$ and $X^{bb}Y^{bb}$
all females wild type — all males bobbed

$F_2$:

| | $X^{bb}$ | $X^{bb+}$ |
|---|---|---|
| $X^{bb}$ | $X^{bb}X^{bb}$ bobbed female | $X^{bb}X^{bb+}$ wild type female |
| $Y^{bb}$ | $X^{bb}Y^{bb}$ bobbed male | $X^{bb+}Y^{bb}$ wild type male |

Thus $\frac{1}{2}$ of the $F_2$ females are bobbed and $\frac{1}{2}$ are wild type; $\frac{1}{2}$ of the $F_2$ males are bobbed and $\frac{1}{2}$ are wild type.

**Second Cross:**

Parents: $X^{bb}X^{bb}$ × $X^{bb}Y^{bb+}$
bobbed female — wild type male

$F_1$: $X^{bb}X^{bb}$ and $X^{bb}Y^{bb+}$
all females bobbed — all males wild type

$F_2$:

| | $X^{bb}$ | $Y^{bb+}$ |
|---|---|---|
| $X^{bb}$ | $X^{bb}X^{bb}$ bobbed females | $X^{bb}Y^{bb+}$ wild type males |

Thus all $F_2$ females are bobbed and all males are wild type.

## SEX-INFLUENCED TRAITS

**5.7.** Let us consider two sex-influenced traits simultaneously, pattern baldness and short index finger, both of which are dominant in men and recessive in women. A heterozygous bald man with long index finger marries a heterozygous long-fingered, bald woman. Determine the phenotypic expectations for their children.

**Solution:**

Let us first select appropriate symbols and define the phenotypic expression of the three genotypes in each sex.

| Genotypes | Males | Females | Genotype | Males | Females |
|---|---|---|---|---|---|
| $B^1B^1$ | bald | bald | $F^1F^1$ | short-finger | short-finger |
| $B^1B^2$ | bald | non-bald | $F^1F^2$ | short-finger | long-finger |
| $B^2B^2$ | non-bald | non-bald | $F^2F^2$ | long-finger | long-finger |

P: $B^1B^2, F^2F^2$ × $B^1B^1, F^1F^2$
bald, long-fingered man — bald, long-fingered woman

F$_1$:

$\frac{1}{2}B^1B^1$ < $\frac{1}{2}F^1F^2$ = $\frac{1}{4}B^1B^1F^1F^2$ bald, short (men) / bald, long (women)
$\frac{1}{2}F^2F^2$ = $\frac{1}{4}B^1B^1F^2F^2$ bald, long (men) / bald, long (women)

$\frac{1}{2}B^1B^2$ < $\frac{1}{2}F^1F^2$ = $\frac{1}{4}B^1B^2F^1F^2$ bald, short (men) / non-bald, long (women)
$\frac{1}{2}F^2F^2$ = $\frac{1}{4}B^1B^2F^2F^2$ bald, long (men) / non-bald, long (women)

F$_1$ summary: **Men:** 1/2 bald, short-fingered : 1/2 bald, long-fingered
**Women:** 1/2 bald, long-fingered : 1/2 non-bald, long-fingered

## SEX-LIMITED TRAITS

**5.8.** Cock-feathering in chickens is a trait limited to expression only in males and determined by the autosomal recessive genotype *hh*. The dominant allele (*H*) produces hen-feathered males. All females are hen-feathered regardless of genotype. A cock-feathered male is mated to three females, each of which produces a dozen chicks. Among the 36 progeny are 15 hen-feathered males, 18 hen-feathered females and 3 cock-feathered males. What are the most probable genotypes of the three parental females?

**Solution:**

In order for both hen-feathered (*H*-) and cock-feathered (*hh*) males to be produced, at least one of the females had to be heterozygous (*Hh*) or recessive (*hh*). The following female genotype possibilities must be explored:

(*a*) 2 *HH*, 1 *Hh* (*c*) 1 *HH*, 1 *Hh*, 1 *hh* (*e*) 2 *Hh*, 1 *hh* (*g*) 2 *HH*, 1 *hh*
(*b*) 1 *HH*, 2 *Hh* (*d*) 3 *Hh* (*f*) 1 *Hh*, 2 *hh* (*h*) 2 *hh*, 1 *HH*

Obviously, the more *hh* or *Hh* hen genotypes, proportionately more cock-feathered males are expected in the progeny. The ratio of 15 hen-feathered males : 3 cock-feathered males is much greater than the 1 : 1 ratio expected when all three females are heterozygous (*Hh*).

P: *hh* (cock-feathered male) × *Hh* (hen-feathered females)

F$_1$: $\frac{1}{2}Hh$ hen-feathered males, $\frac{1}{2}hh$ cock-feathered males

Possibility (*d*) is therefore excluded. Possibilities (*e*) and (*f*), which both contain one or more *hh* genotypes in addition to one or more *Hh* genotypes, must also be eliminated because these matings would produce even more cock-feathered males than possibility (*d*). In possibility (*g*), the 2 *HH* : 1 *hh* hens are expected to produce an equivalent ratio of 2 hen-feathered (*Hh*) : 1 cock-feathered (*hh*) males. This 2 : 1 ratio should be expressed in the 18 male offspring as 12 hen-feathered : 6 cock-feathered. These numbers compare fairly well with the observed 15 : 3, but possibility (*h*) would be even less favorable because even more cock-feathered males would be produced. Let us see if one of the remaining three possibilities will give us expected values closer to our observations.

Possibility (*c*):

P: *hh* (cock-feathered male) × $\frac{1}{3}HH$, $\frac{1}{3}Hh$, $\frac{1}{3}hh$ } hen-feathered females

F$_1$: $\frac{1}{3}(hh \times HH)$ = $\frac{1}{3}Hh$ hen-feathered males

$\frac{1}{3}(hh \times Hh)$ = { $\frac{1}{3} \cdot \frac{1}{2} = \frac{1}{6}Hh$ hen-feathered males; $\frac{1}{3} \cdot \frac{1}{2} = \frac{1}{6}hh$ cock-feathered males }

$\frac{1}{3}(hh \times hh)$ = $\frac{1}{3}hh$ cock-feathered males

Summary: Hen-feathered males $= \frac{1}{3} + \frac{1}{6} = \frac{1}{2}$, Cock-feathered males $= \frac{1}{6} + \frac{1}{3} = \frac{1}{2}$

Again this disagrees with the observations and must be excluded.

Possibility (*b*):

P: *hh* (cock-feathered male) × $\frac{1}{3}HH$, $\frac{2}{3}Hh$ } hen-feathered females

F$_1$: $\frac{1}{3}(hh \times HH) = \frac{1}{3}Hh$ hen-feathered males

$\frac{2}{3}(hh \times Hh) = \begin{cases} \frac{2}{3} \cdot \frac{1}{2} = \frac{2}{6}Hh \text{ hen-feathered males} \\ \frac{2}{3} \cdot \frac{1}{2} = \frac{2}{6}hh \text{ cock-feathered males} \end{cases}$

Summary: Hen-feathered males $= \frac{1}{3} + \frac{2}{6} = \frac{2}{3}$, Cock-feathered males $= \frac{2}{6}$ or $\frac{1}{3}$

These expectations are no closer to the observations than those of possibility (*g*).

Possibility (*a*):

P: $hh$ cock-feathered male $\times$ $\left.\begin{matrix} \frac{2}{3}HH \\ \frac{1}{3}Hh \end{matrix}\right\}$ hen-feathered females

F$_1$: $\frac{2}{3}(hh \times HH) = \frac{2}{3}Hh$ hen-feathered males

$\frac{1}{3}(hh \times Hh) = \begin{cases} \frac{1}{3} \cdot \frac{1}{2} = \frac{1}{6}Hh \text{ hen-feathered males} \\ \frac{1}{3} \cdot \frac{1}{2} = \frac{1}{6}hh \text{ cock-feathered males} \end{cases}$

Summary: Hen-feathered males $= \frac{2}{3} + \frac{1}{6} = \frac{5}{6}$, Cock-feathered males $= \frac{1}{6}$

Set the observation of 3 cock-feathered males equal to the $\frac{1}{6}$, then 5 times 3 or 15 hen-feathered males should represent the $\frac{5}{6}$. These expectations agree perfectly with the observations and therefore it is most probable that two of the females were *HH* and one was *Hh*.

## SEX REVERSAL

**5.9.** Suppose that a hen's ovaries are destroyed by disease, allowing its rudimentary testes to develop. Further suppose that this hen was carrying the dominant sex-linked gene *B* for barred feathers, and upon sex reversal was then crossed to a non-barred female. What phenotypic proportions are expected in the F$_1$ and F$_2$?

**Solution:**

Remember that sex determination in chickens is by the ZW method and that sex reversal does not change this chromosomal constitution. Furthermore at least one sex chromosome (Z) is essential for life.

P: *B*W barred female sex reversed to a functional male $\times$ *b*W normal non-barred female

F$_1$:

| | (*B*) | (W) |
|---|---|---|
| (*b*) | *Bb* barred male | *b*W non-barred female |
| (W) | *B*W barred female | WW lethal |

The proportions are thus $\frac{1}{3}$ males (all barred) : $\frac{2}{3}$ females (half barred and half non-barred).

F$_2$: Two equally frequent kinds of matings are possible among the F$_1$ birds. First mating $= \frac{1}{2}$ of all matings.

*Bb* barred male $\times$ *b*W non-barred female

| Progeny Expectations | | Correction for Frequency of Mating | | Proportion of Total F$_2$ |
|---|---|---|---|---|
| $\frac{1}{4}Bb$ | • | $\frac{1}{2}$ | = | $\frac{1}{8}Bb$ barred males |
| $\frac{1}{4}bb$ | • | $\frac{1}{2}$ | = | $\frac{1}{8}bb$ non-barred males |
| $\frac{1}{4}$ *B*W | • | $\frac{1}{2}$ | = | $\frac{1}{8}$*B*W barred females |
| $\frac{1}{4}$ *b*W | • | $\frac{1}{2}$ | = | $\frac{1}{8}$*b*W non-barred females |

Second mating = $\frac{1}{2}$ of all matings.

$Bb$ barred male × $BW$ barred female

| Progeny Expectations | | Correction for Frequency of Mating | | Proportion of Total $F_2$ |
|---|---|---|---|---|
| $\frac{1}{4}BB$ | · | $\frac{1}{2}$ | = | $\frac{1}{8}BB$ } $\frac{1}{4}$ barred males |
| $\frac{1}{4}Bb$ | · | $\frac{1}{2}$ | = | $\frac{1}{8}Bb$ } |
| $\frac{1}{4}BW$ | · | $\frac{1}{2}$ | = | $\frac{1}{8}BW$ barred females |
| $\frac{1}{4}bW$ | · | $\frac{1}{2}$ | = | $\frac{1}{8}bW$ non-barred females |

Summary of the $F_2$: Barred males = $\frac{1}{8} + \frac{1}{4} = \frac{3}{8}$ Barred females = $\frac{1}{8} + \frac{1}{8} = \frac{1}{4}$

Non-barred males = $\frac{1}{8}$ Non-barred females = $\frac{1}{8} + \frac{1}{8} = \frac{1}{4}$

## SEXUAL PHENOMENA IN PLANTS

**5.10.** A recessive gene in monoecious corn called "tassel-seed" (*ts*), when homozygous, produces only seeds where the staminate inflorescence (tassel) normally appears. No pollen is produced. Thus individuals of genotype *ts/ts* are functionally reduced to a single sex, that of the female. On another chromosome, the recessive gene called "silkless" (*sk*), when homozygous, produces ears with no pistils (silks). Without silks, none of these ears can produce seed and individuals of genotype *sk/sk* are reduced to performing only male functions (production of pollen in the tassel). The recessive gene for tassel-seed is epistatic to the silkless locus. (*a*) What sex ratio is expected in the $F_1$ and $F_2$ from the cross $ts/ts, sk^+/sk^+$ (female) × $ts^+/ts^+, sk/sk$ (male)? (*b*) How could the genes for tassel-seed and silkless be used to establish male and female plants (dioecious) that would continue, generation after generation, to produce progeny in the ratio of 1 male : 1 female?

**Solution:**

(*a*) P: $ts/ts, sk^+/sk^+$ (female) × $ts^+/ts^+, sk/sk$ (male)

$F_1$: $ts^+/ts, sk^+/sk$ monoecious (both male and female flowers)

$F_2$:
- $\frac{9}{16}$ $ts^+/-, sk^+/-$ = $\frac{9}{16}$ monoecious
- $\frac{3}{16}$ $ts^+/-, sk/sk$ = $\frac{3}{16}$ male
- $\frac{3}{16}$ $ts/ts, sk^+/-$ } = $\frac{4}{16}$ female
- $\frac{1}{16}$ $ts/ts, sk/sk$ }

(*b*) P: $ts/ts, sk/sk$ (female) × $ts^+/ts, sk/sk$ (male)

$F_1$:
- $\frac{1}{2}$ $ts^+/ts, sk/sk$ males
- $\frac{1}{2}$ $ts/ts, sk/sk$ females

Subsequent generations would continue to exhibit a 1:1 sex ratio for these dioecious plants.

**5.11.** Pollen tubes containing the same self-incompatibility allele as that found in the diploid tissue of the style grow so slowly that fertilization cannot occur before the flower withers. Pollen produced by a plant of genotype $S^1S^3$ would be of two types, $S^1$ and $S^3$. If this pollen were to land on the stigma of the same plant ($S^1S^3$), none of the pollen tubes would grow. If these pollen grains ($S^1$ and $S^3$) were to alight on a

| | | Male Parent | | | |
|---|---|---|---|---|---|
| | | A | B | C | D |
| Female Parent | A | – | $\frac{1}{4}$C, $\frac{1}{4}$D<br>$\frac{1}{4}$E, $\frac{1}{4}$F | $\frac{1}{2}$C<br>$\frac{1}{2}$D | $\frac{1}{2}$C<br>$\frac{1}{2}$D |
| | B | $\frac{1}{4}$C, $\frac{1}{4}$D<br>$\frac{1}{4}$E, $\frac{1}{4}$F | – | $\frac{1}{2}$C<br>$\frac{1}{2}$E | $\frac{1}{2}$D<br>$\frac{1}{2}$F |
| | C | $\frac{1}{2}$A<br>$\frac{1}{2}$D | $\frac{1}{2}$B<br>$\frac{1}{2}$E | – | $\frac{1}{2}$D<br>$\frac{1}{2}$A |
| | D | $\frac{1}{2}$A<br>$\frac{1}{2}$C | $\frac{1}{2}$B<br>$\frac{1}{2}$F | $\frac{1}{2}$A<br>$\frac{1}{2}$C | – |

stigma of genotype $S^1S^2$, then only the tubes containing the $S^3$ allele would be compatible with the alleles in the tissue of the style. If these pollen grains were to alight on a stigma of genotype $S^2S^4$, all of the pollen tubes would be functional. Four plant varieties (A, B, C and D) are crossed, with the results listed in the table above. Notice that two additional varieties (E and F) appear in the progeny. Determine the genotypes for all six varieties in terms of four self-sterility alleles ($S^1$, $S^2$, $S^3$ and $S^4$).

**Solution:**

None of the genotypes are expected to be homozygous for the self-incompatibility alleles because pollen containing the same allele present in the maternal tissue is not functional and therefore homozygosity is prevented. Thus six genotypes are possible with four self-incompatibility alleles: $S^1S^2$, $S^1S^3$, $S^1S^4$, $S^2S^3$, $S^2S^4$, $S^3S^4$. Crosses between genotypes with both alleles in common produce no progeny (e.g. A × A, B × B, etc.). Crosses between genotypes with only one allele in common produce offspring in the ratio of 1 : 1 (e.g. $S^1S^2$♀ × $S^1S^3$♂ = $\frac{1}{2}S^1S^3 : \frac{1}{2}S^2S^3$). Crosses between genotypes with none of their self-incompatibility alleles in common produce progeny in the ratio 1 : 1 : 1 : 1 (e.g. $S^1S^2 \times S^3S^4 = \frac{1}{4}S^1S^3 : \frac{1}{4}S^1S^4 : \frac{1}{4}S^2S^3 : \frac{1}{4}S^2S^4$). Turning now to the table of results, we find the cross B♀ × A♂ produces offspring in the ratio 1 : 1 : 1 : 1 and therefore neither B nor A contains any alleles in common. If we assume that variety B has the genotype $S^1S^4$, then variety A must have the genotype $S^2S^3$ (the student's solution to this problem may differ from the one presented here in the alleles arbitrarily assigned as a starting point). The cross C♀ × A♂ produces offspring in the ratio 1 : 1, indicating one pair of alleles in common. Since we have already designated variety A to be of genotype $S^2S^3$, let us arbitrarily assign the genotype $S^1S^2$ to variety C. The cross D♀ × A♂ also indicates that one allele is held in common by these two varieties. Let us assign the genotype $S^1S^3$ to variety D. The genotype for variety E may now be determined from the cross C♀ × B♂.

P: $S^1S^2$(C)♀ × $S^1S^4$(B)♂

$F_1$: $\frac{1}{2}S^1S^4$ = variety B, $\frac{1}{2}S^2S^4$ = variety E

Likewise, the genotype for variety F may now be determined from the cross D♀ × B♂.

P: $S^1S^3$(D)♀ × $S^1S^4$(B)♂

$F_1$: $\frac{1}{2}S^1S^4$ = variety B, $\frac{1}{2}S^3S^4$ = variety F

Summary of genotypes for all six varieties:

A = $S^2S^3$ B = $S^1S^4$ C = $S^1S^2$ D = $S^1S^3$ E = $S^2S^4$ F = $S^3S^4$

The student should satisfy him- or herself that the other results shown in the table are compatible with the genotypic assumptions shown above.

# Supplementary Problems

## SEX DETERMINATION AND SEX-LINKED INHERITANCE

### *Heterogametic Males (XY and XO methods)*

**5.12.** A sex-linked recessive gene *c* produces red-green color blindness in humans. A normal woman whose father was color blind marries a color blind man. (*a*) What genotypes are possible for the mother of the color blind man? (*b*) What are the chances that the first child from this marriage will be a color blind boy? (*c*) Of all the girls produced by these parents, what percentage is expected to be color blind? (*d*) Of all the children (sex unspecified) from these parents, what proportion is expected to be normal?

**5.13.** The gene for yellow body color *y* in *Drosophila* is recessive and sex-linked. Its dominant allele $y^+$ produces wild type body color. What phenotypic ratios are expected from the crosses (*a*) yellow male × yellow female, (*b*) yellow female × wild type male, (*c*) wild type female (homozygous) × yellow male, (*d*) wild type (carrier) female × wild type male, (*e*) wild type (carrier) female × yellow male?

**5.14.** A narrow reduced eye called "bar" is a dominant sex-linked condition (*B*) in *Drosophila*, and the full wild type eye is produced by its recessive allele $B^+$. A homozygous wild type female is mated to a bar-eyed male. Determine the $F_1$ and $F_2$ genotypic and phenotypic expectations.

**5.15.** Sex determination in the grasshopper is by the XO method. The somatic cells of a grasshopper are analyzed and found to contain 23 chromosomes. (*a*) What sex is this individual? (*b*) Determine the frequency with which different types of gametes (number of autosomes and sex chromosomes) can be formed in this individual. (*c*) What is the diploid number of the opposite sex?

**5.16.** Male house cats may be black or yellow. Females may be black, tortoise-shell pattern, or yellow. (*a*) If these colors are governed by a sex-linked locus, how can these results be explained? (*b*) Using appropriate symbols, determine the phenotypes expected in the offspring from the cross yellow female × black male. (*c*) Do the same for the reciprocal cross of part (*b*). (*d*) A certain kind of mating produces females, half of which are tortoise-shell and half are black; half the males are yellow and half are black. What colors are the parental males and females in such crosses? (*e*) Another kind of mating produces offspring, $\frac{1}{4}$ of which are yellow males, $\frac{1}{4}$ yellow females, $\frac{1}{4}$ black males and $\frac{1}{4}$ tortoise-shell females. What colors are the parental males and females in such crosses?

**5.17.** In the plant genus *Melandrium*, sex determination is similar to that in humans. A sex-linked gene (*l*) is known to be lethal when homozygous in females. When present in the hemizygous condition in males (*l*Y), it produces blotchy patches of yellow-green color. The homozygous or heterozygous condition of the wild type allele (*LL* or *Ll*) in females, or the hemizygous condition in males (*L*Y) produces normal dark green color. From a cross between heterozygous females and yellow-green males, predict the phenotypic ratio expected in the progeny.

**5.18.** The recessive gene for white eye color in *Drosophila* (*w*) is sex-linked. Another recessive sex-linked gene governing eye color is vermilion (*v*), which when homozygous in females or hemizygous in males together with the autosomal gene for brown eye (*bw*/*bw*), also produces white eye. White genotypes (*w*Y, *ww*) are epistatic to the other loci under consideration. (*a*) What phenotypic results are expected among progeny from mating a white-eyed male of genotype ($bw/bw$, $vw^+/\text{Y}$) with a white-eyed female of genotype ($bw^+/bw$, $vw/v^+w$)? *Hint:* See Problem 5.5. (*b*) What phenotypic proportions are expected in the progeny from the mating of a vermilion female heterozygous at the brown locus but not carrying the white allele with a male which is white due to the *w* allele but heterozygous at the brown locus and hemizygous for the vermilion allele? (*c*) Determine the expected phenotypic ratio in the $F_1$ and $F_2$ from the reciprocal cross of Problem 5.5(*a*).

### *Heterogametic Females (ZW and ZO methods)*

**5.19.** A recessive sex-linked gene (*k*) influences a slower growth rate of the primary feathers of chickens than its dominant allele ($k^+$) for fast feathering. This trait can be used for sexing chicks within a few days after hatching. (*a*) If fast feathering females are crossed to slow feathering males, what phenotypic ratio is expected among the $F_1$ and $F_2$? (*b*) What are the expected $F_1$ and $F_2$ phenotypic ratios from crossing fast feathering males ($k^+/k^+$) to slow feathering females? (*c*) What are the expected $F_1$ and $F_2$ phenotypic ratios from crossing fast feathering males ($k^+/k$) to slow feathering females?

**5.20.** Silver-colored plumage in poultry is due to a dominant sex-linked gene (*S*) and gold-colored plumage to its recessive allele (*s*). List the phenotypic and genotypic expectations of the progeny from the matings (*a*) *s*/W ♀ × *S*/*S*♂, (*b*) *s*/W ♀ × *S*/*s*♂, (*c*) *S*/W♀ × *S*/*s*♂, (*d*) *S*/W♀ × *s*/*s*♂.

**5.21.** In the Rosy Gier variety of carrier pigeon, a cross was made between gray-headed females and creamy-headed males. The $F_1$ ratio was 1 gray-headed female : 1 gray-headed male : 1 creamy-headed male. (*a*) How may these results be explained? (*b*) Diagram this cross using appropriate symbols.

**5.22.** Chickens have an autosomal dominant gene (*C*) that produces a short-legged phenotype called "creeper" in heterozygotes. Normal legs are produced by the recessive genotype (*cc*). The homozygous dominant genotype (*CC*) is lethal. A dominant sex-linked gene (*B*) produces barred plumage, the recessive allele (*b*) produces non-barred plumage. (*a*) Determine the phenotypic expectations among progeny (of both sexes) from the cross of a barred creeper female and a non-barred creeper male. (*b*) Determine the phenotypic ratios within each sex for part (*a*). (*c*) Two chickens were mated and produced progeny in the following proportions: $\frac{1}{12}$ non-barred males, $\frac{1}{6}$ non-barred creeper females, $\frac{1}{12}$ barred males, $\frac{1}{12}$ non-barred females, $\frac{1}{6}$ non-barred creeper males, $\frac{1}{6}$ barred creeper males, $\frac{1}{12}$ barred females and $\frac{1}{6}$ barred creeper females. What are the genotypes and phenotypes of the parents?

**5.23.** A dominant autosomal inhibitor (*I*-) as well as a recessive autosomal inhibitor (*cc*) prevents any color from being produced in chickens. The genotypes *I-C-*, *I-cc* and *iicc* all produce white chickens; only the genotype *iiC-* produces colored birds. A recessive sex-linked gene *k* produces slow growth of the primary wing feathers. Its dominant allele $k^+$ produces fast feathering. A white (*IICC*) slow feathering male is mated to a white (*iicc*) fast feathering female. What are the $F_1$ and $F_2$ phenotypic expectations?

**5.24.** The presence of feathers on the shanks of the Black Langshan breed of chickens is due to the dominant alleles at either or both of two autosomal loci. Non-feathered shanks are the result of the double recessive genotype. A dominant sex-linked gene (*B*) places white bars on a black bird. Its recessive allele (*b*) produces non-barred (black) birds. Trihybrid barred males with feathered shanks are mated to dihybrid non-barred females with feathered shanks. Determine the $F_1$ phenotypic expectations.

***Genic Balance***

**5.25.** In *Drosophila*, the ratio between the number of X chromosomes and the number of sets of autosomes (A) is called the "sex index". Diploid females have a sex index (ratio X/A) = 2/2 = 1.0 Diploid males have a sex ratio of $\frac{1}{2}$ = 0.5. Sex index values between 0.5 and 1.0 give rise to intersexes. Values over 1.0 or under 0.5 produce weak and inviable flies called superfemales (meta-females) and supermales (meta-males) respectively. Calculate the sex index and the sex phenotype in the following individuals: (*a*) AAX, (*b*) AAXXY, (*c*) AAAXX, (*d*) AAXX, (*e*) AAXXX, (*f*) AAAXXX, (*g*) AAY.

***Haplodiploidy***

**5.26.** If the diploid number of the honey bee is 16, (*a*) how many chromosomes will be found in the somatic cells of the drone (male), (*b*) how many bivalents will be seen during the process of gametogenesis in the male, (*c*) how many bivalents will be seen during the process of gametogenesis in the female?

**5.27.** Seven eye colors are known in the honey bee, each produced by a recessive gene at a different locus: brick (*bk*), chartreuse (*ch*), ivory (*i*), cream (*cr*), snow (*s*), pearl (*pe*) and garnet (*g*). Suppose that a wild type queen heterozygous at the brick locus ($bk^+/bk$) was to be artificially inseminated with a mixture of sperm from seven haploid drones each exhibiting a different one of the seven mutant eye colors. Further assume that the semen contribution of each male contains equal concentrations of sperm, that each sperm has an equal opportunity to enter fertilization, and that each zygote thus formed has an equal opportunity to survive. (*a*) What percentage of the drone offspring is expected to be brick-eyed? (*b*) What percentage of worker offspring is expected to be brick-eyed?

***Single Gene Effects***

**5.28.** In the single-celled haploid plant *Chlamydomonas*, there are two mating types, (+) and (−). There is no morphological distinction between the (+) sex and the (−) sex in either the spore stage or the gamete stage (isogametes). The fusion of (+) and (−) gametes produces a 2n zygote which immediately undergoes meiosis producing four haploid spores, two of which are (+) and two are (−). (*a*) Could a pair of genes for sex account for the 1 : 1 sex ratio? (*b*) Does the foregoing information preclude some other form of sex determination? Explain.

**5.29.** Sex determination in the wasp *Bracon* is either by sex-alleles or haplodiploidy. A recessive gene called "veinless" (*v*) is known to assort independently of the sex alleles; the dominant allele $v^+$ results in wild type. For each of the eight crosses listed below, determine the relative frequencies of progeny phenotypes within each of 3 categories: (1) haploid males, (2) diploid males, (3) females. (*a*) $v/v, s^a/s^b \times v^+, s^a$, (*b*) $v/v, s^a/s^b \times v^+, s^c$, (*c*) $v/v, s^a/s^b \times v, s^b$, (*d*) $v/v, s^a/s^b \times v, s^c$, (*e*) $v/v^+, s^a/s^b \times v, s^a$, (*f*) $v/v^+, s^a/s^b \times v, s^c$, (*g*) $v^+/v^+, s^a/s^b \times v, s^a$, (*h*) $v^+/v^+, s^a/s^b \times v, s^c$.

## VARIATIONS OF SEX LINKAGE

**5.30.** An Englishman by the name of Edward Lambert was born in 1717. His skin was like thick bark which had to be shed periodically. The hairs on his body were quill-like and he subsequently has been referred to as the "porcupine man". He had six sons, all of which exhibited the same trait. The trait appeared to be transmitted from father to son through four generations. None of the daughters ever exhibited the trait. In fact, it has never been known to appear in females. (*a*) Could this be an autosomal sex-limited trait? (*b*) How is this trait probably inherited?

**5.31.** Could a recessive mutant gene in humans be located on the X chromosome if a woman exhibiting the recessive trait and a normal man had a normal son? Explain.

**5.32.** A holandric gene is known in humans which causes long hair to grow on the external ears. When men with hairy ears marry normal women, (*a*) what percentage of their sons would be expected to have hairy ears, (*b*) what proportion of the daughters is expected to show the trait, (*c*) what ratio of hairy eared : normal children is expected?

## SEX-INFLUENCED TRAITS

**5.33.** A certain type of white forelock in humans appears to follow the sex-influenced mode of inheritance, being dominant in men and recessive in women. Using the allelic symbols $w$ and $w'$, indicate all possible genotypes and the phenotypes thereby produced in men and women.

**5.34.** The sex-influenced gene governing the presence of horns in sheep exhibits dominance in males but acts recessively in females. When the Dorset breed (both sexes horned) with genotype $hh$ is crossed to the Suffolk breed (both sexes polled or hornless) with the genotype $h'h'$, what phenotypic ratios are expected in the $F_1$ and $F_2$?

**5.35.** The fourth (ring) finger of humans may be longer or shorter than the second (index) finger. The short index finger is thought to be produced by a gene which is dominant in men and recessive in women. What kinds of children and with what frequency would the following marriages be likely to produce: (*a*) heterozygous short-fingered man × short-fingered woman, (*b*) heterozygous long-fingered woman × homozygous short-fingered man, (*c*) heterozygous short-fingered man × heterozygous long-fingered woman, (*d*) long-fingered man × short-fingered woman?

**5.36.** In the Ayrshire breed of dairy cattle, mahogany-and-white color is dependent upon a gene $C^M$ which is dominant in males and recessive in females. Its allele for red-and-white ($C^R$) acts as a dominant in females but recessive in males. (*a*) If a red-and-white male is crossed to a mahogany-and-white female, what phenotypic and genotypic proportions are expected in the $F_1$ and $F_2$? (*b*) If a mahogany-and-white cow has a red-and-white calf, what sex is the calf? (*c*) What genotype is *not* possible for the sire of the calf in part (*b*)?

**5.37.** Long eared goats mated to short eared goats produce an ear of intermediate length in the $F_1$ and an $F_2$ consisting of $\frac{1}{4}$ long, $\frac{1}{2}$ intermediate, and $\frac{1}{4}$ short in both males and females. Non-bearded male goats mated to bearded female goats produce bearded male progeny and non-bearded female progeny. The $F_2$ males have $\frac{3}{4}$ bearded and $\frac{1}{4}$ non-bearded, while the $F_2$ females have $\frac{3}{4}$ non-bearded and $\frac{1}{4}$ bearded. A bearded male with ears of intermediate length whose father and mother were both non-bearded is mated with a non-bearded, intermediate-eared half-sib (sib = sibling = a brother or sister; half-sibs are half-brothers or half-sisters) by the same father but out of a bearded mother. List the phenotypic expectations among the progeny.

**5.38.** A sex-linked recessive gene in humans produces color blind men when hemizygous and color blind women when homozygous. A sex-influenced gene for pattern baldness is dominant in men and recessive in women. A heterozygous bald, color blind man marries a non-bald woman with normal vision whose father was non-bald and color blind and whose mother was bald with normal vision. List the phenotypic expectations for their children.

## SEX-LIMITED TRAITS

**5.39.** A dominant sex-limited gene is known to affect premature baldness in men but is without effect in women. (*a*) What proportion of the male offspring from parents, both of whom are heterozygous, is expected to be bald prematurely? (*b*) What proportion of all their children is expected to be prematurely bald?

**5.40.** The down of baby junglefowl chicks of genotype $S$- is darkly striped, whereas the recessive genotype $ss$ produces in both sexes an unstriped yellowish-white down. In the adult plumage, however, the character behaves as a sex-limited trait. Males, regardless of genotype, develop normal junglefowl plumage. Females of genotype $S$- bear normal junglefowl plumage but the recessive $ss$ is a creamy-buff color. A male bird unstriped at birth is mated to three females, each of which lays 16 eggs. Among the 48 progeny there are 32 unstriped chicks and 16 striped. At maturity there are 16 with creamy-buff and 32 with normal junglefowl plumage. What are the most probable genotypes of the three parental females?

**5.41.** In the clover butterfly, all males are yellow, but females may be yellow if they are of the homozygous recessive genotype $yy$ or white if they possess the dominant allele ($Y$-). What phenotypic proportions, exclusive of sex, are expected in the $F_1$ from the cross $Yy \times Yy$?

**5.42.** The barred plumage pattern in chickens is governed by a dominant sex-linked gene $B$. The gene for cock-feathering $h$ is recessive in males, its dominant allele $H$ producing hen-feathering. Normal females are hen-feathered regardless of genotype (sex-limited trait). Non-barred females heterozygous at the hen-feathered locus are crossed to a barred, hen-feathered male whose father was cock-feathered and non-barred. What phenotypic proportions are expected among the progeny?

**5.43.** Cock-feathering is a sex-limited trait in chickens (see Example 5.10). In the Leghorn breed, all males are cock-feathered and all females are hen-feathered. In the Sebright bantam breed, both males and females are hen-feathered. In the Hamburg breed, males may be either cock-feathered or hen-feathered, but females are always hen-feathered. (*a*) How can these results be explained? (*b*) If the ovaries or testes are removed and the chickens are allowed to molt, they will become cock-feathered regardless of genotype. What kind of chemicals are involved in the expression of genotypes at this locus?

## PEDIGREES

**5.44.** Could the trait represented by the solid symbols in the pedigree on the right be explained on the basis of (*a*) a dominant sex-linked gene, (*b*) a recessive sex-linked gene, (*c*) a holandric gene, (*d*) a sex-limited autosomal dominant, (*e*) a sex-limited autosomal recessive, (*f*) a sex-influenced autosomal gene dominant in males, (*g*) a sex-influenced autosomal gene recessive in males?

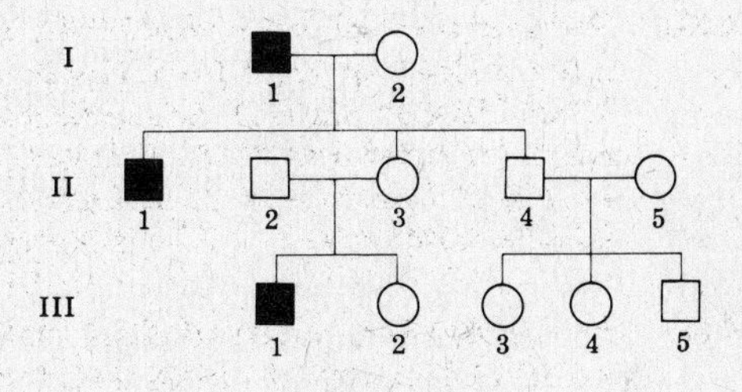

**5.45.**

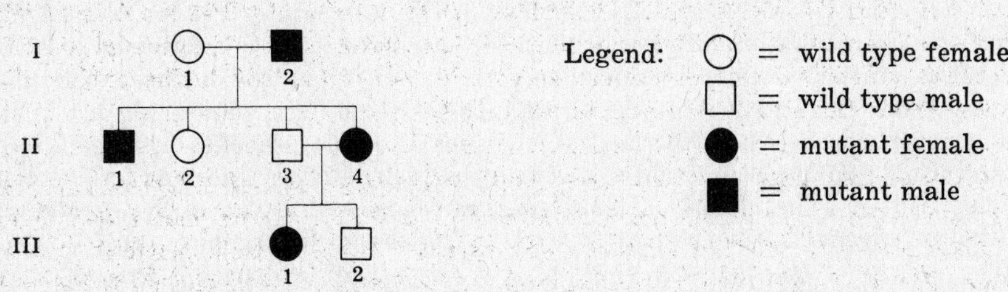


Could the assumption of a sex-linked recessive mutant gene be supported by the above pedigree? Explain?

**5.46.**

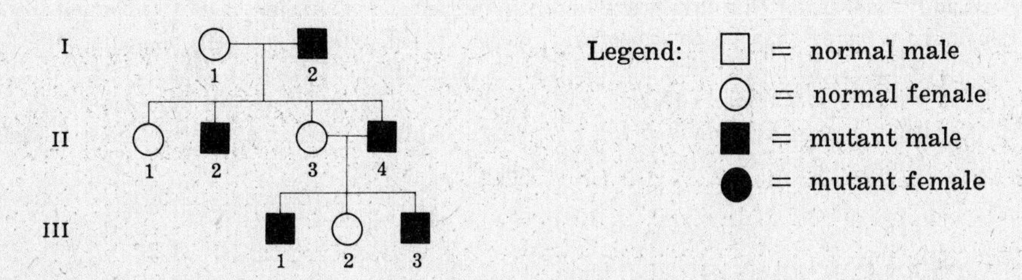


(*a*) Could the above pedigree be used as support for a holandric gene?

(*b*) Does the above pedigree contradict the assumption of a sex-linked recessive gene for the mutant trait?

(*c*) If a mating between III2 and III3 produced a mutant female offspring, which of the above two hypotheses would apply? List the genotype of each individual in the pedigree, using appropriate symbols.

5.47. Could the trait represented by the solid symbols in the pedigree shown below be produced by (*a*) an autosomal dominant, (*b*) an autosomal recessive, (*c*) a sex-linked dominant, (*d*) a sex-linked recessive, (*e*) a sex-limited gene, (*f*) a holandric gene, (*g*) a sex-influenced gene?

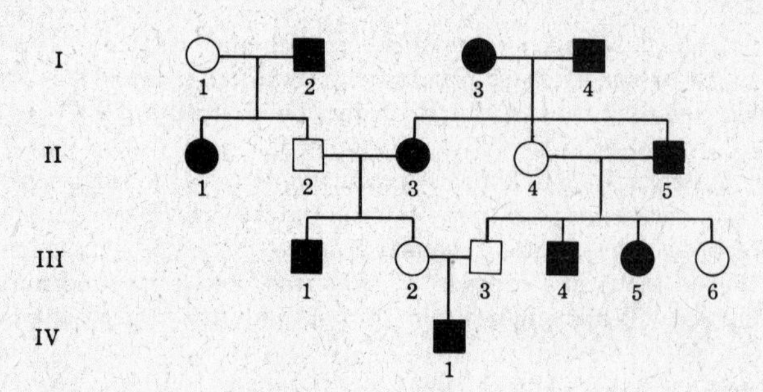


## SEX REVERSALS

5.48. Suppose that a female undergoes sex reversal to become a functional male and is then mated to a normal female. Determine the expected $F_1$ sex ratios from such matings in species with (*a*) ZW method of sex determination, (*b*) XY method of sex determination.

5.49. The hemp plant is dioecious, probably resulting from an XY mechanism of sex determination. Early plantings (May-June) yield the normal 1 : 1 sex ratio. Late plantings in November, however, produce all female plants. If this difference is due to the length of daylight, it should be possible to rear both XY females and XY males under controlled conditions in the greenhouse. What sex ratio would be expected among seedlings grown early in the year from crosses between XY males and XY females?

5.50. Suppose that a hen carrying the recessive sex-linked allele $k$ for slow feathering underwent a sex reversal and sired chicks from hens carrying the dominant allele $k^+$ for fast feathering. What genotypic and phenotypic proportions are expected in the $F_1$ and $F_2$?

5.51. The developing gonad in young larvae of goldfish (*Carassius auratus*) is ambisexual and subject to differentiate into either an ovary or a testis, irrespective of its sex genotype, by exogenous exposure to heterotypic sex hormones. The sex genes are not the direct cause of sex differentiation, but act indirectly by producing sex-inducing hormones. Female hormones (estrogens) and male hormones (androgens) are usually considered to be responsible for the expression of secondary sexual characteristics and for the maintenance of sexual capacities. However, in the case of this species, estrogens can also act as the *gynotermone* (ovary-inducing agent) and androgens can act as the *androtermone* (testis-inducing agent). (*a*) If females are heterogametic (ZW) and males are homogametic (ZZ), predict the offspring expected from a presumptive male (ZZ) converted by estrone (an estrogenic hormone) treatment into a female and mated to a normal male (ZZ). (*b*) If males are heterogametic (XY) and females are homogametic (XX), predict the zygotic expectations from a presumptive male (XY) induced to become a female and mated to a normal male (XY). (*c*) This species produces viable offspring in the ratios predicted in part (*b*). What is so unusual about this finding? (*d*) As an additional proof that males are heterogametic in this species, a methyltestosterone-induced male of a genotypic female is mated to a normal female. What type of progeny is expected? (*e*) An estrone-induced XY female was mated to a normal male and produced 7 sons (1 died). Each of the 6 viable sons was crossed with normal females (XX). Five of the six matings produced both male and female progeny. The sixth mating, however, produced 198 offspring, all males. The male parent lived 8 years. What does this indicate regarding the frequency of such males?

## SEXUAL PHENOMENA IN PLANTS

5.52. A completely pistillate inflorescence (female flower) is produced in the castor bean by the recessive genotype *nn*. Plants of genotype *NN* and *Nn* have mixed pistillate and staminate flowers in the inflorescence. Determine the types of flowers produced in the progeny from the following crosses: (*a*) *NN*♀ × *Nn*♂, (*b*) *Nn*♀ × *Nn*♂, (*c*) *nn*♀ × *Nn*♂.

**5.53.** Asparagus is a dioecious plant in which maleness (staminate plants) is governed by a dominant gene $P$ and femaleness (pistillate plants) by its recessive allele $p$. Sometimes pistillate flowers are found to have small non-functional anthers, and then again some staminate flowers may be found to possess immature pistils. Very rarely a staminate plant may be found to produce seed, most likely by self-fertilization. (*a*) What sex ratio is expected among the $F_1$ from an exceptional staminate-seed plant of genotype $Pp$ when selfed? (*b*) When the staminate $F_1$ plants from part (*a*) are crossed to normal pistillate plants ($pp$), what sex ratio is expected in the progeny? (*c*) What type of mating gives a 1 : 1 sex ratio?

**5.54.** Sex determination in the dioecious plant *Melandrium album* (*Lychnis dioica*) is by the XY method. A sex-linked gene governs leaf size, the dominant allele $B$ producing broad leaves and the recessive allele $b$ producing narrow leaves. Pollen grains bearing the recessive allele are inviable. What phenotypic results are expected from the following crosses?

| | Seed Parent | | Pollen Parent |
|---|---|---|---|
| (*a*) | homozygous broad-leaf | × | narrow-leaf |
| (*b*) | heterozygous broad-leaf | × | narrow-leaf |
| (*c*) | heterozygous broad-leaf | × | broad leaf |

**5.55.** Partial dioecy can be attained in a monoecious plant by the action of a single gene locus which prevents the production of viable gametes in one of the two types of gametangia (organ bearing gametes). The male-sterile condition is ordinarily recessive in most plant species where it has been studied. (*a*) Suppose that we artificially cross a pollen parent ($Ss$) onto a male-sterile (egg) parent of genotype $ss$. Determine the phenotypic ratio in the $F_1$ and $F_2$ (assuming complete randomness of mating, including selfing, among the $F_1$ types). (*b*) Determine the $F_1$ and $F_2$ expectancies in part (*a*) when the parental cross is $ss \times SS$. (*c*) If a locus ($A$), assorting independently of the male sterility locus ($S$), is jointly considered in the cross $ssAA \times Ssaa$, determine the $F_1$ and $F_2$ expectancies for the genotypes $S$-$A$-, $S$-$aa$, $ssA$- and $ssaa$. (*d*) Do likewise for part (*c*) where the parental cross is $ssAA \times SSaa$.

**5.56.** Two or more genes may cooperate to restrict selfing. An example is known in monoecious sorghum where the action of two complementary genes produces an essentially male plant by making the female structures sterile. Plants heterozygous at both loci ($Fs_1/fs_1, Fs_2/fs_2$) result in female sterile plants with no effect on their production of pollen. Whenever three dominant genes are present ($Fs_1/Fs_1, Fs_2/fs_2$ or $Fs_1/fs_1, Fs_2/Fs_2$), dwarf plants are produced which fail to develop a head. Although not yet observed, a genotype with all four dominant alleles would presumably also be dwarf and headless. All other genotypes produce normal plants. If these loci assort independently of one another, determine the $F_1$ phenotypic expectancies from the crosses (*a*) $Fs_1/fs_1, Fs_2/fs_2 \times Fs_1/fs_1, fs_2/fs_2$, (*b*) $Fs_1/fs_1, Fs_2/fs_2 \times Fs_1/Fs_1, fs_2/fs_2$.

**5.57.** In some cases of self-incompatibility, pollen tube growth is so slow that the style withers and dies before fertilization can occur. Sometimes, if pollination is artificially accomplished in the bud stage, the pollen tube can reach the ovary before the style withers. In this case it is possible to produce a genotype homozygous at the self-sterility locus. (*a*) What would be the expected results from natural pollination of such a homozygote ($S^1S^1$) by a heterozygote containing one allele in common ($S^1S^3$)? (*b*) What would be the result of the reciprocal cross of part (*a*)? (*c*) What would be the result of natural pollination of $S^1S^1$ by $S^2S^3$? (*d*) Would the reciprocal cross of part (*c*) make any difference in the progeny expectations?

**5.58.** Two heteromorphic types of flowers are produced in many species of the plant genus *Primula*. One type called "pin" has short anthers and a long style. The other type called "thrum" has highly placed anthers and a short style. Thrum is produced by a dominant gene ($S$) and pin by the recessive allele ($s$). The only pollinations that are compatible are those between styles and anthers of the same height, i.e. between thrum style and pin anther or between thrum anther and pin style. (*a*) What genotype do all thrum plants possess? (*b*) If both the pin and thrum are heterozygous for an independently segregating allelic pair ($Aa$), what genotypic ratio is expected in the next generation?

**5.59.** The self-incompatibility mechanism of many plants probably involves a series of multiple alleles similar to that found in *Nicotiana*. In this species, pollen tubes grow very slowly or not at all down the style that contains the same allele at the self-incompatibility locus ($S$). List the genotypic ratio of progeny sporophytes expected from the following crosses:

| | Seed Parent | | Pollen Parent |
|---|---|---|---|
| (a) | $S_1S_2$ | × | $S_1S_2$ |
| (b) | $S_1S_2$ | × | $S_1S_3$ |
| (c) | $S_1S_2$ | × | $S_3S_4$ |

(d) How much of the pollen is compatible in each of the above three crosses?

**5.60.** A cross is made between two plants of the self-sterile genotype $S^1S^2 \times S^3S^4$. If all the $F_1$ progeny are pollinated only by plants of genotype $S^2S^3$, what genotypic proportions are expected in the $F_2$?

# Answers to Supplementary Problems

**5.12.** (a) *Cc* or *cc* (b) $\frac{1}{4}$ (c) 50% (d) $\frac{1}{2}$

**5.13.** (a) all offspring yellow (b) all females wild type, all males yellow (c) all offspring wild type (d) all females wild type : $\frac{1}{2}$ wild type males : $\frac{1}{2}$ yellow males (e) females and males: $\frac{1}{2}$ wild type : $\frac{1}{2}$ yellow

**5.14.** $F_1$: $B^+/B$ bar eyed females, $B^+/Y$ wild type males; $F_2$ females: $\frac{1}{2}B^+/B^+$ wild type : $\frac{1}{2}B^+/B$ bar eye; $F_2$ males: $\frac{1}{2}B^+/Y$ wild type : $\frac{1}{2}B/Y$ bar eye

**5.15.** (a) male (b) $\frac{1}{2}(11A + 1X) : \frac{1}{2}(11A)$ (c) 24

**5.16.** (a) a pair of codominant sex-linked alleles.

| | Females | Males |
|---|---|---|
| black | $C^BC^B$ | $C^BY$ |
| tortoise-shell | $C^BC^Y$ | – |
| yellow | $C^YC^Y$ | $C^YY$ |

(b) all males yellow, all females tortoise-shell (c) all males black, all females tortoise-shell (d) tortoise-shell female × black male (e) tortoise-shell female × yellow male

**5.17.** $\frac{1}{3}$ dark green females : $\frac{1}{3}$ males with yellow-green patches : $\frac{1}{3}$ males dark green

**5.18.** (a) males all white eyed; females: $\frac{1}{4}$ vermilion : $\frac{1}{4}$ wild type : $\frac{1}{4}$ white : $\frac{1}{4}$ brown (b) males and females: $\frac{3}{4}$ vermilion : $\frac{1}{4}$ white (c) males and females: $\frac{3}{8}$ wild type : $\frac{3}{8}$ vermilion : $\frac{1}{8}$ brown : $\frac{1}{8}$ white

**5.19.** (a) $F_1$: fast males, slow females; $F_2$: males and females: $\frac{1}{2}$ fast : $\frac{1}{2}$ slow (b) $F_1$: all fast; $F_2$ males fast; $F_2$ females: $\frac{1}{2}$ fast : $\frac{1}{2}$ slow (c) $F_1$ both males and females: $\frac{1}{2}$ fast : $\frac{1}{2}$ slow; $F_2$ males: $\frac{5}{8}$ fast : $\frac{3}{8}$ slow; $F_2$ females: $\frac{1}{4}$ fast : $\frac{3}{4}$ slow

**5.20.** (a) silver females (*S*/W), silver males (*S*/*s*) (b) males: $\frac{1}{2}$ silver (*S*/*s*) : $\frac{1}{2}$ gold (*s*/*s*); females: $\frac{1}{2}$ silver (*S*/W) : $\frac{1}{2}$ gold (*s*/W) (c) males: all silver ($\frac{1}{2}S/S : \frac{1}{2}S/s$); females: $\frac{1}{2}$ silver (*S*/W) : $\frac{1}{2}$ gold (*s*/W) (d) all males silver (*S*/*s*), all females gold (*s*/W)

**5.21.** (a) Sex-linked gene with one allele lethal when hemizygous in females or homozygous in males.

| | Male | Female |
|---|---|---|
| gray | $HH$ | $H$W |
| cream | $HH^1$ | – |
| lethal | $H^1H^1$ | $H^1$W |

(b) P: $H\text{W} \times HH^1$; $F_1$: $\frac{1}{3}$ *HH* gray male : $\frac{1}{3}$ $HH^1$ cream male : $\frac{1}{3}$ *H*W gray female

**5.22.** (*a*) $\frac{1}{6}$ non-bar, normal leg females : $\frac{1}{6}$ bar, normal leg male : $\frac{1}{3}$ non-bar, creeper female : $\frac{1}{3}$ bar, creeper male
(*b*) males: $\frac{2}{3}$ bar, creeper : $\frac{1}{3}$ bar, normal leg; females: $\frac{2}{3}$ non-bar, creeper : $\frac{1}{3}$ non-bar, normal leg
(*c*) bar, creeper male ($CcBb$) × non-bar, creeper female ($CcbW$)

**5.23.** $F_1$: males: white, fast; females: white, slow; $F_2$ males and females: $\frac{13}{32}$ white, fast : $\frac{13}{32}$ white, slow : $\frac{3}{32}$ colored, fast : $\frac{3}{32}$ colored, slow

**5.24.** Males and females: $\frac{15}{32}$ bar, feathered : $\frac{15}{32}$ non-bar, feathered : $\frac{1}{32}$ bar, non-feathered : $\frac{1}{32}$ non-bar, non-feathered

**5.25.** (*a*) 0.5 male (*b*) 1.0 female (*c*) 0.67 intersex (*d*) 1.0 female (*e*) 1.5 superfemale (*f*) 1.0 female (triploid) (*g*) lethal

**5.26.** (*a*) 8 (*b*) none; meiosis cannot occur in haploid males (*c*) 8

**5.27.** (*a*) 50% (*b*) 7.14%

**5.28.** (*a*) Yes (*b*) No. A sex chromosome mechanism could be operative without a morphological difference in the chromosomes, gametes, or spores.

**5.29.**

| | Haploid Males | | Diploid Males | | Females | |
|---|---|---|---|---|---|---|
| | Wild type | Veinless | Wild type | Veinless | Wild type | Veinless |
| (*a*) | 0 | all | all | 0 | all | 0 |
| (*b*) | 0 | all | 0 | 0 | all | 0 |
| (*c*) | 0 | all | 0 | all | 0 | all |
| (*d*) | 0 | all | 0 | 0 | 0 | all |
| (*e*) | $\frac{1}{2}$ | $\frac{1}{2}$ | $\frac{1}{2}$ | $\frac{1}{2}$ | $\frac{1}{2}$ | $\frac{1}{2}$ |
| (*f*) | $\frac{1}{2}$ | $\frac{1}{2}$ | 0 | 0 | $\frac{1}{2}$ | $\frac{1}{2}$ |
| (*g*) | all | 0 | all | 0 | all | 0 |
| (*h*) | all | 0 | 0 | 0 | all | 0 |

**5.30.** (*a*) No. It is highly unlikely that a mutant autosomal sex-limited gene would be transmitted to all his sons through four generations without showing segregation. (*b*) holandric gene (Y-linked)

**5.31.** Yes, if it was incompletely sex-linked and the father carried the dominant normal gene on the homologous portion of his Y chromosome.

**5.32.** (*a*) 100% (*b*) none (*c*) 1 hairy : 1 normal

**5.33.**

| Genotypes | Men | Women | | Genotypes |
|---|---|---|---|---|
| $w$ dominant in men: | | | | $w'$ dominant in men: |
| $ww$ | forelock | forelock | | $w'w'$ |
| $ww'$ | forelock | normal | (or) | $w'w$ |
| $w'w'$ | normal | normal | | $ww$ |

**5.34.** $F_1$: all males horned, all females polled; $F_2$ males: $\frac{3}{4}$ horned : $\frac{1}{4}$ polled; $F_2$ females: $\frac{3}{4}$ polled : $\frac{1}{4}$ horned

**5.35.** (*a*) all males short; females: $\frac{1}{2}$ short : $\frac{1}{2}$ long (*b*) same as (*a*) (*c*) males: $\frac{3}{4}$ short : $\frac{1}{4}$ long; females: $\frac{1}{4}$ short : $\frac{3}{4}$ long (*d*) all males short, all females long

**5.36.** (*a*) $F_1$: $C^MC^R$ mahogany males, $C^MC^R$ red females; $F_2$ males and females: $\frac{1}{4}C^MC^M$ : $\frac{1}{2}C^MC^R$ : $\frac{1}{4}C^RC^R$; $F_2$ males: $\frac{3}{4}$ mahogany : $\frac{1}{4}$ red; $F_2$ females: $\frac{1}{4}$ mahogany : $\frac{3}{4}$ red (*b*) female (*c*) $C^MC^M$

5.37.

| Phenotype | Males | Females |
|---|---|---|
| bearded, long eared | 3/16 | 1/16 |
| bearded, intermediate eared | 3/8 | 1/8 |
| bearded, short eared | 3/16 | 1/16 |
| non-bearded, long eared | 1/16 | 3/16 |
| non-bearded, intermediate eared | 1/8 | 3/8 |
| non-bearded, short eared | 1/16 | 3/16 |

5.38.

| Phenotype | Daughters | Sons |
|---|---|---|
| bald, normal vision | 1/8 | 3/8 |
| bald, color blind | 1/8 | 3/8 |
| non-bald, normal vision | 3/8 | 1/8 |
| non-bald, color blind | 3/8 | 1/8 |

**5.39.** (*a*) $\frac{3}{4}$ (*b*) $\frac{3}{8}$

**5.40.** (2*ss*:1*SS*) or (2*Ss*:1*ss*)

**5.41.** $\frac{5}{8}$ yellow : $\frac{3}{8}$ white

**5.42.** Males: $\frac{3}{8}$ barred, hen-feathered : $\frac{1}{8}$ barred, cock-feathered : $\frac{3}{8}$ non-barred, hen-feathered : $\frac{1}{8}$ non-barred, cock-feathered; females: $\frac{1}{2}$ barred, hen-feathered : $\frac{1}{2}$ non-barred, hen-feathered

**5.43.** (*a*) Leghorns are homozygous *hh*. Sebright bantams are homozygous *HH*. Hamburgs are segregating at this locus; one or the other allele has not been "fixed" in the breed. (*b*) Gonads are the source of steroid sex hormones as well as of reproductive cells. The action of these genes is dependent upon the presence or absence of these sex hormones.

**5.44.** (*a*) no (*b*) yes (*c*) no (*d*) yes (*e*) yes (*f*) yes (*g*) no

**5.45.** No. Under the assumption, III1 must be of heterozygous genotype and therefore should be phenotypically normal; III2 must carry the recessive mutant in hemizygous condition and therefore should be phenotypically mutant.

**5.46.** (*a*) yes (*b*) no (*c*) sex-linked recessive gene; *Aa* (I1, II1, 3, III2), *a*Y (I2, II2, 4, III1, 3)

**5.47.** (*a*) through (*f*) no (*g*) yes (if black is dominant in males and recessive in females)

**5.48.** (*a*) 2 females : 1 male (*b*) all females

**5.49.** 2 males : 1 female

**5.50.** $F_1$: $\frac{1}{3}k^+/k$ fast males : $\frac{1}{3}k^+$/W fast females : $\frac{1}{3}k$/W slow females; $F_2$: $(\frac{1}{8}k^+/k^+ + \frac{1}{4}k^+/k) = \frac{3}{8}$ fast males : $\frac{1}{8}k/k$ slow males : $\frac{1}{4}k^+$/W fast females : $\frac{1}{4}k$/W slow females

**5.51.** (*a*) all ZZ males (*b*) 1 XX female : 2 XY males : 1 YY male (*c*) In most other organisms with XY sex determination, at least one X chromosome is essential for survival. (*d*) all XX females (*e*) $\frac{1}{6}$ tested sons proved YY; therefore YY males are not rare in this species and they appear to be as viable as normal XY males.

**5.52.** (*a*) all mixed (*b*) $\frac{3}{4}$ mixed : $\frac{1}{4}$ pistillate (*c*) $\frac{1}{2}$ mixed : $\frac{1}{2}$ pistillate

**5.53.** (*a*) $\frac{3}{4}$ staminate : $\frac{1}{4}$ pistillate (*b*) $\frac{2}{3}$ staminate : $\frac{1}{3}$ pistillate (*c*) $Pp \times pp$

**5.54.** (*a*) only broad-leaved males (*b*) $\frac{1}{2}$ broad-leaved males : $\frac{1}{2}$ narrow-leaved males (*c*) all females broad-leaved; $\frac{1}{2}$ males broad-leaved : $\frac{1}{2}$ males narrow-leaved

**5.55.** (*a*) $F_1$: $\frac{1}{2}$ male sterile : $\frac{1}{2}$ male fertile (normal = monoecious); $F_2$: $\frac{3}{8}$ male sterile : $\frac{5}{8}$ normal (*b*) $F_1$: all normal; $F_2$: $\frac{1}{4}$ male sterile : $\frac{3}{4}$ normal (*c*) $F_1$: $\frac{1}{2}SsAa$ normal : $\frac{1}{2}ssAa$ male sterile; $F_2$: $\frac{15}{32}S\text{-}A\text{-}$ : $\frac{9}{32}ssA\text{-}$ : $\frac{5}{32}S\text{-}aa$ : $\frac{3}{32}ssaa$ (*d*) $F_1$: $SsAa$ normal; $F_2$: $\frac{9}{16}S\text{-}A\text{-}$ : $\frac{3}{16}S\text{-}aa$ : $\frac{3}{16}ssA\text{-}$ : $\frac{1}{16}ssaa$

**5.56.** (*a*) $\frac{1}{8}$ dwarf : $\frac{1}{4}$ female sterile : $\frac{5}{8}$ normal (monoecious) (*b*) $\frac{1}{2}$ normal : $\frac{1}{4}$ female sterile : $\frac{1}{4}$ dwarf

**5.57.** (*a*) all $S^1S^3$ (*b*) no progeny (*c*) $\frac{1}{2}S^1S^2$ : $\frac{1}{2}S^1S^3$ (*d*) no

**5.58.** (*a*) $Ss$ (*b*) $\frac{1}{8}AASs$ : $\frac{1}{4}AaSs$ : $\frac{1}{8}aaSs$ : $\frac{1}{8}AAss$ : $\frac{1}{4}Aass$ : $\frac{1}{8}aass$

**5.59.** (*a*) none (*b*) $\frac{1}{2}S_1S_3$ : $\frac{1}{2}S_2S_3$ (*c*) $\frac{1}{4}S_1S_3$ : $\frac{1}{4}S_1S_4$ : $\frac{1}{4}S_2S_3$ : $\frac{1}{4}S_2S_4$ (*d*) $a$ = none, $b = \frac{1}{2}$, $c$ = all

**5.60.** $\frac{1}{4}S_1S_2$ : $\frac{1}{3}S_2S_3$ : $\frac{1}{12}S_1S_3$ : $\frac{1}{12}S_2S_4$ : $\frac{1}{4}S_3S_4$

# Chapter 6

## Linkage and Chromosome Mapping

### RECOMBINATION AMONG LINKED GENES

**1. Linkage.**

When two or more genes reside in the same chromosome, they are said to be *linked*. They may be linked together on one of the autosomes or connected together on the sex chromosome (Chapter 5). Genes on different chromosomes are distributed into gametes independently of one another (Mendel's Law of Independent Assortment). Genes on the same chromosome, however, tend to stay together during the formation of gametes. Thus the results of testcrossing dihybrid individuals will yield different results, depending upon whether the genes are linked or on different chromosomes.

**Example 6.1.** Genes on different chromosomes assort independently, giving a 1:1:1:1 testcross ratio.

Parents: $AaBb \times aabb$

Gametes: (*AB*) (*Ab*) (*aB*) (*ab*) (*ab*)

$F_1$: $\frac{1}{4}\,AaBb : \frac{1}{4}\,Aabb : \frac{1}{4}\,aaBb : \frac{1}{4}\,aabb$

**Example 6.2.** Linked genes do not assort independently, but tend to stay together in the same combinations as they were in the parents. Genes to the left of the slash line (/) are on one chromosome and those to the right are on the homologous chromosome.

Parents: $AB/ab \times ab/ab$

Gametes: (*AB*) (*ab*) (*ab*)

$F_1$: $\frac{1}{2}\,AB/ab : \frac{1}{2}\,ab/ab$

Large deviations from a 1:1:1:1 ratio in the testcross progeny of a dihybrid could be used as evidence for linkage. Linked genes do not always stay together, however, because homologous non-sister chromatids may exchange segments of varying length with one another during meiotic prophase. Recall from Chapter 1 that homologous chromosomes pair with one another in a process called "synapsis" and that the points of genetic exchange, called "chiasmata", produce recombinant gametes through crossing over.

**2. Crossing Over.**

During meiosis each chromosome replicates, forming two identical sister chromatids; homologous chromosomes pair (synapse) and crossing over occurs between non-sister chromatids. This latter process involves the breakage and reunion of only two of the four strands at any given point on the chromosomes. In the diagram below, a crossover occurs in the region between the A and B loci.

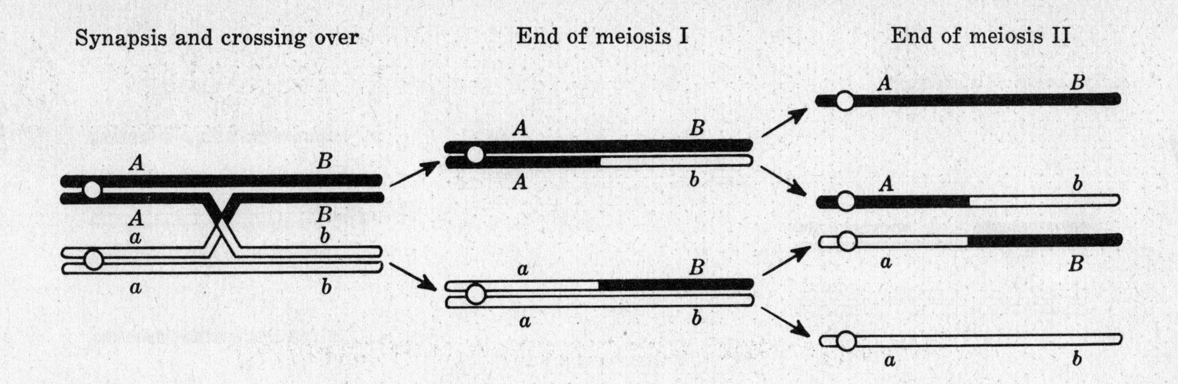


Notice that two of the meiotic products ($AB$ and $ab$) have the genes linked in the same way as they were in the parental chromosomes. These products are produced from chromatids that were not involved in crossing over and are referred to as *non-crossover* or *parental* types. The other two meiotic products ($Ab$ and $aB$) produced by crossing over have recombined the original linkage relationships of the parent into two new forms called *recombinant* or *crossover types*.

The alleles of double heterozygotes (dihybrids) at two linked loci may appear in either of two positions relative to one another. If the two dominant (or wild type) alleles are on one chromosome and the two recessives (or mutants) on the other ($AB/ab$), the linkage relationship is called *coupling phase*. When the dominant allele of one locus and the recessive allele of the other occupy the same chromosome ($Ab/aB$), the relationship is termed *repulsion phase*. Parental and recombinant gametes will be of different types, depending upon how these genes are linked in the parent.

**Example 6.3.** Coupling Parent: $AB/ab$

Gametes:
- Parental: ($AB$) ($ab$)
- Recombinant: ($Ab$) ($aB$)

**Example 6.4.** Repulsion Parent: $Ab/aB$

Gametes:
- Non-crossover: ($Ab$) ($aB$)
- Crossover: ($AB$) ($ab$)

### 3. Chiasma Frequency.

A pair of synapsed chromosomes (bivalent) consists of four chromatids called a *tetrad*. Every tetrad usually experiences at least one chiasma somewhere along its length. Generally speaking, the longer the chromosome the greater the number of chiasmata. Each type of chromosome within a species has a characteristic (or average) number of chiasmata. The frequency with which a chiasma occurs between any two genetic loci also has a characteristic or average probability. The further apart two genes are located on a chromosome, the greater the opportunity for a chiasma to occur between them. The closer two genes are linked, the smaller the chance for a chiasma occurring between them. These chiasmata probabilities are useful in predicting the proportions of parental and recombinant gametes expected to be formed from a given genotype. The percentage of crossover (recombinant) gametes formed by a given genotype is a direct reflection of the frequency with which a chiasma forms between the genes in question. Only when a crossover forms *between* the gene loci under consideration will recombination be detected.

**Example 6.5.** Crossing over outside the A-B region fails to recombine these markers.

Synapsis and crossing over — End of meiosis I — End of meiosis II

A B
a b

When a chiasma forms between two gene loci, only half of the meiotic products will be of crossover type. Therefore chiasma frequency is twice the frequency of crossover products.

$$\text{Chiasma } \% = 2(\text{crossover } \%) \quad \text{or} \quad \text{Crossover } \% = \tfrac{1}{2}(\text{chiasma } \%)$$

**Example 6.6.** If a chiasma forms between the loci of genes $A$ and $B$ in 30% of the tetrads of an individual of genotype $AB/ab$, then 15% of the gametes will be recombinant ($Ab$ or $aB$) and 85% will be parental ($AB$ or $ab$).

**Example 6.7.** Suppose progeny from the testcross $Ab/aB \times ab/ab$ were found in the proportions 40% $Ab/ab$, 40% $aB/ab$, 10% $AB/ab$ and 10% $ab/ab$. The genotypes $AB/ab$ and $ab/ab$ were produced from crossover gametes. Thus 20% of all gametes formed by the dihybrid parent were crossover types. This means that a chiasma occurs between these two loci in 40% of all tetrads.

### 4. Multiple Crossovers.

When two strand double crossovers occur between two genetic markers, the products, as detected through the progeny phenotypes, are only parental types.

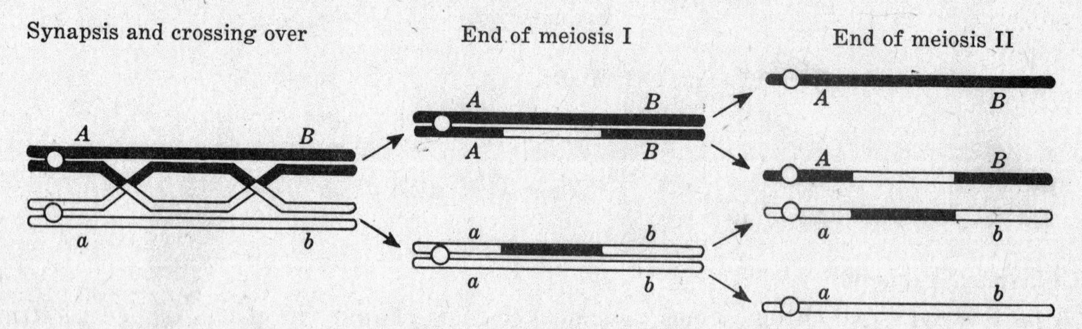


In order to detect these double crossovers, a third gene locus ($C$) between the outside markers must be used.

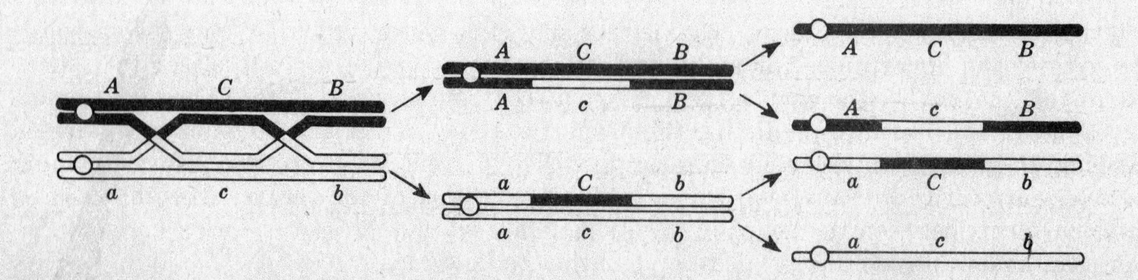

If there is a certain probability that a crossover will form between the $A$ and $C$ loci and another independent probability of a crossover forming between the $C$ and $B$ loci, then the probability of a double crossover is the product of the two independent probabilities.

**Example 6.8.** If a crossover between the $A$ and $C$ loci occurs in 20% of the tetrads and between $C$ and $B$ loci in 10% of the tetrads in an individual of genotype $ACB/acb$, then 2% ($0.2 \times 0.1$) of the gametes are expected to be of double crossover types $AcB$ and $aCb$.

Odd numbers of two strand crossovers (1, 3, 5, etc.) between two gene loci produce detectable recombinations between the outer markers, but even numbers of two strand crossovers (2, 4, 6, etc.) do not.

### 5. Limits of Recombination.

If two gene loci are so far apart in the chromosome that the probability of a chiasma forming between them is 100%, then 50% of the gametes will be parental type (non-crossover) and 50% recombinant (crossover) type. When such dihybrid individuals are testcrossed, they are expected to produce progeny in a 1:1:1:1 ratio as would be expected for genes on different chromosomes. Recombination between two linked genes cannot exceed 50% even when multiple crossovers occur between them.

## GENETIC MAPPING

### 1. Map Distance.

The places where genes reside in the chromosome (loci) are positioned in linear order analogous to beads on a string. There are two major aspects to genetic mapping: (i) the determination of the linear order with which the genetic units are arranged with respect to one another (gene order) and (ii) the determination of the relative distances between the genetic units (gene distance). The unit of distance which has the greatest utility in predicting the outcome of certain types of matings is an expression of the probability that crossing over will occur between the two genes under consideration. One unit of map distance (centimorgan) is therefore equivalent to 1% crossing over.

**Example 6.9.** If the genotype $Ab/aB$ produces 8% each of the crossover gametes $AB$ and $ab$, then the distance between $A$ and $B$ is estimated to be 16 map units.

**Example 6.10.** If the map distance between the loci $B$ and $C$ is 12 units, then 12% of the gametes of genotype $BC/bc$ should be crossover types; i.e. 6% $Bc$ and 6% $bC$.

Each chiasma produces 50% crossover products. Fifty percent crossing over is equivalent to 50 map units. If the average (mean) number of chiasmata is known for a chromosome pair, the total length of the map for that linkage group may be predicted:

$$\text{Total length} = \text{mean number of chiasmata} \times 50$$

### 2. Two-Point Testcross.

The easiest way to detect crossover gametes in a dihybrid is through the testcross progeny. Suppose we testcross dihybrid individuals in coupling phase ($AC/ac$) and find in the progeny phenotypes 37% dominant at both loci, 37% recessive at both loci, 13% dominant at the first locus and recessive at the second, and 13% dominant at the second locus and recessive at the first. Obviously the last two groups (genotypically $Ac/ac$ and $aC/ac$) were produced by crossover gametes from the dihybrid parent. Thus 26% of all gametes (13 + 13) were of crossover types and the distance between the loci $A$ and $C$ is estimated to be 26 map units.

### 3. Three-Point Testcross.

Double crossovers usually do not occur between genes less than 5 map units apart. For genes further apart, it is advisable to use a third marker between the other two in order to detect any double crossovers. Suppose that we testcross trihybrid individuals of genotype $ABC/abc$ and find in the progeny the following:

| | | | |
|---|---|---|---|
| 36% $ABC/abc$ | 9% $Abc/abc$ | 4% $ABc/abc$ | 1% $AbC/abc$ |
| 36% $abc/abc$ | 9% $aBC/abc$ | 4% $abC/abc$ | 1% $aBc/abc$ |
| 72% Parental type : | 18% Single crossovers between A and B (region I) : | 8% Single crossovers between B and C (region II) : | 2% Double crossovers |

To find the distance *A-B* we must count all crossovers (both singles and doubles) that occurred in region I = 18% + 2% = 20% or 20 map units between the loci *A* and *B*. To find the distance *B-C* we must again count all crossovers (both singles and doubles) that occurred in region II = 8% + 2% = 10% or 10 map units between the loci *B* and *C*. The *A-C* distance is therefore 30 map units when double crossovers are detected in a three-point linkage experiment and 26 map units when double crossovers are undetected in the two-point linkage experiment above.

Without the middle marker (*B*), double crossovers would appear as parental types and hence we *underestimate* the true map distance (crossover percentage). In this case the 2% double crossovers would appear with the 72% parental types, making a total of 74% parental types and 26% recombinant types. Therefore for any three linked genes whose distances are known, the amount of *detectable* crossovers (recombinants) between the two outer markers *A* and *C* when the middle marker *B* is missing is: (*A-B* crossover percentage) plus (*B-C* crossover percentage) minus (2 × double crossover percentage).

**Example 6.11.** Given distances *A-B* = 20, *B-C* = 10, *A-C* = 30 map units, the percentage of detectable crossovers from the dihybrid testcross $AC/ac \times ac/ac$ = 0.20 + 0.10 − 2(0.20)(0.10) = 0.30 − 2(0.02) = 0.30 − 0.04 = 0.26 or 26% (13% $Ac/ac$ and 13% $aC/ac$).

### 4. Gene Order.

The additivity of map distances allows us to place genes in their proper linear order. Three linked genes may be in any one of three different orders, depending upon which gene is in the middle. We will ignore left and right end alternatives for the present. If double crossovers do not occur, map distances may be treated as completely additive units. When we are given the distances *A-B* = 12, *B-C* = 7, *A-C* = 5, we should be able to determine the correct order.

**Case 1.** Let us assume that *A* is in the middle.

| B 12 A | A 5 C |
|---|---|
| B 7 C | |

The distances *B-C* are not equitable. Therefore *A* cannot be in the middle.

**Case 2.** Let us assume that *B* is in the middle.

| A 12 B | B 7 C |
|---|---|
| A 5 C | |

The distances *A-C* are not equitable. Therefore *B* cannot be in the middle.

**Case 3.** Let us assume that $C$ is in the middle.

| $A$ 5 $C$ | $C$ 7 $B$ |
|---|---|
| $A$ 12 $B$ | |

The distances $A$-$B$ are equitable. Therefore $C$ must be in the middle. Most students should be able to perceive the proper relationships intuitively.

### (a) *Linkage Relationships from a Two-Point Testcross.*

Parental combinations will tend to stay together in the majority of the progeny and the crossover types will always be the least frequent classes. From this information, the mode of linkage (coupling or repulsion) may be determined for the dihybrid parent.

**Example 6.12.** P: Dihybrid Parent $Aa, Bb$ (linkage relationships unknown) × Testcross Parent $ab/ab$

$F_1$: 42% $AaBb$, 42% $aabb$ } Parental types; 8% $Aabb$, 8% $aaBb$ } Recombinant types

The testcross parent contributes $ab$ to each progeny. The remaining genes come from the dihybrid parent. Thus $A$ and $B$ must have been on one chromosome of the dihybrid parent and $a$ and $b$ on the other, i.e. in coupling phase ($AB/ab$), because these were the combinations that appeared with greatest frequency in the progeny.

**Example 6.13.** P: Dihybrid Parent $Aa, Bb$ (linkage relationships unknown) × Testcross Parent $ab/ab$

$F_1$: 42% $Aabb$, 42% $aaBb$ } Parental types; 8% $AaBb$, 8% $aabb$ } Recombinant types

By reasoning similar to that in Example 6.12, $A$ and $b$ must have been on one chromosome of the dihybrid parent and $a$ and $B$ on the other, i.e. in repulsion phase ($Ab/aB$).

### (b) *Linkage Relationships from a Three-Point Testcross.*

In a testcross involving three linked genes, the parental types are expected to be most frequent and the double crossovers to be the least frequent. The gene order is determined by manipulating the parental combinations into the proper order for the production of double crossover types.

**Example 6.14.** P: Trihybrid Parent $Aa, Bb, Cc$ (linkage relationships unknown) × Testcross Parent $abc/abc$

| $F_1$: | | | | |
|---|---|---|---|---|
| | 36% $Aabbcc$ | 9% $aabbCc$ | 4% $AabbCc$ | 1% $AaBbCc$ |
| | 36% $aaBbCc$ | 9% $AaBbcc$ | 4% $aaBbcc$ | 1% $aabbcc$ |
| | 72% | 18% | 8% | 2% |

The 72% group is composed of parental types because non-crossover gametes are always produced in the highest frequency. Obviously the only contribution the testcross parent makes to all the progeny is $abc$. Thus the trihybrid parent must have had $A$, $b$ and $c$ on one chromosome and $a$, $B$ and $C$ on the other. But which locus is in the middle? Again, three cases can be considered.

Case 1. Can we produce the least frequent double crossover types (2% of the $F_1$) if the $B$ locus is in the middle?

*A* *b* *c*

*a* *B* *C*

× *abc*/*abc* = *ABc*/*abc* and *abC*/*abc*

These are not double crossover types and therefore the *B* locus is not in the middle.

Case 2. Can we produce the double crossover types if the *C* locus is in the middle? Remember to keep *A*, *b* and *c* on one chromosome and *a*, *B* and *C* on the other when switching different loci to the middle position.

*A* *c* *b*

*a* *C* *B*

× *acb*/*acb* = *ACb*/*acb* and *acB*/*acb*

These are not double crossover types and therefore the *C* locus is not in the middle.

Case 3. Can we produce the double crossover types if the *A* locus is in the middle?

*b* *A* *c*

*B* *a* *C*

× *bac*/*bac* = *bac*/*bac* and *BAC*/*bac*

These are the double crossover types and we conclude that the *A* locus is in the middle.

Now that we know the gene order and the parental linkage relationships, we can deduce the single crossovers. Let us designate the distance *B*-*A* as region I, and the *A*-*C* distance as region II. Single crossovers in region I:

*b* *A* *c*

*B* *a* *C*

× *bac*/*bac* = *baC*/*bac* and *BAc*/*bac*

Single crossovers in region II:

*b* *A* *c*

*B* *a* *C*

× *bac*/*bac* = *bAC*/*bac* and *Bac*/*bac*

### 5. Recombination Percentage vs. Map Distance.

In a two-point linkage experiment, the greater the unmarked distance (i.e. without segregating loci) between two genes, the greater the chance of double crossovers (and other even numbers of crossovers) occurring without detection. Therefore the most reliable estimates of the amount of crossing over will be gained from closely linked genes. Double crossovers do not occur within a distance of 10 to 12 map units in *Drosophila*. The minimum double crossover distance will vary between different species. Within this minimum distance, recombination percentage is equivalent to map distance. Outside this minimum distance, the relationship between recombination percentage and map distance becomes nonlinear as illustrated in Fig. 6-1. The true map distance will thus be underestimated by the recombination fraction, and at large distances they virtually become independent of each other.

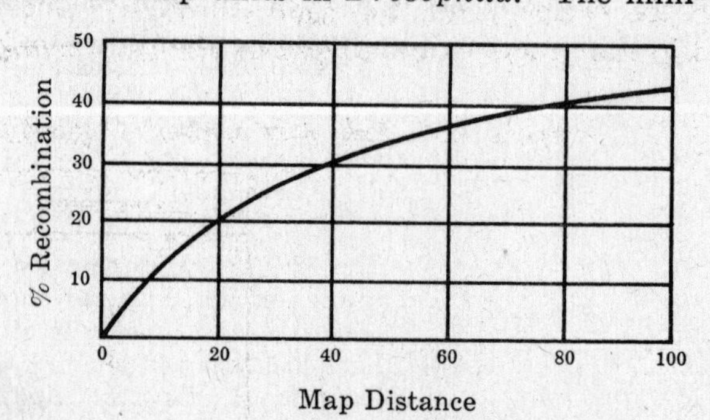


Fig. 6-1. Relationship between map distance and recombination percentage.

### 6. Genetic vs. Physical Maps.

The frequency of crossing over usually varies in different segments of the chromosome, but is a highly predictable event between any two gene loci. Therefore the actual physical distances between linked genes bears no direct relationship to the map distances calculated on the basis of crossover percentages. The linear order, however, is identical in both cases.

### 7. Combining Map Segments.

Segments of map determined from three-point linkage experiments may be combined whenever two of the three genes are held in common.

**Example 6.15.** Consider three map segments.

(1) $a$ — 8 — $b$ — 10 — $c$

(2) $c$ — 10 — $b$ — 22 — $d$

(3) $c$ — 30 — $e$ — 2 — $d$

Superimpose each of these segments by aligning the genes shared in common.

(1) $a$ — 8 — $b$ — 10 — $c$

(2) $d$ — 22 — $b$ — 10 — $c$

(3) $d$ — 2 — $e$ — 30 — $c$

Then combine the three segments into one map.

The $a$ to $d$ distance $=$ ($d$ to $b$) $-$ ($a$ to $b$) $= 22 - 8 = 14$.

The $a$ to $e$ distance $=$ ($a$ to $d$) $-$ ($d$ to $e$) $= 14 - 2 = 12$.

$d$ — 2 — $e$ — 12 — $a$ — 8 — $b$ — 10 — $c$

Additional segments of map added in this manner can produce a total linkage map over 100 map units long. However, as explained previously, the maximum recombination between any two linked genes is 50%. That is, genes very far apart on the same chromosome may behave as though they were on different chromosomes (assorting independently).

All other factors being equal, the greater the number of individuals in an experiment, the more accurate the linkage estimates should be. Therefore in averaging the distances from two or more replicate experiments, the linkage estimates may be *weighted* according to the sample size. For each experiment, multiply the sample size by the linkage estimate. Add the products and divide by the total number of individuals from all experiments.

**Example 6.16.** Let $n$ = number of individuals, $d$ = map distance.

| Experiment | $n$ | $d$ | $nd$ |
|---|---|---|---|
| 1 | 239 | 12.3 | 2,940 |
| 2 | 652 | 11.1 | 7,237 |
| 3 | 966 | 12.9 | 12,461 |
| | 1857 | | 22,638 |

22,638/1857 = 12.2 map units (weighted average)

**8. Interference and Coincidence.**

In most of the higher organisms, the formation of one chiasma actually reduces the probability of another chiasma forming in an immediately adjacent region of the chromosome. This reduction in chiasma formation may be thought of as being due to a physical inability of the chromatids to bend back upon themselves within certain minimum distances. The net result of this *interference* is the observation of fewer double crossover types than would be expected according to map distances. The strength of interference varies in different segments of the chromosome and is usually expressed in terms of a *coefficient of coincidence,* or the ratio between the observed and the expected double crossovers.

$$\text{Coefficient of Coincidence} = \frac{\%\text{ observed double crossovers}}{\%\text{ expected double crossovers}}$$

Coincidence is the complement of interference.

$$\text{Coincidence} + \text{Interference} = 1.0$$

When interference is complete (1.0), no double crossovers will be observed and coincidence becomes zero. When we observe all the double crossovers expected, coincidence is unity and interference becomes zero. When interference is 30% operative, coincidence becomes 70%, etc.

**Example 6.17.** Given the map distances $A$-$B = 10$ and $B$-$C = 20$, then $0.1 \times 0.2 = 0.02$ or 2% double crossovers are expected if there is no interference. Suppose we observe 1.6% double crossovers in a testcross experiment.

$$\text{Coincidence} = 1.6/2.0 = 0.8$$

This simply means that we observed only 80% of the double crossovers that were expected on the basis of combining independent probabilities (map distances).

$$\text{Interference} = 1.0 - 0.8 = 0.2$$

Thus 20% of the expected double crossovers did not form due to interference.

The percentage of double crossovers that will probably be observed can be predicted by multiplying the expected double crossovers by the coefficient of coincidence.

**Example 6.18.** Given a segment of map, $a \overset{10}{\text{———}} b \overset{20}{\text{——————}} c$, with 40% interference, we expect $0.1 \times 0.2 = 0.02$ or 2% double crossovers on the basis of combining independent probabilities. However, we will only observe 60% of those expected because of the interference. Therefore we should observe $0.02 \times 0.6 = 0.012$ or 1.2% double crossover types.

## LINKAGE ESTIMATES FROM $F_2$ DATA

**1. Sex-Linked Traits.**

In organisms where the male is XY or XO, the male receives only the Y chromosome from the paternal parent (or no chromosome homologous with the X in the case of XO sex determination). The Y contains, on its differential segment, no alleles homologous to those on the X chromosome received from the maternal parent. Thus for completely sex-linked traits the parental and recombinant gametes formed by the female can be observed directly in the $F_2$ males, regardless of the genotype of the $F_1$ males.

**Example 6.19.** Consider in *Drosophila* the recessive sex-linked bristle mutant scute (*sc*), and on the same chromosome the gene for vermilion eye color (*v*).

P: $\frac{+\,+}{+\,+}$ ♀♀ wild type females × $\frac{sc\ v}{\rightharpoondown}$ ♂♂ scute, vermilion males (⇁ = Y chromosome)

$F_1$: $\frac{++}{sc\ v}$ ♀♀ wild type females and $\frac{++}{\rightharpoondown}$ ♂♂ wild type males

$F_2$:

| | ♀ \ ♂ | $++$ | Y |
|---|---|---|---|
| Parental Gametes | $++$ | $++/++$ wild type | $++/Y$ wild type |
| | $sc\ v$ | $++/sc\ v$ wild type | $sc\ v/Y$ scute, vermilion |
| Crossover Gametes | $+v$ | $++/+v$ wild type | $+v/Y$ vermilion |
| | $sc+$ | $++/sc+$ wild type | $sc+/Y$ scute |
| | | Females | Males |

**Example 6.20.** Let us consider the same two sex-linked genes as in Example 6.19, using scute parental males and vermilion parental females.

P: $\frac{+v}{+v}$ ♀♀ vermilion females × $\frac{sc+}{\rightharpoondown}$ ♂♂ scute males

$F_1$: $\frac{+v}{sc+}$ ♀♀ wild type females × $\frac{+v}{\rightharpoondown}$ ♂♂ vermilion males

$F_2$:

| | ♀ \ ♂ | $+v$ | Y |
|---|---|---|---|
| Parental Gametes | $+v$ | $+v/+v$ vermilion | $+v/Y$ vermilion |
| | $sc+$ | $+v/sc+$ wild type | $sc+/Y$ scute |
| Crossover Gametes | $++$ | $+v/++$ wild type | $++/Y$ wild type |
| | $sc\ v$ | $+v/sc\ v$ vermilion | $sc\ v/Y$ scute, vermilion |
| | | Females | Males |

If the original parental females are double recessive (testcross parent), then both male and female progeny of the $F_2$ can be used to estimate the percentage of crossing over.

**Example 6.21.** P: $\frac{sc\ v}{sc\ v}$ ♀♀ scute, vermilion females × $\frac{++}{\rightharpoondown}$ ♂♂ wild type males

$F_1$: $\frac{sc\ v}{++}$ ♀♀ wild type females × $\frac{sc\ v}{\rightharpoondown}$ ♂♂ scute, vermilion males

$F_2$:

| | ♀ \ ♂ | *sc v* | Y |
|---|---|---|---|
| Parental Gametes | *sc v* | *sc v/sc v* scute, vermilion | *sc v*/Y scute, vermilion |
| | + + | + +/*sc v* wild type | + +/Y wild type |
| Crossover Gametes | *sc* + | *sc* +/*sc v* scute | *sc* +/Y scute |
| | + *v* | + *v*/*sc v* vermilion | + *v*/Y vermilion |
| | | Females | Males |

In organisms where the female is the heterogametic sex (ZW or ZO methods of sex determination), the $F_2$ females can be used for detection of crossing over between sex-linked genes. If the male is used as a testcross parent, both males and females of the $F_2$ can be used to estimate the strength of the linkage.

### 2. Autosomal Traits.

A poor alternative to the testcross method for determining linkage and estimating distances is by allowing dihybrid $F_1$ progeny to produce an $F_2$ either by random mating among the $F_1$ or, in the case of plants, by selfing the $F_1$. Such an $F_2$ which obviously does not conform to the 9:3:3:1 ratio expected for genes assorting independently may be considered evidence for linkage. Two methods for estimating the degree of linkage from $F_2$ data are presented below.

#### (*a*) ***Square Root Method.***

The frequency of double recessive phenotypes in the $F_2$ may be used as an estimator of the frequency of non-crossover gametes when the $F_1$ is in coupling phase, and as an estimator of the frequency of crossover gametes when the $F_1$ is in repulsion phase.

**Example 6.22.** $F_1$ in coupling phase. *AB/ab*

$F_2$: The frequency of *ab* gametes = $\frac{1}{2}$ of the frequency of all non-crossover gametes. If the crossover percentage is 20%, we would expect 80% non-crossover gametes (40% *AB* and 40% *ab*). The probability of two *ab* gametes uniting to form the double recessive $ab/ab = (0.4)^2 = 0.16$ or 16%. Now, if we do not know the crossover percentage, but the $F_2$ data tells us that 16% are double recessive, then the percentage of non-crossover gametes = $2\sqrt{\text{freq. of double recessives}} = 2\sqrt{0.16} = 2(0.4) = 0.8$ or 80%. If 80% are non-crossovers, the other 20% must be crossover types. Therefore the map distance between *A* and *B* is estimated at 20 units.

**Example 6.23.** $F_1$ in repulsion phase. *Ab/aB*

$F_2$: The reasoning is similar to that in Example 6.22. With 20% crossing over we expect 10% of the gametes to be *ab*. The probability of two of these gametes uniting to form the double recessive $(ab/ab) = (0.1)^2 = 0.01$ or 1%. Now, if we do not know the crossover percentage, but the $F_2$ data tells us that 1% are double recessives, then the percentage of crossover gametes = $2\sqrt{\text{freq. of double recessives}} = 2\sqrt{0.01} = 2(0.1) = 0.2$ or 20%.

#### (*b*) ***Product-Ratio Method.***

An estimate of the frequency of recombination from double heterozygous (dihybrid) $F_1$ parents can be ascertained from $F_2$ phenotypes *R-S-*, *R-ss*, *rrS-* and *rrss* appearing in the frequencies *a*, *b*, *c* and *d* respectively. The ratio of crossover to parental types, called the *product ratio*, is a function of recombination.

For coupling data: $x = bc/ad$

For repulsion data: $x = ad/bc$

The recombination fraction represented by the value of x may be read directly from a product ratio table (Table 6.1). The product-ratio method utilizes all of the $F_2$ data available and not just the double recessive class as in the square root method. The product-ratio method should therefore yield more accurate estimates of recombination than the square root method.

**Table 6.1. Recombination Fraction Estimated by the Product-Ratio Method**

| Recombination Fraction | Ratio of Products | | Recombination Fraction | Ratio of Products | |
|---|---|---|---|---|---|
| | *ad/bc* (Repulsion) | *bc/ad* (Coupling) | | *ad/bc* (Repulsion) | *bc/ad* (Coupling) |
| .00 | .000000 | .000000 | .26 | .1608 | .1467 |
| .01 | .000200 | .000136 | .27 | .1758 | .1616 |
| .02 | .000801 | .000552 | .28 | .1919 | .1777 |
| .03 | .001804 | .001262 | .29 | .2089 | .1948 |
| .04 | .003213 | .002283 | .30 | .2271 | .2132 |
| .05 | .005031 | .003629 | | | |
| .06 | .007265 | .005318 | .31 | .2465 | .2328 |
| .07 | .009921 | .007366 | .32 | .2672 | .2538 |
| .08 | .01301 | .009793 | .33 | .2892 | .2763 |
| .09 | .01653 | .01262 | .34 | .3127 | .3003 |
| .10 | .02051 | .01586 | .35 | .3377 | .3259 |
| .11 | .02495 | .01954 | .36 | .3643 | .3532 |
| .12 | .02986 | .02369 | .37 | .3927 | .3823 |
| .13 | .03527 | .02832 | .38 | .4230 | .4135 |
| .14 | .04118 | .03347 | .39 | .4553 | .4467 |
| .15 | .04763 | .03915 | .40 | .4898 | .4821 |
| .16 | .05462 | .04540 | .41 | .5266 | .5199 |
| .17 | .06218 | .05225 | .42 | .5660 | .5603 |
| .18 | .07033 | .05973 | .43 | .6081 | .6034 |
| .19 | .07911 | .06787 | .44 | .6531 | .6494 |
| .20 | .08854 | .07671 | .45 | .7013 | .6985 |
| .21 | .09865 | .08628 | .46 | .7529 | .7510 |
| .22 | .1095 | .09663 | .47 | .8082 | .8071 |
| .23 | .1211 | .1078 | .48 | .8676 | .8671 |
| .24 | .1334 | .1198 | .49 | .9314 | .9313 |
| .25 | .1467 | .1328 | .50 | 1.0000 | 1.0000 |

*Source*: F. R. Immer and M. T. Henderson, "Linkage studies in barley," *Genetics*, 28: 419-440, 1943.

**Example 6.24.** Coupling Data.

P: *RS/RS* × *rs/rs*

$F_1$: *RS/rs* (coupling phase)

$F_2$:

| | Phenotypes | Numbers |
|---|---|---|
| (*a*) | *R-S-* | 1221 |
| (*b*) | *R-ss* | 219 |
| (*c*) | *rrS-* | 246 |
| (*d*) | *rrss* | 243 |

$$x \text{ (for coupling data)} = \frac{bc}{ad} = \frac{(219)(246)}{(1221)(243)} = \frac{53,874}{296,703} = 0.1816$$

Locating the value of x in the body of the coupling column (Table 6.1), we find that 0.1816 lies between the values 0.1777 and 0.1948 which corresponds to recombination fractions of 0.28 and 0.29 respectively. Therefore, without interpolation, recombination is approximately 28%.

**Example 6.25.** Repulsion Data.

P: $Ve/Ve \times vE/vE$

$F_1$: $Ve/vE$ (repulsion phase)

| | Phenotypes | Numbers |
|---|---|---|
| (*a*) | *V-E-* | 36 |
| (*b*) | *V-ee* | 12 |
| (*c*) | *vvE-* | 16 |
| (*d*) | *vvee* | 2 |

$$\text{x (for repulsion data)} = \frac{ad}{bc} = \frac{(36)(2)}{(12)(16)} = \frac{72}{192} = 0.3750$$

Locating the value of x in the body of the repulsion column, we find that 0.3750 lies between the values 0.3643 and 0.3927 which corresponds to recombination fractions of 0.36 and 0.37 respectively. Therefore recombination is approximately 36%.

## USE OF GENETIC MAPS

### 1. Predicting Results of a Dihybrid Cross.

If the map distance between any two linked genes is known, the expectations from any type of mating may be predicted by use of the gametic checkerboard.

**Example 6.26.** Given genes *A* and *B* 10 map units apart and parents *AB*/*AB* ♂♂ × *ab*/*ab* ♀♀, the $F_1$ will all be heterozygous in coupling phase (*AB*/*ab*). Ten percent of the $F_1$ gametes are expected to be of crossover types (5% *Ab* and 5% *aB*). Ninety percent of the $F_1$ gametes are expected to be parental types (45% *AB* and 45% *ab*). The $F_2$ can be derived by use of the gametic checkerboard, combining independent probabilities by multiplication.

| | | Parental Types | | Crossover Types | |
|---|---|---|---|---|---|
| | | 0.45 *AB* | 0.45 *ab* | 0.05 *Ab* | 0.05 *aB* |
| Parental Types | 0.45 *AB* | .2025 *AB*/*AB* | .2025 *AB*/*ab* | .0225 *AB*/*Ab* | .0225 *AB*/*aB* |
| | 0.45 *ab* | .2025 *ab*/*AB* | .2025 *ab*/*ab* | .0225 *ab*/*Ab* | .0225 *ab*/*aB* |
| Crossover Types | 0.05 *Ab* | .0225 *Ab*/*AB* | .0225 *Ab*/*ab* | .0025 *Ab*/*Ab* | .0025 *Ab*/*aB* |
| | 0.05 *aB* | .0225 *aB*/*AB* | .0225 *aB*/*ab* | .0025 *aB*/*Ab* | .0025 *aB*/*aB* |

Summary of Phenotypes:
0.7025 or $70\frac{1}{4}$% *A-B-*
0.0475 or $4\frac{3}{4}$% *A-bb*
0.0475 or $4\frac{3}{4}$% *aaB-*
0.2025 or $20\frac{1}{4}$% *aabb*

### 2. Predicting Results of a Trihybrid Testcross.

Map distances or crossover percentages may be treated as any other probability estimates. Given a particular kind of mating, the map distances involved, and either the coincidence or interference for this region of the chromosome, we should be able to predict the results in the offspring generation.

**Example 6.27.** Parents: $AbC/aBc \times abc/abc$

Map: $a$ —10— $b$ —20— $c$

Interference: 40%

**Step 1.** For gametes produced by the trihybrid parent determine the parental types, single crossovers in each of the two regions, and the double crossover types.

$F_1$:

| | Step 1 | Steps 2 thru 5 | |
|---|---|---|---|
| Parental Types | $AbC$ | 35.6% | 71.2% |
| | $aBc$ | 35.6% | |
| Singles in Region I | $ABc$ | 4.4% | 8.8% |
| | $abC$ | 4.4% | |
| Singles in Region II | $Abc$ | 9.4% | 18.8% |
| | $aBC$ | 9.4% | |
| Double Crossovers | $ABC$ | 0.6% | 1.2% |
| | $abc$ | 0.6% | |
| | | 100.0% | |

**Step 2.** The frequency of double crossovers expected to be observed is calculated by multiplying the two decimal equivalents of the map distances by the coefficient of coincidence.

$$0.1 \times 0.2 \times 0.6 = 0.012 \text{ or } 1.2\%$$

This percentage is expected to be equally divided (0.6% each) between the two double crossover types.

**Step 3.** Calculate the single crossovers in region II (between $b$ and $c$) and correct it for the double crossovers which also occurred in this region:

$$20\% - 1.2\% = 18.8\%$$

equally divided into two classes = 9.4% each.

**Step 4.** The single crossovers in region I (between $a$ and $b$) are calculated in the same manner as step 3:

$$10\% - 1.2\% = 8.8\%$$

divided equally among the two classes = 4.4% each.

**Step 5.** Total all the single crossovers and all the double crossovers and subtract from 100% to obtain the percentage of parental types:

$$100 - (8.8 + 18.8 + 1.2) = 71.2\%$$

to be equally divided among the two parental classes = 35.6% each.

For convenience, we need not write out the entire genotype or phenotype of the progeny because, for example, when the gamete $AbC$ from the trihybrid parent unites with the gamete produced by the testcross parent ($abc$), obviously the genotype is $AbC/abc$. Phenotypically it will exhibit the dominant trait at the $A$ locus, the recessive trait at the $B$ locus and the dominant trait at the $C$ locus. All this could be predicted directly from the gamete $AbC$.

An alternative method for predicting $F_1$ progeny types is by combining the probabilities of crossovers and/or non-crossovers in appropriate combinations. This method can be used only when there is no interference.

**Example 6.28.** Parents: $ABC/abc \times abc/abc$

Coincidence: 1.0

Map: $a$ —— I 10 —— $b$ —— II 20 —— $c$

No. of progeny: 2000

**Step 1.** Determine the parental, single crossover and double crossover progeny types expected.

$F_1$:

| | Step 1 | Steps 2 thru 5 |
|---|---|---|
| Parental Types | $ABC$ | 720 |
| | $abc$ | 720 |
| Singles in Region I | $Abc$ | 80 |
| | $aBC$ | 80 |
| Singles in Region II | $ABc$ | 180 |
| | $abC$ | 180 |
| Double Crossovers | $AbC$ | 20 |
| | $aBc$ | 20 |
| | | 2000 |

**Step 2.** The number of double crossovers expected to appear in the progeny is $0.1 \times 0.2 \times 2000 = 40$, equally divided between the two double crossover types (20 each).

**Step 3.** The probability of a single crossover occurring in region I is 10%. Hence there is a 90% chance that a crossover will not occur in that region. The combined probability that a crossover will not occur in region I and will occur in region II is $(0.9)(0.2) = 0.18$ and the number of region II single crossover progeny expected is $0.18(2000) = 360$, equally divided between the two classes (180 each).

**Step 4.** Likewise the probability of a crossover occurring in region I and not in region II is $0.1(0.8) = 0.08$ and the number of region I single crossover progeny expected is $0.08(2000) = 160$, equally divided among the two classes (80 each).

**Step 5.** The probability that a crossover will not occur in region I and region II is $0.9(0.8) = 0.72$ and the number of parental type progeny expected is $0.72(2000) = 1440$, equally divided among the two parental types (720 each).

## CROSSOVER SUPPRESSION

Many extrinsic and intrinsic factors are known to contribute to the crossover rate. Among these are the effects of sex, age, temperature, proximity to the centromere or heterochromatic regions (darkly staining regions presumed to carry little genetic information), chromosomal aberrations such as inversions, and many more. Two specific cases of crossover suppression are presented in this section: (1) complete absence of crossing over in male *Drosophila* and (2) the maintenance of balanced lethal systems as permanent trans heterozygotes through the prevention of crossing over.

### 1. Absence of Crossing Over in Male Drosophila.

One of the unusual characteristics of *Drosophila* is the apparent absence of crossing over in males. This fact is shown clearly by the non-equivalent results of reciprocal crosses.

**Example 6.29.** Testcross of heterozygous females.

Consider two genes on the third chromosome of *Drosophila*, hairy ($h$) and scarlet ($st$), approximately 20 map units apart.

P: $h+/+st$ ♀♀ × $h\,st/h\,st$ ♂♂
wild type females — hairy, scarlet males

F₁:

| | ♀ \ ♂ | (h st) |
|---|---|---|
| 80% Parental Types | 40% (h +) | h +/h st = 40% hairy |
| | 40% (+ st) | + st/h st = 40% scarlet |
| 20% Recombinant Types | 10% (h st) | h st/h st = 10% hairy and scarlet |
| | 10% (+ +) | + +/h st = 10% wild type |

**Example 6.30.** Testcross of heterozygous males (reciprocal cross of Example 6.29).

P: *h st/h st* ♀♀ × *h +/+ st* ♂♂
hairy, scarlet females × wild type males

F₁:

| | ♂ \ ♀ | h st |
|---|---|---|
| Only Parental Types | (h +) | h +/h st = 50% hairy |
| | (+ st) | + st/h st = 50% scarlet |

When dihybrid males are crossed to dihybrid females (both in repulsion phase) the progeny will always appear in the ratio 2 : 1 : 1 regardless of the degree of linkage between the genes. The double recessive class never appears.

**Example 6.31.** P: *h +/+ st* ♀♀ × *h +/+ st* ♂♂
wild type females × wild type males

F₁:

| | | 50% (h +) | 50% (+ st) |
|---|---|---|---|
| 80% Parental Types | 40% (h +) | h +/h +<br>20% hairy | h +/+ st<br>20% wild type |
| | 40% (+ st) | + st/h +<br>20% wild type | + st/+ st<br>20% scarlet |
| 20% Recombinant Types | 10% (h st) | h st/h +<br>5% hairy | h st/+ st<br>5% scarlet |
| | 10% (+ +) | + +/h +<br>5% wild type | + +/+ st<br>5% wild type |

Summary: 50% wild type, 25% hairy, 25% scarlet } 2 : 1 : 1

*Drosophila* is not unique in this respect. For example, crossing over is completely suppressed in female silkworms. Other examples of complete and partial suppression of crossing over are common in genetic literature.

**2. Balanced Lethal Systems.**

A gene which is lethal when homozygous and linked to another lethal with the same mode of action can be maintained in permanent dihybrid condition in repulsion phase when associated with a genetic condition which prevents crossing over (see "inversions" in Chapter 9). Balanced lethals breed true and their behavior simulates that of a homozygous genotype. These systems are commonly used to maintain laboratory cultures of lethal, semilethal, or sterile mutants.

**Example 6.32.** Two dominant genetic conditions, curly wings (*Cy*) and plum eye color (*Pm*), are linked on chromosome 2 of *Drosophila* and associated with a chromosomal inversion which prevents crossing over. *Cy* or *Pm* are lethal when homozygous. Half the progeny from repulsion heterozygotes die, and the viable half are repulsion heterozygotes just like the parents.

P: $Cy\ Pm^+/Cy^+\ Pm$ ♀♀ × $Cy\ Pm^+/Cy^+\ Pm$ ♂♂
curly, plum females — curly, plum males

$F_1$:

| | $Cy\ Pm^+$ | $Cy^+\ Pm$ |
|---|---|---|
| $Cy\ Pm^+$ | $Cy\ Pm^+/Cy\ Pm^+$ dies | $Cy\ Pm^+/Cy^+\ Pm$ curly, plum |
| $Cy^+\ Pm$ | $Cy^+\ Pm/Cy\ Pm^+$ curly, plum | $Cy^+\ Pm/Cy^+\ Pm$ dies |

Balanced lethals may be used to determine on which chromosome an unknown genetic unit resides (see Problem 6.12). Sex-linked genes make themselves known through the non-equivalence of progeny from reciprocal matings (Chapter 5). Without the aid of a balanced lethal system, the assignment of an autosomal gene to a particular linkage group may be made through observation of the peculiar genetic ratios obtained from abnormal individuals possessing an extra chromosome (trisomic) bearing the gene under study (Chapter 9).

## TETRAD ANALYSIS IN ASCOMYCETES

Fungi which produce sexual spores (*ascospores*) housed in a common sac (*ascus*) are called *ascomycetes*. One of the simplest ascomycetes is the unicellular baker's yeast *Saccharomyces cerevisiae* (Fig. 6-2 below). Asexual reproduction is by budding, a mitotic process usually with unequal cytokinesis. The sexual cycle involves the union of entire cells of opposite mating type, forming a diploid zygote. The diploid cell may reproduce diploid progeny asexually by budding or haploid progeny by meiosis. The four haploid nuclei form ascospores enclosed by the ascus. Rupture of the ascus releases the haploid spores which then germinate into new yeast cells.

Another ascomycete of interest to geneticists is the bread mold *Neurospora crassa* (Fig. 6-3 below). The fungal mat or *mycelium* is composed of intertwined filaments called *hyphae*. The tips of hyphae may pinch off asexual spores called *conidia*, which germinate into more hyphae. The vegetative hyphae are segmented, with several haploid nuclei in each segment. Hyphae from one mycelium may anastomose with hyphae of another mycelium to form a mixture of nuclei in a common cytoplasm called a *heterokaryon*. A pair of alleles, *A* and *a*, governs the two mating types. Sexual reproduction occurs only when cells of opposite mating type unite. Specialized regions of the mycelium produce immature female fruiting bodies (*protoperithecia*) from which extrude receptive filaments called *trichogynes*. A conidium or hyphae from the opposite mating type fuses with the trichogyne, undergoes several karyokineses, and fertilizes many female nuclei. Each of the resulting diploid zygotes lies within an elongated sac called the *ascus* (*asci*, plural). The zygote divides by meiosis to form four nuclei, followed by a mitotic division which yields four pairs of nuclei, maturing into eight *ascospores*. A mature fruiting body (*perithecium*) may contain over 100 asci, each containing eight ascospores. The confines of the ascus force the polar organization of division to orient lengthwise in the ascus and also prevent the meiotic or mitotic products from slipping past each other. Each of the four chromatids of first meiotic prophase are now represented by a pair of ascospores in tandem order within the ascus.

In the case of yeast the ascospores representing the four chromatids of meiosis are in no special order, but in the bread mold *Neurospora* the ascospores are linearly ordered in the ascus in the same sequence as the chromatids were on the meiotic metaphase plate. The

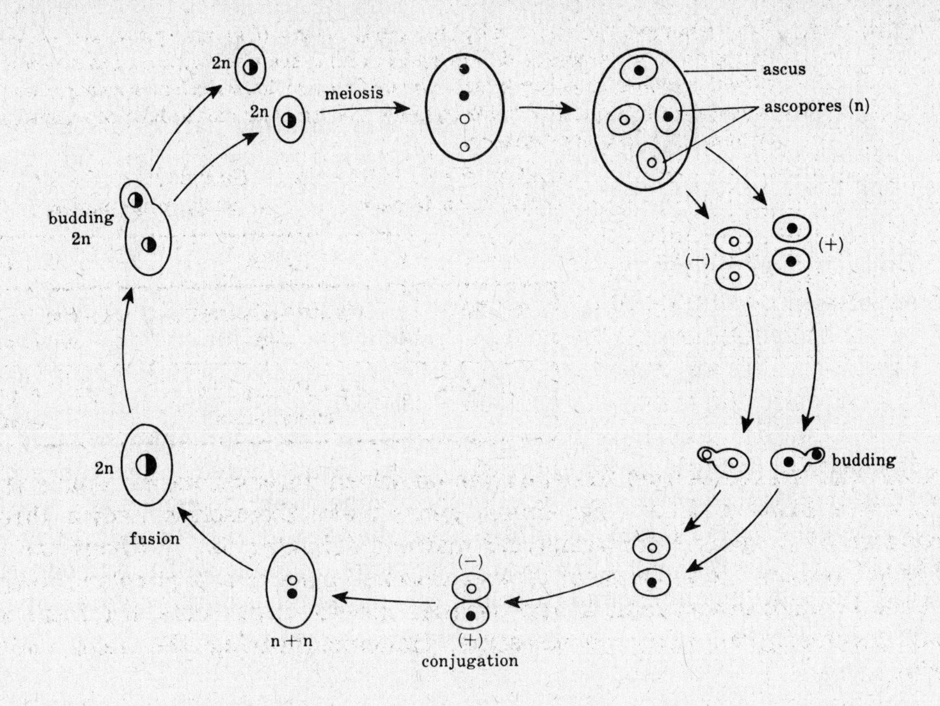


Fig. 6-2. Life cycle of *Saccharomyces cerevisiae.*

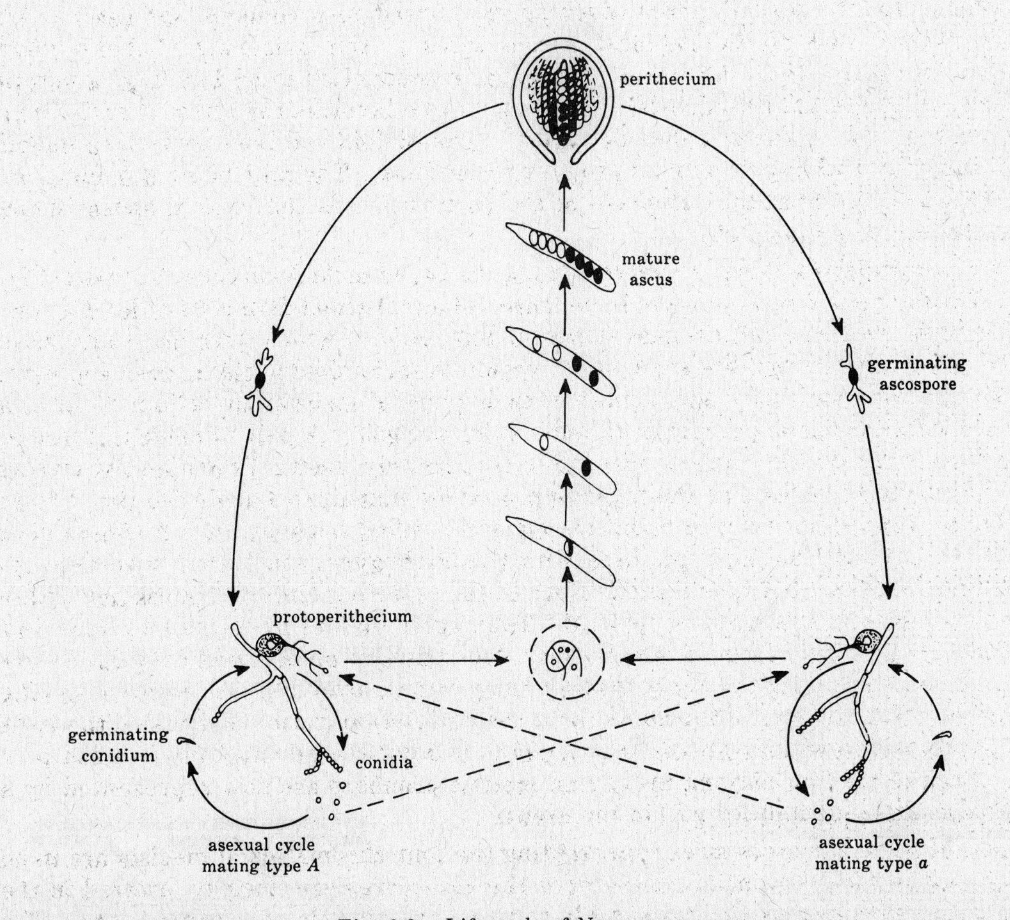


Fig. 6-3. Life cycle of *Neurospora crassa.*

recovery and investigation of all of the products from a single meiotic event is called *tetrad analysis*.

Each ascus of *Neurospora*, when analyzed for a segregating pair of alleles, reveals one of two linear ratios: (1) 4:4 ratio, attributed to *first division segregation* and (2) 2:2:2:2 ratio resulting from *second division segregation*.

**1. First Division Segregation.**

A cross between a culture with a wild type ($c^+$) spreading form of mycelial growth and one with a restricted form of growth called colonial ($c$) is diagrammed in Fig. 6-4($a$) below. If the ascospores are removed one by one from the ascus in linear order and each is grown as a separate culture, a linear ratio of 4 colonial : 4 wild type indicates that a first division segregation has occurred. That is, during first meiotic anaphase both of the $c^+$ chromatids moved to one pole and both of the $c$ chromatids moved to the other pole. The 4:4 ratio indicates that *no crossing over* has occurred between the gene and its centromere. The further the gene locus is from the centromere, the greater is the opportunity for crossing over to occur in this region. Therefore if the meiotic products of a number of asci are analyzed and most of them are found to exhibit a 4:4 pattern, then the locus of $c$ must be close to the centromere.

**2. Second Division Segregation.**

Let us now investigate the results of a crossover between the centromere and the $c$ locus (Fig. 6-4($b$)). Note that crossing over in meiotic prophase results in a $c^+$ chromatid and a $c$ chromatid being attached to the same centromere. Hence $c^+$ and $c$ fail to separate from each other during first anaphase. During second anaphase, sister chromatids move to opposite poles, thus affecting segregation of $c^+$ from $c$. The 2:2:2:2 linear pattern is indicative of a second division segregation ascus produced by *crossing over* between the gene and its centromere.

## RECOMBINATION MAPPING WITH TETRADS

**1. Ordered Tetrads.**

The frequency of crossing over between the centromere and the gene in question is a reflection of its map distance from the centromere. Thus the percentage of asci showing second division segregation is a measure of linkage intensity. It must be remembered, however, that one crossover event gives one second division ascus, but that only half of the *ascospores* in that ascus are recombinant type. Therefore to convert second division asci frequency to crossover frequency, we divide the former by two.

**2. Unordered Tetrads.**

The meiotic products of most ascomycetes are usually not in a linear order as in the ascus of *Neurospora*. Let us analyze unordered tetrads involving two linked genes from the cross ++ × $ab$. The fusion nucleus is diploid (++/$ab$) and immediately undergoes meiosis. If a crossover does not occur between these two loci or if a two-strand double crossover occurs between them, the resulting meiotic products will be of two kinds, equally frequent, resembling the parental combinations. Such a tetrad is referred to as a *parental ditype* (PD).

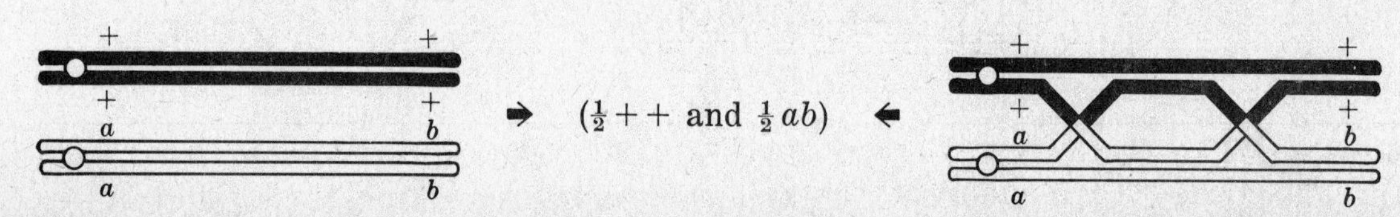

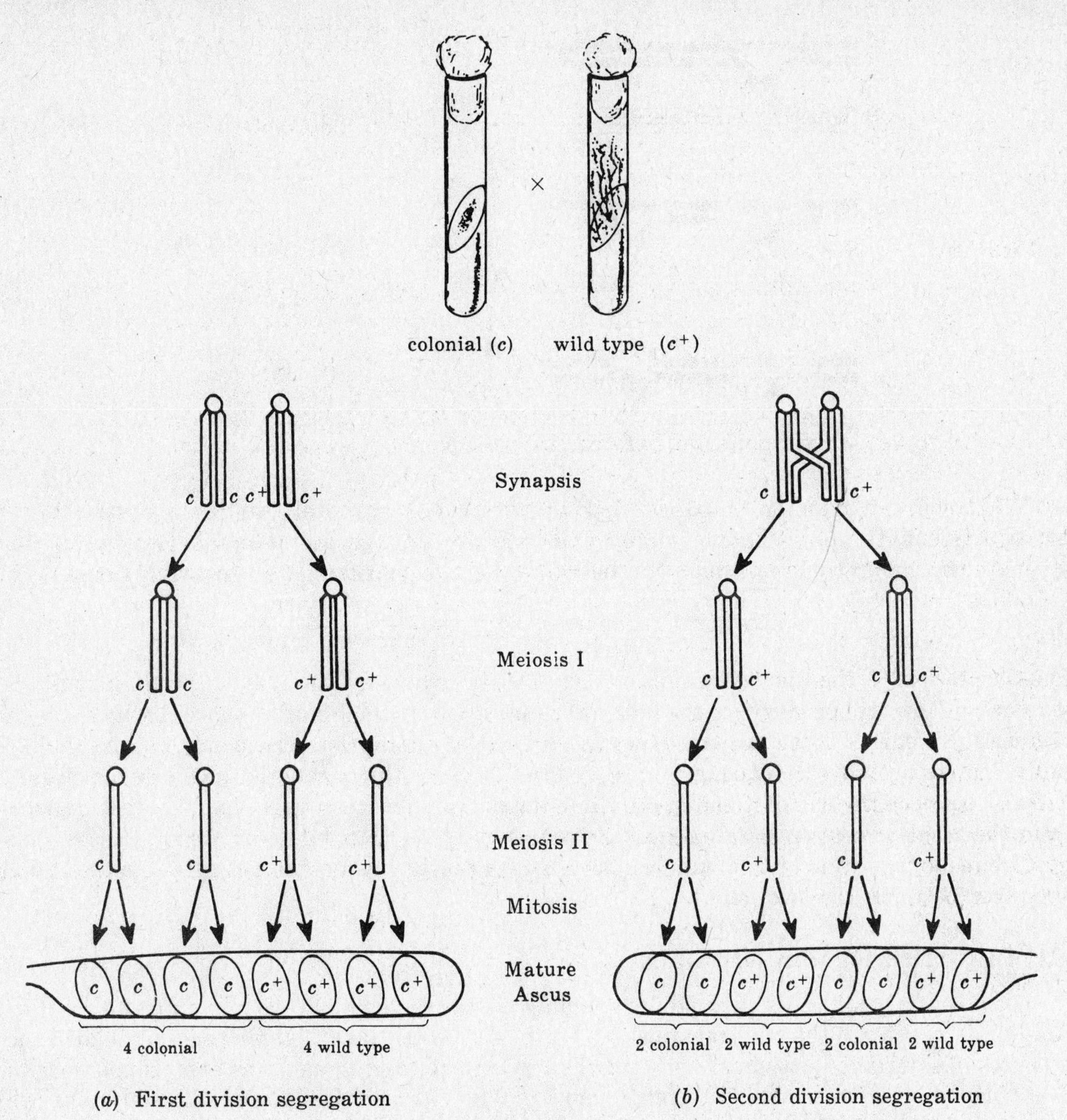


(*a*) First division segregation (*b*) Second division segregation

**Fig. 6-4.** *Neurospora* spore patterns.

A four-strand double crossover between the two genes results in two kinds of products, neither of which are parental combinations. This tetrad is called a *non-parental ditype* (NPD), and is the rarest of the tetrad double crossovers.

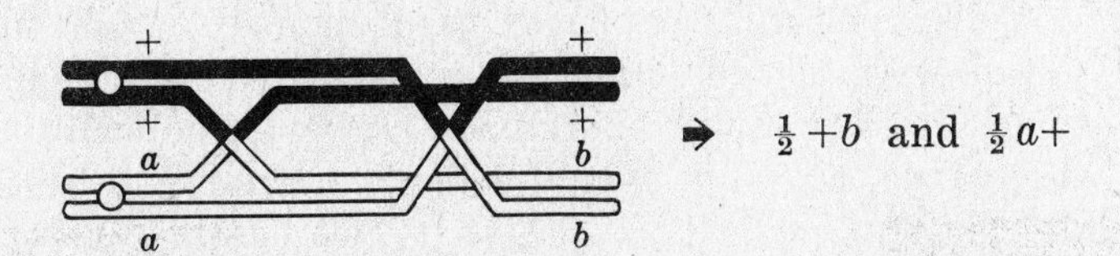


$\Rightarrow$ $\frac{1}{2}+b$ and $\frac{1}{2}a+$

A *tetratype* (TT) is produced by either a single crossover or a three-strand double crossover (of two types) between the two genes.

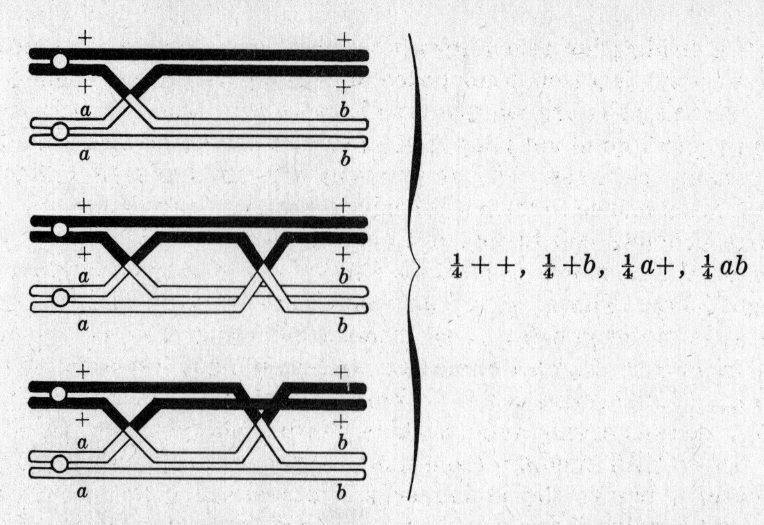


Whenever the number of parental ditypes and non-parental ditypes are statistically non-equivalent, this may be considered evidence for linkage between the two genes. To estimate the amount of recombination between the two markers, we use the formula:

$$\text{Recombination Frequency} = \frac{\text{NPD} + \frac{1}{2}\text{TT}}{\text{total number of tetrads}}$$

The derivation of the above formula becomes clear when we analyze these diagrams and see that all of the products from a NPD tetrad are recombinant, but only half of the products from a TT tetrad are recombinant. Recombination frequency is not always equivalent to crossover frequency (map distance). If a third genetic marker was present midway between the loci of $a$ and $b$, the three-strand double crossovers could be distinguished from the single crossovers and crossover frequency could thus be determined. Recombination frequency analysis of two widely spaced genes thereby can establish only minimum map distances between the two genes.

## MAPPING THE HUMAN GENOME

Until recently, the only method for mapping human genes was through pedigree analysis. Sex-linked genes are the ones most easily distinguished because of their peculiar inheritance patterns. Assigning autosomal genes to their specific chromosomes was sometimes possible if a chromosomal abnormality (e.g. reciprocal translocation or segmental deletion) was involved. Closely linked genes could occasionally be discovered in large family pedigrees, but loosely linked genes often mimic independent assortment. Now, however, rapid progress is being made in human gene mapping through a variety of techniques (most of which are beyond the scope of this book), including somatic cell hybridization, radiation-induced gene segregation, chromosome-mediated gene transfer, DNA-mediated gene transfer, amino acid sequencing, and linkage disequilibrium. For an excellent review of these procedures, see "The anatomy of the human genome," by V. A. McKusick, *Journal of Heredity* 71(6): 370–391, Nov.–Dec. 1980.

*Somatic cell hybridization* (SCH) techniques can be used to fuse human fibroblast cells in tissue culture with cells from another species (e.g. mouse). Cell fusion is promoted by adding inactivated ("dead"), irradiated Sendai virus. Such hybrid cells tend to progressively lose human chromosomes at random. If one human chromosome contains an essential gene missing in the mouse genome, the only hybrid cells to survive will be the ones that retain the essential human chromosome. Evidence that two human genes are on the same chromosome (*syntenic*) exists when these loci are retained together as various human chromosomes are lost from hybrid cells. Mouse chromosomes are easily distinguished from human chromosomes, and special staining techniques reveal distinctive bands that allow identification of the selected human chromosome.

Somatic cell hybridization techniques have also been used to mass produce monoclonal antibodies. A myeloma is a tumor composed of cells derived from hemopoietic tissues of the bone marrow. Plasma cells are mature lymphocytes that secrete antibodies. Each plasma cell produces only one kind of antibody, but such cells cannot be maintained in tissue culture the way that myeloma cells can. Mouse myeloma cells can be fused with spleen cells from a specifically immunized mouse to form a hybrid myeloma or "hybridoma". Polyethylene glycol (PEG) is used to promote cell fusion, but even so very few hybridomas are formed. The potentially immortal hybridoma produces a single kind of antibody (immunoglobulin) called a monoclonal antibody. These "pure" antibodies are in great demand because they react with only one kind of antigen (i.e., are monospecific) and they can be economically mass produced for a variety of diagnostic, medical, industrial, and research purposes. Myelomas can be induced in certain strains of mice. Variant (mutant) myeloma cells have been isolated that are deficient in the enzyme hypoxanthine phosphoribosyl transferase ($HPRT^-$). These cells cannot grow in HAT medium (containing *h*ypoxanthine, *a*minopterin, and *t*hymidine) because aminopterin blocks the endogenous synthesis of both purines and pyrimidines; however, they will survive if fused with normal ($HPRT^+$) spleen cells. The latter cells can survive by utilizing the exogenous hypoxanthine and thymidine, but they grow so poorly in HAT medium that they either die or are rapidly outgrown by the hybrids. The clones that survive in HAT medium are then assayed for the antibodies specific to the immunizing antigen. Once the desired clone is found, it can be frozen and stored for later use or propagated indefinitely in tissue culture or injected into syngeneic mice (genetically identical to the source of the cells) to induce antibody-secreting tumors. To date, no human myeloma line has been fused with human lymphocytes to make an intraspecific hybridoma. Because human chromosomes are rapidly lost from human × mouse hybrids, the prospects for making a useful interspecific hybridoma is poor.

# Solved Problems

## RECOMBINATION AMONG LINKED GENES

**6.1.** In the adjacent human pedigree where the male parent does not appear, it is assumed that he is phenotypically normal. Both hemophilia (*h*) and color blindness (*c*) are sex-linked recessives. Insofar as possible, determine the genotypes for each individual in the pedigree.

Legend:

- □ ○ Non-hemophilic, normal vision
- Color blind male
- Hemophilic male
- Hemophilic and color blind male

**Solution:**

Let us begin with males first because, being hemizygous for sex-linked genes, the linkage relationship on their single X chromosome is obvious from their phenotype. Thus I1, I2 and III3 are all hemophilic with normal color vision and therefore must be *hC*/Y. Non-hemophilic, color blind males II1 and II3 must be *Hc*/Y. Normal males II2, II6 and III1 must possess both dominant alleles *HC*/Y. III2 is both hemophilic and color blind and therefore must possess both recessives *hc*/Y. Now let us determine the female genotypes. I3 is normal but produces sons, half of which are color blind and half normal. The X chromosome contributed by I3 to her color blind sons II1 and II3 must have been *Hc*; the X chromosome she contributed to her normal sons II2 and II6 must have been *HC*. Therefore the genotype for I3 is *Hc*/*HC*.

Normal females II4, II5 and II7 each receive *hC* from their father (I2), but could have received either *Hc* or *HC* on the X chromosome they received from their mother (I3). II4 has a normal son (III1) to which she gives *HC*; therefore II4 is probably *hC*/*HC*, although it is possible for II4 to be *hC*/*Hc* and produce a *HC* gamete by crossing over. II5, however, could not be *hC*/*HC* and produce a son with both hemophilia and color blindness (III2); therefore II5 must be *hC*/*Hc*, in order to give the crossover gamete *hc* to her son.

## GENETIC MAPPING

**6.2.** Two dominant mutants in the first linkage group of the guinea pig govern the traits pollex (*Px*), which is the ativistic return of thumb and little toe, and rough fur (*R*). When dihybrid pollex, rough pigs (with identical linkage relationships) were crossed to normal pigs, their progeny fell into four phenotypes: 79 rough, 103 normal, 95 rough, pollex and 75 pollex. (*a*) Determine the genotypes of the parents. (*b*) Calculate the amount of recombination between *Px* and *R*.

**Solution:**

(*a*) The parental gametes always appear with greatest frequency, in this case 103 normal and 95 rough, pollex. This means that the two normal genes were on one chromosome of the dihybrid parent and the two dominant mutations on the other (i.e., coupling linkage).

P: $Px\,R/px\,r$ × $px\,r/px\,r$
pollex, rough — normal

(*b*) The 79 rough and 75 pollex types are recombinants, constituting 154 out of 352 individuals = 0.4375 or approximately 43.8% recombination.

**6.3.** A kidney-bean shaped eye is produced by a recessive gene *k* on the third chromosome of *Drosophila*. Orange eye color, called "cardinal", is produced by the recessive gene *cd* on the same chromosome. Between these two loci is a third locus with a recessive allele *e* producing ebony body color. Homozygous kidney, cardinal females are mated to homozygous ebony males. The trihybrid $F_1$ females are then testcrossed to produce the $F_2$. Among 4000 $F_2$ progeny are the following:

| | |
|---|---|
| 1761 kidney, cardinal | 97 kidney |
| 1773 ebony | 89 ebony, cardinal |
| 128 kidney, ebony | 6 kidney, ebony, cardinal |
| 138 cardinal | 8 wild type |

(*a*) Determine the linkage relationships in the parents and $F_1$ trihybrids.
(*b*) Estimate the map distances.

**Solution:**

(*a*) The parents are homozygous lines:

$k\,e^+\,cd/k\,e^+\,cd$ ♀♀ × $k^+\,e\,cd^+/k^+\,e\,cd^+$ ♂♂
kidney, cardinal — ebony

The $F_1$ is then trihybrid:

$k\,e^+\,cd/k^+\,e\,cd^+$
wild type

The linkage relationships in the trihybrid $F_1$ can also be determined directly from the $F_2$. By far the most frequent $F_2$ phenotypes are kidney, cardinal (1761) and ebony (1773), indicating that kidney and cardinal were on one chromosome in the $F_1$ and ebony on the other.

(*b*) Crossing over between the loci *k* and *e* produces the kidney, ebony (128) and cardinal (138) offspring. Double crossovers are the triple mutants (6) and wild type (8). Altogether there are $128 + 138 + 6 + 8 = 280$ crossovers between *k* and *e*:

$$280/4000 = 0.07 \text{ or } 7\% \text{ crossing over} = 7 \text{ map units}$$

Crossovers between *e* and *cd* produced the single crossover types kidney (97) and ebony, cardinal (89). Double crossovers again must be counted in this region.

$$97 + 89 + 6 + 8 = 200 \text{ crossovers between } e \text{ and } cd$$
$$200/4000 = 0.05 \text{ or } 5\% \text{ crossing over} = 5 \text{ map units}$$

**6.4.** The map distances for six genes in the second linkage group of the silkworm *Bombyx mori* are shown in the table on the right. Construct a genetic map which includes all of these genes.

| | *Gr* | *Rc* | *S* | *Y* | *P* | *oa* |
|---|---|---|---|---|---|---|
| *Gr* | – | 25 | 1 | 19 | 7 | 20 |
| *Rc* | 25 | – | 26 | 6 | 32 | 5 |
| *S* | 1 | 26 | – | 20 | 6 | 21 |
| *Y* | 19 | 6 | 20 | – | 26 | 1 |
| *P* | 7 | 32 | 6 | 26 | – | 27 |
| *oa* | 20 | 5 | 21 | 1 | 27 | – |

**Solution:**

**Step 1.** It makes little difference where one begins to solve this kind of problem, so we shall begin at the top. The *Gr-Rc* distance is 25 map units, and the *Gr-S* distance is 1 unit. Therefore the relationship of these three genes may be either

(a) $S \overset{1}{} Gr$ —— 25 —— $Rc$

or

(b) $Gr \overset{1}{} S$ —— 24 —— $Rc$

The table, however, tells us that the distance *S-Rc* is 26 units. Therefore alternative (*a*) must be correct, i.e., *Gr* is between *S* and *Rc*.

**Step 2.** The *Gr-Y* distance is 19 units. Again two alternatives are possible:

(c) $S \overset{1}{} Gr$ —— 19 —— $Y$ —— 6 —— $Rc$

or

(d) $Y$ —— 18 —— $S \overset{1}{} Gr$ —— 25 —— $Rc$

In the table we find that the distance $Y\text{-}Rc = 6$. Hence possibility (*c*) must be correct, i.e., *Y* lies between the loci of *Gr* and *Rc*.

**Step 3.** The distance *Gr-P* is 7 map units. Two alternatives for these loci are

(e) $S \overset{1}{} Gr$ —— 7 —— $P$ —— 12 —— $Y$ —— 6 —— $Rc$

or

(f) $P$ —— 6 —— $S \overset{1}{} Gr$ —— 19 —— $Y$ —— 6 —— $Rc$

The distance *P-S* is read from the table, and thus alternative (*f*) must be correct.

**Step 4.** There are 20 units between *Gr* and *oa*. These two genes may be in one of two possible relationships:

(g) $P$ —— 6 —— $S \overset{1}{} Gr$ —— 19 —— $Y \overset{1}{} oa$ —— 5 —— $Rc$

or

(h) $oa$ —— 13 —— $P$ —— 6 —— $S \overset{1}{} Gr$ —— 19 —— $Y$ —— 6 —— $Rc$

The table indicates that *Y* and *oa* are 1 map unit apart. Therefore (*g*) is the completed map.

**6.5.** Three recessive genes in linkage group V of the tomato are *a* producing absence of anthocyanin pigment, *hl* producing hairless plants, and *j* producing jointless fruit stems (pedicels). Among 3000 progeny from a trihybrid testcross, the following phenotypes were observed:

| | |
|---|---|
| 259 hairless | 268 anthocyaninless, jointless, hairless |
| 40 jointless, hairless | 941 anthocyaninless, hairless |
| 931 jointless | 32 anthocyaninless |
| 260 normal | 269 anthocyaninless, jointless |

(*a*) How were the genes originally linked in the trihybrid parent? (*b*) Estimate the distance between the genes.

**Solution:**

(*a*) The most frequent phenotypes observed among the offspring are the jointless (931) and anthocyaninless, hairless (941). Hence *j* was on one chromosome of the trihybrid parent, *a* and *hl* on the other. The double crossover (DCO) types are the least frequent phenotypes: jointless, hairless (40) and anthocyaninless (32).

**Case 1.** If jointless is in the middle, we could not obtain the double crossover types as given:

P: $\dfrac{A\ j\ Hl}{a\ J\ hl}$

$F_1$: DCO $\begin{cases} A\ J\ Hl = \text{normal} \\ a\ j\ hl = \text{triple mutant} \end{cases}$

**Case 2.** If *h* was in the middle, the double crossover types could be formed. Therefore the parental genotype is as shown below:

P: $\dfrac{J\ hl\ a}{j\ Hl\ A}$

$F_1$: DCO $\begin{cases} J\ Hl\ a = \text{anthocyaninless} \\ j\ hl\ A = \text{jointless, hairless} \end{cases}$

(*b*) Now that the genotype of the trihybrid parent is known, we can predict the single crossover types.

P: $\dfrac{J\,hl\,a}{j\,Hl\,A}$

$F_1$: Single crossovers (SCO) between *j* and *hl* (region I) yield:

SCO(I) < $J\,Hl\,A$ = normal (260)
$j\,hl\,a$ = jointless, hairless, anthocyaninless (268)

Therefore the percentage of *all* crossovers (singles and doubles) that occurred between *j* and *hl* is $260 + 268 + 32 + 40 = 600/3000 = 0.2 = 20\%$ or 20 map units.

Similarly the single crossovers between *hl* and *a* (region II) may be obtained.

P: $\dfrac{J\,hl\,a}{j\,Hl\,A}$

$F_1$: SCO(II) < $J\,hl\,A$ = hairless (259)
$j\,Hl\,a$ = jointless, anthocyaninless (269)

$259 + 269 + 32 + 40 = 600/3000 = 0.2 = 20\%$ or 20 map units

Note the similarity of numbers between the SCO(II) jointless, anthocyaninless (269) and the SCO(I) triple mutant (268). Attempts to obtain map distances by matching pairs with similar numbers could, as this case proves, lead to erroneous estimates. The single crossover types in each region must first be determined in order to avoid such errors.

**6.6.** The recessive mutation called "lemon" (*le*) produces a pale yellow body color in the parasitic wasp *Bracon hebetor*. This locus exhibits 12% recombination with a recessive eye mutation called "canteloupe" (*c*). Canteloupe shows 14% recombination with a recessive mutation called "long" (*l*), causing antennal and leg segments to elongate. Canteloupe is the locus in the middle. A homozygous lemon female is crossed with a hemizygous long male (males are haploid). The $F_1$ females are then testcrossed to produce the $F_2$. (*a*) Diagram the crosses and the expected $F_1$ and $F_2$ female genotypes and phenotypes. (*b*) Calculate the amount of wild types expected among the $F_2$ females.

**Solution:**

(*a*) P: $le\,l^+/le\,l^+$ ♀ × $le^+\,l$ ♂
lemon — long

$F_1$: $le\,l^+/le^+\,l$ ♀♀ × $le\,l$ ♂♂
wild type — lemon, long

$F_2$:

| | ♀ \ ♂ | *le l* |
|---|---|---|
| Parental Types | $le\,l^+$ | $le\,l^+/le\,l$ lemon |
| | $le^+\,l$ | $le^+\,l/le\,l$ long |
| Recombinant Types | $le\,l$ | $le\,l/le\,l$ lemon, long |
| | $le^+\,l^+$ | $le^+\,l^+/le\,l$ wild type |

(*b*) Since the canteloupe locus is not segregating in this cross, double crossovers will appear as parental types. The percentage of recombination expected to be observed is $0.12 + 0.14 - 2(0.12)(0.14) = 0.2264 = 22.64\%$. Half of the recombinants are expected to be wild type: $22.64\%/2 = 11.32\%$ wild type.

**6.7.** Several three-point testcrosses were made in maize utilizing the genes booster ($B$, a dominant plant color intensifier), liguleless leaf ($lg_1$), virescent seedling ($v_4$, yellowish-green), silkless ($sk$, abortive pistils), glossy seedling ($gl_2$), and tassel seed ($ts_1$, pistillate terminal inflorescence). Using the information from the following testcrosses, map this region of the chromosome.

*Testcross #1.* Trihybrid parent is heterozygous for booster, liguleless, tassel seed.

Testcross Progeny

71 booster, liguleless, tassel seed
111 wild type
48 liguleless
35 booster, tassel seed
17 tassel seed
24 booster, liguleless
6 booster
3 liguleless, tassel seed

*Testcross #2.* Trihybrid parent is heterozygous for booster, liguleless, tassel seed.

Testcross Progeny

57 tassel seed
57 booster, liguleless
20 wild type
31 booster
21 liguleless, tassel seed
21 booster, liguleless, tassel seed
8 booster, tassel seed
7 liguleless

*Testcross #3.* Trihybrid parent is heterozygous for booster, liguleless, silkless.

Testcross Progeny

52 silkless
8 booster, silkless
2 booster, liguleless, silkless
148 booster
56 booster, liguleless
13 liguleless
131 liguleless, silkless

*Testcross #4.* Trihybrid parent is heterozygous for booster, liguleless, silkless.

Testcross Progeny

6 booster
137 booster, silkless
291 booster, liguleless, silkless
142 liguleless
3 liguleless, silkless
30 silkless
34 booster, liguleless
339 wild type

*Testcross #5.* Trihybrid parent is heterozygous for liguleless, virescent, glossy.

Testcross Progeny

431 wild type
399 liguleless, virescent, glossy
256 virescent
310 liguleless, glossy
128 virescent, glossy
153 liguleless
44 glossy
51 virescent, liguleless

*Testcross #6.* Trihybrid parent is heterozygous for booster, liguleless, virescent.

Testcross Progeny

60 wild type
37 liguleless, booster, virescent
32 virescent, booster
34 liguleless
18 virescent
23 liguleless, booster
11 booster
12 virescent, liguleless

*Testcross #7.* Trihybrid parent is heterozygous for virescent, liguleless, booster.

Testcross Progeny

25 booster
11 booster, liguleless
8 booster, virescent
2 liguleless, booster, virescent

**Solution:**

Following the procedures established in this chapter, we determine from each of the testcrosses the gene order (which gene is in the middle) and the percent crossing over in each region. Note that the results of testcrosses #1 and #2 may be combined, recognizing that the linkage relationships are different in the trihybrid parents. Likewise, the results of #3 and #4 may be combined, as well as #6 with #7. The analysis of these seven testcrosses are summarized below in tabular form.

| Testcross No. | Trihybrid Parent | Parental Type Progeny | Recombinant Progeny | | | Total |
|---|---|---|---|---|---|---|
| | | | Region I | Region II | DCO | |
| 1 | $\frac{+++}{lg_1\,B\,ts_1}$ | 111 71 | 35 48 | 17 24 | 6 3 | 315 |
| 2 | $\frac{++ts_1}{lg_1\,B+}$ | 57 57 | 31 21 | 20 21 | 8 7 | 222 |
| | | 296 | 135 | 82 | 24 | 537 |
| | | | 25.1% | 15.3% | 4.5% | |
| 3 | $\frac{+B+}{lg_1+sk}$ | 148 131 | 52 56 | 8 13 | 0 2 | 410 |
| 4 | $\frac{+++}{lg_1\,B\,sk}$ | 339 291 | 137 142 | 30 34 | 6 3 | 982 |
| | | 909 | 387 | 85 | 11 | 1392 |
| | | | 27.8% | 6.1% | 0.8% | |
| 5 | $\frac{+++}{lg_1\,gl_2\,v_4}$ | 431 399 | 128 153 | 256 310 | 44 51 | |
| | | 830 | 281 | 566 | 95 | 1772 |
| | | | 15.9% | 31.9% | 5.4% | |
| 6 | $\frac{+++}{lg_1\,B\,v_4}$ | 60 37 | 32 34 | 18 23 | 11 12 | 227 |
| 7 | $\frac{+B+}{lg_1+v_4}$ | 25 0 | 0 11 | 8 0 | 0 2 | 46 |
| | | 122 | 77 | 49 | 25 | 273 |
| | | | 28.2% | 17.9% | 9.2% | |

To find the map distances between $lg_1$ and $B$ in the first two testcrosses, we add the double crossovers (4.5%) to the region I single crossovers (25.1%) = 29.6% or 29.6 map units. Likewise, to find the map distance between $B$ and $ts_1$, we add 4.5% to the region II single crossovers (15.3%) = 19.8% or 19.8 map units. Thus this segment of map becomes

$lg_1$ —— 29.6 —— $B$ —— 19.8 —— $ts_1$

Three other map segments are similarly derived from testcross data:

#3 and #4 $lg_1$ —— 28.6 —— $B$ —— 6.9 —— $sk$

#5 $lg_1$ —— 21.3 —— $gl_2$ —— 37.3 —— $v_4$

#6 and #7 $lg_1$ —— 37.4 —— $B$ —— 27.1 —— $v_4$

Now let us combine all four maps into one:

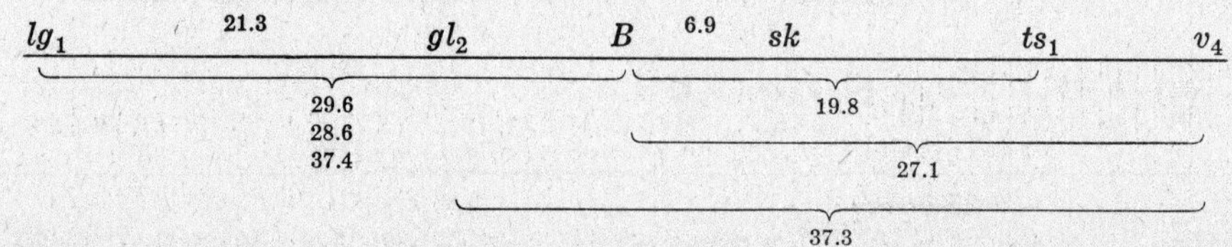

The weighted average distance $lg_1$-$B$ is determined next:

| Expt. | No. in Expt. $x$ | Distance $y$ | $xy$ |
|---|---|---|---|
| 1, 2 | 537 | 29.6 | 15,895.2 |
| 3, 4 | 1392 | 28.6 | 39,811.2 |
| 6, 7 | 273 | 37.4 | 10,210.2 |
| | 2202 | | 65,916.6 |

$$65{,}916.6/2202 = 29.9 \text{ map units (weighted average)}$$

The $sk$-$ts_1$ distance $= (B\text{-}ts_1) - (B\text{-}sk) = 19.8 - 6.9 = 12.9$ map units.

The $gl_2$-$B$ distance has two estimators:

(1) $(lg_1\text{-}B) - (lg_1\text{-}gl_2) = 29.9 - 21.3 = 8.6$

(2) $(gl_2\text{-}v_4) - (B\text{-}v_4) = 37.3 - 27.1 = 10.2$

All other factors being equal, the second estimate is likely to be less accurate because of the greater distances involved. There is no easy way to accurately average these two values. We will arbitrarily use the estimate of 8.6 map units until more definitive experimental results are obtained.

Likewise, the $ts_1$-$v_4$ distance has two estimators:

(1) $(B\text{-}v_4) - (B\text{-}ts_1) = 27.1 - 19.8 = 7.3$

(2) $(gl_2\text{-}v_4) - [(gl_2\text{-}B) + (B\text{-}ts_1)] = 37.3 - (8.6 + 19.8) = 8.9$

Again, the second estimate is likely to be less accurate because of the distances involved, and we will use 7.3 map units as our estimate of the $ts_1$-$v_4$ distance. The unified map now appears as follows:

$lg_1$ — 21.3 — $gl_2$ — 8.6 — $B$ — 6.9 — $sk$ — 12.9 — $ts_1$ — 7.3 — $v_4$

Additional experimental data may considerably modify certain portions of this genetic map. It should always be remembered that these maps are only estimates, and as such are continually subject to refinements.

## LINKAGE ESTIMATES FROM $F_2$ DATA

**6.8.** Two dominant sex-linked traits are known in the mouse (*Mus musculus*): bent (*Bn*), appearing as a short crooked tail, and tabby (*Ta*) with dark transverse stripes. Homozygous bent, tabby females are mated to normal (wild type) males. All of the $F_1$ offspring are mated together to produce an $F_2$. Inadvertently, the male-female data from the $F_2$ was not recorded. Among 200 $F_2$ offspring were found 141 bent, tabby, 47 wild type, 7 tabby and 5 bent.

(*a*) Estimate the amount of recombination between bent and tabby assuming that the male : female ratio is 1 : 1.

(*b*) Estimate the amount of recombination when the male : female ratio is variable and unreliable in this colony.

**Solution:**

(*a*) P: *Bn Ta*/*Bn Ta* ♀♀ × *bn ta*/Y ♂♂
bent, tabby females — normal males

$F_1$: *Bn Ta*/*bn ta* ♀♀ × *Bn Ta*/Y ♂♂
bent, tabby females — bent, tabby males

$F_2$:

| | ♀ \ ♂ | *Bn Ta* | Y |
|---|---|---|---|
| Parental Types | *Bn Ta* | *Bn Ta/Bn Ta* bent, tabby | *Bn Ta*/Y bent, tabby |
| | *bn ta* | *bn ta/Bn Ta* bent, tabby | *bn ta*/Y wild type (47) |
| Recombinant Types | *Bn ta* | *Bn ta/Bn Ta* bent, tabby | *Bn ta*/Y bent (5) |
| | *bn Ta* | *bn Ta/Bn Ta* bent, tabby | *bn Ta*/Y tabby (7) |
| | | Females | Males |

If we assume that the male : female ratio is 1 : 1, then out of 200 offspring, half of them or (100) should have been males. The difference $100 - (47 + 5 + 7) = 41$ estimates the probable number of bent, tabby males. The estimate of recombination from the male data $= (5 + 7)/100 = 0.12$ or 12%.

(*b*) If the male : female ratio is unreliable, then we might better use the 47 wild type males as an estimator of the number of bent, tabby males. The estimated total number of males $= 47 + 47 + 5 + 7 = 106$. The amount of recombination $= (5 + 7)/106 = 0.113$ or 11.3%.

**6.9.** White eyes (*w/w* females; *w*/Y males) in *Drosophila* can be produced by the action of a sex-linked recessive gene. White eyes can also be produced through the interaction of two other genes; the recessive sex-linked gene *v* for vermilion eye color, and the autosomal recessive gene *bw* for brown eye color (see Problem 5.5). Consider the parental cross: $bw/bw,\ w^+v^+/w\ v$ ♀♀ (brown-eyed females) × $bw/bw,\ w\ v$/Y ♂♂ (white-eyed males), where the $F_1$ progeny consists of 70 brown-eyed and 130 white-eyed individuals. Estimate the distance between the sex-linked genes *w* and *v*.

**Solution:**

$F_1$:

| | ♀ \ ♂ | $bw/,\ w\ v/$ | $bw/$, Y/ |
|---|---|---|---|
| Parental Types | $bw/,\ w^+v^+/$ | $bw/bw,\ w^+v^+/w\ v$ brown | $bw/bw,\ w^+v^+$/Y brown |
| | $bw/,\ w\ v/$ | $bw/bw,\ w\ v/w\ v$ white | $bw/bw,\ w\ v$/Y white |
| Recombinant Types | $bw/,\ w^+v/$ | $bw/bw,\ w^+v/w\ v$ white | $bw/bw,\ w^+v$/Y white |
| | $bw/,\ w\ v^+/$ | $bw/bw,\ w\ v^+/w\ v$ white | $bw/bw,\ w\ v^+$/Y white |

Only the genotypes of the brown offspring are known for certain. The 70 brown offspring constitute only one-half of the offspring produced by non-crossover maternal gametes. Therefore we *estimate* that 70 of the white individuals were also produced by non-crossover maternal gametes. Thus 140 out of 200 $F_1$ flies are estimated to be parental type offspring = 70%. The other 30% must be crossover types. The best estimate of linkage between the white and vermilion loci would be 30 map units.

**6.10.** Elongate tomato fruit is produced by plants homozygous for a recessive gene *o*, round fruit shape is produced by the dominant allele at this locus (*O*). A compound inflorescence is the result of another recessive gene *s*, simple inflorescence is produced by the dominant allele at this locus (*S*). A Yellow Pear variety (with elongate fruit and simple inflorescence) is crossed to a Grape Cluster variety (with round fruit and compound inflorescence). The $F_1$ plants are randomly crossed to produce the $F_2$. Among 259 $F_2$ are found 126 round, simple: 63 round, compound: 66 long, simple: 4 long, compound. Estimate the amount of recombination by the "square root method."

**Solution:**

P: *oS/oS* (Yellow Pear Variety, long, simple) × *Os/Os* (Grape Cluster Variety, round, compound)

$F_1$: *oS/Os* (round, simple)

$F_2$:

| | | Parental Gametes | | Crossover Gametes | |
|---|---|---|---|---|---|
| | | *o S* | *O s* | *o s* | *O S* |
| Parental Gametes | *o S* | *o S/o S* long, simple | *o S/O s* round, simple | *o S/o s* long, simple | *o S/O S* round, simple |
| | *O s* | *O s/o S* round, simple | *O s/O s* round, compound | *O s/o s* round, compound | *O s/O S* round, simple |
| Crossover Gametes | *o s* | *o s/o S* long, simple | *o s/O s* round, compound | *o s/o s* long, compound | *o s/O S* round, simple |
| | *O S* | *O S/o S* round, simple | *O S/O s* round, simple | *O S/o s* round, simple | *O S/O S* round, simple |

Notice that the double recessive phenotype (long, compound) occupies only one of the 16 frames in the gametic checkerboard. This genotype is produced by the union of two identical double recessive gametes (*o s*). If we let x = the frequency of formation of *os* gametes, then $x^2$ = frequency of occurrence of the *o s/o s* genotype (long, compound phenotype) = 4/259 = 0.0154. Thus, $x = \sqrt{0.0154} = 0.124$. But x estimates only half of the crossover gametes. Therefore 2x estimates all of the crossover gametes = 2(0.124) = 0.248 or 24.8% recombination.

## USE OF GENETIC MAPS

**6.11.** The genes for two nervous disorders, waltzer (*v*) and jittery (*ji*) are 18 map units apart on chromosome 10 in mice. A phenotypically normal $F_1$ group of mice carrying these two genes in coupling phase is being maintained by a commercial firm. An order arrives for two dozen young mice each of waltzer, jittery, and waltzer plus jittery. Assuming that the average litter size is seven offspring, and including a 10% safety factor to ensure the recovery of the needed number of offspring, calculate the minimum number of females that need to be bred.

**Solution:**

$F_1$: *v ji*/+ +

$F_2$: If 18% are crossover types, then 82% should be parental types.

| | | .41 $v\ ji$ | .41 $+\ +$ | .09 $v\ +$ | .09 $+\ ji$ |
|---|---|---|---|---|---|
| 82% Parental Type | .41 $v\ ji$ | .1681<br>$v\ ji/v\ ji$<br>waltzer and jittery | .1681<br>$v\ ji/+\ +$<br>wild type | .0369<br>$v\ ji/v\ +$<br>waltzer | .0369<br>$v\ ji/+\ ji$<br>jittery |
| | .41 $+\ +$ | .1681<br>$+\ +/v\ ji$<br>wild type | .1681<br>$+\ +/+\ +$<br>wild type | .0369<br>$+\ +/v\ +$<br>wild type | .0369<br>$+\ +/+\ ji$<br>wild type |
| 18% Crossover Types | .09 $v\ +$ | .0369<br>$v\ +/v\ ji$<br>waltzer | .0369<br>$v\ +/+\ +$<br>wild type | .0081<br>$v\ +/v\ +$<br>waltzer | .0081<br>$v\ +/+\ ji$<br>wild type |
| | .09 $+\ ji$ | .0369<br>$+\ ji/v\ ji$<br>jittery | .0369<br>$+\ ji/+\ +$<br>wild type | .0081<br>$+\ ji/v\ +$<br>wild type | .0081<br>$+\ ji/+\ ji$<br>jittery |

Summary: wild type = .6681 or 66.81%
waltzer and jittery = .1681 or 16.81%
waltzer = .0819 or 8.19%
jittery = .0819 or 8.19%

Waltzer or jittery phenotypes are least frequent and hence are the limiting factors. If 8.19% of all the progeny are expected to be waltzers, how many offspring need to be raised to produce 24 waltzers? $.0819x = 24$, $x = 24/.0819 = 293.04$ or approximately 293 progeny. Adding the 10% safety factor, $293 + 29.3 = 322.3$ or approximately 322 progeny.

If each female has 7 per litter, how many females need to be bred? $322/7 = 46$ females.

## CROSSOVER SUPPRESSION

**6.12.** Suppose you are given a strain of *Drosophila* exhibiting an unknown abnormal genetic trait (mutation). We mate the mutant females to males from a balanced lethal strain ($Cy\,Pm^+/Cy^+\,Pm$, $D\,Sb^+/D^+\,Sb$) where curly wings ($Cy$) and plum eye ($Pm$) are on chromosome 2 and dichaete wing ($D$) and stubble bristles ($Sb$) are on chromosome 3. Homozygosity for either curly, plum, dichaete or stubble is lethal. The trait does not appear in the $F_1$. The $F_1$ males with curly wings and stubby bristles are then backcrossed to the original mutant females. In the progeny the mutation appears in equal association with curly and stubble. *Drosophila melanogaster* has a haploid number of 4 including an X, 2, 3 and 4 chromosome. (*a*) Determine whether the mutation is a dominant or a recessive. (*b*) To which linkage group (on which chromosome) does the mutation belong?

**Solution:**

(*a*) If the mutation was a dominant (let us designate it $M$), then each member of the strain (pure line) would be of genotype $MM$. Since the trait does not appear in our balanced lethal stock, they must be homozygous recessive ($M^+M^+$). Crosses between these two lines would be expected to produce only heterozygous genotypes ($M^+M$) and would be phenotypically of the mutant type. But since the mutant type did not appear in the $F_1$, the mutation must be a recessive (now properly redesignated $m$). The dominant wild type allele may now be designated $m^+$.

(*b*) Let us assume that this is a sex-linked recessive mutation. The $F_1$ males receive their single X chromosome from the mutant female ($mm$). Therefore all males of the $F_1$ should exhibit the mutant trait because males would be hemizygous for all sex-linked genes ($mY$). Since the mutant type did not appear in the $F_1$, our recessive mutation could not be sex-linked.

Let us assume that our recessive mutation is on the second chromosome. The curly, stubble $F_1$ males carry the recessive in the heterozygous condition ($Cy\,m^+/Cy^+\,m$, $Sb/Sb^+$). Notice that we omit the designation of loci with which we are not concerned. When these carrier males are then backcrossed to the original mutant females ($Cy^+\,m/Cy^+\,m$, $Sb^+/Sb^+$), the $F_2$ expectations are as follows:

| | | | |
|---|---|---|---|
| $Cy\,m^+/Cy^+\,m,\ Sb/Sb^+$ | curly, stubble | $Cy^+\,m/Cy^+\,m,\ Sb/Sb^+$ | *mutant*, stubble |
| $Cy\,m^+/Cy^+\,m,\ Sb^+/Sb^+$ | curly | $Cy^+\,m/Cy^+\,m,\ Sb^+/Sb^+$ | *mutant* |

Note that the mutant cannot appear with curly. Therefore our recessive mutation is not on chromosome 2.

Let us then assume that our mutant gene is on the third chromosome. When $F_1$ carrier males ($Cy/Cy^+,\ Sb\,m^+/Sb^+\,m$) are backcrossed to the original mutant females ($Cy^+/Cy^+, Sb^+\,m/Sb^+\,m$), the $F_2$ expectations are as follows:

| | | | |
|---|---|---|---|
| $Cy/Cy^+,\ Sb\,m^+/Sb^+\,m$ | curly, stubble | $Cy^+/Cy^+,\ Sb\,m^+/Sb^+\,m$ | stubble |
| $Cy/Cy^+,\ Sb^+\,m/Sb^+\,m$ | curly, *mutant* | $Cy^+/Cy^+,\ Sb^+\,m/Sb^+\,m$ | *mutant* |

Note that the mutant cannot appear with stubble. Hence our recessive mutation is not on chromosome 3.

If the mutant is not sex-linked, not on 2 nor on 3, then obviously it must be on the fourth chromosome. Let us prove this. When $F_1$ carrier males ($Cy/Cy^+, Sb/Sb^+, m^+/m$) are backcrossed to the original mutant females ($Cy^+/Cy^+, Sb^+/Sb^+, m/m$), the $F_2$ expectations are as follows:

| | | | |
|---|---|---|---|
| $Cy/Cy^+$ | $Sb/Sb^+$ | $m^+/m$ | curly, stubble |
| | | $m/m$ | curly, stubble, *mutant* |
| | $Sb^+/Sb^+$ | $m^+/m$ | curly |
| | | $m/m$ | curly, *mutant* |
| $Cy^+/Cy^+$ | $Sb/Sb^+$ | $m^+/m$ | stubble |
| | | $m/m$ | stubble, *mutant* |
| | $Sb^+/Sb^+$ | $m^+/m$ | wild type |
| | | $m/m$ | *mutant* |

Note that our recessive mutant occurs in equal association with curly and stubble which satisfies the conditions of the problem. We conclude that this mutation is on the fourth chromosome.

## RECOMBINATION MAPPING WITH TETRADS

**6.13.** A strain of *Neurospora* requiring methionine ($m$) was crossed to a wild type ($m^+$) strain with the results shown below. How far is this gene from its centromere?

| No. of Asci | Spores | | | |
|---|---|---|---|---|
| | 1+2 | 3+4 | 5+6 | 7+8 |
| 6 | + | $m$ | + | $m$ |
| 5 | $m$ | + | + | $m$ |
| 6 | $m$ | + | $m$ | + |
| 7 | + | $m$ | $m$ | + |
| 40 | $m$ | $m$ | + | + |
| 36 | + | + | $m$ | $m$ |

**Solution:**

Non-crossover asci are those which appear with greatest frequency = 40 + 36 = 76 out of 100 total asci. The other 24/100 of these asci are crossover types. While 24% of the asci are crossover types, only half of the spores in these asci are recombinant. Therefore the distance from the gene to the centromere is 12 map units. The origin of each of the crossover type asci is as follows:

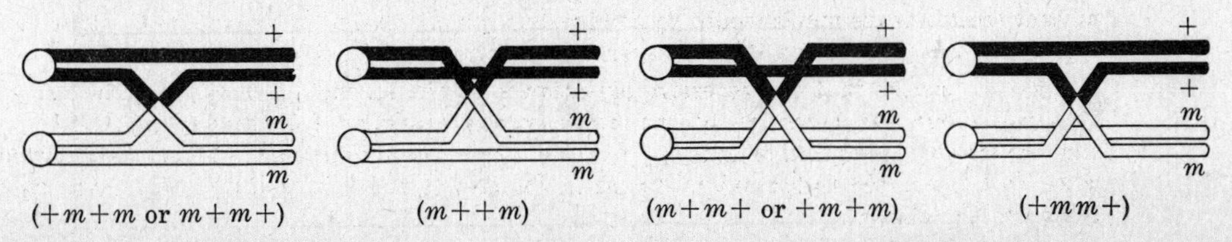

**6.14.** Two linked genes are involved in a *Neurospora* cross: $(a\,+) \times (+\,b)$, where $a$ is closest to the centromere. Diagram the simplest explanation to account for the following spore patterns: (*a*) $(a\,b)(+\,b)(a\,+)(+\,+)$, (*b*) $(+\,b)(a\,+)(+\,b)(a\,+)$, (*c*) $(a\,+)(+\,b)(a\,b)(+\,+)$.

**Solution:**

(*a*) Two kinds of double crossover events can account for this spore pattern.

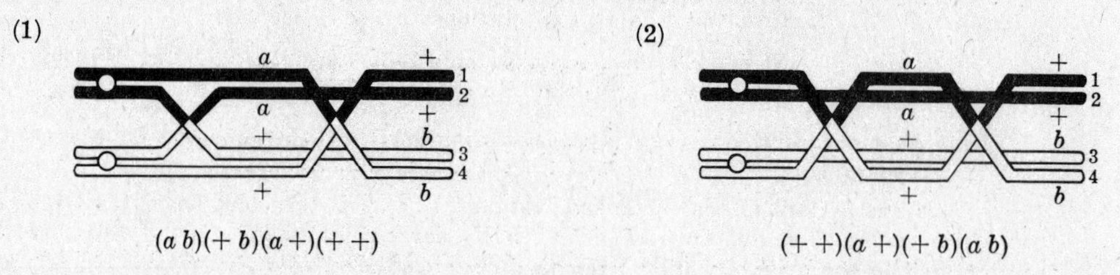


(1) This four-strand double crossover can be symbolized as (2, 3)(1, 4) which indicates that the first crossover involves strands 2 and 3 and the second crossover involves strands 1 and 4.

(2) This two-strand double crossover can be symbolized (1, 4)(1, 4). Note that the readings of the two spore patterns are reversed left to right, but otherwise are in the same linear order. Both of these patterns are possible for either kind of crossover depending upon the first meiotic metaphase orientation of the chromosomes.

(*b*) A single crossover (2, 3) gives the right to left pattern; a single crossover (1,4) gives the left to right pattern).

(*c*) A three-strand double crossover (2, 3)(2, 4) gives the right to left spore pattern; a three-strand double (1, 4)(2, 4) reverses the pattern.

**6.15.** Two strains of *Neurospora*, one mutant for gene $a$, the other mutant for gene $b$, are crossed. Results are shown below. Determine the linkage relationships between these two genes.

| | Percent of Asci | Spores | | | |
|---|---|---|---|---|---|
| | | 1 + 2 | 3 + 4 | 5 + 6 | 7 + 8 |
| (1) | 79 | $a\,+$ | $a\,+$ | $+\,b$ | $+\,b$ |
| (2) | 14 | $a\,+$ | $+\,+$ | $ab$ | $+\,b$ |
| (3) | 6 | $a\,+$ | $ab$ | $+\,+$ | $+\,b$ |
| (4) | 1 | $a\,+$ | $+\,b$ | $a\,+$ | $+\,b$ |

**Solution:**

Pattern (1) represents the non-crossover types showing first division segregation (4 : 4) for both *a* and *b*. Pattern (2) shows second division segregation (2 : 2 : 2 : 2) for *a*, but first division segregation for *b*. Genes which show high frequency of second division segregation are usually further from the centromere than genes with low frequency of second division segregation. Judging by the relatively high frequency of pattern (2) these are probably single crossovers and we suspect that *a* is more distal from its centromere than *b*.

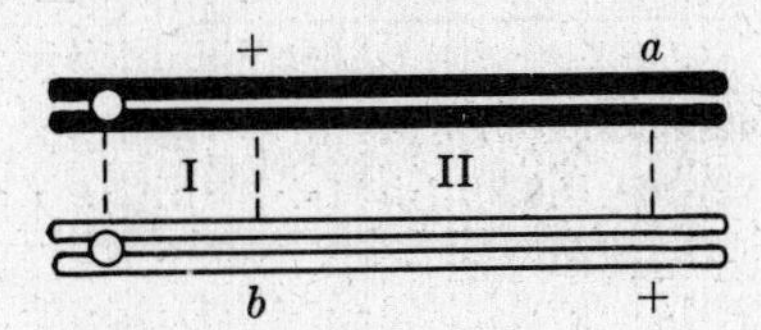


Pattern (3), indicating first division segregation for *a* and second division segregation for *b*, cannot be generated by a single crossover if *a* and *b* are linked as shown above, but requires a double crossover.

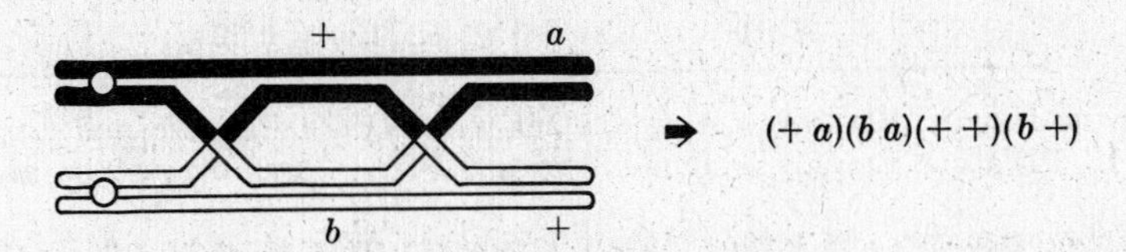


Furthermore, pattern (4) could be produced from the linkage relationships as assumed above by a single crossover in region I.

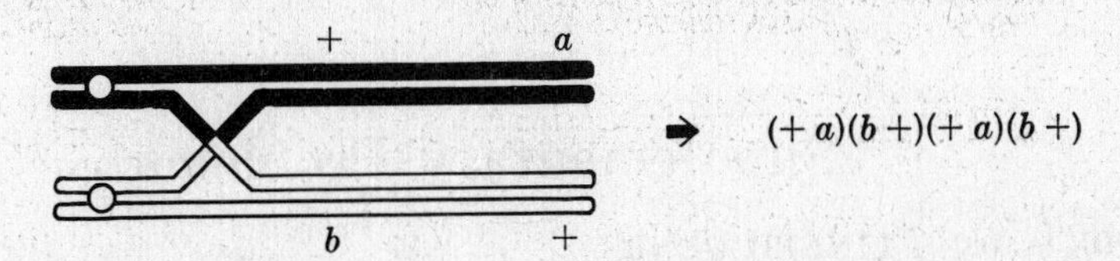


Double crossovers are expected to be much less frequent than single crossovers. Under the above assumptions, double crossover pattern (3) is more frequent than one of the single crossover patterns (4). This does not make sense, and thus our assumption must be wrong. The locus of *a* must be further from the centromere than *b*, but it need not be on the same side of the centromere with *b*. Let us place *a* on the other side of the centromere.

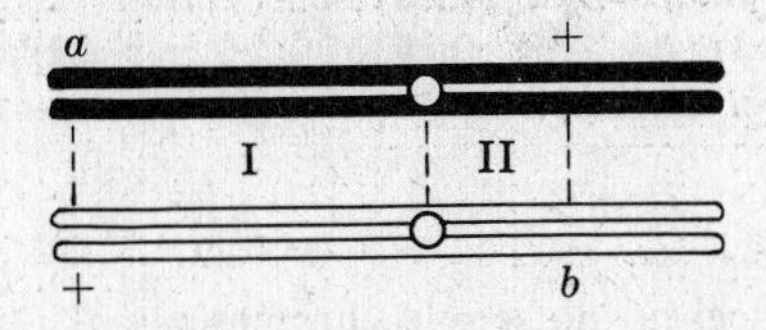


Now a single crossover in region I produces pattern (2), a single crossover in region II produces pattern (3), and a two-strand double crossover (I, II) produces pattern (4). The percentage of asci are numerically acceptable under this assumption.

The distance *a*-centromere $= \frac{1}{2}(\text{SCO I} + \text{DCO}) = \frac{1}{2}(14+1) = 7.5$ map units.

The distance centromere-*b* $= \frac{1}{2}(\text{SCO II} + \text{DCO}) = \frac{1}{2}(6+1) = 3.5$ map units.

**6.16.** The cross (*abc*) × (+++) is made in an ascomycete with unordered tetrads. From the analysis of 100 asci, determine the linkage relationships between these three loci as completely as the data allow.

(1) 40 (*abc*)(*abc*)(+ + +)(+ + +)

(2) 42 (*ab* +)(*ab* +)(+ + *c*)(+ + *c*)

(3) 10 (*a* + *c*)(+ + *c*)(*ab* +)(+ *b* +)

(4) 8 (*a* + +)(+ + +)(*abc*)(+ *bc*)

**Solution:**

Pattern (1) is parental ditype (PD) for *ab*, *ac* and *bc*. Pattern (2) is PD for *ab*, nonparental ditype (NPD) for *ac* and *bc*. Pattern (3) is tetratype (TT) for *ab* and *ac*, NPD for *bc*. Pattern (4) is TT for *ab* and *ac*, PD for *bc*. For each pair of markers the relative frequencies of each type of tetrad are as follows:

| | PD | NPD | TT |
|---|---|---|---|
| *ab* | 40<br>42<br>82/100 = 0.82 | 0 | 10<br>8<br>18/100 = 0.18 |
| *ac* | 40/100 = 0.40 | 42/100 = 0.42 | 10<br>8<br>18/100 = 0.18 |
| *bc* | 40<br>8<br>48/100 = 0.48 | 42<br>10<br>52/100 = 0.52 | 0 |

For the *ab* pair, PD's are not equivalent with NPD's. Thus *a* and *b* must be linked. For pairs *ac* and *bc*, PD's are roughly equivalent with NPD's. Thus *c* must be assorting independently on another chromosome. The recombination frequency between *a* and *b* $= \dfrac{\text{NPD} + \frac{1}{2}\text{TT}}{\text{total number of tetrads}} = \dfrac{0 + \frac{1}{2}(18)}{100} = 9\%$ or 9 map units. Single crossovers between either *b* and its centromere or *c* and its centromere or both would produce TT's for *bc*. Since none occurred we can assume that the locus of *b* and *c* are both very near their respective centromeres.

# Supplementary Problems

## RECOMBINATION AMONG LINKED GENES

**6.17.** There is 21% crossing over between the locus of *p* and that of *c* in the rat. Suppose that 150 primary oocytes could be scored for chiasmata within this region of the chromosome. How many of these oocytes would be expected to have a chiasma between these two genes?

**6.18.** The longest chromosome in the sweet pea has a minimum uncorrected map length (based on known genetic markers) of 118 units. Cytological observations of the longest chromosome in meiotic cells revealed an average chiasmata frequency of 2.96 per tetrad. Calculate the maximum number of crossover units remaining in this chromosome for the mapping of new genes outside the range already known.

## GENETIC MAPPING

**6.19.** The distances between 8 loci in the second chromosome of *Drosophila* are presented in the following table. Construct a genetic map to include these eight loci. The table is symmetrical above and below the diagonal.

| | *d* | *dp* | *net* | *J* | *ed* | *ft* | *cl* | *ho* |
|---|---|---|---|---|---|---|---|---|
| *d* | — | 18 | 31 | 10 | 20 | 19 | 14.5 | 27 |
| *dp* | | — | 13 | 28 | 2 | 1 | 3.5 | 9 |
| *net* | | | — | 41 | 11 | 12 | 16.5 | 4 |
| *J* | | | | — | 30 | 29 | 24.5 | 37 |
| *ed* | | | | | — | 1 | 5.5 | 7 |
| *ft* | | | | | | — | 4.5 | 8 |
| *cl* | | | | | | | — | 12.5 |
| *ho* | | | | | | | | — |

**6.20.** The recessive gene *sh* produces shrunken endosperm in corn kernels and its dominant allele $sh^+$ produces full, plump kernels. The recessive gene *c* produces colorless endosperm and its dominant allele $c^+$ produces colored endosperm. Two homozygous plants are crossed, producing an $F_1$ all phenotypically plump and colored. The $F_1$ plants are testcrossed and produce 149 shrunken, colored : 4,035 shrunken, colorless : 152 plump, colorless : 4,032 plump, colored. (*a*) What were the phenotypes and genotypes of the original parents? (*b*) How are the genes linked in the $F_1$? (*c*) Estimate the map distance between *sh* and *c*.

**6.21.** The presence of one of the Rh antigens on the surface of the red blood cells (Rh-positive) in humans is produced by a dominant gene *R*, Rh-negative cells are produced by the recessive genotype *rr*. Oval shaped erythrocytes (elliptocytosis or ovalocytosis) are caused by a dominant gene *E*, its recessive allele *e* producing normal red blood cells. Both of these genes are linked approximately 20 map units apart on one of the autosomes. A man with elliptocytosis, whose mother had normally shaped erythrocytes and a homozygous Rh-positive genotype and whose father was Rh-negative and heterozygous for elliptocytosis, marries a normal Rh-negative woman. (*a*) What is the probability of their first child being Rh-negative and elliptocytotic? (*b*) If their first child is Rh-positive, what is the chance that it will also be elliptocytotic?

**6.22.** The Rh genotypes, as discussed in Problem 6.21, are given for each individual in the pedigree shown below. Solid symbols represent elliptocytotic individuals. (*a*) List the *E* locus genotypes for each individual in the pedigree. (*b*) List the *gametic contribution* (for both loci) of the elliptocytotic individuals (of genotype *Rr*) beside each of their offspring in which it can be detected. (*c*) How often in part (*b*) did *R* segregate with *E* and *r* with *e*? (*d*) On the basis of random assortment, in how many of the offspring of part (*b*) would we expect to find *R* segregating with *e* or *r* with *E*? (*e*) If these genes assort independently, calculate the probability of *R* segregating with *E* and *r* with *e* in all 10 cases. (*f*) Is the solution to part (*c*) suggestive of linkage between these two loci? (*g*) Calculate part (*e*) if the siblings II1 and II2 were identical twins (developed from a single egg). (*h*) How are these genes probably linked in I1?

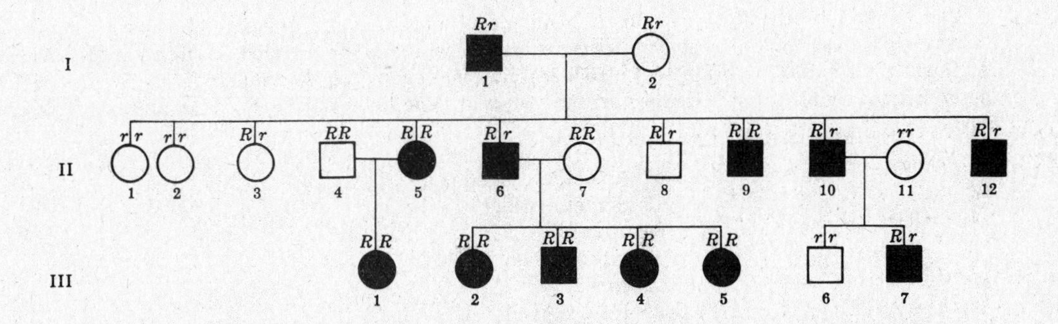


**6.23.** A hereditary disease called "retinitis pigmentosa" in humans causes excessive pigmentation of the retina with consequent partial or complete blindness. It is caused by a dominant incompletely sex-linked gene *R*. Afflicted individuals appear as solid symbols in the pedigree.

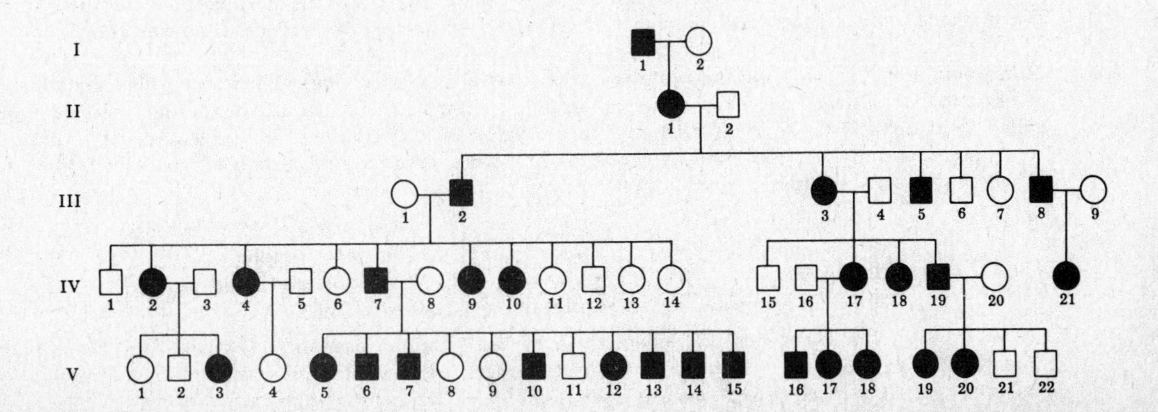

(*a*) Determine the genotypes, insofar as possible, for each individual in the pedigree, designating the chromosome (X or Y) in which the mutant gene probably resides in the afflicted individuals. Label each individual produced by a crossover gamete "CO". (*b*) Only in the gametes of afflicted males does the opportunity exist for the mutant gene to crossover from the X to the Y chromosome. Beginning with afflicted males, wherein the location of the mutant gene is known for certain, determine the number of opportunities (progeny) for the detection of such crossovers. (*c*) How many crossovers were actually observed? (*d*) What percentage crossing over does this represent?

**6.24.** Two recessive genes in *Drosophila* (*b* and *vg*) produce black body and vestigial wings respectively. When wild type flies are testcrossed, the $F_1$ are all dihybrid in coupling phase. Testcrossing the female $F_1$ produced 1930 wild type : 1888 black and vestigial : 412 black : 370 vestigial. (*a*) Calculate the distance between *b* and *vg*. (*b*) Another recessive gene *cn* lies between the loci of *b* and *vg*, producing cinnabar eye color. When wild type flies are testcrossed, the $F_1$ are all trihybrid. Testcrossing the $F_1$ females produced 664 wild type : 652 black, cinnabar, vestigial : 72 black, cinnabar : 68 vestigial : 70 black : 61 cinnabar, vestigial : 4 black, vestigial : 8 cinnabar. Calculate the map distances. (*c*) Do the *b-vg* distances calculated in parts (a) and (*b*) coincide? Explain. (*d*) What is the coefficient of coincidence?

**6.25.** In corn, a dominant gene *C* produces colored aleurone; its recessive allele *c* produces colorless. Another dominant gene *Sh* produces full, plump kernels; its recessive allele *sh* produces shrunken kernels due to collapsing of the endosperm. A third dominant gene *Wx* produces normal starchy endosperm and its recessive allele *wx* produces waxy starch. A homozygous plant from a seed with colorless, plump and waxy endosperm is crossed to a homozygous plant from a seed with colored, shrunken and starchy endosperm. The $F_1$ is testcrossed to a colorless, shrunken, waxy strain. The progeny seed exhibits the following phenotypes: 113 colorless, shrunken, starchy : 4 colored, plump, starchy : 2,708 colorless, plump, waxy : 626 colorless, plump, starchy : 2 colorless, shrunken, waxy : 116 colored, plump, waxy : 2,538 colored, shrunken, starchy : 601 colored, shrunken, waxy. (*a*) Construct a genetic map for this region of the chromosome. Round all calculations to the nearest tenth of a percent. (*b*) Calculate the interference in this region.

**6.26.** A gene called "forked" (*f*) produces shortened, bent or split bristles and hairs in *Drosophila*. Another gene called "outstretched" (*od*) results in wings being carried at right angles to the body. A third gene called "garnet" (*g*) produces pinkish eye color in young flies. Wild type females heterozygous at all three loci were crossed to wild type males. The $F_1$ data appears below.

$F_1$: Females: all wild type

Males:
57 garnet, outstretched
419 garnet, forked
60 forked
1 outstretched, forked
2 garnet
439 outstretched
13 wild type
9 outstretched, garnet, forked
1000

(*a*) Which gene is in the middle? (*b*) What was the linkage relationship between alleles at the forked and outstretched loci in the maternal parent? (*c*) What was the linkage relationship between alleles at the forked and garnet loci in the maternal parent? (*d*) On what chromosome do these three genes reside? (*e*) Calculate the map distances. (*f*) How much interference is operative?

**6.27.** Maize plants homozygous for the recessive gene "variable sterile" (*va*) exhibit irregular distribution of chromosomes during meiosis. Yellowish-green seedlings are the result of another recessive gene called "virescent" (*v*). A third recessive called "glossy" (*gl*) produces shiny leaves. All three of these genes are linked. Two homozygous plants were crossed and produced an all normal $F_1$. When the $F_1$ was testcrossed, progeny phenotypes appeared as follows:

60 virescent
48 virescent, glossy
7 glossy
270 variable sterile, virescent, glossy
4 variable sterile, virescent
40 variable sterile
62 variable sterile, glossy
235 wild type

(*a*) What were the genotypes and phenotypes of the original parents? (*b*) Diagram the linkage relationships in the $F_1$. (*c*) Determine the gene order. (*d*) Calculate the amount of recombination observed. (*e*) How much interference is operative?

**6.28.** Five sex-linked recessive genes of *Drosophila* (*ec*, *sc*, *v*, *cv* and *ct*) produce traits called echinus, scute, vermilion, crossveinless and cut respectively. Echinus is a mutant producing rough eyes with large facets. Scute manifests itself by the absence or reduction in the number of bristles on certain parts of the body. Vermilion is a bright orange-red eye color. Crossveinless prevents the development of supporting structures in the wings. Cut produces scalloped and pointed wings with manifold (pleiotropic) effects in other parts of the body. At the beginning of our experiments we do not know the gene order. From the results of the following three experiments, construct a genetic map for this region of the X chromosome. Whenever possible use weighted averages.

*Experiment 1.* Echinus females crossed to scute, crossveinless males produced all wild type females and all echinus males in the $F_1$. When the $F_1$ females were testcrossed, the results (including both male and female progeny) were as follows:

| | |
|---|---|
| 810 echinus | 89 scute |
| 828 scute, crossveinless | 62 echinus, scute |
| 88 crossveinless | 103 echinus, crossveinless |

*Experiment 2.* Crossveinless females crossed to echinus, cut males produced all wild type females and all crossveinless males in the $F_1$. When the $F_1$ females were testcrossed, the results (including both male and female progeny) were as follows:

| | |
|---|---|
| 2207 crossveinless | 223 crossveinless, cut |
| 2125 echinus, cut | 217 echinus |
| 273 echinus, crossveinless | 5 wild type |
| 265 cut | 3 echinus, crossveinless, cut |

*Experiment 3.* Cut females crossed to vermilion, crossveinless males produced all wild type females and cut males in the $F_1$. When the $F_1$ females were testcrossed, the results (including both male and female progeny) were as follows:

| | |
|---|---|
| 766 vermilion, crossveinless | 73 vermilion |
| 759 cut | 85 crossveinless, cut |
| 140 vermilion, cut | 2 wild type |
| 158 crossveinless | 2 vermilion, crossveinless, cut |

## LINKAGE ESTIMATES FROM $F_2$ DATA

**6.29.** Two recessive sex-linked genes are known in chickens (ZW method of sex determination), rapid feathering (*sl*) and gold plumage (*s*). The dominant alleles produce slow feathering (*Sl*) and silver plumage (*S*) respectively. Females of the Silver Penciled Rock breed, with slow feathering and silver plumage, are crossed to males of the Brown Leghorn breed, with rapid feathering and gold plumage. The $F_2$ progeny data appear below:

| | Slow, silver | Rapid, silver | Slow, gold | Rapid, gold |
|---|---|---|---|---|
| Males | 94 | 40 | 7 | 127 |
| Females | 117 | 28 | 7 | 156 |

(*a*) Determine the $F_1$ genotypes and phenotypes. (*b*) In what linkage phase are the $F_1$ males? (*c*) Calculate the amount of recombination expected to occur between these two loci in males.

**6.30.** Assume the genotype *AB*/*AB* is testcrossed and produces an $F_2$ consisting of 37 *A-B-*, 11 *A-bb*, 12 *aaB-* and 4 *aabb*. Estimate the percentage recombination between *A* and *B* by the square root method.

**6.31.** Two recessive genes in the third linkage group of corn produce crinkly leaves and dwarf plants respectively. A pure crinkly plant is pollinated by a pure dwarf plant. The $F_2$ progeny consist of 104 normal : 43 dwarf : 51 crinkly : 2 dwarf, crinkly. Using the square root method, estimate the amount of recombination between these two loci.

**6.32.** Colored kernel is dominant to colorless in corn; full kernel is dominant to shrunken. A pure colored, full variety is crossed to a colorless, shrunken variety. In the $F_2$ there was 73% colored, full : 2% colored, shrunken : 2% colorless, full : 23% colorless, shrunken. Estimate, by square root method, the crossover percentage between these two genes.

**6.33.** Several sex-linked genes in chickens are known, among which are the $S$ locus governing plumage color, the $K$ locus controlling the rate of feather development, and the $Id$ locus which determines whether or not melanin pigment can develop in the dermis. $F_2$ data is shown below, where $X$ and $Y$ represent dominant phenotypes and $x$ and $y$ represent recessive phenotypes.

| | Coupling | | | | Repulsion | | | |
|---|---|---|---|---|---|---|---|---|
| | $XY$ | $Xy$ | $xY$ | $xy$ | $XY$ | $Xy$ | $xY$ | $xy$ |
| $S$ and $Id$ | 157 | 171 | 145 | 177 | 83 | 108 | 91 | 75 |
| $K$ and $S$ | 603 | 94 | 159 | 648 | 12 | 24 | 26 | 9 |

Using the product-ratio method, estimate the amount of recombination between $S$ and $Id$ and between $K$ and $S$ using a weighted average of the estimates derived from coupling and repulsion data.

**6.34.** The results of selfing dihybrid sweet peas have been reported by several investigators. Let us consider five loci on the "D chromosome":

$D_1, d_1$ = tendril vs. acacia leaves
$D_2, d_2$ = bright vs. dull flower color
$D_3, d_3$ = presence vs. absence of flake modifier
$D_4, d_4$ = hairy vs. glabrous (smooth)
$D_5, d_5$ = full vs. picotee flower pattern

Ten experiments were carried through the $F_2$ generation with the following results (let $X$ and $Y$ represent dominant phenotypes and $x$ and $y$ represent recessive phenotypes):

| | | Coupling Series | | | | Repulsion Series | | | |
|---|---|---|---|---|---|---|---|---|---|
| | | $XY$ | $Xy$ | $xY$ | $xy$ | $XY$ | $Xy$ | $xY$ | $xy$ |
| (1) | $D_1D_2$ | 634 | 192 | 163 | 101 | 2263 | 809 | 882 | 124 |
| (2) | $D_1D_3$ | 286 | 11 | 19 | 79 | — | — | — | — |
| (3) | $D_1D_4$ | 3217 | 790 | 784 | 544 | 692 | 278 | 300 | 58 |
| (4) | $D_1D_5$ | — | — | — | — | 3112 | 1126 | 1114 | 277 |
| (5) | $D_2D_3$ | 992 | 273 | 265 | 124 | 296 | 97 | 88 | 20 |
| (6) | $D_2D_4$ | 1580 | 482 | 421 | 148 | 1844 | 592 | 603 | 171 |
| (7) | $D_2D_5$ | 1438 | 443 | 424 | 162 | 1337 | 465 | 415 | 128 |
| (8) | $D_3D_4$ | 200 | 52 | 44 | 35 | — | — | — | — |
| (9) | $D_3D_5$ | 373 | 112 | 103 | 42 | — | — | — | — |
| (10) | $D_4D_5$ | 1076 | 298 | 283 | 188 | 2537 | 1003 | 993 | 131 |

(*a*) Using the product-ratio method (Table 6.1) determine the crossover values for each experiment accurate to two decimal places. (*b*) Where the estimates are available in both coupling and repulsion, take a weighted average of the two and express all 10 crossover values to two decimal places on the same map. (*c*) The distance calculated from experiments $(D_1D_2)$ plus $(D_1D_3)$ is not the same as the distance from the experiment involving $(D_2D_3)$. Why? (*d*) What is the total map distance represented by the map in part (*b*)? (*e*) To what average chiasma frequency does the answer to part (*d*) correspond?

**6.35.** The duplicate recessive genes ($r_1$ and $r_2$) produce a short, velvetlike fur called "rex". Two rex rabbits of different homozygous genotypes were mated and produced an $F_1$ which was then testcrossed to produce 64 rex and 6 normal testcross progeny. (*a*) Assuming independent assortment, how many normal and rex phenotypes would be expected among 70 progeny? (*b*) Do the data indicate linkage? (*c*) What is the genotype and the phenotype of the $F_1$? (*d*) What is the genotype and phenotype of the testcross individuals? (*e*) Calculate the map distance.

**6.36.** A dominant gene $C$ is necessary for any pigment to be developed in rabbits. Its recessive allele $c$ produces albino. Black pigment is produced by another dominant gene $B$ and brown pigment by its recessive allele $b$. The C locus exhibits recessive epistasis over the B locus. Homozygous brown rabbits mated to albinos of genotype *ccBB* produce an $F_1$ which is then testcrossed to produce the $F_2$. Among the $F_2$ progeny were found 17 black, 33 brown and 50 albino. (*a*) Assuming independent assortment, what $F_2$ ratio is expected? (*b*) Do the $F_2$ results approximate the expectations in part (*a*)? (*c*) What are the genotypes and phenotypes of the $F_1$? (*d*) What are the genotypes and phenotypes of the testcross individuals? (*e*) Estimate the map distance between these two genes.

## USE OF GENETIC MAPS

**6.37.** Two loci are known to be in linkage group IV of the rat. Kinky hairs in the coat and vibrissae (long nose "whiskers") are produced in response to the recessive genotype *kk* and a short, stubby tail is produced by the recessive genotype *st/st*. The dominant alleles at these loci produce normal hairs and tails, respectively. Given 30 map units between the loci of *k* and *st*, determine the expected $F_1$ phenotypic proportions from heterozygous parents which are (*a*) both in coupling phase, (*b*) both in repulsion phase, (*c*) one in coupling and the other in repulsion phase.

**6.38.** In mice, the genes frizzy (*fr*) and albino (*c*) are linked on chromosome 1 at a distance of 20 map units. Dihybrid wild type females in repulsion phase are mated to dihybrid wild type males in coupling phase. Predict the offspring phenotypic expectations.

**6.39.** Deep yellow hemolymph (blood) in silkworm larvae is the result of a dominant gene *Y* at locus 25.6 (i.e., 25.6 crossover units from the end of the chromosome). Another dominant mutation *Rc*, 6.2 map units from the *Y* locus, produces a yellowish-brown cocoon (rusty). Between these two loci is a recessive mutant *oa* governing mottled translucency in the larval skin, mapping at locus 26.7. *Rc* and *oa* are separated by 5.1 crossover units. An individual which is homozygous for yellow blood, mottled translucent larval skin and wild type cocoon color is crossed to an individual of genotype $Y^+\ oa^+\ Rc/Y^+\ oa^+\ Rc$ which spins a rusty cocoon. The $F_1$ males are then testcrossed to produce 3000 $F_2$ progeny. Coincidence is assumed to be 10%. (*a*) Predict the numbers within each phenotypic class that will appear in the $F_2$ (to the nearest whole numbers). (*b*) On the basis of probabilities, how many *more* $F_2$ progeny would need to be produced in order to recover one each of the DCO phenotypes?

**6.40.** The eyes of certain mutant *Drosophila* have a rough texture due to abnormal facet structure. Three of the mutants which produce approximately the same phenotype (mimics) are sex-linked recessives: roughest (*rst*), rugose (*rg*) and roughex (*rux*). The loci of these genes in terms of their distances from the end of the X chromosome are 2, 11, and 15 map units respectively. (*a*) From testcrossing wild type females of genotype $\frac{rst\ +\ rux}{+\ rg\ +}$, predict the number of wild type and rough-eyed flies expected among 20,000 progeny. Assume no interference. (*b*) Approximately how many rough-eyed progeny flies are expected for every wild type individual? (*c*) If the females of part (*a*) were of genotype $\frac{rst\ rg\ rux}{+\ +\ +}$, what would be the approximate ratio of wild type : rough-eyed progeny?

**6.41.** In Asiatic cotton, a pair of factors (*R* and *r*) controls the presence or absence respectively of anthocyanin pigmentation. Another gene, about ten map units away from the *R* locus, controls chlorophyll production. The homozygous recessive genotype at this locus (*yy*) produces a yellow (chlorophyll deficient) plant which dies early in the seedling stage. The heterozygote *Yy* is phenotypically green and indistinguishable from the dominant homozygote *YY*. Obviously, testcrosses are not possible for the *Y* locus. When dihybrids are crossed together, calculate the expected phenotypic proportions among the seedlings and among the mature $F_1$ when parents are (*a*) both in coupling phase, (*b*) both in repulsion phase, (*c*) one in coupling and one in repulsion position. (*d*) Which method (in parts (*a*), (*b*) or (*c*)) is expected to produce the greatest mortality?

**6.42.** The "agouti" hair pattern, commonly found in many wild animals, is characterized by a small band of light pigment near the tip of the hair. In the rabbit, the agouti locus has a dominant allele *A* which puts the narrow band of light pigment on the hair (agouti pattern). Its recessive allele *a* results in non-agouti (black coat). Approximately 28 map units from the agouti locus is a gene which regulates the size of the agouti band; wide-banded agouti being produced by the recessive allele *w* and normal (narrow) agouti band by the dominant allele *W*. Midway between these two loci is a third locus controlling body size. The dominant allele at this locus (*Dw*) produces normal growth, but the recessive allele *dw* produces a dwarf which dies shortly after birth. Assume the genotype of a group of trihybrid females has the *A* and *Dw* loci in coupling, and the *A* and *W* loci in repulsion. These females are mated to a group of wide-banded agouti males of genotype *A dw w*/*a Dw w*. (*a*) Predict the phenotypic percentages (to the nearest decimal) expected in the $F_1$ at birth. (*b*) How much genetic death loss is anticipated? (*c*) Predict the phenotypic percentages expected in the $F_1$ when mature.

## CROSSOVER SUPPRESSION

**6.43.** A black bodied *Drosophila* is produced by a recessive gene *b* and vestigial wings by another recessive gene *vg* on the same chromosome. These two loci are approximately 20 map units apart. Predict the progeny phenotypic expectations from (*a*) the mating of repulsion phase females × coupling phase males, (*b*) the reciprocal cross of part (*a*), (*c*) the mating where both parents are in repulsion phase.

**6.44.** Poorly developed mucous glands in the female silkworm *Bombyx mori* cause eggs to be easily separated from the papers on which they are laid. This is a dominant genetic condition; its wild type recessive allele $Ng^+$ produces normally "glued" eggs. Another dominant gene *C*, 14 map units from *Ng*, produces a golden-yellow color on the outside of the cocoon and nearly white inside. Its recessive wild type allele $C^+$ produces normally pigmented or wild type cocoon color. A pure "glueless" strain is crossed to a pure golden strain. The $F_1$ females are then mated to their brothers to produce the $F_2$. Predict the number of individuals of different phenotypes expected to be observed in a total of 500 $F_2$ offspring. *Hint.* Crossing over does not occur in female silkworms.

**6.45.** Two autosomal recessive genes, "dumpy" (*dp*, a reduction in wing size) and "net" (*net*, extra veins in the wing) are linked on chromosome 2 of *Drosophila.* Homozygous wild type females are crossed to net, dumpy males. Among 800 $F_2$ offspring were found: 574 wild type : 174 net, dumpy : 25 dumpy : 27 net. Estimate the map distance.

**6.46.** Suppose that an abnormal genetic trait (mutation) appeared suddenly in a female of a pure culture of *Drosophila melanogaster.* We mate the mutant female to a male from a balanced lethal strain (*Cy/Pm, D/Sb*, where curly (*Cy*) and plum (*Pm*) are on chromosome 2 and dichaete (*D*) and stubble (*Sb*) are on chromosome 3). About half of the $F_1$ progeny (both males and females) exhibit the mutant phenotype. The $F_1$ mutant males with curly wings and stubble bristles are then mated to unrelated virgin wild type females. In the $F_2$ the mutant trait never appears with stubble. Recall that this species of *Drosophila* has chromosomes X, 2, 3 and 4. Could the mutation be (*a*) an autosomal recessive, (*b*) a sex-linked recessive, (*c*) an autosomal dominant, (*d*) a sex-linked dominant? (*e*) In which chromosome does the mutant gene reside? (*f*) Suppose the mutant trait in the $F_2$ appeared in equal association with curly and stubble. In which chromosome would the mutant gene reside? (*g*) Suppose the mutant trait in the $F_2$ appeared only in females. In which chromosome would the mutant gene reside? (*h*) Suppose the mutant trait in the $F_2$ never appeared with curly. In which chromosome would the mutant gene reside?

## RECOMBINATION MAPPING WITH TETRADS

**6.47.** Given the adjoining meiotic metaphase orientation in *Neurospora*, determine the simplest explanation to account for the following spore patterns. *Hint.* See Problem 6.14.

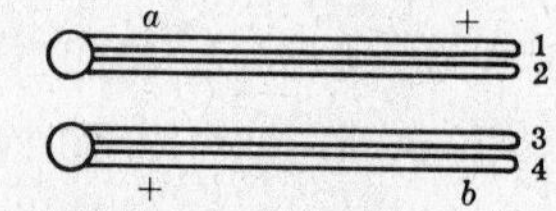


(*a*) (+ *b*)(*a* +)(*a* +)(+ *b*)
(*b*) (*a* +)(+ *b*)(+ *b*)(*a* +)
(*c*) (+ +)(*a* +)(*ab*)(+ *b*)
(*d*) (+ +)(*ab*)(*a* +)(+ *b*)
(*e*) (*ab*)(+ +)(*a* +)(+ *b*)
(*f*) (*a* +)(+ +)(*ab*)(+ *b*)
(*g*) (*ab*)(+ *b*)(+ +)(*a* +)
(*h*) (*a* +)(+ *b*)(+ +)(*ab*)
(*i*) (*ab*)(*a* +)(+ +)(+ *b*)
(*j*) (*a* +)(*ab*)(+ +)(+ *b*)
(*k*) (*a* +)(*ab*)(+ *b*)(+ +)
(*l*) (*ab*)(*a* +)(+ *b*)(+ +)

**6.48.** A certain strain of *Neurospora* cannot grow unless adenine is in the culture medium. Adenineless is a recessive mutation (*ad*). Another strain produces yellow conidia (*ylo*). Below are shown the results from crossing these two strains. Calculate the map distance between these two genes.

| No. of Asci | Meiotic Products |
|---|---|
| 106 | (*ad* +)(*ad* +)(+ *ylo*)(+ *ylo*) |
| 14 | (*ad* +)(*ad ylo*)(+ +)(+ *ylo*) |

**6.49.** A strain of *Neurospora*, unable to synthesize the vitamin thiamine (*t*), is crossed to a strain unable to synthesize the amino acid arginine (*a*).

| No. of Asci | Ascospores | | | |
|---|---|---|---|---|
| | Pair 1 | Pair 2 | Pair 3 | Pair 4 |
| 37 | + + | + + | *at* | *at* |
| 33 | + *t* | + *t* | *a* + | *a* + |
| 34 | *at* | *at* | + + | + + |
| 36 | *a* + | *a* + | + *t* | + *t* |

What information can be gleaned from these results for use in mapping these two loci?

**6.50.** Two strains of *Neurospora*, one mutant for gene $x$ and the other for gene $y$, were crossed with results as shown below. Determine the gene order and calculate the map distance of each gene from its centromere.

| No. of Asci | Meiotic Products |
|---|---|
| 52 | $(+\,y)(+\,y)(x\,+)(x\,+)$ |
| 2 | $(+\,+)(xy)(+\,y)(x\,+)$ |
| 17 | $(x\,+)(+\,y)(x\,+)(+\,y)$ |
| 1 | $(xy)(x\,+)(+\,+)(+\,y)$ |
| 6 | $(+\,y)(xy)(+\,+)(x\,+)$ |
| 1 | $(+\,y)(x\,+)(+\,+)(xy)$ |
| 15 | $(x\,+)(+\,y)(+\,y)(x\,+)$ |
| 1 | $(+\,+)(+\,y)(xy)(x\,+)$ |
| 5 | $(xy)(+\,y)(+\,+)(x\,+)$ |

**6.51.** A riboflavineless strain ($r$) of *Neurospora* is crossed with a tryptophaneless strain ($t$) to give

| No. of Asci | Meiotic Products | No. of Asci | Meiotic Products |
|---|---|---|---|
| 129 | $(r+)(r+)(+t)(+t)$ | 13 | $(rt)(r+)(+t)(++)$ |
| 1 | $(+t)(r+)(r+)(+t)$ | 17 | $(r+)(rt)(+\,+)(+t)$ |
| 2 | $(r+)(+t)(r+)(+t)$ | 17 | $(r+)(rt)(+t)(+\,+)$ |
| 1 | $(r+)(+t)(+t)(r+)$ | 2 | $(r+)(+\,+)(rt)(+t)$ |
| 15 | $(rt)(r+)(+\,+)(+t)$ | 1 | $(rt)(+\,+)(r+)(+t)$ |

Construct a map which includes these two genes.

**6.52.** Two of the three genes ($s$, $t$ or $u$) are linked, the third assorts independently and is very tightly linked to its centromere. Analyze the unordered tetrads produced by the cross $(stu) \times (+\,+\,+)$. *Hint.* See Problem 6.16.

| No. of Tetrads | Tetrads |
|---|---|
| 59 | $(stu)(stu)(+\,+\,+)(+\,+\,+)$ |
| 53 | $(s\,+\,u)(s\,+\,u)(+\,t\,+)(+\,t\,+)$ |
| 26 | $(st\,+)(+\,t\,+)(s\,+\,u)(+\,+\,u)$ |
| 30 | $(s\,+\,+)(+\,+\,+)(stu)(+\,tu)$ |
| 32 | $(stu)(+\,t\,+)(s\,+\,u)(+\,+\,+)$ |

# Answers to Supplementary Problems

**6.17.** 63

**6.18.** 30 units

**6.19.** *net* 4 *ho* 7 *ed* 1 *ft* 1 *dp* 3.5 *cl* 14.5 *d* 10 *J*

**6.20.** (*a*) $sh\,c/sh\,c$ (shrunken, colorless) $\times$ $sh^+\,c^+/sh^+\,c^+$ (plump, colored) (*b*) $sh^+\,c^+/sh\,c$ (coupling phase) (*c*) 3.6 map units

**6.21.** (*a*) $\frac{2}{5}$ (*b*) $\frac{1}{5}$

**6.22.** (*a*) all open symbols = *ee*, all solid symbols = *Ee* (*b*) *re* (II1, II2, III6), *RE* (II5, 9, III2-5, 7) (*c*) all 10 cases (*d*) 5 each (*e*) 1/1024 (*f*) yes (*g*) 1/512 (*h*) *RE*/*re* (coupling phase)

**6.23.** (*a*) all open symbols *rr*; all afflicted individuals *Rr*; location of mutant gene is on X chromosome in II1, III2, 3, 5, 8, IV2, 4, 9, 10, 17, 18, 19, 21, V3, 5, 12, 16, 17, 18, 19, 20; location of mutant gene is on Y chromosome in IV7, V6, 7, 10, 13, 14, 15; crossover individuals are IV6, 7, 11, 13, 14, V5, 11, 12; location of mutant gene in I1 is unknown. (*b*) 27 opportunities; they are IV1, 2, 4, 6, 7, 9-14, 21, V5-15, 19-22 (*c*) 8 (*d*) 29.63%

**6.24.** (*a*) 17 map units (*b*) *b-cn* = 8.9 map units; *cn-vg* = 9.5 map units (*c*) No. In part (*a*), two-point testcross cannot detect double crossovers which then appear as parental types, thus underestimating the true map distance. (*d*) 0.89

**6.25.** (*a*) *c* —3.5— *sh* —18.4— *wx* (*b*) 86.1%

**6.26.** (*a*) *f* (*b*) repulsion phase (*c*) coupling phase (*d*) X chromosome (sex-linked) (*e*) *g-f* = 12.0, *f-od* = 2.5 (*f*) none

**6.27.** (*a*) $va^+ \, gl^+ \, v^+/va^+ \, gl^+ \, v^+$ (wild type) × $va\,gl\,v/\,va\,gl\,v$ (variable sterile, glossy, virescent) (*b*) $va^+ \, gl^+ \, v^+/va\,gl\,v$ (*c*) *gl* is in the middle (*d*) *va-gl* = 13.6%, *gl-v* = 18.3% (*e*) 39%

**6.28.** *Experiment 1*: *sc* —7.6— *ec* —9.7— *cv*

*Experiment 2*: *ec* —10.3— *cv* —8.4— *ct*

*Experiment 3*: *cv* —8.2— *ct* —15.2— *v*

Combined Map: *sc* —7.6— *ec* —10.1— *cv* —8.3— *ct* —15.2— *v*

**6.29.** (*a*) *Sl S/sl s* (slow, silver males) × *sl s*/W (rapid, gold females) (*b*) coupling phase (*c*) 14.2%

**6.30.** 50%

**6.31.** 20%

**6.32.** 4%

**6.33.**

| Genes | Coupling | Repulsion | Weighted Average |
|---|---|---|---|
| *S* and *Id* | 0.48 | 0.44 | 0.47 |
| *K* and *S* | 0.15 | 0.27 | 0.16 |

**6.34.** (*a*)

| Genes | $D_1D_2$ | $D_1D_3$ | $D_1D_4$ | $D_1D_5$ | $D_2D_3$ | $D_2D_4$ | $D_2D_5$ | $D_3D_4$ | $D_3D_5$ | $D_4D_5$ |
|---|---|---|---|---|---|---|---|---|---|---|
| Coupling | 0.40 | 0.08 | 0.36 | — | 0.42 | 0.48 | 0.47 | 0.35 | 0.46 | 0.38 |
| Repulsion | 0.37 | — | 0.40 | 0.45 | 0.45 | 0.48 | 0.48 | — | — | 0.35 |

(*b*)

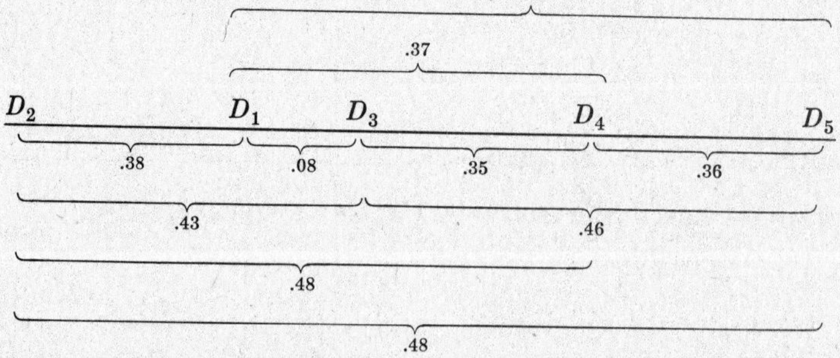

$D_2$ —.38— $D_1$ —.08— $D_3$ —.35— $D_4$ —.36— $D_5$

($D_1$ to $D_5$: .45; $D_1$ to $D_4$: .37; $D_2$ to $D_3$: .43; $D_3$ to $D_5$: .46; $D_2$ to $D_4$: .48; $D_2$ to $D_5$: .48)

(*c*) (1) Undetected double crossovers underestimate the true distance in the $D_2D_3$ experiment. (2) Sampling error in different experiments yields slightly different estimates of the true values. (*d*) 117 map units (*e*) 2.34 chiasmata per tetrad

**6.35.** (*a*) 52.5 rex : 17.5 normal (*b*) yes (*c*) $R_1r_2/r_1R_2$, normal (*d*) $r_1r_2/r_1r_2$, rex (*e*) 17.14 map units

**6.36.** (*a*) 2 albino : 1 black : 1 brown (*b*) no (*c*) *Cb/cB*, black (*d*) *cb/cb*, albino (*e*) 34 map units

**6.37.**

| Phenotype | (*a*) | (*b*) | (*c*) |
|---|---|---|---|
| normal | .6225 | .5225 | .5525 |
| kinky hair | .1275 | .2275 | .1975 |
| stub tail | .1275 | .2275 | .1975 |
| kinky hair, stub tail | .1225 | .0225 | .0525 |

**6.38.** 54% wild type : 21% frizzy : 21% albino : 4% frizzy, albino

**6.39.** (*a*) 1408 yellow, mottled : 1408 rusty : 16 yellow, rusty : 16 mottled : 76 yellow, mottled, rusty : 76 wild type. Note: rounding errors may allow one whole individual difference in each of these phenotypic classes. (*b*) 32,651

**6.40.** (*a*) 19,964 rough-eyed : 36 wild type (*b*) approximately 1 : 555 (*c*) approximately 1 : 1.289

**6.41.** (*a*)(*b*)(*c*)

| | Seedlings | | | Adult | | |
|---|---|---|---|---|---|---|
| | (*a*) | (*b*) | (*c*) | (*a*) | (*b*) | (*c*) |
| normal chlorophyll, anthocyanin present | .7025 | .5025 | .5225 | .9367 | .6700 | .6967 |
| normal chlorophyll, anthocyanin absent | .0475 | .2475 | .2275 | .0633 | .3300 | .3033 |
| yellow, anthocyanin present | .0475 | .2475 | .2275 | | | |
| yellow, anthocyanin absent | .2025 | .0025 | .0225 | | | |

(*d*) All three crosses = 25% mortality.

**6.42.** (*a*) 40.9% wide agouti : 22.0% black : 18.9% narrow agouti, dwarf : 12.1% narrow agouti : 3.1% wide agouti, dwarf : 3.0% black, dwarf (*b*) 25% (*c*) 54.5% wide agouti : 29.3% black : 16.1% narrow agouti

**6.43.** (*a*) 55% wild type : 20% black : 20% vestigial : 5% black, vestigial
(*b*), (*c*) 50% wild type : 25% black : 25% vestigial

**6.44.** 250 glueless, golden : 125 glueless : 125 golden

**6.45.** 13 map units

**6.46.** A dominant trait may appear suddenly in a population by mutation of a recessive wild type gene to a dominant allelic form. Such a mutant individual would be heterozygous. (*a*) No (*b*) No (*c*) Yes (*d*) No (*e*) Chromosome 3 (*f*) Chromosome 4 (*g*) X chromosome (sex-linked) (*h*) Chromosome 2

**6.47.** (*a*) (1, 3) (*b*) (2, 4) (*c*) (1, 3)(1, 3) or (1, 3)(2, 4) (*d*) (1, 3)(2, 3) or (1, 3)(1, 4) (*e*) (2, 3)(1, 3) or (1, 4)(1, 3) (*f*) (2, 3)(2, 3) or (1, 4)(2, 3) (*g*) (2, 4)(1, 3) or (2, 4)(2, 4) (*h*) (2, 4)(2, 3) or (2, 4)(1, 4) (*i*) (1, 3) (*j*) (2, 3) (*k*) (2, 4) (*l*) (1, 4)

**6.48.** 5.83 map units

**6.49.** Gene $a$ is segregating independently of $t$, and both loci are closely linked to their respective centromeres.

**6.50.** Centromere to $y$ = 18.5 map units; centromere to $x$ = 26.5 map units

**6.51.** ●—1.8—$r$—16.4—$t$

**6.52.** Genes $u$ and $s$ are linked on the same side of the centromere with $u$ closest to the centromere. Centromere to $u$ = 16 units, $u$ to $s$ = 14 units. Gene $t$ is on another chromosome and very closely linked to its centromere.

# Chapter 7

## Statistical Distributions

### THE BINOMIAL EXPANSION

In $(p+q)^n$, the $p$ and $q$ represent the probabilities of alternative independent events, and the power $n$ to which the binomial is raised represents the number of trials. The sum of the factors in the binomial must add to unity; thus

$$p + q = 1$$

Recall from Chapter 2 that when two independent events are occurring with the probabilities $p$ and $q$, then the probability of their joint occurrence is $pq$. That is, the combined probability is the product of the independent events. When alternative possibilities exist for the satisfaction of the conditions of the problem, the probabilities are combined by addition.

**Example 7.1.** In two tosses of a coin, with $p = \text{heads} = \frac{1}{2}$ and $q = \text{tails} = \frac{1}{2}$, there are four possibilities.

| First Toss | | Second Toss | | Probability |
|---|---|---|---|---|
| heads ($p$) | (and) | heads ($p$) | = | $p^2$ |
| heads ($p$) | (and) | tails ($q$) | = | $pq$ |
| tails ($q$) | (and) | heads ($p$) | = | $pq$ |
| tails ($q$) | (and) | tails ($q$) | = | $q^2$ |
| | | | | 1.0 |

which may be expressed as follows:

$$\underset{\text{(two heads)}}{p^2} + \underset{\text{(one head : one tail)}}{2pq} + \underset{\text{(two tails)}}{q^2} = 1.0$$

**Example 7.2.** Expanding the binomial $(p+q)^2$ produces the same expression as in the previous example. Thus $(p+q)^2 = p^2 + 2pq + q^2$.

**Example 7.3.** When a coin is tossed three times, the probabilities for any combination of heads and/or tails can be found from $(p+q)^3 = p^3 + 3p^2q + 3pq^2 + q^3$.

Let $p$ = probability of heads = $\frac{1}{2}$ and $q$ = probability of tails = $\frac{1}{2}$.

| No. of Heads | No. of Tails | Term | Probability |
|---|---|---|---|
| 3 | 0 | $p^3$ | $(\frac{1}{2})^3 = \frac{1}{8}$ |
| 2 | 1 | $3p^2q$ | $3(\frac{1}{2})^2(\frac{1}{2}) = \frac{3}{8}$ |
| 1 | 2 | $3pq^2$ | $3(\frac{1}{2})(\frac{1}{2})^2 = \frac{3}{8}$ |
| 0 | 3 | $q^3$ | $(\frac{1}{2})^3 = \frac{1}{8}$ |

The expansion of $(p+q)^3$ is found by multiplying $(p^2 + 2pq + q^2)$ by $(p+q)$. This process can be extended for higher powers, but obviously becomes increasingly laborious. A short method for expanding $(p+q)$ to any power $(n)$ may be performed by following these rules. (1) The coefficient of the first term is one. The power of the first factor $(p)$ is $n$, and that of $(q)$ is zero (note: any factor to the zero power is unity). (2) Thereafter in each term, multiply the coefficient by the power of $p$ and divide by the number of that term in the expansion. The result is the coefficient of the *next* term. (3) Also thereafter, the power of $p$ will decrease by one and the power of $q$ will increase by one in each term of the expansion. (4) The fully expanded binomial will have $(n+1)$ terms. The coefficients are symmetrical about the middle term(s) of the expansion.

Summary:

| Term | Coefficient | Powers | |
|---|---|---|---|
| | | $p$ | $q$ |
| 1 | 1 | $n$ | 0 |
| 2 | $n(1)/1$ | $n-1$ | 1 |
| 3 | $n(n-1)/1 \cdot 2$ | $n-2$ | 2 |
| 4 | $n(n-1)(n-2)/1 \cdot 2 \cdot 3$ | $n-3$ | 3 |
| . | . | . | . |
| . | . | . | . |
| . | . | . | . |
| $n+1$ | 1 | 0 | $n$ |

**1. Single Terms of the Expansion.**

The coefficients of the binomial expansion represent the number of ways in which the conditions of each term may be satisfied. The number of combinations $(C)$ of $n$ different things taken $k$ at a time is expressed by

$$_nC_k = \frac{n!}{(n-k)!\,k!} \tag{7.1}$$

where $n!$ (called "factorial $n$") $= n(n-1)(n-2)\ldots 1$. ($0! = 1$ by definition.)

**Example 7.4.** If $n = 4$, then $n! = 4(4-1)(4-2)(4-3) = 4 \cdot 3 \cdot 2 \cdot 1 = 24$.

**Example 7.5.** The number of ways to obtain two heads in three tosses of a coin is

$$_3C_2 = \frac{3!}{(3-2)!\,2!} = \frac{3 \times 2 \times 1}{(1) \times 2 \times 1} = 3$$

These three combinations are HHT, HTH, THH.

Formula (*7.1*) can be used for calculating the coefficients in a binomial expansion,

$$(p+q)^n = \sum_{k=0}^{n} {_nC_k}\, p^{n-k} q^k = \sum_{k=0}^{n} \frac{n!}{(n-k)!\,k!} p^{n-k} q^k \tag{7.2}$$

where $\sum_{k=0}^{n}$ means to sum what follows as $k$ increases by one unit in each term of the expansion from zero to $n$. This method is obviously much more laborious than the short method presented previously. However, it does have utility in the calculation of one or a few specific terms of a large binomial expansion. To represent this formula in another way, we can let $p =$ probability of the occurrence of one event (e.g. a success) and $q =$ probability of the occurrence of the alternative event (e.g. a failure); then the probability that in $n$ trials a success will occur $s$ times and a failure will occur $f$ times is given by

$$\left(\frac{n!}{s!\,f!}\right)(p^s)(q^f) \tag{7.3}$$

**2. The Multinomial Distribution.**

The binomial distribution may be generalized to accommodate any number of variables. If events $e_1, e_2, \ldots, e_k$ can occur with probabilities $p_1, p_2, \ldots, p_k$ respectively, then the probability that $e_1, e_2, \ldots, e_k$ will occur $k_1, k_2, \ldots, k_n$ times respectively is

$$\frac{N!}{k_1!\,k_2!\ldots k_n!} p_1^{k_1} p_2^{k_2} \ldots p_n^{k_n} \tag{7.4}$$

where $k_1 + k_2 + \cdots + k_n = N$.

## THE POISSON DISTRIBUTION

When the probability ($p$) of a rare event (e.g. a specific mutation) is relatively small and the sample size ($n$) is relatively large, the binomial distribution is essentially the same as a *Poisson distribution*, but is much easier to solve by the latter. Another advantage in using the Poisson distribution instead of the binomial distribution is that it allows analysis of data where $pn$ is known, but neither $p$ nor $n$ alone is known. Under these conditions, any term of the binomial expansion is closely approximated by the point Poisson formula

$$\frac{n!}{k!(n-k)!}\,p^k q^{(n-k)} = \frac{e^{-np}\,(np)^{k}}{k!}$$

where $k$ is the number of the rare events, $q$ is the probability of the common event (e.g. no mutation), and $e$ is the base of the Napierian or natural system of logarithms $\left(\frac{1}{1!} + \frac{1}{2!} + \frac{1}{3!} + \cdots = 2.71828 + \cdots\right)$. The mean ($\mu$) of rare events is equivalent to $np$. The probabilities that such events happen 0, 1, 2, 3, . . . times is given by the series of terms

$$\frac{e^{-\mu}}{0!},\quad \frac{\mu e^{-\mu}}{1!},\quad \frac{\mu^2 e^{-\mu}}{2!},\quad \frac{\mu^3 e^{-\mu}}{3!},\quad \cdots$$

Table 7.1 displays some values of $e^{-np}$ or $e^{-\mu}$ that can be helpful in solving certain problems.

The variance (see Chapter 11) of the binomial distribution is $npq$ or $np(1-p)$. The Poisson distribution has the same variance, but since $p$ is very small the probability of the common event $(1-p)$ is almost 1.0. Therefore, in a Poisson distribution the mean ($np$) is essentially the same as the variance.

Knowledge of these aspects of the Poisson distribution will be essential in Chapter 15 for understanding the Luria-Delbrück fluctuation test, a classic experiment that initiated modern bacterial genetics.

**Table 7.1. Values of $e^{-\mu}$**

($0 < \mu < 1$)

| $\mu$ | 0 | 1 | 2 | 3 | 4 | 5 | 6 | 7 | 8 | 9 |
|---|---|---|---|---|---|---|---|---|---|---|
| 0.0 | 1.0000 | .9900 | .9802 | .9704 | .9608 | .9512 | .9418 | .9324 | .9231 | .9139 |
| 0.1 | .9048 | .8958 | .8869 | .8781 | .8694 | .8607 | .8521 | .8437 | .8353 | .8270 |
| 0.2 | .8187 | .8106 | .8025 | .7945 | .7866 | .7788 | .7711 | .7634 | .7558 | .7483 |
| 0.3 | .7408 | .7334 | .7261 | .7189 | .7118 | .7047 | .6977 | .6907 | .6839 | .6771 |
| 0.4 | .6703 | .6636 | .6570 | .6505 | .6440 | .6376 | .6313 | .6250 | .6188 | .6126 |
| 0.5 | .6065 | .6005 | .5945 | .5886 | .5827 | .5770 | .5712 | .5655 | .5599 | .5543 |
| 0.6 | .5488 | .5434 | .5379 | .5326 | .5273 | .5220 | .5169 | .5117 | .5066 | .5016 |
| 0.7 | .4966 | .4916 | .4868 | .4819 | .4771 | .4724 | .4677 | .4630 | .4584 | .4538 |
| 0.8 | .4493 | .4449 | .4404 | .4360 | .4317 | .4274 | .4232 | .4190 | .4148 | .4107 |
| 0.9 | .4066 | .4025 | .3985 | .3946 | .3906 | .3867 | .3829 | .3791 | .3753 | .3716 |

($\mu$ = 1, 2, 3, . . ., 10)

| $\mu$ | 1 | 2 | 3 | 4 | 5 | 6 | 7 | 8 | 9 | 10 |
|---|---|---|---|---|---|---|---|---|---|---|
| $e^{-\mu}$ | .36788 | .13534 | .04979 | .01832 | .006738 | .002479 | .000912 | .000335 | .000123 | .000045 |

*Note*: To obtain values of $e^{-\mu}$ for other values of $\mu$ use the laws of exponents.
Example: $e^{-3.48} = (e^{-3.00})(e^{-0.48}) = (.04979)(.6188) = .03081$.

*Source*: Murray R. Spiegel, *Schaum's Outline of Theory and Problems of Statistics*, McGraw-Hill Book Company, New York, 1961, p. 348.

## TESTING GENETIC RATIOS

### 1. Sampling Theory.

If we toss a coin, we expect that half of the time it will land heads up and half of the time tails up. This hypothesized probability is based upon an infinite number of coin tossings wherein the effects of chance deviations from 0.5 in favor of either heads or tails cancel one another. All actual experiments, however, involve finite numbers of observations and therefore some deviation from the expected numbers (sampling error) is to be anticipated. Let us assume that there is no difference between the observed results of a coin-tossing experiment and the expected results that cannot be accounted for by chance alone (*null hypothesis*). How great a deviation from the expected 50-50 ratio in a given experiment should be allowed before the null hypothesis is rejected? Conventionally, the null hypothesis in most biological experiments is rejected when the deviation is so large that it could be accounted for by chance less than 5% of the time. Such results are said to be *significant*. When the null hypothesis is rejected at the 5% level, we take one chance in twenty of discarding a valid hypothesis. It must be remembered that statistics can never render absolute proof of the hypothesis, but merely sets limits to our uncertainty. If we wish to be even more certain that the rejection of the hypothesis is warranted we could use the 1% level, often called *highly significant*, in which case the experimenter would be taking only one chance in a hundred of rejecting a valid hypothesis.

### 2. Sample Size.

If our coin-tossing experiment is based on small numbers, we might anticipate relatively large deviations from the expected values to occur quite often by chance alone. However, as the sample size increases the deviation should become proportionately less, so that in a sample of infinite size the plus and minus chance deviations cancel each other completely to produce the 50-50 ratio.

### 3. Degrees of Freedom.

Assume a coin is tossed 100 times. We may arbitrarily assign any number of heads from 0 to 100 as appearing in this hypothetical experiment. However, once the number of heads is established, the remainder is tails and must add to 100. In other words, we have $n-1$ *degrees of freedom* (df) in assigning numbers at random to the $n$ classes within an experiment.

**Example 7.6.** In an experiment involving three phenotypes ($n = 3$), we can fill two of the classes at random, but the number in the third class must constitute the remainder of the total number of individuals observed. Therefore we have $3-1 = 2$ degrees of freedom.

**Example 7.7.** A 9:3:3:1 dihybrid ratio has four phenotypes ($n = 4$). There are $4-1 = 3$ degrees of freedom.

The number of degrees of freedom in these kinds of problems is the number of variables ($n$) under consideration minus one. For most genetic problems, the degrees of freedom will be one less than the number of phenotypic classes. Obviously the more variables involved in an experiment the greater the total deviation may be by chance.

### 4. Chi-Square Test.

In order to evaluate a genetic hypothesis, we need a test which can convert deviations from expected values into the probability of such inequalities occurring by chance. Furthermore, this test must also take into consideration the size of the sample and the number of variables (degrees of freedom). The chi-square test (pronounced ki-; symbolized $\chi^2$) includes all of these factors.

$$\chi^2 = \sum_{i=1}^{n} \frac{(o_i - e_i)^2}{e_i} = \frac{(o_1 - e_1)^2}{e_1} + \frac{(o_2 - e_2)^2}{e_2} + \cdots + \frac{(o_n - e_n)^2}{e_n} \tag{7.5}$$

where $\sum_{i=1}^{n}$ means to sum what follows it as the $i$ classes increase from 1 to $n$, o represents the number of observations within a class, e represents the number expected in the class according to the hypothesis under test and $n$ is the number of classes. The value of chi-square may then be converted into the probability that the deviation is due to chance by entering Table 7.2 at the appropriate number of degrees of freedom.

**Table 7.2. Chi-Square Distribution**

| Degrees of Freedom | Probability | | | | | | | | | | |
|---|---|---|---|---|---|---|---|---|---|---|---|
| | 0.95 | 0.90 | 0.80 | 0.70 | 0.50 | 0.30 | 0.20 | 0.10 | 0.05 | 0.01 | 0.001 |
| 1 | 0.004 | 0.02 | 0.06 | 0.15 | 0.46 | 1.07 | 1.64 | 2.71 | 3.84 | 6.64 | 10.83 |
| 2 | 0.10 | 0.21 | 0.45 | 0.71 | 1.39 | 2.41 | 3.22 | 4.60 | 5.99 | 9.21 | 13.82 |
| 3 | 0.35 | 0.58 | 1.01 | 1.42 | 2.37 | 3.66 | 4.64 | 6.25 | 7.82 | 11.34 | 16.27 |
| 4 | 0.71 | 1.06 | 1.65 | 2.20 | 3.36 | 4.88 | 5.99 | 7.78 | 9.49 | 13.28 | 18.47 |
| 5 | 1.14 | 1.61 | 2.34 | 3.00 | 4.35 | 6.06 | 7.29 | 9.24 | 11.07 | 15.09 | 20.52 |
| 6 | 1.63 | 2.20 | 3.07 | 3.83 | 5.35 | 7.23 | 8.56 | 10.64 | 12.59 | 16.81 | 22.46 |
| 7 | 2.17 | 2.83 | 3.82 | 4.67 | 6.35 | 8.38 | 9.80 | 12.02 | 14.07 | 18.48 | 24.32 |
| 8 | 2.73 | 3.49 | 4.59 | 5.53 | 7.34 | 9.52 | 11.03 | 13.36 | 15.51 | 20.09 | 26.12 |
| 9 | 3.32 | 4.17 | 5.38 | 6.39 | 8.34 | 10.66 | 12.24 | 14.68 | 16.92 | 21.67 | 27.88 |
| 10 | 3.94 | 4.86 | 6.18 | 7.27 | 9.34 | 11.78 | 13.44 | 15.99 | 18.31 | 23.21 | 29.59 |
| | Nonsignificant | | | | | | | | Significant | | |

*Source:* R. A. Fisher and F. Yates, *Statistical Tables for Biological, Agricultural and Medical Research* (6th edition), Table IV, Oliver & Boyd, Ltd., Edinburgh, by permission of the authors and publishers.

An alternative method for computing chi-square in problems involving *only two phenotypes* will give the same result as the conventional method and often makes computation easier,

$$\chi^2 = \frac{(a - rb)^2}{r(a + b)} \tag{7.6}$$

where $a$ and $b$ are the numbers in the two phenotypic classes and $r$ is the expected ratio of $a$ to $b$.

(a) ***Chi-Square Limitations.***

The chi-square test as used for analyzing the results of genetic experiments has two important limitations. (1) It must be used only on the numerical data itself, never on percentages or ratios derived from the data. (2) It cannot properly be used for experiments wherein the expected frequency within any phenotypic class is less than 5.

(b) ***Correction for Small Samples.***

The formula from which the chi-square table is derived is based upon a continuous distribution, namely that of the "normal" curve (see Chapter 11). Such a distribution might be expected when we plot the heights of a group of people. The most frequent class would be the average height and successively fewer people would be in the taller or shorter phenotypes. All sizes are possible from the shortest to the tallest, i.e. heights form a *continuous distribution*. However, the kinds of genetic problems in the previous chapters of this book involve separate or *discrete phenotypic classes* such as blue eyes vs. brown eyes. A correction should therefore be applied in the calculation of chi-square to correct for this lack of continuity. The "Yates' Correction for Continuity" is applied as follows:

$$\chi^2 \text{ (corrected)} = \sum_{i=1}^{n} \frac{[|o_i - e_i| - 0.5]^2}{e_i} = \frac{[|o_1 - e_1| - 0.5]^2}{e_1} + \frac{[|o_2 - e_2| - 0.5]^2}{e_2} + \cdots + \frac{[|o_n - e_n| - 0.5]^2}{e_n} \tag{7.7}$$

This correction usually makes little difference in the chi-square of most problems, but may become an important factor near the critical values. Yates' correction should be routinely applied whenever only one degree of freedom exists, or in small samples where each expected frequency is between 5 and 10. If the corrected and uncorrected methods each lead to the same conclusion, there is no difficulty. However, if these methods do not lead to the same conclusion, then either more data needs to be collected or a more sophisticated statistical test should be employed. Most genetic texts make no mention of this correction. Therefore problems in this book requiring the application of Yates' correction will be limited to those in the section entitled "Correction for Small Samples" in both the Solved and Supplementary Problems.

# Solved Problems

## THE BINOMIAL EXPANSION

**7.1.** Expand the binomial $(p+q)^5$.

**Solution:**

1st term: $1p^5q^0$; coefficient of 2nd term $= (5 \cdot 1)/1$

2nd term: $5p^4q^1$; coefficient of 3rd term $= (4 \cdot 5)/2$

3rd term: $10p^3q^2$; coefficient of 4th term $= (3 \cdot 10)/3$

4th term: $10p^2q^3$; coefficient of 5th term $= (2 \cdot 10)/4$

5th term: $5p^1q^4$; coefficient of 6th term $= (1 \cdot 5)/5$

6th term: $1p^0q^5$

Summary: $(p+q)^5 = p^5 + 5p^4q + 10p^3q^2 + 10p^2q^3 + 5pq^4 + q^5$

**7.2.** Expand $(p+q)^2$ by formula *(7.2)*.

**Solution:**

$$(p+q)^n = \sum_{k=0}^{n} \frac{n!}{(n-k)!\,k!} p^{n-k}q^k$$

$$(p+q)^2 = \frac{2!}{(2-0)!\,0!}p^{2-0}q^0 + \frac{2!}{(2-1)!\,1!}p^{2-1}q^1 + \frac{2!}{(2-2)!\,2!}p^{2-2}q^2$$

$$= p^2 + 2pq + q^2$$

Note that $0! = 1$, and $x^0 = 1$ where $x$ is any number.

**7.3.** Find the middle term of the expansion $(p+q)^{10}$ by application of formula *(7.2)*.

**Solution:**

The middle term of the expansion $(p+q)^{10}$ is the 6th term since there are $(n+1)$ terms in the expansion. The power of $q$ starts at zero in the first term and increases by one in each successive term so that the sixth term would have $q^5$, and so $k = 5$. Then the 6th term is

$$\frac{n!}{(n-k)!\,k!}p^{n-k}q^k = \frac{10 \cdot 9 \cdot 8 \cdot 7 \cdot 6 \cdot 5!}{(10-5)!\,5!}p^{10-5}q^5 = \frac{10 \cdot 9 \cdot 8 \cdot 7 \cdot 6}{5 \cdot 4 \cdot 3 \cdot 2 \cdot 1}p^5q^5 = 252p^5q^5$$

**7.4.** A multiple allelic series is known with seven alleles. How many kinds of matings are possible?

**Solution:**

$$\text{No. of genotypes possible} = \text{No. of different allelic combinations (heterozygotes)} + \text{No. of genotypes with two of the same alleles (homozygotes)}$$

$$= \frac{n!}{(n-k)!\,k!} + n = \frac{7!}{5!\,2!} + 7 = \frac{7 \cdot 6 \cdot 5!}{2 \cdot 5!} + 7$$

$$= 21 + 7 = 28 \text{ genotypes}$$

$$\text{No. of different matings} = \text{No. of matings between unlike genotypes} + \text{No. of matings between identical genotypes}$$

$$= \frac{28!}{26!\,2!} + 28 = \frac{28 \cdot 27 \cdot 26!}{2 \cdot 26!} + 28 = 406$$

**7.5.** Determine the probability of obtaining 6 heads and 3 tails in 9 tosses of a coin by applying formula (*7.3*).

**Solution:**

In tossing a coin, we may consider heads to be a success ($s$) and tails a failure ($f$). The probability of obtaining 6 successes and 3 failures in 9 trials is

$$\frac{n!}{s!\,f!}p^s q^f = \frac{9!}{6!\,3!}p^6q^3 = \frac{9 \cdot 8 \cdot 7 \cdot 6!}{1 \cdot 2 \cdot 3 \cdot 6!}p^6q^3 = 84p^6q^3$$

If the probability of obtaining a head ($p$) is equal to the probability of obtaining a tail ($q$) $= \frac{1}{2}$, then $84p^6q^3 = 84(\frac{1}{2})^9 = 84/512 = 21/128$.

**7.6.** The M-N blood types of humans are under the genetic control of a pair of codominant alleles as explained in Example 2.9, page 21. In families of size six where both parents are blood type MN, what is the chance of finding 3 children of type M, 2 of type MN, and 1 N?

**Solution:**

P: $L^ML^N \times L^ML^N$

$F_1$: $\frac{1}{4}L^ML^M$ = type M
$\frac{1}{2}L^ML^N$ = type MN
$\frac{1}{4}L^NL^N$ = type N

Let $p_1$ = probability of child being type M $= \frac{1}{4}$
$p_2$ = probability of child being type MN $= \frac{1}{2}$
$p_3$ = probability of child being type N $= \frac{1}{4}$

Let $k_1$ = number of children of type M required = 3
$k_2$ = number of children of type MN required = 2
$k_3$ = number of children of type N required = 1

$$N = \sum k_i = 6$$

$$(p_1+p_2+p_3)^N = \frac{N!}{k_1!\,k_2!\,k_3!}p_1^{k_1}p_2^{k_2}p_3^{k_3} \qquad \text{Formula } (7.4)$$

$$(p_1+p_2+p_3)^6 = \frac{6!}{3!\,2!\,1!}(\tfrac{1}{4})^3(\tfrac{1}{2})^2(\tfrac{1}{4}) = \frac{6 \cdot 5 \cdot 4 \cdot 3!}{2 \cdot 3!}(\tfrac{1}{4})^4(\tfrac{1}{2})^2 = \frac{15}{256}$$

## THE POISSON DISTRIBUTION

**7.7** Suppose that only one out of a thousand individuals in a population is an albino; all the rest are normally pigmented. If a sample of 100 individuals is drawn at random from this population, calculate, by Poisson distribution, the probabilities that it contains (*a*) no albinos, (*b*) one albino, (*c*) two albinos, (*d*) three or more albinos.

**Solution:**

Let $n$ = sample size = 100
$p$ = probability of albino = 0.001
$np$ = probable number of albinos in a population of size $n$ = (100)(0.001) = 0.1
$k$ = number of rare events (albinos) = 1, 2, 3 or more

(*a*) If $k = 0$, then referring to Table 7.1 we have

$$\frac{e^{-np}(np)^k}{k!} = \frac{e^{-0.1}(0.1)^0}{0!} = \frac{e^{-0.1}(1)}{1} = 0.9048$$

(Note: any number to the zero power is one.)

(*b*) If $k = 1$

$$\frac{e^{-0.1}(0.1)^1}{1!} = (0.9048)(0.1) = 0.09048$$

(*c*) If $k = 2$

$$\frac{e^{-0.1}(0.1)^2}{2!} = \frac{(0.9048)(0.01)}{2} = 0.004524$$

(*d*) The probability of finding 3 or more albinos is one minus the sum of probabilities for 0, 1, and 2 albinos.

$$1 - (0.9048 + 0.09048 + 0.004524) = 1 - 0.999804 = 0.000196$$

Note that trying to solve this problem by applying the binomial distribution

$$1 - [{}_{100}C_0(0.001)^0(0.999)^{100} + {}_{100}C_1(0.001)^1(0.999)^{99} + {}_{100}C_2(0.001)^2(0.999)^{98}]$$

would be a very difficult approach.

**7.8.** Sex-linked female-sterile mutations can be induced by the chemical mutagen EMS (ethyl methane sulfonate) in *Drosophila melanogaster*. Apparently mutations in many different X-linked genes can result in female sterility. In one experiment involving 131 female-sterile mutations, 30 such EMS-induced genes were each found to have one mutation and 12 genes were found to have two mutations. Estimate the total number of X-linked genes that, if mutated, could produce female sterility.

**Solution:**

According to the Poisson distribution, we expect $x$ genes to have one mutation with a frequency

$$\frac{(np)\,e^{-np}}{1!}$$

where $n$ = the total number of female-sterile genes on the X chromosome
$p$ = the probablity of a mutation per such gene
$np$ = the mean number of mutations per such gene

We expect $y$ genes to have two mutations with a frequency

$$\frac{(np)^2\,e^{-np}}{2!}$$

We make the ratio

$$\frac{y}{x} = \frac{(np)^2\,e^{-np}}{2(np)\,e^{-np}} = \frac{np}{2}$$

From the data given

$$\frac{y}{x} = \frac{12/n}{30/n} = 0.4$$

Thus, $np/2 = 0.4$ or $np = 0.8$.

Since the mean number of mutations per mutable female-sterile gene ($np$) was observed to be $131/n$, and $np = 0.8$, then

$$0.8 = \frac{131}{n} \quad \text{or} \quad n = 164$$

**7.9.** A bacterial culture susceptible to infection by a virus was inoculated into 20 tubes each containing 0.2 milliliter of broth and allowed to multiply to a concentration of about

$10^9$ cells/milliliter. After this multiplication, 9 of the tubes were found to contain some cells that had mutated to viral resistance; 11 of the tubes had no resistant cells. Estimate the mutation rate from viral sensitivity to viral resistance in this organism under these experimental conditions.

**Solution:**

The fraction of cultures with no mutants ($x$) is 11/20 = 0.55. The zero term of the Poisson distribution is $x = e^{-np}$, where $n$ is the number of cells per culture = $(0.2)(10^9) = 2 \times 10^8$ and $p$ is the probability of mutation from viral sensitivity to viral resistance (mutation rate).

$$0.55 = e^{-p(2\times 10^8)}$$

From Table 7.1 it is seen that the value of $e^{-np}$ that approximates 0.55 is about 0.6. Therefore,

$$p(2 \times 10^8) = 0.6$$

and

$$p = \frac{0.6}{2 \times 10^8} = 3 \times 10^{-9}$$

Alternatively, taking natural logarithms (obtained from a reference source), from the equation

$$x = e^{-np}$$

we obtain

$$\ln x = -np$$

$$p = \frac{-\ln x}{n} = \frac{-\ln 0.55}{2 \times 10^8} = \frac{0.598}{2 \times 10^8} \approx 3 \times 10^{-9}$$

**7.10.** Genes that are tightly linked (very close together on the same chromosome) are usually unlikely to experience more than one crossover. Hence, small map distances are likely to reflect the actual amount of crossing over. However, as the gene distances increase, the opportunity for multiple crossovers increases, tending to follow a Poisson distribution. Meioses involving any finite number of crossovers per meiosis produce an observed recombination frequency (RF) of 50% among their products. No matter how far apart two loci may be, they theoretically cannot give a RF greater than 50%. The further apart two genes are, the greater the error of estimate of true linkage distance provided by the recombinant progeny. The proportion of meioses in which at least one crossover occurs is predicted by RF $= \frac{1}{2}(1 - e^{-\mu})$, where $\mu$ = actual mean number of zero crossover events (not the observed number). If there are no zero crossover events, then RF $= \frac{1}{2}(1 - 0) = \frac{1}{2}$, because the expected zero class frequency is $e^{-\mu}\mu^0/0! = e^{-\mu}$. Given any observed RF, $\mu$ can be calculated from the above formula. The following formula can be used to convert observed RF to the most likely amount of crossing over (as for closely linked genes): RF$' = \mu/2$, because when $\mu = 0.02$, $e^{-\mu} = 0.98$, and RF $= \frac{1}{2}(1 - 0.98) = \frac{1}{2}(0.02) = \mu/2$. Likewise, when $\mu = 0.05$, $e^{-\mu} = 0.95$, and RF $= \frac{1}{2}(1 - 0.95) = \frac{1}{2}(0.05) = \mu/2$. Thus, RF$' = \mu/2$. A mean ($\mu$) of one crossover between the two loci equates with 50 map units. If the observed RF between two linked loci is 0.33, estimate the true amount of crossovers (map units) this probably represents, and calculate the error of estimate in this experiment.

**Solution:**

$$0.33 = \tfrac{1}{2}(1 - e^{-\mu})$$
$$0.66 = 1 - e^{-\mu}$$
$$e^{-\mu} = 1 - 0.66 = 0.34$$

In Table 7.1 we find that $e^{-0.34}$ lies between $\mu = 1$ (0.36788) and $\mu = 2$ (0.13534). Thus, 0.34/0.36788 = 0.9242; and 0.9242 corresponds to $\mu = 0.08$. Therefore $\mu = 1.08$, and RF$' = 1.08/2 = 0.54$ or 54 map units. The error of underestimate of actual crossing over = 54/33 = 1.64 or about 64%.

## CHI-SQUARE TEST

**7.11.** (*a*) A coin is tossed 10 times and lands heads up 6 times and tails up 4 times. Are these results consistent with the expected 50-50 ratio? (*b*) If the coin is tossed 100 times with the same relative magnitude of deviation from the expected ratio, is the hypothesis still acceptable? (*c*) What conclusion can be drawn from the results of parts (*a*) and (*b*)?

**Solution:**

(*a*)

| Phenotypic Classes | Observed (o) | Expected (e) | Deviations (o − e) | $(o-e)^2$ | $(o-e)^2/e$ |
|---|---|---|---|---|---|
| Heads | 6 | $\frac{1}{2}(10) = 5$ | 1 | 1 | $1/5 = 0.2$ |
| Tails | 4 | $\frac{1}{2}(10) = 5$ | −1 | 1 | $1/5 = 0.2$ |
| | 10 | 10 | 0 | | $\chi^2 = 0.4$ |

Two mathematical check points are always present in the chi-square calculations. (1) The total of the expected column must equal the total observations. (2) The sum of the deviations should equal zero. The squaring of negative deviations converts all values to a positive scale. The number of degrees of freedom is the number of phenotypes minus one $(2-1=1)$. We enter Table 7.2 on the first line (df $=$ 1) and find the computed value of 0.4 lying in the body of the table between the values 0.15 and 0.46 corresponding to the probabilities 0.7 and 0.5 shown at the top of the respective columns. This implies that the magnitude of the deviation in our experimental results could be anticipated by chance alone in more than 50% but less than 70% of an infinite number of experiments of comparable size. This range of values is far above the critical probability value of 0.05 or 5%. Therefore we accept the null hypothesis and conclude that our coin is conforming to the expected probabilities of heads $= \frac{1}{2}$ and tails $= \frac{1}{2}$.

(*b*) In (*a*), heads appeared in 60% and tails in 40% of the tosses. The same relative magnitude of deviations will now be considered in a sample of size 100. In problems such as this, where expected values are equivalent in all the phenotypic classes, chi-square may be calculated more rapidly by adding the squared deviations and making a single division by the expected number.

| Phenotypes | o | e | (o − e) | $(o-e)^2$ |
|---|---|---|---|---|
| Heads | 60 | $\frac{1}{2}(100) = 50$ | 10 | 100 |
| Tails | 40 | $\frac{1}{2}(100) = 50$ | −10 | 100 |
| | 100 | 100 | 0 | $\chi^2 = 200/50 = 4.0$ |

With $df = 1$, this $\chi^2$ value lies between 6.64 and 3.84 corresponding to the probabilities 0.01 and 0.05 respectively. This means that a deviation as large as or larger than the one observed in this experiment is to be anticipated by chance alone in less than 5% of an infinite number of trials of similar size. This is in the "critical region" and we are therefore obliged to reject the null hypothesis and conclude that our coin is not conforming to the expected 50-50 ratio. Either of two explanations may be involved: (1) this is not a normal well-balanced coin or (2) our experiment is among the one in twenty (5%) expected to have a large deviation produced by chance alone.

(*c*) The results of parts (*a*) and (*b*) demonstrate the fact that large samples provide a more critical test of a hypothesis than small samples. Proportionately larger deviations have a greater probability of occurring by chance in small samples than in large samples.

**7.12.** In the garden pea, yellow cotyledon color is dominant to green, and inflated pod shape is dominant to the constricted form. When both of these traits were considered jointly in self-fertilized dihybrids, the progeny appeared in the following numbers: 193 green, inflated: 184 yellow, constricted: 556 yellow, inflated: 61 green, constricted. Test the data for independent assortment.

**Solution:**

P: *Gg Cc* (yellow, inflated) × *Gg Cc* (yellow inflated)

$F_1$: (expectations)

$\frac{9}{16}$ *G- C-* yellow, inflated
$\frac{3}{16}$ *G- cc* yellow, constricted
$\frac{3}{16}$ *gg C-* green, inflated
$\frac{1}{16}$ *gg cc* green, constricted

| Phenotypes | Observed | Expected | Deviation d | $d^2$ | $d^2/e$ |
|---|---|---|---|---|---|
| yellow, inflated | 556 | $\frac{9}{16}(994) = 559.1$ | $-3.1$ | 9.61 | 0.017 |
| yellow, constricted | 184 | $\frac{3}{16}(994) = 186.4$ | $-2.4$ | 5.76 | 0.031 |
| green, inflated | 193 | $\frac{3}{16}(994) = 186.4$ | 6.6 | 43.56 | 0.234 |
| green, constricted | 61 | $\frac{1}{16}(994) = 62.1$ | $-1.1$ | 1.21 | 0.019 |
| | 994 | 994.0 | 0 | | $\chi^2 = 0.301$ |

$$\text{df} = 4 - 1 = 3, \quad p > 0.95$$

This is not a significant chi-square value, and thus we accept the null hypothesis, i.e. the magnitude of the deviation $(o - e)$ is to be expected by chance alone in greater than 95% of an infinite number of experiments of comparable size. This is far above the critical value of 5% necessary for acceptance of the hypothesis. We may therefore accept the data as being in conformity with a 9 : 3 : 3 : 1 ratio, indicating that the gene for cotyledon color assorts independently of the gene for pod form.

**7.13.** Pure red fleshed tomatoes crossed with yellow fleshed tomatoes produced an all red $F_1$. Among 400 $F_2$ plants, 90 were yellow. It is hypothesized that a single pair of alleles is involved such that $Y\text{-} = $ red and $yy = $ yellow. Test this hypothesis by use of formula (*7.6*).

**Solution:**

P: $YY$ (red) $\times$ $yy$ (yellow)

$F_1$: $Yy$ (red)

$F_2$: (expectations) $\frac{3}{4}$ $Y$- (red)
$\frac{1}{4}$ $yy$ (yellow)

Let $a$ = number of yellow fleshed fruits = 90, $b$ = number of red fleshed fruits = 400 − 90 = 310, $r$ = expected ratio of $a$ to $b$ = $\frac{1}{3}$. Then

$$\chi^2 = \frac{(a - rb)^2}{r(a+b)} = \frac{[90 - \frac{1}{3}(310)]^2}{\frac{1}{3}(90 + 310)} = 1.33$$

$$\text{df} = 1, \quad p = 0.2\text{-}0.3$$

This is not a significant value and hence we may accept the hypothesis.

Some work can be saved in this method by always letting the larger number = $a$. Let $a$ = number of red fleshed fruits = 310, $b$ = number of yellow fleshed fruits = 90, $r$ = expected ratio of $a$ to $b$ = 3. Then

$$\chi^2 = \frac{[310 - 3(90)]^2}{3(400)} = \frac{(40)^2}{1200} = 1.33$$

**7.14.** A total of 160 families with four children each were surveyed with the following results:

| Girls | 4 | 3 | 2 | 1 | 0 |
|---|---|---|---|---|---|
| Boys | 0 | 1 | 2 | 3 | 4 |
| Families | 7 | 50 | 55 | 32 | 16 |

Is the family distribution consistent with the hypothesis of equal numbers of boys and girls?

**Solution:**

Let $a$ = probability of a girl = $\frac{1}{2}$, $b$ = probability of a boy = $\frac{1}{2}$.

| $(a+b)^4$ = | $a^4$ | + | $4a^3b$ | + | $6a^2b^2$ | + | $4ab^3$ | + | $b^4$ |
|---|---|---|---|---|---|---|---|---|---|
| | 4 girls 0 boys | | 3 girls 1 boy | | 2 girls 2 boys | | 1 girl 3 boys | | 0 girls 4 boys |
| = | $(\frac{1}{2})^4$ | + | $4(\frac{1}{2})^3(\frac{1}{2})$ | + | $6(\frac{1}{2})^2(\frac{1}{2})^2$ | + | $4(\frac{1}{2})(\frac{1}{2})^3$ | + | $(\frac{1}{2})^4$ |
| = | $\frac{1}{16}$ | + | $\frac{4}{16}$ | + | $\frac{6}{16}$ | + | $\frac{4}{16}$ | + | $\frac{1}{16}$ |

Expected number with 4 girls and 0 boys $= \frac{1}{16}(160) = 10$

3 girls and 1 boy $= \frac{4}{16}(160) = 40$

2 girls and 2 boys $= \frac{6}{16}(160) = 60$

1 girl and 3 boys $= \frac{4}{16}(160) = 40$

0 girls and 4 boys $= \frac{1}{16}(160) = 10.$ Then

$$\chi^2 = \frac{(7-10)^2}{10} + \frac{(50-40)^2}{40} + \frac{(55-60)^2}{60} + \frac{(32-40)^2}{40} + \frac{(16-10)^2}{10} = 9.02$$

$$\text{df} = 5-1 = 4, \quad p = 0.05\text{–}0.10$$

This value is close to, but less than, the critical value 9.49. We may therefore accept the hypothesis, but the test would be more definitive if it could be run on a larger sample. It is a well-known fact that a greater mortality occurs in males than in females and therefore an attempt should be made to ascertain family composition on the basis of sex of *all* children *at birth* including prematures, aborted fetuses, etc.

### *Correction for Small Samples.*

**7.15.** In Problem 7.11(*b*), it was shown that observations of 60:40 produced a significant chi-square at the 5% level when uncorrected for continuity. Apply Yates' correction for continuity and retest the data.

**Solution:**

| o | e | $[\lvert o-e\rvert - 0.5]$ | $[\lvert o-e\rvert - 0.5]^2$ |
|---|---|---|---|
| 60 | 50 | $10-0.5 = 9.5$ | 90.25 |
| 40 | 50 | $10-0.5 = 9.5$ | 90.25 |
| | | | $\chi^2 = 180.50/50 = 3.61$ |

Notice that the correction 0.5 is always applied to the *absolute value* $\lvert o-e\rvert$ of the deviation of expected from observed numbers. This is not a significant chi-square value. Because the data is discrete (jumping from unit to unit) there is a tendency to underestimate the probability, causing too many rejections of the null hypothesis. Yates' correction removes this bias and produces a more accurate test near the critical values (column headed by a probability of 0.05 in Table 7.2).

# Supplementary Problems

## THE BINOMIAL EXPANSION

7.16. Black hair in the guinea pig is dominant to white hair. In families of five offspring where both parents are heterozygous black, with what frequency would we expect to find (*a*) 3 whites and 2 blacks, (*b*) 2 whites and 3 blacks, (*c*) 1 white and 4 blacks, (*d*) all whites?

7.17. In families of size three, what is the probability of finding the oldest child to be a girl and the youngest a boy?

7.18. In families of five children, what is the probability of finding (*a*) 3 or more boys, (*b*) 3 or more boys or 3 or more girls?

7.19. A dozen strains of corn are available for a cross-pollination experiment. How many different ways can these strains be paired?

7.20. Five coat colors in mice are agouti, cinnamon, black, chocolate and albino. (*a*) List all of the possible crosses between different phenotypes. (*b*) Verify the number of different crosses by applying formula (*7.1*), page 141.

7.21. In mice litters of size 8, determine (*a*) the most frequently expected number of males and females, (*b*) the term of the binomial part (*a*) represents, (*c*) the percentage of all litters of size 8 expected to have 4 males and 4 females.

7.22. There are at least 12 alleles at the sex-linked "white" locus in *Drosophila.* Find the number of possible (*a*) genotypes, (*b*) types of matings.

7.23. White plumage in chickens can result from the action of a recessive genotype *cc* or from the action of an independently assorting dominant gene *I*. The White Plymouth Rock breed has genotype *ccii* and the White Leghorn breed has genotype *CCII*. The hybrid bird produced by crossing these two breeds is also white. How frequently in clutches (a nest of eggs; a brood of chicks) of 10 chicks produced by hybrid birds would you expect to find 5 colored and 5 white chicks?

7.24. A case of dominant interaction among coat colors has been discovered in the dog: *B*- results in black, *bb* in brown; *I*- inhibits color development, *ii* allows color to be produced. White dogs of genotype *BBII* testcrossed to brown dogs produce all white puppies in the $F_1$. In the $F_2$, determine the fraction of all litters of size 6 which is expected to contain 3 white, 1 black and 2 brown puppies.

7.25. A pair of alleles in the rat, *C* and *c*, act on coat color in such a way that the genotypes *CC* and *Cc* allow pigment to be produced, but the genotype *cc* prevents any pigment from being produced (albinos). Black rats possess the dominant gene *R* of an independently assorting locus. Cream rats are produced by the recessive genotype *rr*. When black rats of genotype *RRCC* are testcrossed to albino rats, the $F_1$ is all black. Determine the percentage of $F_2$ litters of size seven which is expected to have 4 black, 2 cream and 1 albino.

7.26. Two independently assorting loci, each with codominant allelic pairs, are involved in the shape and color characteristics of radishes. The shape may be long or round, due to different homozygous genotypes, or oval due to the heterozygous genotype. The color may be red or white due to different homozygous genotypes, or purple due to the heterozygous genotype. A long white variety is crossed to a round red variety. The $F_1$ is all oval purple. A dozen seeds are saved from each self-pollinated $F_1$ plant and grown out the next season in sibling groups. Assuming 100% germination, determine the proportion of plants in each group of a dozen progeny expected to exhibit 5 oval, purple: 3 oval, white: 2 round, purple: 1 long, purple: 1 round, red.

## THE POISSON DISTRIBUTION

7.27. A bacterial suspension contains 5 million cells per milliliter. This culture is serially diluted tenfold in six successive tubes by adding 1 milliliter from the previous tube into 9 milliliters of diluent fluid. (*a*) How many cells/milliliter are expected in the sixth dilution tube? (*b*) If $\mu = 5$ and $e^{-5} = 0.006738$ (Table 7.1), calculate the probabilities of finding in the sixth tube 0, 1, 2, 3, 4 cells. (*c*) Suppose that 5 cells are found in the sixth tube. Give the 95% confidence limits for this estimate of the sixth-tube mean. *Hint:* Confidence limits of 95% are within $\pm 2$ standard deviations of the mean. The variance is the square of the standard deviation. (*d*) Suppose that each tube contains the expected number of cells. Which tube contains the greatest dilution from which the estimate of the mean number of cells in the tube is accurate to within about 10% of the mean itself?

7.28. Radioactive elements are extensively used in molecular genetic research. Radioactive disintegration of these atoms follows a Poisson distribution. What is the error of estimating the mean number of radioactive disintegrations per minute (dpm) if 100 dpm are actually detected when counted for (*a*) 1 minute? (*b*) 100 minutes? *Hint:* The 95% confidence limits are approximated by the mean $\pm$ 2 standard deviations. The variance is the square of the standard deviation.

7.29. If two linked genes produce 27.5% recombinants, estimate the probable amount of crossing over that actually occurred between these loci, assuming that the probability of 0, 1, 2, . . ., *n* exchanges occur during meiosis according to a Poisson distribution.

## TESTING GENETIC RATIOS

**7.30.** Determine the number of degrees of freedom when testing the ratios (*a*) 3:1, (*b*) 9:3:3:1, (*c*) 1:2:1, (*d*) 9:3:4. Find the number of degrees of freedom in applying a chi-square test to the results from (*e*) testcrossing a dihybrid, (*f*) testcrossing a trihybrid, (*g*) trihybrid × trihybrid cross, (*h*) mating repulsion dihybrid *Drosophila* males and females.

**7.31.** Two phenotypes appear in an experiment in the numbers 4:16. (*a*) How well does this sample fit a 3:1 ratio? Would a sample with the same proportional deviation fit a 3:1 ratio if it were (*b*) 10 times larger than (*a*), (*c*) 20 times larger than (*a*)?

**7.32.** The flowers of four o'clock plants may be red, pink, or white. Reds crossed to whites produced only pink offspring. When pink flowered plants were crossed they produced 113 red, 129 white and 242 pink. It is hypothesized that these colors are produced by a single gene locus with codominant alleles. Is this hypothesis acceptable on the basis of a chi-square test?

**7.33.** A heterozygous genetic condition called "creeper" in chickens produces shortened and deformed legs and wings, giving the bird a squatty appearance. Matings between creepers produced 775 creeper : 388 normal progeny. (*a*) Is the hypothesis of a 3:1 ratio acceptable? (*b*) Does a 2:1 ratio fit the data better? (*c*) What phenotype is probably produced by the gene for creeper when in homozygous condition?

**7.34.** Among fraternal (non-identical, dizygotic) twins, the expected sex ratio is 1 MM : 2 MF : 1 FF (M = male, F = female). A sample from a sheep population contained 50 MM, 142 MF, and 61 FF twin pairs. (*a*) Do the data conform within statistically acceptable limits to the expectations? (*b*) If identical (monozygotic) twin pairs = total pairs − (2 × MF pairs), what do the data indicate concerning the frequency of monozygotic sheep twins?

**7.35.** Genetically pure white dogs, when testcrossed to brown dogs, produce an all white $F_1$. Data on 190 $F_2$ progeny: 136 white, 41 black, 13 brown. These coat colors are postulated to be under the genetic control of two loci exhibiting dominant epistasis (12:3:1 ratio expected). (*a*) Test this hypothesis by chi-square. (*b*) When the $F_1$ was backcrossed to the brown parental type, the following numbers of phenotypes appeared among 70 progeny: 39 white, 19 black, 12 brown. Are these results consistent with the hypothesis?

**7.36.** Pure black rats, when testcrossed to albinos, produce only black $F_1$ offspring. The $F_2$ in one experiment was found to consist of 43 black, 14 cream and 22 albino. The genetic control of these coat colors is postulated to involve two gene loci with recessive epistasis (9:3:4 ratio expected). Is the genetic hypothesis consistent with the data?

**7.37.** Colored aleurone in corn is hypothesized to be produced by the interaction of two dominant genes in the genotype *A-C-*; all other genotypes at these two loci produce colorless aleurone. A homozygous colored strain is testcrossed to a pure colorless strain. The $F_1$ exhibits only kernels with colored aleurone. The $F_2$ exhibits 3300 colored : 2460 colorless. Analyze the data by chi-square test.

**7.38.** The results of phenotypic analysis of 96 $F_2$ progeny in two replicate experiments is shown below.

| Experiment | Phenotype 1 | Phenotype 2 |
|---|---|---|
| 1 | 70 | 26 |
| 2 | 76 | 20 |

Calculate chi-square for each experiment assuming a (*a*) 3:1 ratio, (*b*) 13:3 ratio. (*c*) Which hypothesis is most consistent with the data?

**7.39.** A total of 320 families with six children each were surveyed with the results shown below. Does this distribution indicate that boys and girls are occurring with equal frequency?

| No. of girls | 6 | 5 | 4 | 3 | 2 | 1 | 0 |
|---|---|---|---|---|---|---|---|
| No. of boys | 0 | 1 | 2 | 3 | 4 | 5 | 6 |
| No. of families | 6 | 33 | 71 | 99 | 69 | 37 | 5 |

**7.40.** Yellowish-green corn seedlings are produced by a gene called "virescent-4" ($v_4$). A dark brown color of the outer seed coat called "chocolate pericarp" is governed by a dominant gene $Ch$. A virescent-4 strain is crossed to a strain homogyzous for chocolate pericarp. The $F_1$ is then testcrossed. The resulting progeny are scored for phenotype with the following results: 216 green seedling, light pericarp : 287 green seedling, chocolate pericarp : 293 virescent seedling, light pericarp : 204 virescent seedling, chocolate pericarp. (*a*) Are these results compatible with the hypothesis of independent assortment? (*b*) Perform a genetic analysis of the data in the light of the results of part (*a*).

**7.41.** Purple anthocyanin pigment in tomato stems is governed by a dominant gene $A$, and its recessive allele $a$ produces green stem. Hairy stem is governed by a dominant gene $Hl$, and hairless stem by its recessive allele $hl$. A dihybrid purple, hairy plant is testcrossed and produces 73 purple, hairy : 12 purple, hairless : 75 green, hairless : 9 green, hairy. (*a*) Is the $F_1$ purple : green ratio compatible with the expectation for alleles (i.e. a 1 : 1 ratio)? (*b*) Is the $F_1$ hairy : hairless ratio compatible with the expectation for alleles (i.e. a 1 : 1 ratio)? (*c*) Test the $F_1$ data for independent assortment (i.e. a 1 : 1 : 1 : 1 ratio). What conclusion do you reach?

**7.42.** In guinea pigs, it is hypothesized that a dominant allele $L$ governs short hair and its recessive allele $l$ governs long hair. Codominant alleles at an independently assorting locus are assumed to govern hair color, such that $C^yC^y$ = yellow, $C^yC^w$ = cream and $C^wC^w$ = white. From the cross $Ll\,C^yC^w \times Ll\,C^yC^w$, the following progeny were obtained: 50 short cream : 21 short yellow : 23 short white : 21 long cream : 7 long yellow : 6 long white. Are the data consistent with the genetic hypothesis?

***Correction for Small Samples***

**7.43.** A dominant gene in corn ($Kn$) results in the proliferation of vascular tissues in a trait called "knotted leaf". A heterozygous knotted leaf plant is testcrossed to a normal plant producing 153 knotted leaf and 178 normal progeny. Apply Yates' correction in the calculation of chi-square testing a 1 : 1 ratio. Are these results consistent with the hypothesis?

**7.44.** Observations of 30 : 3 in a genetic experiment are postulated to be in conformity with a 3 : 1 ratio. Is a 3 : 1 ratio acceptable at the 5% level on the basis of (*a*) an uncorrected chi-square test, (*b*) a corrected chi-square test?

# Answers to Supplementary Problems

**7.16.** (*a*) 90/1024 (*b*) 270/1024 (*c*) 405/1024 (*d*) 1/1024

**7.17.** $\frac{1}{4}$

**7.18.** (*a*) $\frac{1}{2}$ (*b*) 1.0

**7.19.** 66

**7.20.** (*a*) (1) agouti × cinnamon (2) agouti × black (3) agouti × chocolate (4) agouti × albino (5) cinnamon × black (6) cinnamon × chocolate (7) cinnamon × albino (8) black × chocolate (9) black × albino (10) chocolate × albino

**7.21.** (*a*) 4 males : 4 females (*b*) 5th (*c*) 27.34%

**7.22.** (*a*) Males = 12; females = 78 (*b*) 936

**7.23.** $252(13/16)^5(3/16)^5 \cong 2.1\%$ by 4 place logarithms

**7.24.** 1215/65,536

**7.25.** 9.25%

**7.26.** $332{,}640/4{,}294{,}967{,}296 \cong 7.745 \times 10^{-5}$

**7.27.** (*a*) 5 cells in 10 milliliters = 0.5 cell/milliliter
(*b*) 0 = 0.006738; 1 = 0.033690; 2 = 0.084225; 3 = 0.140375; 4 = 0.175469
(*c*) 0.03 to 9.97
(*d*) The fourth tube contains 500 cells. The 95% confidence limits $= 500 \pm 2\sqrt{500} = 500 \pm 2(22.36) = 500 \pm 44.72$

**7.28.** (*a*) $100 \pm 2\sqrt{100} = 100 \pm 2(10) = 100 \pm 20$ or 20% error
(*b*) $10{,}000 \pm 2\sqrt{10{,}000} = 10{,}000 \pm 2(100) = 10{,}000 \pm 200 = 2\%$ error

**7.29.** 40%

**7.30.** **(*a*) 1 (*b*) 3 (*c*) 2 (*d*) 2 (*e*) 3 (*f*) 7 (*g*) 7 (*h*) 2 (2 : 1 : 1 ratio expected; see "crossover suppression" in Chapter 6)**

**7.31.** (*a*) $\chi^2 = 0.27$; $p = 0.5$–$0.7$; acceptable (*b*) $\chi^2 = 2.67$; $p = 0.1$–$0.2$; acceptable (*c*) $\chi^2 = 5.33$; $p = 0.01$–$0.05$; not acceptable.

**7.32.** Yes. $\chi^2 = 1.06$; $p = 0.5$–$0.7$; acceptable

**7.33.** (*a*) $\chi^2 = 43.37$; $p < 0.001$; not acceptable (*b*) $\chi^2 = 0.000421$; $p > 0.95$; a 2:1 ratio fits the data almost perfectly (*c*) lethal

**7.34.** (*a*) $\chi^2 = 4.76$; $0.10 > p > 0.05$; hypothesis acceptable (*b*) Monozygotic twins are estimated to be $-31$; the negative estimate indicates that identical sheep twins are rare provided that unlike-sex twins do not have a survival advantage over like-sex twins.

**7.35.** (*a*) $\chi^2 = 1.22$; $p = 0.5$–$0.7$; acceptable (*b*) Yes; expected 2:1:1 ratio; $\chi^2 = 2.32$; $p = 0.3$–$0.5$

**7.36.** Yes; $\chi^2 = 0.35$; $p = 0.8$–$0.9$

**7.37.** 9:7 ratio expected; $\chi^2 = 2.54$; $p = 0.1$–$0.2$; genetic hypothesis is acceptable.

**7.38.** (*a*) *Experiment 1:* $\chi^2 = 0.22$; $p = 0.5$–$0.7$. *Experiment 2:* $\chi^2 = 0.88$; $p = 0.3$–$0.5$.
(*b*) *Experiment 1:* $\chi^2 = 4.38$; $p = 0.05$–$0.01$. *Experiment 2:* $\chi^2 = 0.27$; $p = 0.5$–$0.7$.
(*c*) 3:1 ratio

**7.39.** Yes. $\chi^2 = 2.83$; $p = 0.8$–$0.9$; the distribution is consistent with the assumption that boys and girls occur with equal frequency.

**7.40.** (*a*) No. $\chi^2 = 25.96$; $p < 0.001$ (*b*) $v_4$ is linked to $Ch$, exhibiting approximately 42% recombination

**7.41.** (*a*) Yes. $\chi^2 = 0.006$; $p = 0.90$–$0.95$ (*b*) Yes. $\chi^2 = 0.077$; $p = 0.7$–$0.8$ (*c*) $\chi^2 = 95.6$; $p < 0.001$; the observations do not conform to a 1:1:1:1 ratio, therefore genes $a$ and $hl$ are probably linked.

**7.42.** Yes; $\chi^2 = 2.69$; $p = 0.7$–$0.8$

**7.43.** Yes; $\chi^2 = 1.74$; $p = 0.1$–$0.2$

**7.44.** (*a*) No. $\chi^2 = 4.45$; $p < 0.05$ (*b*) Yes. $\chi^2 = 3.64$; $p > 0.05$

# Chapter 8

## Compound Genetic Analyses

The problems in this chapter are designed to test the student's accumulated understanding of the basic Mendelian principles as presented in all seven of the preceding chapters. Therefore, no attempt should be made to solve the problems of this chapter until the concepts of independent assortment, multifactorial crosses, sex-linkage, gene interaction, autosomal linkage and testing of genetic ratios have been thoroughly mastered.

The problems encountered in genetic analysis become compounded whenever two or more traits are considered simultaneously and especially so whenever different modes of transmission and control are also involved. Regardless of the number of traits under consideration in a problem, the analysis can be greatly simplified by investigating only two traits at a time. Whenever the data do not appear to be quite close to the values expected on the basis of the hypothesis formulated during the analysis, a statistical test of the data (using the 5% level of rejection) should be employed to substantiate the conclusions.

Let us consider some of the combinations involving sex-linkage, autosomal linkage and independent assortment that might be generated by three gene loci.

| | |
|---|---|
| (1) Three autosomally linked loci | (*A B C*) |
| (2) Three sex-linked loci | [*A B C*] |
| (3) Three autosomal loci, two of which are linked | (*A B*) (*C*) |
| (4) Two autosomally linked loci, one sex-linked | (*A B*) [*C*] |
| (5) Two sex-linked loci, one autosomal | [*A B*] (*C*) |
| (6) Three autosomal loci assorting independently | (*A*) (*B*) (*C*) |
| (7) Two independently assorting autosomal loci, one sex-linked | (*A*) (*B*) [*C*] |

Using the same symbols of parentheses ( ) for autosomal loci and brackets [ ] for sex-linked loci, we can generate many more combinations with four gene loci.

| | |
|---|---|
| (1) (*A B C D*) | (7) (*A B*) (*C D*) |
| (2) [*A B C D*] | (8) (*A B*) (*C*) (*D*) |
| (3) (*A B C*) [*D*] | (9) [*A B*] (*C*) (*D*) |
| (4) [*A B C*] (*D*) | (10) (*A B*) (*C*) [*D*] |
| (5) (*A B C*) (*D*) | (11) (*A*) (*B*) (*C*) (*D*) |
| (6) (*A B*) [*C D*] | (12) (*A*) (*B*) (*C*) [*D*] |

The imposition of gene interactions, multiple allelic loci, codominance relationships, lethal genes, etc. on any of the above systems could considerably complicate the analysis of such problems.

## Solved Problems

**8.1.** Black body color, scarlet eye color and hooked bristles are three autosomal recessive mutations in *Drosophila* governed by the genes *b*, *st* and *hk* respectively. The results of a trihybrid testcross are given below:

245 black
239 black, scarlet
13 black, hook
12 black, hook, scarlet
228 hook
233 hook, scarlet
16 scarlet
14 wild type

Determine the mode of inheritance for these three genes.

**Solution:**

The failure of the progeny to approximate a 1:1:1:1:1:1:1:1 ratio indicates that these three loci are not assorting independently. Therefore, two or all three of these loci may be linked. In order to determine the linkage relationships, let us make three comparisons between two genes, ignoring the third locus.

(1) Black vs. Scarlet:

228 + 14 = 242 neither black nor scarlet
245 + 13 = 258 black, but not scarlet
233 + 16 = 249 scarlet, but not black
239 + 12 = 251 black and scarlet

(2) Black vs. Hook:

16 + 14 = 30 neither black nor hook
239 + 245 = 484 black, but not hook
233 + 228 = 461 hook, but not black
12 + 13 = 25 black and hook

(3) Hook vs. Scarlet:

245 + 14 = 259 neither hook nor scarlet
228 + 13 = 241 hook, but not scarlet
239 + 16 = 255 scarlet, but not hook
233 + 12 = 245 hook and scarlet

Notice that in comparisons (1) and (3) a 1:1:1:1 ratio is approximated by the data, indicating that black and scarlet are assorting independently, as are the hook and scarlet loci. Comparison (2) however indicates a gross deviation from a 1:1:1:1 ratio, indicating that black and hook are on the same chromosome. Comparison (3) need not have been made with the knowledge that black and scarlet are assorting independently, and that black and hook are linked. It follows logically that scarlet and hook should assort independently.

From comparison (2), the parental types black (484) and hook (461) indicate that the trihybrid parent was ($b+/+hk$, $+/st$). The crossover types (neither black nor hook, 30; black plus hook, 25) indicate that $(30 + 25)/1000 = 0.055$ or 5.5 map units separate the black and hook loci.

**8.2.** The color of eyes, the morphology of feathers and the color of down are three genetically determined traits in chickens. Matings between pure brown-eyed silkies with full down and pure non-brown, non-silkie individuals with light down produce both male and female offspring which are all non-brown, non-silkie with light down. The $F_1$ males are testcrossed to brown-eyed silky females with full-colored down. The reciprocal testcross is also made between $F_1$ females and brown-eyed silky males with full-colored down. The first 100 males and females were scored for phenotype with the following results:

| Phenotypes | Original Testcross | | Reciprocal Testcross | |
|---|---|---|---|---|
| | Males | Females | Males | Females |
| light down | 20 | 22 | 51 | 0 |
| light down, brown-eyed | 1 | 2 | 0 | 0 |
| light down, silkie | 23 | 21 | 49 | 0 |
| light down, brown-eyed, silkie | 2 | 2 | 0 | 0 |
| brown-eyed | 26 | 30 | 0 | 55 |
| brown-eyed, silkie | 21 | 17 | 0 | 45 |
| silkie | 2 | 2 | 0 | 0 |
| normal | 5 | 4 | 0 | 0 |

**How are these traits inherited? Reconstruct these crosses using appropriate genetic symbols.**

**Solution:**

P : pure brown, silkie, full-colored down × pure non-brown, non-silkie, light down
$F_1$: all non-brown, non-silkie, light down

This tells us that the allele for non-brown eye color (*Br*) is dominant to the one for brown eye color (*br*), that non-silkie (*H*) is dominant to silkie (*h*), and that light down (*Li*) is dominant to full-colored down (*li*).

The difference between the original and the reciprocal testcrosses indicates that one or more genes are sex-linked. Silkie segregates approximately equally with non-silkie in both males and females of both testcrosses. This is characteristic of an autosomal gene. Therefore, either the genes for eye color or down color, or both, could be sex-linked. Recall from Chapter 5 that in birds the female (ZW) is the heterogametic sex. When the heterogametic parent exhibits all recessive traits, both male and female offspring can be used to indicate the crossover gametes produced by the male parent (Chapter 6). Note the similarity of numbers between males and females in the original testcross. Let us make three comparisons combining the male and female data.

(1) Eye color vs. Feather morphology:

brown-eyed and silkie = 2 + 2 + 21 + 17 = 42
brown-eyed, not silkie = 1 + 2 + 26 + 30 = 59
silkie, not brown-eyed = 23 + 21 + 2 + 2 = 48
neither silkie nor brown-eyed = 20 + 22 + 5 + 4 = 51
200

(2) Eye color vs. Down color:

brown-eyed and light down = 1 + 2 + 2 + 2 = 7
brown-eyed, not light down = 26 + 30 + 21 + 17 = 94
light down, not brown-eyed = 20 + 22 + 23 + 21 = 86
neither light down nor brown-eyed = 2 + 2 + 5 + 4 = 13
200

(3) Down color vs. Feather morphology:

light down and silkie = 23 + 21 + 2 + 2 = 48
light down, not silkie = 20 + 22 + 1 + 2 = 45
silkie, not light down = 21 + 17 + 2 + 2 = 42
neither light down nor silkie = 26 + 30 + 5 + 4 = 65
200

Note that comparison (2) fails to approximate a 1:1:1:1 ratio, indicating that eye color and down color are linked. We have previously established that at least one of these genes is sex-linked. It thus follows that being linked, they must both be sex-linked. The amount of recombination between *br* and *li* may be determined from the data in comparison (2):

$$7 + \tfrac{13}{200} = \tfrac{20}{200} = 0.10 \text{ or } 10\% \text{ recombination}$$

We may now reconstruct these crosses:

P: *brli*/W, *h*/*h* ♀♀ × *BrLi*/*BrLi*, *H*/*H* ♂♂
pure brown-eyed, full-colored down, silkie females — pure non-brown-eyed, light down, non-silkie males

$F_1$: *BrLi*/W, *H*/*h* females
*BrLi*/*brli*, *H*/*h* males — non-brown-eyed, light down, non-silkie

Original $F_1$ testcross:

*BrLi*/*brli*, *H*/*h*♂♂ × *brli*/W, *h*/*h*♀♀
non-brown-eyed, light down, non-silkie males — brown-eyed, full-colored down, silkie females

Original testcross offspring:

| | | ♂ \ ♀ | *brli*/, *h*/ | W/, *h*/ |
|---|---|---|---|---|
| 0.5 *H*/ | 0.9 Non-Crossover Types | 0.45 *BrLi*/ | 0.225 *BrLi*/*brli*, *H*/*h* light down | 0.225 *BrLi*/W, *H*/*h* light down |
| | | 0.45 *brli*/ | 0.225 *brli*/*brli*, *H*/*h* brown-eyed | 0.225 *brli*/W, *H*/*h* brown-eyed |
| | 0.1 Crossover Types | 0.05 *Brli*/ | 0.025 *Brli*/*brli*, *H*/*h* normal | 0.025 *Brli*/W, *H*/*h* normal |
| | | 0.05 *brLi*/ | 0.025 *brLi*/*brli*, *H*/*h* brown-eyed, light down | 0.025 *brLi*/W, *H*/*h* brown-eyed, light down |
| 0.5 *h*/ | 0.9 Non-Crossover Types | 0.45 *BrLi*/ | 0.225 *BrLi*/*brli*, *h*/*h* light down, silkie | 0.225 *BrLi*/W, *h*/*h* light down, silkie |
| | | 0.45 *brli*/ | 0.225 *brli*/*brli*, *h*/*h* brown-eyed, silkie | 0.225 *brli*/W, *h*/*h* brown-eyed, silkie |
| | 0.1 Crossover Types | 0.05 *Brli*/ | 0.025 *Brli*/*brli*, *h*/*h* silkie | 0.025 *Brli*/W, *h*/*h* silkie |
| | | 0.05 *brLi*/ | 0.025 *brLi*/*brli*, *h*/*h* brown-eyed, light down, silkie | 0.025 *brLi*/W, *h*/*h* brown-eyed, light down, silkie |
| | | | Males | Females |

Reciprocal $F_1$ testcross:

*BrLi*/W, *H*/*h*♀♀ × *brli*/*brli*, *h*/*h*♂♂
non-brown-eyed, light down, silkie females — brown-eyed, full-colored down, silkie males

Reciprocal testcross offspring:

| ♀ \ ♂ | *brli*/, *h*/ | |
|---|---|---|
| ½ *BrLi*/, *H*/ | *BrLi*/*brli*, *H*/*h* light down | Males |
| ½ *BrLi*/, *h*/ | *BrLi*/*brli*, *h*/*h* light down, silkie | |
| ½ W/, *H*/ | *brli*/W, *H*/*h* brown-eyed | Females |
| ½ W/, *h*/ | *brli*/W, *h*/*h* brown-eyed, silkie | |

**8.3.** Three genes are involved in the inheritance of the coat colors of mice in this problem. Brown female mice crossed to albino males produce blotchy (patches of light fur), black $F_1$ females and black $F_1$ males. Information on three types of matings are available:

(1) $F_1 \times F_1 = F_2$
(2) Testcross: $F_1$ trihybrid (blotchy, black) females × testcross albino males
(3) Reciprocal testcross: blotchy, black males × testcross albino females

The results of these three crosses are tabulated below.

| Phenotypes | (1) | | (2) | | (3) | |
|---|---|---|---|---|---|---|
| | Males | Females | Males | Females | Males | Females |
| blotchy, black | 35 | 29 | 17 | 13 | 0 | 7 |
| blotchy, brown | 13 | 15 | 15 | 16 | 0 | 9 |
| black | 36 | 40 | 19 | 14 | 6 | 0 |
| brown | 11 | 12 | 14 | 18 | 8 | 0 |
| albino | 33 | 36 | 61 | 59 | 17 | 16 |
| | 128 | 132 | 126 | 120 | 31 | 32 |

Analyze these results to determine the mode of inheritance for these coat colors.

**Solution:**

On the assumption of independent gene action, three loci with two alleles each is expected to produce $2^3 = 8$ phenotypes. The data, however, indicate only 5 phenotypes. Therefore gene interaction is indicated between two or all three of these genes.

In the testcrosses (2) and (3), black, brown and albino colors appear in both male and female offspring, indicating autosomal loci, but in testcross (3) blotchy appears only in females, indicating that the gene for blotchy (*Blo*) is both sex-linked and dominant.

It appears that the male-female data in crosses (1) and (2) are so similar that combining the results within each cross is warranted. By ignoring the blotchy vs. non-blotchy phenotypes in cross (1), we might be able to ascertain the type of interaction that is operative.

Black males and females = 35 + 29 + 36 + 40 = 140
Brown males and females = 13 + 15 + 11 + 12 = 51
Albino males and females = 33 + 36 = 69
260

There are approximately three times as many black as brown offspring, and a few more albinos than brown. There are three modified two-factor ratios with three phenotypes each: 12:3:1, 9:3:4, and 9:6:1. Obviously, in this problem, the 9:3:4 ratio fits best, i.e. 9 blacks : 3 brown : 4 albino. This is probably a case of recessive interaction, wherein the dominant allele of one locus ($C$) allows pigment to be developed and the recessive allele ($c$) prevents any pigment from forming. The alleles of the hypostatic locus determine the type of pigment which can be produced ($B$ = black, $b$ = brown).

Let us combine the interacting autosomal loci with the sex-linked locus and compare the expected with the observed results.

P: *C/C, b/b, blo/blo* ♀♀ × *c/c, B/B, Blo/*Y ♂♂
brown, non-blotchy females — albino males (carrying the gene for blotchy)

$F_1$: *C/c, B/b, Blo/blo* = black, blotchy females
*C/c, B/b, blo/*Y = black males

(1) $F_2$:

| | | *blo* | Y |
|---|---|---|---|
| 1/2 *Blo* | 9/16 *C-B-* | 9/32 blotchy black | 9/32 blotchy, black |
| | 3/16 *C-bb* | 3/32 blotchy, brown | 3/32 blotchy, brown |
| | 4/16 *cc-* | 4/32 albino | 4/32 albino |
| 1/2 *blo* | 9/16 *C-B-* | 9/32 black | 9/32 black |
| | 3/16 *C-bb* | 3/32 brown | 3/32 brown |
| | 4/16 *cc-* | 4/32 albino | 4/32 albino |
| | | Females | Males |

Summary: 128 males + 132 females = 260 total

9/32 blotchy, black × 260 ≅ 73, or about 36 of each sex
3/32 blotchy, brown × 260 ≅ 24, or about 12 of each sex
9/32 black × 260 ≅ 73, or about 36 of each sex
3/32 brown × 260 ≅ 24, or about 12 of each sex
8/32 albino × 260 ≅ 65, or about 32 of each sex

(2) Testcross: *c/c, B/b, Blo/blo* ♀♀ × *c/c, b/b, blo/Y* ♂♂
trihybrid blotchy, black females    albino testcross males

Testcross progeny:

| | 1/2 *Blo/blo* | 1/2 *blo/blo* | 1/2 *Blo/Y* | 1/2 *Blo/Y* |
|---|---|---|---|---|
| 1/4 *C/c, B/b* | 1/8 blotchy, black | 1/8 black | 1/8 blotchy, black | 1/8 black |
| 1/4 *C/c, b/b* | 1/8 blotchy, brown | 1/8 brown | 1/8 blotchy, brown | 1/8 brown |
| 1/4 *c/c, B/b* | 1/8 albino | 1/8 albino | 1/8 albino | 1/8 albino |
| 1/4 *c/c, b/b* | 1/8 albino | 1/8 albino | 1/8 albino | 1/8 albino |
| | Females | | Males | |

Summary: 126 males + 120 females = 246 total

1/8 blotchy, black × 246 ≅ 30, or about 15 of each sex
1/8 blotchy, brown × 246 ≅ 30, or about 15 of each sex
1/8 black × 246 ≅ 30, or about 15 of each sex
1/8 brown × 246 ≅ 30, or about 15 of each sex
1/2 albino × 246 ≅ 123, or about 61 of each sex

(3) Reciprocal testcross: *C/c, B/b, Blo/Y* × *c/c, b/b, blo/blo*
hybrid blotchy, black, males    albino testcross males

Testcross progeny:

| | *Blo/blo* | Y/*blo* |
|---|---|---|
| 1/4 *C/c, B/b* | 1/4 blotchy, black | 1/4 black |
| 1/4 *C/c, b/b* | 1/4 blotchy, brown | 1/4 brown |
| 1/4 *c/c, B/b* | 1/4 albino | 1/4 albino |
| 1/4 *c/c, b/b* | 1/4 albino | 1/4 albino |
| | Females | Males |

Summary:

1/4 blotchy, black females × 32 = 8
1/4 blotchy, brown females × 32 = 8
1/2 albino females × 32 = 16
1/4 black males × 31 ≅ 8
1/4 brown males × 31 ≅ 8
1/2 albino males × 31 ≅ 16

All of these expected values correspond quite well with the observed data, supporting the hypothesis that three independently assorting but interacting loci are operative, two autosomal and one sex-linked.

**8.4.** In *Drosophila*, genes *a*, *b*, *c*, *d* and *e* are recessive to their wild alleles. Females (all genetically alike) known to be heterozygous for all above mentioned genes are mated to males which have the phenotype *a*, *c*, *d* and *e*. The following offspring are obtained:

| Phenotypes | ♂♂ | ♀♀ |
|---|---|---|
| + + + + + | 0 | 0 |
| + + + + *e* | 24 | 44 |
| + + + *d* + | 5 | 14 |
| + + *c* + + | 0 | 0 |
| + *b* + + + | 0 | 0 |
| *a* + + + + | 224 | 458 |
| + + + *d* *e* | 229 | 466 |
| + + *c* + *e* | 19 | 45 |
| + *b* + + *e* | 24 | 0 |
| *a* + + + *e* | 3 | 7 |
| + + *c* *d* + | 5 | 11 |
| + *b* + *d* + | 7 | 0 |
| *a* + + *d* + | 18 | 39 |
| + *b* *c* + + | 0 | 0 |
| *a* + *c* + + | 220 | 439 |
| *a* *b* + + + | 213 | 0 |
| + + *c* *d* *e* | 227 | 463 |
| + *b* + *d* *e* | 221 | 0 |
| *a* + + *d* *e* | 0 | 0 |
| + *b* *c* + *e* | 23 | 0 |
| *a* + *c* + *e* | 4 | 9 |
| *a* *b* + + *e* | 6 | 0 |
| + *b* *c* *d* + | 5 | 0 |
| *a* + *c* *d* + | 19 | 35 |
| *a* *b* + *d* + | 20 | 0 |
| *a* *b* *c* + + | 219 | 0 |
| + *b* *c* *d* *e* | 214 | 0 |
| *a* + *c* *d* *e* | 0 | 0 |
| *a* *b* + *d* *e* | 0 | 0 |
| *a* *b* *c* + *e* | 4 | 0 |
| *a* *b* *c* *d* + | 17 | 0 |
| *a* *b* *c* *d* *e* | 0 | 0 |
| Totals | 1970 | 2030 |

Make as complete a genetic analysis as the data allow.

**Solution:**

Perhaps one of the first things noted on preliminary survey of the data is a significant difference between the numbers of males and females within some genotypes. Chi-square test of the overall male:female ratio is not significant, however.

| Phenotypes | o | e | (o − e) | $(o-e)^2$ | $(o-e)^2/e$ |
|---|---|---|---|---|---|
| Males | 1970 | 2000 | −30 | 900 | 0.45 |
| Females | 2030 | 2000 | +30 | 900 | 0.45 |
| | | | | | $\chi^2 = 0.90$ |
| | | | | | $p = 0.3 - 0.5$ |

Let us rearrange the data and see if we can determine which locus (or loci) is responsible for this difference.

**Table A. Phenotypes in Which the Number of Females Exceed the Number of Males**

| Phenotypes | | | | | ♂♂ | ♀♀ |
|---|---|---|---|---|---|---|
| + | + | + | + | $e$ | 24 | 44 |
| + | + | + | $d$ | + | 5 | 14 |
| $a$ | + | + | + | + | 224 | 458 |
| + | + | + | $d$ | $e$ | 229 | 466 |
| + | + | $c$ | + | $e$ | 19 | 45 |
| $a$ | + | + | + | $e$ | 3 | 7 |
| + | + | $c$ | $d$ | + | 5 | 11 |
| $a$ | + | + | $d$ | + | 18 | 39 |
| $a$ | + | $c$ | + | + | 220 | 439 |
| + | + | $c$ | $d$ | $e$ | 227 | 463 |
| $a$ | + | $c$ | + | $e$ | 4 | 9 |
| $a$ | + | $c$ | $d$ | + | 19 | 35 |
| Totals | | | | | 997 | 2030 |

**Table B. Phenotypes in Which the Number of Males Exceed the Number of Females**

| Phenotypes | | | | | ♂♂ | ♀♀ |
|---|---|---|---|---|---|---|
| + | $b$ | + | + | $e$ | 24 | 0 |
| + | $b$ | + | $d$ | + | 7 | 0 |
| $a$ | $b$ | + | + | + | 213 | 0 |
| + | $b$ | + | $d$ | $e$ | 221 | 0 |
| + | $b$ | $c$ | + | $e$ | 23 | 0 |
| $a$ | $b$ | + | + | $e$ | 6 | 0 |
| + | $b$ | $c$ | $d$ | + | 5 | 0 |
| $a$ | $b$ | + | $d$ | + | 20 | 0 |
| $a$ | $b$ | $c$ | + | + | 219 | 0 |
| + | $b$ | $c$ | $d$ | $e$ | 214 | 0 |
| $a$ | $b$ | $c$ | + | $e$ | 4 | 0 |
| $a$ | $b$ | $c$ | $d$ | + | 17 | 0 |
| Totals | | | | | 973 | 0 |

It is obvious that only the $b$ locus is associated with these male-female differences. Table A shows that the dominant wild type phenotype at this locus can be correlated with approximately twice as many females as males. In Table B, the recessive phenotype fails to appear in females. In light of these observations, we might suspect that the $b$ locus is sex-linked; thus

Parents: $b/+$ ♀♀ × $+/Y$ ♂♂

F$_1$:

| ♀ \ ♂ | (+) | (Y) |
|---|---|---|
| (b) | $b/+$ | $b/Y$ |
| (+) | $+/+$ | $+/Y$ |

According to this hypothesis, there would be twice as many wild type females as males. Furthermore, the $b$ phenotype shows only in males. This hypothesis is consistent with the observations at the $b$ locus.

Since segregation of the sex-linked alleles at the $b$ locus fails to produce different phenotypes in the female progeny, we are obliged to use only the male data in further analysis of the problem.

This mating is a testcross as far as the $a$, $c$, $d$ and $e$ loci are concerned. The female parents are heterozygous at all these loci. If these four loci were assorting independently of each other, we would expect equal numbers of offspring within these four phenotypic classes. Ignoring the male-female differences produced by the $b$ locus, it is obvious that such expectations are not realized in the male data, and therefore two or more of the genes under consideration must be linked. Let us then compare the male data for two phenotypes at a time and look for deviations from a 1:1:1:1 ratio as an indication of linkage.

**Table C. Comparison of *a* and *c***

| + + | + *c* | *a* + | *a c* |
|---|---|---|---|
| 24 | 19 | 224 | 220 |
| 5 | 5 | 3 | 4 |
| 229 | 227 | 18 | 19 |
| 24 | 23 | 213 | 219 |
| 7 | 5 | 6 | 4 |
| 221 | 214 | 20 | 17 |
| 510 | 493 | 484 | 483 |

**Table D. Comparison of *a* and *d***

| + + | + *d* | *a* + | *a d* |
|---|---|---|---|
| 24 | 5 | 224 | 18 |
| 19 | 229 | 3 | 19 |
| 24 | 5 | 220 | 20 |
| 23 | 7 | 213 | 17 |
| 90 | 227 | 4 | 74 |
| | 221 | 6 | |
| | 5 | 219 | |
| | 214 | 4 | |
| | 913 | 893 | |

**Table E. Comparison of *a* and *e***

| + + | + *e* | *a* + | *a e* |
|---|---|---|---|
| 5 | 24 | 224 | 3 |
| 5 | 229 | 18 | 4 |
| 7 | 19 | 220 | 6 |
| 5 | 24 | 213 | 4 |
| 22 | 227 | 19 | 17 |
| | 221 | 20 | |
| | 23 | 219 | |
| | 214 | 17 | |
| | 981 | 950 | |

The totals of Table C approximate a 1:1:1:1 ratio. Therefore, *a* and *c* are assorting independently. The comparisons in Tables D and E, however, indicate that *a*, *d* and *e* are linked to each other. The next step is to determine the gene order and map distances for the loci *a*, *d* and *e*.

**Table F. Linkage Data for *a*, *d* and *e***

| | + + + | *ade* | *a* + *e* | + *d* + | + + *e* | *ad* + | *a* + + | + *de* |
|---|---|---|---|---|---|---|---|---|
| | | | 3 | 5 | 24 | 18 | 224 | 229 |
| | | | 4 | 5 | 19 | 19 | 220 | 227 |
| | | | 6 | 7 | 24 | 20 | 213 | 221 |
| | | | 4 | 5 | 23 | 17 | 219 | 214 |
| Totals | | | 17 | 22 | 90 | 74 | 876 | 891 |

The parental types are those which appear in greatest numbers. Thus the alleles ($a + +$) were on one chromosome of the maternal parent and alleles ($+ de$) were on the other, though not necessarily in that order. The double-crossover types are those which appear in lowest numbers ($+ + +$) and ($ade$), again not necessarily in the order shown. In order for parental combinations to produce the double-crossover combinations, the gene order must have the $a$ locus in the middle. The maternal genotype must then be $+ a + / d + e$.

$$\text{The distance } d \text{ to } a = \frac{90 + 74}{1970} = 0.083 \text{ or } 8.3 \text{ map units}$$

$$\text{The distance } a \text{ to } e = \frac{22 + 17}{1970} = 0.020 \text{ or } 2.0 \text{ map units}$$

The distance $d$ to $e$ = $8.3 + 2.0 = 10.3$ map units

The original mating can now be symbolized as follows:

$$\left(\frac{+a+}{d+e}, \frac{+}{b}, \frac{c}{+}\right) ♀♀ \quad \times \quad \left(\frac{dae}{dae}, \frac{+}{Y}, \frac{c}{c}\right) ♂♂$$

heterozygous wild type females — males exhibiting $a$, $c$, $d$ and $e$ phenotypes

The male progeny from this cross may now be calculated. As an example, the probability of the male phenotype (+ + + + +) appearing in the progeny is calculated as follows:

(1) Calculate the probability of the formation of the double-crossover gamete

$$(+ \ + \ +) = (0.083)(0.02)(0.5) = 0.00083$$

*Note.* The probability of all double-crossover gametes = (0.083)(0.02), but only half of these double crossovers will be of the type (+ + +); therefore, we multiply by 0.5.

(2) The probability of the sex-linked allele (+) being transmitted = 0.5.

(3) The probability of the (+) allele at the $c$ locus being transmitted in a material gamete = 0.5.

(4) The combined probability of producing an individual of the phenotype (+ + + + +) among 1970 male progeny is (0.00083)(0.5)(0.5)(1970) = 0.41. This is considerably less than one whole individual, and thus we are not surprised to find that no males of this phenotype were observed. The same type of calculation is made for all 32 possible male phenotypes in Table G.

**Table G. Calculation of the Expected Numbers of Male Phenotypes**

| Phenotypes | Gametic Probability at Loci *d, a, e* | *b* | *c* | Total Males | Number Expected | Number Observed |
|---|---|---|---|---|---|---|
| (1) + + + + + | (0.083)(0.02)(0.5) | (0.5) | (0.5) | (1970) = | 0.41 | 0 |
| (2) + + + + *e* | (0.083)(0.5) | (0.5) | (0.5) | (1970) = | 20.44 | 24 |
| (3) + + + *d* + | (0.02)(0.5) | (0.5) | (0.5) | (1970) = | 4.92 | 5 |
| (4) + + *c* + + | as for (1) | | | = | 0.41 | 0 |
| (5) + *b* + + + | as for (1) | | | = | 0.41 | 0 |
| (6) *a* + + + + | (0.897)(0.5) | (0.5) | (0.5) | (1970) = | 220.89 | 224 |
| (7) + + + *d e* | as for (6) | | | = | 220.89 | 229 |
| (8) + + *c* + *e* | as for (2) | | | = | 20.44 | 19 |
| (9) + *b* + + *e* | as for (2) | | | = | 20.44 | 24 |
| (10) *a* + + + *e* | as for (3) | | | = | 4.92 | 3 |
| (11) + + *c d* + | as for (3) | | | = | 4.92 | 5 |
| (12) + *b* + *d* + | as for (3) | | | = | 4.92 | 7 |
| (13) *a* + + *d* + | as for (2) | | | = | 20.44 | 18 |
| (14) + *b c* + + | as for (1) | | | = | 0.41 | 0 |
| (15) *a* + *c* + + | as for (6) | | | = | 220.89 | 220 |
| (16) *a b* + + + | as for (6) | | | = | 220.89 | 213 |
| (17) + + *c d e* | as for (6) | | | = | 220.89 | 227 |
| (18) + *b* + *d e* | as for (6) | | | = | 220.89 | 221 |
| (19) *a* + + *d e* | as for (1) | | | = | 0.41 | 0 |
| (20) + *b c* + *e* | as for (2) | | | = | 20.44 | 23 |
| (21) *a* + *c* + *e* | as for (3) | | | = | 4.92 | 4 |
| (22) *a b* + + *e* | as for (3) | | | = | 4.92 | 6 |
| (23) + *b c d* + | as for (3) | | | = | 4.92 | 5 |
| (24) *a* + *c d* + | as for (2) | | | = | 20.44 | 19 |
| (25) *a b* + *d* + | as for (2) | | | = | 20.44 | 20 |
| (26) *a b c* + + | as for (6) | | | = | 220.89 | 219 |
| (27) + *b c d e* | as for (6) | | | = | 220.89 | 214 |
| (28) *a* + *c d e* | as for (1) | | | = | 0.41 | 0 |
| (29) *a b* + *d e* | as for (1) | | | = | 0.41 | 0 |
| (30) *a b c* + *e* | as for (3) | | | = | 4.92 | 4 |
| (31) *a b c d* + | as for (2) | | | = | 20.44 | 17 |
| (32) *a b c d e* | as for (1) | | | = | 0.41 | 0 |
| | | | | | 1973.28 | 1970 |

The expected values fail to add exactly to 1970 because of accumulated rounding errors, especially those of the map distances. The observed male data correspond quite closely to the expected numbers. The female expectations would be twice that of the males for all genotypes bearing the wild type allele (+) at the $b$ locus, and no $b$ phenotypes would be expected to appear. Thus, our hypothesis as proposed above is supported by the congruity between the expected and observed numbers.

## Supplementary Problems

**8.5.** Four recessive mutations in *Drosophila* are listed in the following table:

| Name | Symbol | Description |
|---|---|---|
| curley | *cu* | wings curved upward and forward |
| grooveless | *gvl* | transverse groove between thorax and scutellum almost eliminated |
| rosy | *ry* | deep ruby eye color |
| spineless | *ss* | reduced bristle size |

Tetrahybrid females are testcrossed, producing a large progeny, the first 1000 of which were scored for phenotypes. The results are tabulated below:

6 wild type
5 grooveless
240 curley
13 rosy
1 spineless
230 curley, grooveless
228 rosy, spinelsss
16 rosy, grooveless
1 spineless, grooveless
0 curley, rosy
15 curley, spineless
218 rosy, spineless, grooveless
6 curley, rosy, spineless
14 curley, spineless, grooveless
0 curley, grooveless, rosy
7 curley, spineless, grooveless, rosy

How are these traits inherited? If linkage is involved, calculate the distances.

**8.6.** Dwarf rabbits appear in approximately one-quarter of the offspring from matings between heterozygotes. These dwarfs die shortly after birth. Some rabbits exhibit a short, plushlike coat called "rex" instead of the normal length of hair. The homozygous recessive genotype at either of two loci can produce the rex phenotype. A group of genetically identical trihybrid females were mated to a doubly heterozygous rex male carrying the dwarf gene. The following offspring were observed at maturity (no records were available on the offspring phenotypes at birth):

| Phenotype | Males | Females |
|---|---|---|
| normal | 10 | 8 |
| rex | 89 | 93 |

Analyze the genetic relationships between these three loci as fully as the data allow.

**8.7.** Four recessive mutants in *Drosophila* are blistery and warped wings (*by*), pale body color called "straw" (*stw*), threadlike arista (*th*), and bright red eye color called "cinnabar" (*cn*). Tetrahybrid wild type females were testcrossed to blister, straw, thread, cinnabar males and produced a very large group of progeny, the first 10,000 of which were scored for phenotypes with the results shown in the following table. Determine the inheritance relationships of these traits and represent the parental genotypes by using appropriate genetic symbols.

| Phenotype | Males | Females |
|---|---|---|
| wild type | 51 | 49 |
| blister | 3 | 2 |
| thread | 3 | 4 |
| straw | 1130 | 1138 |
| cinnabar | 1140 | 1136 |
| blister, straw | 65 | 75 |
| blister, cinnabar | 77 | 73 |
| blister, thread | 50 | 38 |
| thread, straw | 69 | 79 |
| thread, cinnabar | 70 | 68 |
| straw, cinnabar | 54 | 60 |
| blister, thread, straw | 1110 | 1134 |
| blister, thread, cinnabar | 1121 | 1115 |
| blister, straw, cinnabar | 3 | 3 |
| thread, straw, cinnabar | 4 | 2 |
| blister, thread, straw, cinnabar | 42 | 32 |

**8.8.** The ability of mice to develop coat pigment is dependent upon a dominant autosomal gene ($C$), the recessive genotype ($cc$) producing no pigment (albino). Fourteen percent recombination occurs between the $C$ locus and the hypostatic $P$ locus, where the dominant allele ($P$) produces full color and the recessive allele ($p$) reduces the amount of pigment (dilute). A sex-linked mutation is known to produce a short, crooked tail called "bent." A pure full-colored, bent strain is testcrossed, resulting in $F_1$ progeny which are all phenotypically full-colored with bent tails. The $F_1$ females are then crossed to albino, bent males homozygous for the dilution factor to produce the $F_2$. Predict the expected numbers in the different phenotypic classes among 100 male and 200 female $F_2$ progeny.

**8.9.** Recall from Chapter 5 that sex in the parasitic wasp *Habrobracon juglandis* is regulated by a single gene locus with multiple alleles. Heterozygotes ($X^aX^b$) are females, homozygotes ($X^aX^a$ or $X^bX^b$) and haploids ($X^a$ or $X^b$) are males. Two recessive mutations called "fused" (*fu*) and "white" (*wh*) produced fused antennal and tarsal segments and white eyes respectively. A pure white, fused male is crossed to a pure wild type female. The wild type female offspring are then backcrossed to the parental males. Five-hundred each of male and female diploid backcross offspring were scored to phenotypes with the following results:

| Phenotypes | Males | Females |
|---|---|---|
| wild type | 30 | 236 |
| fused | 231 | 22 |
| white-eyed | 20 | 214 |
| white-eyed, fused | 219 | 28 |

(*a*) How are these traits genetically regulated? Is linkage involved? If so, determine the amount of recombination between the linked genes.
(*b*) Diagram the genotypes of the parents and the $F_1$ using appropriate genetic symbols.

**8.10.** In addition to the genes $a$, $b$, $c$, $d$ and $e$ considered in Problem 8.4, suppose that another recessive ($f$) is found, and a female heterozygous for $c$, $d$ and $f$ is mated to a male with the phenotype $c$, $d$ and $f$. The following offspring are obtained:

| Phenotype | Males | Females |
|---|---|---|
| $+\ +\ +$ | 23 | 48 |
| $+\ +\ f$ | 5 | 11 |
| $+\ d\ +$ | 9 | 19 |
| $c\ +\ +$ | 11 | 22 |
| $+\ d\ f$ | 12 | 24 |
| $c\ +\ f$ | 11 | 23 |
| $c\ d\ +$ | 5 | 12 |
| $c\ d\ f$ | 22 | 43 |

Analyze the data fully. How do the results here affect the hypothesis proposed in Problem 8.4?

**8.11.** Five recessive traits in *Drosophila* ($a$, $b$, $c$, $d$ and $e$) are involved in matings between trihybrid females (all genetically identical) exhibiting the $a$ and $e$ phenotypes and a dihybrid male exhibiting the $b$ and $d$ phenotypes. The data on offspring phenotypes appear below. Analyze the results as completely as the data allow.

| Phenotypes | Females | Males |
|---|---|---|
| + + + + $e$ | 140 | 100 |
| + + + $d$ $e$ | 360 | 356 |
| + + $c$ + $e$ | 0 | 42 |
| + + $c$ $d$ $e$ | 0 | 7 |
| + $b$ + + $e$ | 359 | 8 |
| + $b$ + $d$ $e$ | 138 | 43 |
| $a$ + + + + | 131 | 82 |
| $a$ + + $d$ + | 368 | 360 |
| + $b$ $c$ + $e$ | 0 | 355 |
| + $b$ $c$ $d$ $e$ | 0 | 84 |
| $a$ + $c$ + + | 0 | 42 |
| $a$ + $c$ $d$ + | 0 | 8 |
| $a$ $b$ + + + | 377 | 9 |
| $a$ $b$ + $d$ + | 127 | 41 |
| $a$ $b$ $c$ + + | 0 | 361 |
| $a$ $b$ $c$ $d$ + | 0 | 102 |

## Answers to the Supplementary Problems

**8.5.** The genes *cu*, *ry* and *ss* are autosomally linked with *ry* in the middle; *gvl* segregates independently on another autosome. The distance *cu* to *ry* is 2.6 map units; *ry* to *ss* is 6.0 map units.

**8.6.** The two genes governing the rex trait exhibit duplicate recessive interaction and are autosomally linked in the trans position in the maternal parent, approximately 18 map units apart. The dwarf trait is produced by a recessive lethal assorting independently on another autosome.

P: $\dfrac{R_1 \quad r_2}{r_1 \quad R_2}, \dfrac{Dw}{dw} \times \dfrac{r_1 \quad r_2}{r_1 \quad r_2}, \dfrac{Dw}{dw}$

normal females × rex males

**8.7.** Blister and thread are autosomally linked loci approximately 6 map units apart. The straw and cinnabar loci are linked on another autosome approximately 4 map units apart.

P: $\dfrac{by \quad th}{+ \quad +}, \dfrac{stw \quad +}{+ \quad cn} \times \dfrac{by \quad th}{by \quad th}, \dfrac{stw \quad cn}{stw \quad cn}$

tetrahybrid wild type females × blister, straw, thread, cinnabar testcross males

**8.8**

| Phenotypes | Males | Females |
|---|---|---|
| colored, bent | 43 | 43 |
| albino, bent | 50 | 50 |
| dilute, bent | 7 | 7 |
| colored | | 43 |
| albino | | 50 |
| dilute | | 7 |

**8.9.** (*a*) The sex locus is linked to the fused locus with 10% recombination occurring between them.

(*b*) P: $\dfrac{X^a \quad fu}{X^a \quad fu}, \dfrac{wh}{wh}♂ \times \dfrac{X^a \quad +}{X^b \quad +}, \dfrac{+}{+}♀$

pure white, fused male × pure wild type female

$F_1$: $\dfrac{X^a \quad fu}{X^a \quad +}, \dfrac{+}{wh}$ wild type males

$\dfrac{X^b \quad +}{X^a \quad fu}, \dfrac{+}{wh}$ wild type females

**8.10.** The significant difference between the numbers of male and female progeny can be explained by the presence of a sex-linked recessive lethal ($l$) in the female parent; thus

P: $\frac{+}{l}$ ♀ × $\frac{+}{\text{Y}}$ ♂

F$_1$:

| | (+) | (Y) |
|---|---|---|
| (+) | + + | + Y |
| ($l$) | + $l$ | $l$Y dies |

The loci $c$ and $d$ appear to be assorting independently, but $c$ is linked to $f$, and $d$ is also linked to $f$. Therefore $c$, $f$ and $d$ are all on the same chromosome. The distance $c$ to $f$ = 33.7 units, $f$ to $d$ = 30.6 units, and $c$ to $d$ = 64.3 units.

**8.11.** The loci $b$, $c$ and $d$ are sex-linked with $c$ in the middle. The distance $b$ to $c$ is 10 map units, and the distance $c$ to $d$ is 20 map units. Twenty percent interference is operative in this region of the chromosome. The loci $a$ and $e$ are linked on an autosome, but because crossing over does not occur in male *Drosophila*, no estimate of the linkage distance can be obtained from this cross. The genotypes of the parents are

$$\left(\frac{d\ +\ +}{+\ c\ b}, \frac{ae}{ae}\right) ♀♀ \times \left(\frac{d\ +\ b}{\text{Y}}, \frac{a\ +}{+\ e}\right) ♂$$

# Chapter 9

## Cytogenetics

### THE UNION OF CYTOLOGY WITH GENETICS

Perhaps one reason Mendel's discoveries were not appreciated by the scientific community of his day (1865) was that the mechanics of mitosis and meiosis had not yet been discovered. During the years 1870-1900 rapid advances were made in the study of cells (*cytology*). At the turn of the century, when Mendel's laws were rediscovered, the cytological basis was available to render the statistical laws of genetics intelligible in terms of physical units. *Cytogenetics* is the hybrid science which attempts to correlate cellular events, especially those of the chromosomes, with genetic phenomena.

### VARIATION IN CHROMOSOME NUMBERS

Each species has a characteristic number of chromosomes. Most higher organisms are diploid, with two sets of homologous chromosomes: one set donated by the father, the other set by the mother. Variation in the number of sets of chromosomes (*ploidy*) is commonly encountered in nature. It is estimated that one-third of the angiosperms (flowering plants) have more than two sets of chromosomes (polyploid). The term *euploidy* is applied by organisms with chromosomes that are multiples of some basic number ($n$).

**1. Euploidy.**

(*a*) ***Monoploid.*** One set of chromosomes ($n$) is characteristically found in the nuclei of some lower organisms such as fungi. Monoploids in higher organisms are usually smaller and less vigorous than the normal diploids. Few monoploid animals survive. A notable exception exists in male bees and wasps. Monoploid plants are known but are usually sterile.

(*b*) ***Triploid.*** Three sets of chromosomes ($3n$) can originate by the union of a monoploid gamete ($n$) with a diploid gamete ($2n$). The extra set of chromosomes of the triploid is distributed in various combinations to the germ cells, resulting in genetically unbalanced gametes. Because of the sterility which characterizes triploids, they are not commonly found in natural populations.

(*c*) ***Tetraploid.*** Four sets of chromosomes ($4n$) can arise in body cells by the somatic doubling of the chromosome number. Doubling is accomplished either spontaneously or it can be induced in high frequency by exposure to chemicals such as the alkaloid *colchicine*. Tetraploids are also produced by the union of unreduced diploid ($2n$) gametes.

(i) *Autotetraploid.* The prefix "auto" indicates that the ploidy involves only homologous chromosome sets. Somatic doubling of a diploid produces four sets of homologous chromosomes (autotetraploid). Union of unreduced diploid gametes from the same species would accomplish the same result. Meiotic chromosome pairing usually produces quadrivalents (four synapsing chromosomes) which can produce genetically balanced gametes if *disjunction* is by two's, i.e. two chromosomes of the quadrivalent going to one pole and the other two to the opposite pole. If disjunction is not stabilized in this fashion for all quadrivalents, the gametes will be genetically unbalanced. Sterility will be expressed in proportion to the production of unbalanced gametes.

(ii) *Allotetraploid.* The prefix "allo" indicates that non-homologous sets of chromosomes are involved. The union of unreduced ($2n$) gametes from different diploid species could produce, in one step, an allotetraploid which appears and behaves like a new species. Alternatively, two diploid plant species may hybridize to produce a sterile diploid $F_1$. The sterility results from the failure of each set of chromosomes to provide sufficient genetic homology to affect pairing. The sterile diploid can become fertile if it undergoes doubling of the chromosome number. The allotetraploid thus produced has two matched sets of chromosomes which can pair just as effectively as in the diploid. Double diploids of this kind, found only in plants, are called *amphidiploids.*

**Example 9.1.** Let the two sets of chromosomes of one diploid species be $AA$ and the other $BB$.

P: $AA \times BB$

$F_1$: $AB$ (sterile hybrid)

$\downarrow$ (chromosome doubling)

Amphidiploid: $AABB$ (fertile)

(*d*) ***Polyploid.*** This term can be applied to any organism with more than $2n$ chromosomes. Ploidy levels higher than tetraploid are not commonly encountered in natural populations, but some of our most important crops are polyploid. For example, common bread wheat is hexaploid ($6n$), some strawberries are octaploid ($8n$), etc. Some triploids as well as tetraploids exhibit a more robust phenotype than their diploid counterparts, often having larger leaves, flowers and fruits (*gigantism*). Many commercial fruits and ornamentals are polyploids. Sometimes a specialized tissue within a diploid organism will be polyploid. For example, some liver cells of humans are polyploid. A common polyploid with which the reader should already be familiar is the triploid endosperm tissue of corn and other grains. Polyploids offer an opportunity for studying *dosage effects,* i.e. how two or more alleles of one locus behave in the presence of a single dose of an alternative allele. Dominance refers to the masking effect which one allele has over another allele. When one allele in the pollen is able to mask the effect of a double dose of another allele in the resulting endosperm, the former is said to exhibit *xenia* over the latter.

**Example 9.2.** In corn, starchy endosperm is governed by a gene $S$ which shows xenia with respect to its allele for sugary endosperm ($s$). Four genotypes are possible for these triploid cells: starchy = $SSS$, $SSs$, $Sss$; sugary = $sss$.

The term *haploid,* strictly applied, refers to the *gametic* chromosome number. For diploids ($2n$) the haploid number is $n$; for an allotetraploid ($4n$) the haploid (reduced) number is $2n$; for an allohexaploid ($6n$) the haploid number is $3n$; etc. Lower organisms such as bacteria and viruses are called haploids because they have a single set of genetic elements. However, since they do not form gametes comparable to those of higher organisms, the term monoploid would seem to be more appropriate.

## 2. Aneuploidy.

Variations in chromosome number may occur which do not involve whole sets of chromosomes, but only parts of a set. The term *aneuploidy* is given to variations of this nature, and the suffix "-somic" is a part of their nomenclature.

(*a*) ***Monosomic.*** Diploid organisms which are missing one chromosome of a single pair are monosomics with the genomic formula $2n-1$. The single chromosome without a pairing partner may go to either pole during meiosis, but more frequently will lag at anaphase and fails to be included in either nucleus. Monosomics can thus form two kinds of gametes, $(n)$ and $(n-1)$. In plants, the $n-1$ gametes seldom function. In animals, loss of one whole chromosome often results in genetic unbalance which is manifested by high mortality or reduced fertility.

(*b*) ***Trisomic.*** Diploids which have one extra chromosome are represented by the chromosomal formula $2n+1$. One of the pairs of chromosomes has an extra member, so that a trivalent structure may be formed during meiotic prophase. If two chromosomes of the trivalent go to one pole and the third goes to the opposite pole, then gametes will be $(n+1)$ and $(n)$ respectively. Trisomy can produce different phenotypes, depending upon which chromosome of the complement is present in triplicate. In humans, the presence of one small extra chromosome (autosome 21) has a very deleterious effect resulting in Down's syndrome, formerly called mongolism.

(*c*) ***Tetrasomic.*** When one chromosome of an otherwise diploid organism is present in quadruplicate, this is expressed as $2n+2$. A quadrivalent may form for this particular chromosome during meiosis which then has the same problem as that discussed for autotetraploids.

(*d*) ***Double Trisomic.*** If two different chromosomes are each represented in triplicate, the double trisomic can be symbolized as $2n + 1 + 1$.

(*e*) ***Nullosomic.*** An organism which has lost a chromosome pair is a nullosomic. The result is usually lethal to diploids $(2n-2)$. Some polyploids, however, can lose two homologues of a set and still survive. For example, several nullosomics of hexaploid wheat $(6n-2)$ exhibit reduced vigor and fertility but can survive to maturity because of the genetic redundancy in polyploids.

## VARIATION IN CHROMOSOME SIZE

In general, chromosomes of most organisms are too small and too numerous to be considered as good subjects for cytological investigation. *Drosophila* was considered to be a favorable organism for genetic studies because it produces large numbers of progeny within the confines of a small bottle in a short interval of time. Many distinctive phenotypes can be recognized in laboratory strains. It was soon discovered that crossing over does not occur in male fruit flies, thereby making it especially useful for genetic analyses. Later, its unusual sex mechanism was found to be a balance between male determiners on the autosomes and female determiners on the sex chromosomes. Although it had been known for over 30 years that some species of dipterans had extra large chromosomes in certain organs of the body, their utility in cytogenetic studies of *Drosophila* was not recognized until about 1934. There are only four pairs of chromosomes in the diploid complement of *D. melanogaster*, but their size in reproductive cells and most body cells is quite small. Unusually large chromosomes, 100 times as large as those in other parts of the body, are found in the larval salivary gland cells. These giant chromosomes (Fig. 9-1) are thought to be composed of many (100-1000) chromatin strands (*chromonemata*) fused together (*polytene*). The chromosome pairs are in a continual prophase condition with close somatic pairing between homologues throughout their length. Distinctive crossbanding is presumed to be groups of *chromomeres* (highly coiled areas) of the chromonema replicates. The crossbanding pattern of each chromosome is a constant characteristic within a species. Abnormal genetic behavior can sometimes be correlated with aberrations which are easily observed in the giant chromosome.

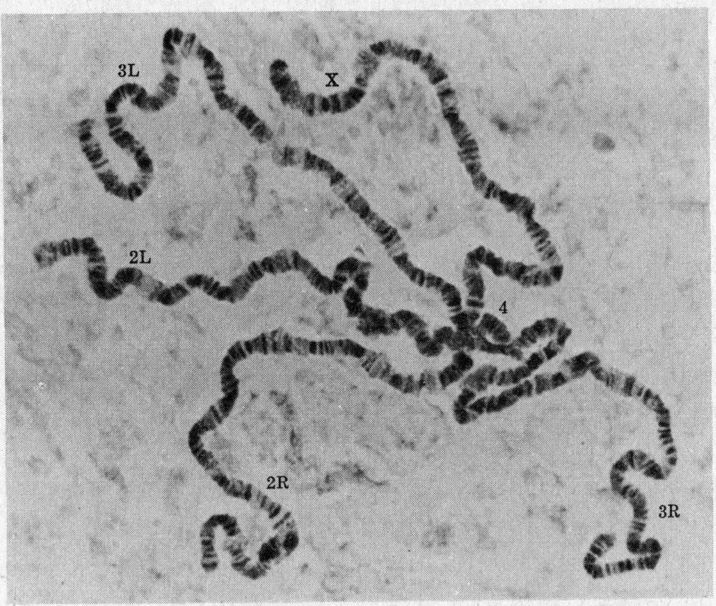


Courtesy of B. P. Kaufmann

**Fig. 9-1.** Salivary gland chromosomes of *Drosophila melanogaster.*

## VARIATION IN THE ARRANGEMENT OF CHROMOSOME SEGMENTS

### 1. Translocations.

Chromosomes occasionally undergo spontaneous rupture, or can be induced to rupture in high frequency by ionizing radiations. The broken ends of such chromosomes behave as though they were "sticky" and may rejoin into non-homologous combinations (*translocations*). A reciprocal translocation involves the exchange of segments between two non-homologous chromosomes. During meiosis, an individual which is heterozygous for a reciprocal translocation must form a cross-shaped configuration in order to affect pairing of all homologous segments.

**Example 9.3.** Assume that a reciprocal translocation occurs between chromosomes 1-2 and 3-4.

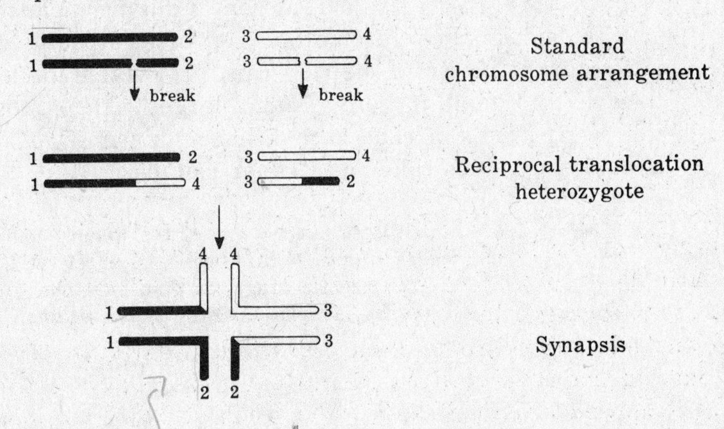


The only way that functional gametes can be formed from a translocation heterozygote is by the alternate disjunction of chromosomes.

**Example 9.4.** At the end of the meiotic prophase begun in Example 9.3, a ring of four chromosomes is formed. If the adjacent chromosomes move to the poles as indicated in the diagram below, all of the gametes will contain some extra segments (duplications) and some pieces will be missing (deficiencies).

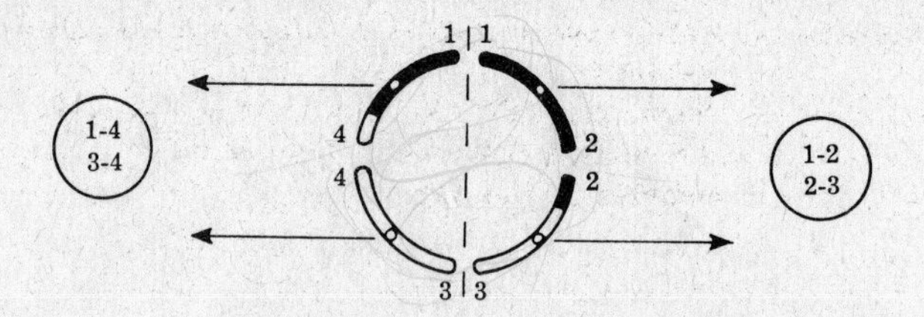


**Example 9.5.** By forming a "figure-8", alternate disjunction produces functional gametes.

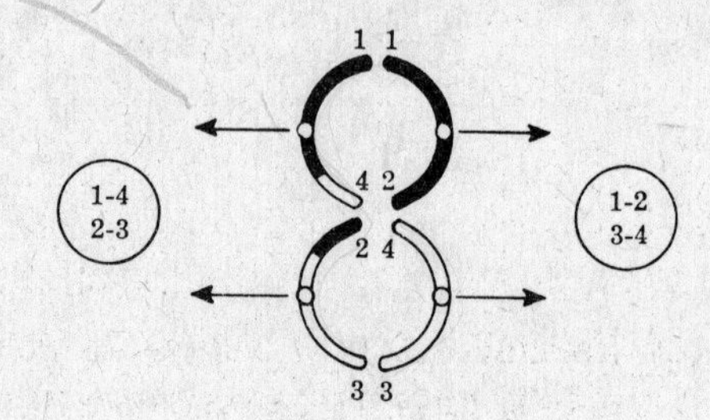


Translocation heterozygotes have several distinctive manifestations. (1) Semi-sterility is usually characteristic of translocation heterozygotes because of adjacent disjunctions as shown in Example 9.4. (2) Some genes which formerly were on non-homologous chromosomes will no longer appear to be assorting independently. (3) The phenotypic expression of a gene may be modified when it is translocated to a new position in the genome. *Position effects* are particularly evident when genes in euchromatin (lightly staining areas usually containing genetic elements) are shifted near heterochromatic regions (darker staining areas presumably devoid of active genes).

(*a*) ***Translocation Complexes.*** In the evening primrose of the genus *Oenothera*, an unusual series of reciprocal translocations has occurred involving all 7 of its chromosome pairs. If we label each chromosome end with a different number, the normal set of 7 chromosomes would be 1-2, 3-4, 5-6, 7-8, 9-10, 11-12 and 13-14; a translocation set would be 2-3, 4-5, 6-7, 8-9, 10-11, 12-13 and 14-1. A multiple translocation heterozygote like this would form a ring of 14 chromosomes during meiosis. Different lethals in each of the two haploid sets of 7 chromosomes enforces structural heterozygosity. Since only alternate segregation from the ring can form viable gametes, each group of 7 chromosomes behaves as though it were a single large linkage group with recombination confined to the pairing ends of each chromosome. Each set of 7 chromosomes which is inherited as a single unit is called a "Renner complex".

**Example 9.6.** In *O. lamarckiana*, one of the Renner complexes is called *gaudens* and the other is called *velans*. This species is largely self-pollinated. The lethals become effective in the zygotic stage so that only the *gaudens-velans* (G-V) zygotes are viable. *Gaudens-gaudens* (G-G) or *velans-velans* (V-V) zygotes are lethal.

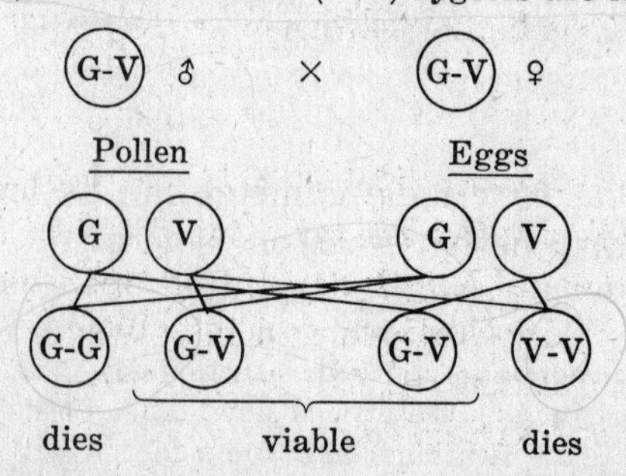

**Example 9.7.** The two complexes in *O. muricata* are called *rigens* (R) and *curvans* (C). Gametic lethals in each complex act differentially in the gametophytes. Pollen with the *rigens* complex are inactive; eggs with the *curvans* complex are inhibited. Only the *curvans* pollen and the *rigens* eggs are functional to give the *rigens-curvans* complex in the zygote.

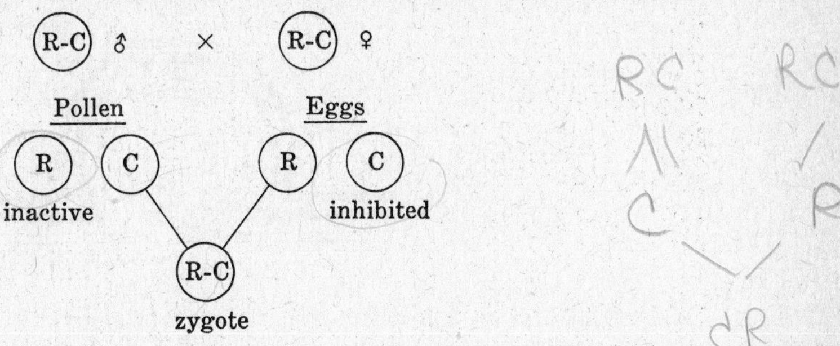


## 2. Inversions.

Assume that the normal order of segments within a chromosome is (1-2-3-4-5-6) and that breaks occur in regions 2-3 and 5-6, and that the broken piece is reinserted in reverse order. The inverted chromosome now has segments (1-2-5-4-3-6). One way in which inversions might arise is shown in Fig. 9-2. An inversion heterozygote has one chromosome in the inverted order and its homologue in the normal order. During meiosis the synaptic configuration attempts to maximize the pairing between homologous regions in the two chromosomes. This is usually accomplished by a loop in one of the chromosomes. Crossing over within the inverted segment gives rise to crossover gametes which are inviable because of duplications and deficiencies. Chromatids which are not involved in crossing over will be viable. Thus as we have seen with translocations, inversions produce semisterility and altered linkage relationships. Inversions are sometimes called "crossover suppressors". Actually they do not prevent crossovers from occurring but they do prevent the crossover products from functioning. Genes within the inverted segment are thus held together and transmitted as one large linked group. Balanced lethal systems (Chapter 6) involve either a translocation or an inversion to prevent the recovery of crossover products and thus maintain heterozygosity generation after generation. In some organisms, these "inversions" have a selective advantage under certain environmental conditions and become more prevalent in the population than the standard chromosome order. Two types of inversion heterozygotes will be considered in which crossing over occurs within the inverted segment.

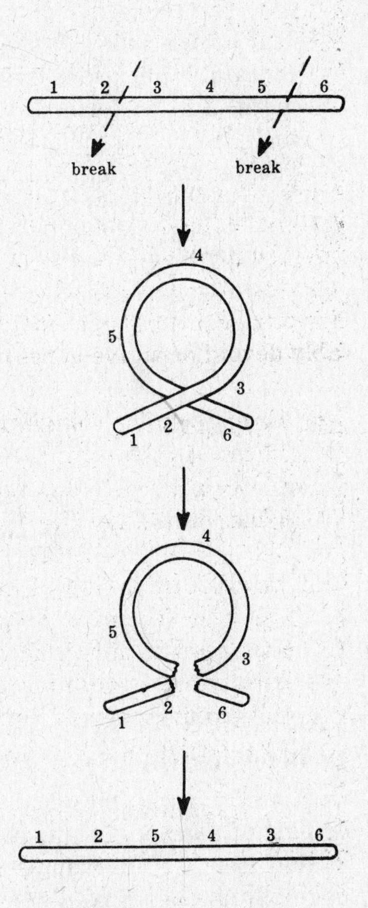


**Fig. 9-2.** Origin of an inversion.

(*a*) ***Pericentric Inversion.*** The centromere lies within the inverted region. First meiotic anaphase figures appear normal, but the two chromatids of each chromosome usually have arms of unequal length depending upon where the crossover occurred. Half of the products contain duplications and deficiencies and do not function. The other half of the gametes are functional: one-quarter have the normal chromosome order, one-quarter have the inverted arrangement.

**Example 9.8.** Assume an inversion heterozygote as shown below with crossing over in region 3-4.

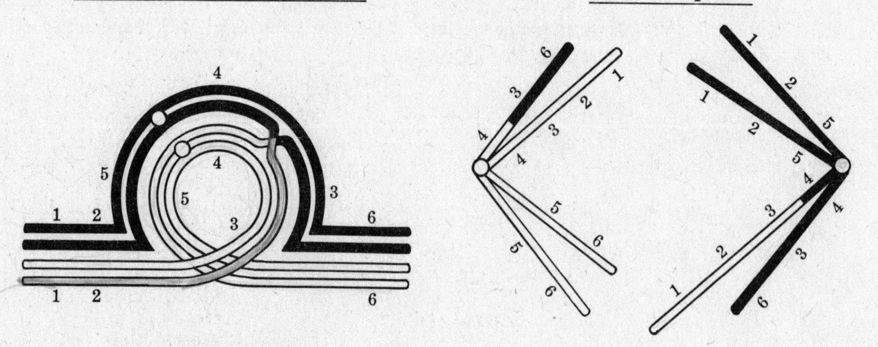


(*b*) ***Paracentric Inversion.*** The centromere lies outside the inverted segment. Crossing over within the inverted segment produces a *dicentric* chromosome (possessing two centromeres) which forms a bridge from one pole to the other during first anaphase. The bridge will rupture somewhere along its length and the resulting fragments will contain duplications and/or deficiencies. Also, an *acentric* fragment (without a centromere) will be formed; and since it usually fails to move to either pole, it will not be included in the meiotic products. Again, half of the products are non-functional, one-quarter are functional with a normal chromosome, and one-quarter are functional with an inverted chromosome.

**Example 9.9.** Assume an inversion heterozygote as shown below with crossing over in region 4-5.

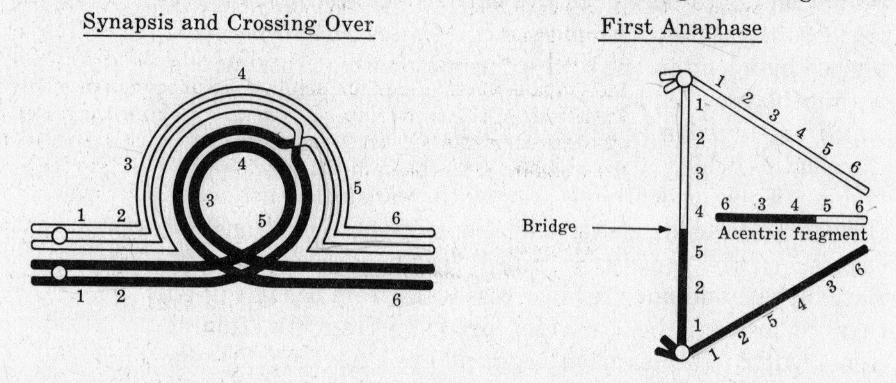


## VARIATION IN THE NUMBER OF CHROMOSOMAL SEGMENTS

### 1. Deletions (Deficiencies).

Loss of a chromosomal segment may be so small that it includes only a single gene or part of a gene. In this case the phenotypic effects may resemble those of a mutant allele at that locus. For example, the "notch" phenotype of *Drosophila* discussed in Problem 5.4, page 79, is a sex-linked deletion which acts like a dominant mutation; a deletion at another sex-linked locus behaves as a recessive mutation, producing yellow body color when homozygous. Deletions never back mutate to the normal condition, because a lost piece of chromosome cannot be replaced. In this way, as well as others to be explained in subsequent chapters, a deletion can be distinguished from a gene mutation. A loss of any considerable portion of a chromosome is usually lethal to a diploid organism because of genetic unbalance. When an organism heterozygous for a pair of alleles, *A* and *a*, loses a small portion of the chromosome bearing the dominant allele, the recessive allele on the other chromosome will become expressed phenotypically. This is called *pseudodominance*, but it is a misnomer because the condition is hemizygous rather than dizygous at this locus.

**Example 9.10.** A deficiency in the segment of chromosome bearing the dominant gene $A$ allows the recessive allele $a$ to become phenotypically expressed.

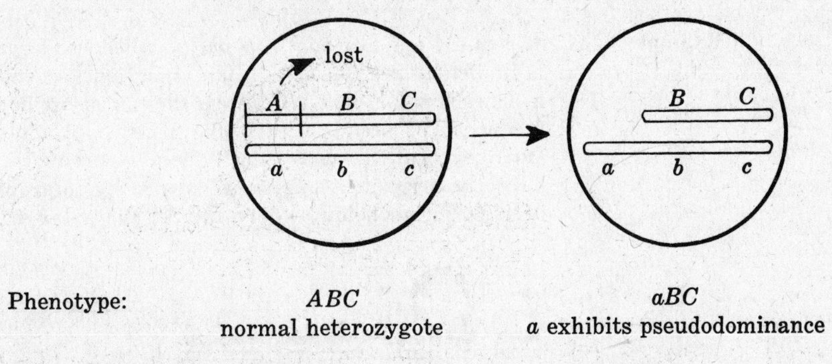


Phenotype: $ABC$ normal heterozygote → $aBC$ $a$ exhibits pseudodominance

A deletion heterozygote may be detected cytologically during meiotic prophase when the forces of pairing cause the normal chromosome segment to bulge away from the region in which the deletion occurs (Fig. 9-3).

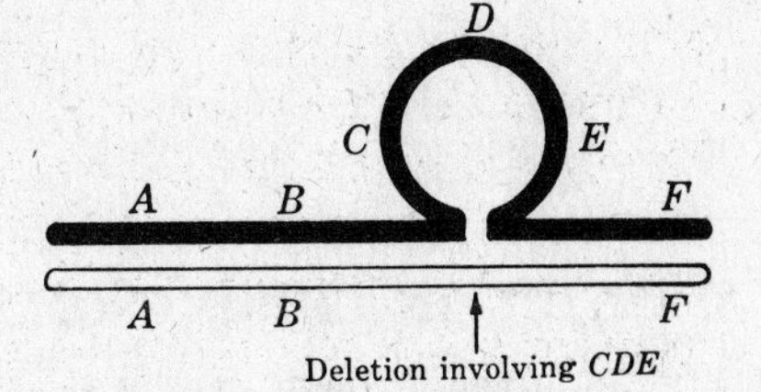


**Fig. 9-3.** Synapsis in a deletion heterozygote.

Overlapping deletions have been extensively used to locate the physical position of genes in the chromosome (cytological mapping).

**Example 9.11.** A laboratory stock of *Drosophila* females is heterozygous in coupling phase for two linked genes at the tip of the X chromosome, *ac* (achaete) and *sc* (scute). A deletion in one chromosome shows pseudodominance for both achaete and scute. In other individuals, another deletion displays pseudodominance only for achaete. Obviously, these two deletions overlap. In the giant chromosomes of *Drosophila*, the absence of these segments of chromosome is easily seen. The actual location of the scute gene resides in the band or bands which differentiate the two overlapping deletions.

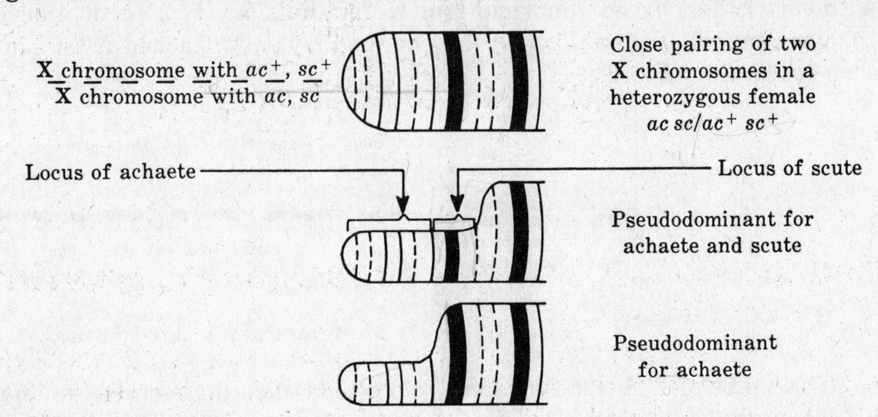


## 2. Duplications (Additions).

Extra segments in a chromosome may arise in a variety of ways. Generally speaking, their presence is not as deleterious to the organism as a deficiency. It is assumed that some duplications are useful in the evolution of new genetic material. Because the old genes can continue to provide for the present requirements of the organism, the superfluous genes may be free to mutate to new forms without a loss in immediate adaptability. Genetic redundancy, of which this is one type, may protect the organism from the effects of a deleterious recessive gene or from an otherwise lethal deletion. During meiotic pairing the chromosome bearing the duplicated segment forms a loop to maximize the juxtaposition of homologous regions. In some cases, extra genetic material is known to cause a distinct phenotypic effect. Relocation of chromosomal material without altering its quantity may result in an altered phenotype (position effect).

**Example 9.12.** A reduced eye size in *Drosophila,* called "bar eye", is known to be associated with a duplicated region on the X chromosome. Genetically the duplication behaves as a dominant factor. Wild type flies arise in homozygous bar-eye cultures with a frequency of about 1 in 1600. With approximately the same frequency, a very small eye called "double-bar" is also produced. These unusual phenotypes apparently arise in a pure bar culture by improper synapsis and unequal crossing over as shown below, where the region (*a-b-c-d*) is a duplication.

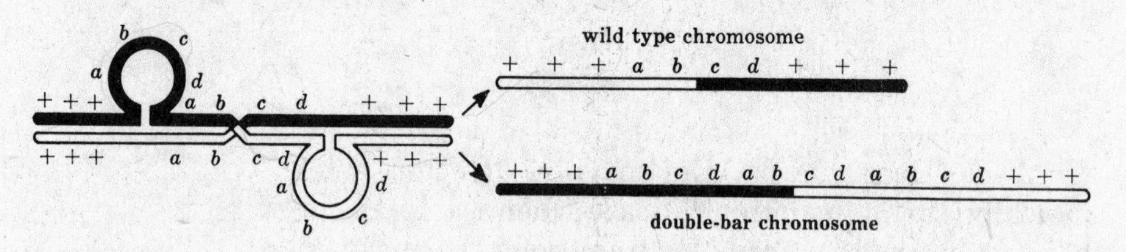


## VARIATION IN CHROMOSOME MORPHOLOGY

It has already been shown that a translocation can change the structure of the chromosome both genetically and morphologically. The length of the chromosome may be longer or shorter, depending upon the size of the translocated piece. An inversion does not normally change the length of the chromosome, but if the inversion includes the centromere (pericentric) the position of the centromere may be changed considerably. Deletions or duplications, if viable, may sometimes be detected cytologically by a change in the size of the chromosome (or banding pattern in the case of the giant chromosomes of *Drosophila*), or by the presence of "bulges" in the pairing figure. Chromosomes with unequal arm lengths may be changed to *isochromosomes* having arms of equal length and genetically homologous with each other, by an abnormal transverse division of the centromere. The telocentric X chromosome of *Drosophila* may be changed to an "attached-X" form by a misdivision of the centromere (Fig. 9-4).

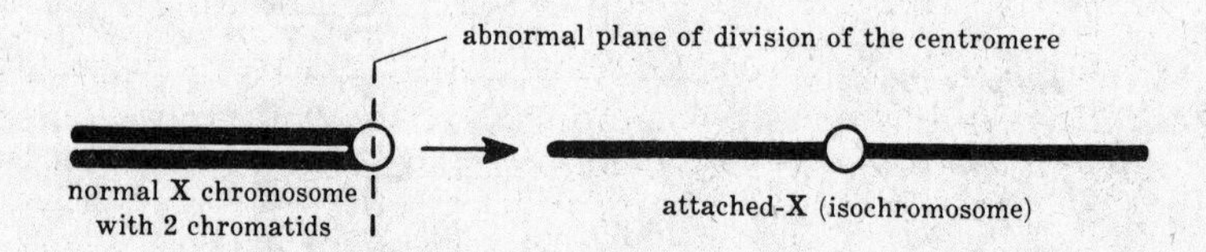


Fig. 9-4. Origin of an attached-X chromosome.

The shape of a chromosome may change at each division once it has been broken. Following reduplication of a broken chromosome, the "sticky" ends of the sister chromatids fuse. When the chromatids move to opposite poles, a bridge is formed. The bridge will break somewhere along its length and the cycle repeats at the next division. This sequence of events is called a *breakage-fusion-bridge cycle.* Mosaic tissue can be produced by a breakage-fusion-bridge cycle appearing as irregular patches of an unexpected phenotype on a background of normal tissue (*variegation*). The size of the unusual trait generally bears an inverse relationship to the period of development at which the original break occurred. That is, the earlier a break occurs, the larger will be the size of the abnormal tissue. Chromosomes are not always rod shaped. Occasionally ring chromosomes are encountered in higher organisms. If breaks occur at each end of a chromosome, the broken ends may become joined to form a ring chromosome. The deleted fragments do not contain centromeres and will be lost. The phenotypic consequences of these deletions vary, depending upon the specific genes involved. Crossing over between ring chromosomes can lead to bizarre anaphase figures.

**Example 9.13.** **A single exchange in a ring homozygote produces a double bridge at first anaphase.**

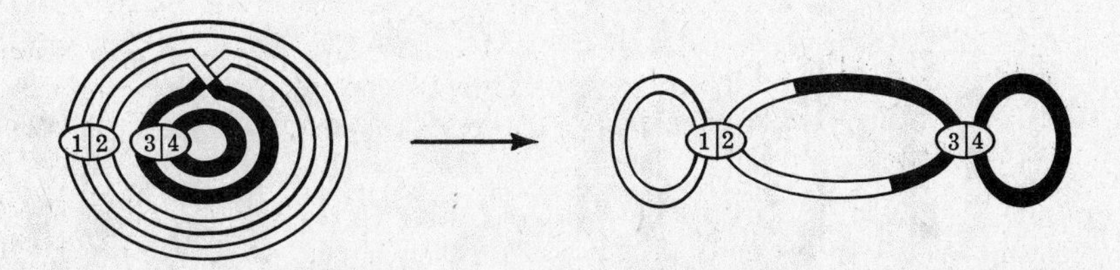


A whole arm fusion (*Robertsonian translocation*) is an eucentric, reciprocal translocation between two acrocentric chromosomes where the break in one chromosome is near the front of the centromere and the break in the other chromosome is immediately behind its centromere. The smaller chromosome thus formed consists of largely inert heterochromatic material near the centromeres; it usually carries no essential genes and tends to become lost. A Robertsonian translocation thus results in a reduction of the chromosome number (Fig. 9-5).

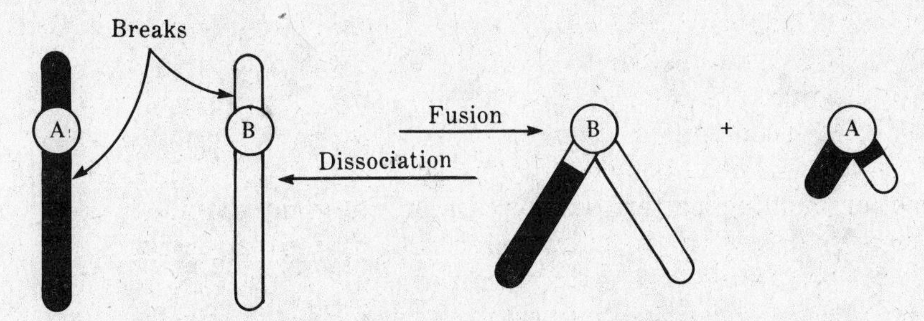


**Fig. 9-5.** Formation of a metacentric chromosome by fusion of two acrocentric chromosomes (Robertsonian translocation). (Idea from Monroe W. Strickberger, *Genetics*, 2nd ed., Macmillan, 1976, p. 524.)

**Example 9.14.** Humans have 46 chromosomes whereas the great apes (chimpanzees, gorillas and orangutans) have 48. It seems likely that humans evolved from a common human/ape ancestor by (among other structural changes) centric fusion of two acrocentrics to produce a single large chromosome (2) containing the combined genetic content of the two acrocentrics. Structural rearrangements of chromosomes may lead to reproductive isolation and the formation of new species. The mule is a hybrid from crossing the horse ($2n = 64$) and the ass or donkey ($2n = 62$). The mule is sterile because there is insufficient homology between the two sets of chromosomes to pair successfully at meiosis.

## HUMAN CYTOGENETICS

The diploid human chromosome number of 46 (23 pairs) was established by Tjio and Levan in 1956. When grouped as homologous pairs, the somatic chromosome complement (*karyotype*) of a cell becomes an *idiogram*. Formerly, a chromosome could only be distinguished by its length and the position of its centromere at the time of maximum condensation (metaphase). No single autosome could be easily identified, but a chromosome could be assigned to one of seven groups (A through G) according to the "Denver system" of classification (Fig. 9-6). Group A consists of large, metacentric chromosomes (1–3); group B contains submedian chromosomes (4–5); group C has medium-sized chromosomes with submedian centromeres (pairs 6–12); group D consists of medium-sized chromosomes (pairs 13–15) with one very short arm (acrocentric); chromosomes in group E (16–18) are a little shorter than in group D with median or submedian centromeres; group F (19–20) contains short, metacentric chromosomes; and group G has the smallest acrocentric chromosomes (21, 22). The X and Y sex chromosomes are not members of the autosome groups, but are usually placed together in one part of the idiogram. The Y chromosome may vary in size from one individual to another but usually has the appearance of G-group autosomes. The X chromosome has the appearance of a group C autosome.

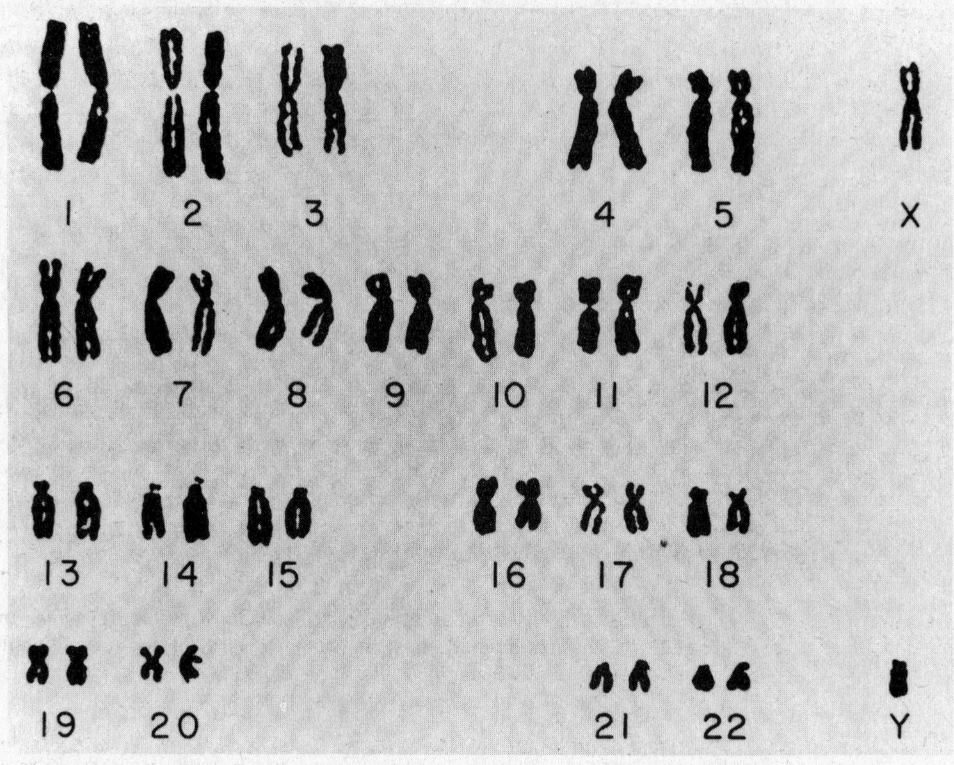


**Fig. 9-6.** Karyotype of the chromosomes of a normal human male. (From L. P. Wisniewski and K. Hirschhorn (eds.), *A Guide to Human Chromosome Defects*, 2nd ed., The March of Dimes Birth Defects Foundation, White Plains, N.Y.; BD:OAS XVI(6), 1980, with permission.)

More recently, special staining techniques (e.g. Giemsa, quinacrine) have revealed specific banding patterns (G bands, Q bands, etc.) for each chromosome, allowing individual identification of each chromosome in the karyotype (Fig. 9-7).

The X chromosome can be identified in many non-dividing (interphase) cells of females as a dark-staining mass called *sex chromatin* or *Barr body* (after Dr. Murray L. Barr) attached to the nuclear membrane. The analogue of sex chromatin in certain white blood cells is a "drumstick" appendage attached to the multilobed nucleus of neutrophilic leukocytes. Dr. Mary Lyon theorizes that sex chromatin results from condensation (*heterochromatinization,* (darkly stained)) and inactivation of any X chromosomes in excess of one per cell. Sex-linked traits are not expressed more intensely in females with two doses of X-linked genes than in males with only one X chromosome. At a particular stage early in development of females, one of the two X chromosomes in a cell becomes inactivated as a *dosage compensation* mechanism. Different cells inactivate one of the two chromosomes in an apparently random manner, but sub-

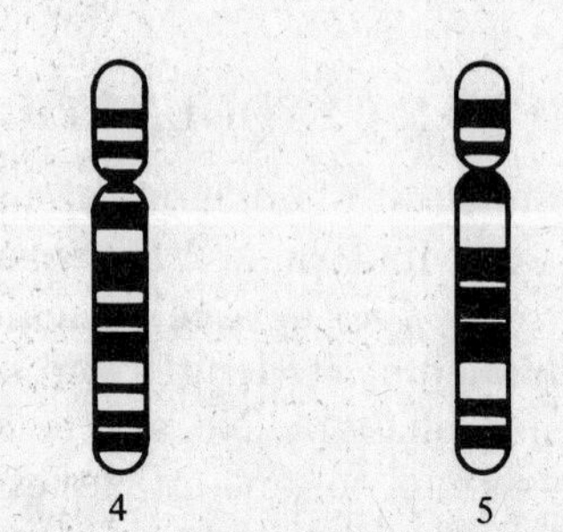


**Fig. 9-7.** Diagrams of the banding patterns that distinguish human chromosomes 4 and 5.

sequently all derived cells retain the same functional chromosome. Females are thus a mixture of two kinds of cells; in some cells one X chromosome is active, and in different cells the other X chromosome is active. The same principle applies to mammals other than humans.

Prenatal screening of babies for gross chromosomal aberrations (polyploidy, aneuploidy, deletions, translocations, etc.), as well as sex prediction, is now possible. A fluid sample can be taken from the "bag of water" (*amnion*) surrounding the fetus in utero, a process termed *amniocentesis*. The cells found in amniotic fluid are of fetal origin. Such cells can be cultured in vitro in a highly nutritive solution, treated with colchicine to stop division at metaphase, subjected to a hypotonic salt solution to cause the cells to swell and scatter the chromosomes, placed on a slide, stained and photographed under a microscope. Individual chromosomes are then cut from the resulting photograph and matched as homologous pairs to form an idiogram.

**Example 9.15.** Aneuploid females with only one X chromosome (XO) have a karyotype with $2n - 1 = 45$. They are called *Turner females* (after Henry Turner, who first described them), and they exhibit a group of characteristics that together define *Turner's syndrome*: short stature, webbing of neck skin, underdeveloped gonads, shieldlike chest and impaired intelligence. Turner females are sex-chromatin negative.

**Example 9.16.** Abnormal males possessing an extra X chromosome (XXY) have a karyotype with $2n + 1 = 47$. They are called *Klinefelter males* (after Harry Klinefelter, who first described them), and they exhibit *Klinefelter's syndrome*: sterility, long limbs, feminine breast development (gynecomastia), sparse body hair and mental deficiency. Klinefelter males are sex-chromatin positive. If some portion of the extra X chromosome is not inactivated, this could account for the phenotypic differences not only between XXY Klinefelter males and XY normal males, but also between XO Turner females and XX normal females.

**Example 9.17.** XXX "superfemales" (*metafemales*) are karyotyped as $2n + 1 = 47$ trisomics and exhibit two Barr bodies. These individuals may range phenotypically from normal fertile females to nearly like those with Turner's syndrome. They have a high incidence of mental retardation.

**Example 9.18.** Tall, trisomic XYY males were first discovered in relatively high frequencies in penal and mental institutions. The presence of an extra Y chromosome was thought to predispose such a male to antisocial behavior, hence the name "tall-aggressive syndrome". Subsequently, more XYY males have been found among the noninstitutionalized population, casting doubt upon the validity of the above hypothesis. XYY males do tend to have subnormal IQs, however, and this may contribute to impulsive behavior.

**Example 9.19.** Down's syndrome (named after the physician Langdon Down and formerly called mongolism or mongolian idiocy) is usually associated with a trisomic condition for one of the smallest human autosomes (21). It is the most common chromosomal abnormality in live births (1/600 births). These unfortunate individuals are mentally retarded, short, possessing eye folds resembling those of Mongolian races, have stubby fingers and a swollen tongue. Women over 45 years of age are about twenty times more likely to give birth to a child with Down's syndrome than women aged 20. Nondisjunction of chromosome pair 21 during spermatogenesis can also produce a child with Down's syndrome, but paternal age does not seem to be associated with its incidence. In about 2–5% of the cases, the normal chromosome number is present ($2n = 46$), but the extra chromosome 21 is attached (translocated) to one of the larger autosomes.

**Example 9.20.** Human autosomal monosomics are rarer than trisomics, possibly because harmful recessive mutants on the remaining homologue are hemizygous and can be expressed. Most cases of autosomal monosomy are mosaics of normal diploid ($2n$) and monosomic ($2n - 1$) cells resulting from *mitotic* non-disjunctions. Mosaics involving sex chromosomes are also known: e.g. XO:XX, XO:XY, XXY:XX, as well as autosomal mosaics such as 21-21:21-21-21, etc.

**Example 9.21.** Babies missing a portion of the short arm of chromosome 5 have a distinctive catlike cry; hence the French name "*cri du chat*" (cry of the cat) syndrome. They also are mentally retarded, have moon faces, saddle noses, small mandibles (micrognathia) and malformed, low-set ears.

**Example 9.22.** Deletion of part of the long arm of chromosome 22 produces an abnormality known as a *Philadelphia chromosome* (so named because it was discovered in that city). It is found only in the bone marrow (along with chromosomally normal cells) in approximately 90% of patients with chronic myelocytic leukemia (a kind of cancer). Usually the missing piece of chromosome 22 can be found translocated to one of the larger autosomes (most frequently chromosome 9).

# Solved Problems

## VARIATION IN CHROMOSOME NUMBERS

**9.1.** An autotetraploid of genotype *AAaa* forms only diploid gametes by random assortment from the quadrivalents. The locus of *A* is very close to the centromere so that crossing over in that area is negligible. (*a*) Determine the expected frequencies of zygotes produced by selfing. (*b*) What reduction in the frequency of the recessive phenotype is anticipated by this autotetraploid in comparison with a selfed diploid of genotype *Aa*?

**Solution:**

(*a*) For genes which are tightly linked to their centromeres, the distribution of alleles into gametes follows the same pattern as chromosomal assortment. Let us first use a checkerboard to determine the kinds and frequencies of different combinations of alleles by two's expected in the gametes of the autotetraploid. Note that the diagonal represents the nonexistent possibility of a given chromosome (or allele) with itself in a gamete. The table is symmetrical above and below the diagonal.

| | *A* | *A* | *a* | *a* |
|---|---|---|---|---|
| *A* | – | *AA* | *Aa* | *Aa* |
| *A* | | – | *Aa* | *Aa* |
| *a* | | | – | *aa* |
| *a* | | | | – |

Ratio of possible gametes: 1 *AA* : 4 *Aa* : 1 *aa* = 1/6 *AA* : 2/3 *Aa* : 1/6 *aa*. Using these gametic expectations, let us now construct a zygotic checkerboard for the prediction of progeny genotypes.

| | 1/6 *AA* | 2/3 *Aa* | 1/6 *aa* |
|---|---|---|---|
| 1/6 *AA* | 1/36 *AAAA* | 2/18 *AAAa* | 1/36 *AAaa* |
| 2/3 *Aa* | 2/18 *AAAa* | 4/9 *AAaa* | 2/18 *Aaaa* |
| 1/6 *aa* | 1/36 *AAaa* | 2/18 *Aaaa* | 1/36 *aaaa* |

Ratio of offspring genotypes: 1/36 *AAAA* (quadruplex) : 8/36 *AAAa* (triplex) : 18/36 *AAaa* (duplex) : 8/36 *Aaaa* (simplex) : 1/36 *aaaa* (nulliplex).

(*b*) If one dose of the dominant allele is sufficient to phenotypically mask one or more doses of the recessive allele, then the phenotypic ratio is expected to be 35 *A* : 1*a*. One-quarter of the offspring of a selfed diploid heterozygote (*Aa*) is expected to be of the recessive phenotype. The reduction in the frequency of the recessive trait is from 1/4 to 1/36, or ninefold. When homozygous genotypes produce a less desirable phenotype than heterozygotes, polyploidy can act as a buffer to reduce the incidence of homozygotes.

**9.2.** Assume an autopolyploid of genotype $AAaa$ with the locus of $A$ 50 map units or more from the centromere so that a crossover always occurs in this area. In this case, the *chromatids* will assort independently. Assuming random assortment of chromatids to the gametes by two's, determine (*a*) the expected genotypic ratio of the progeny which results from selfing this autopolyploid, (*b*) the increase in the incidence of heterozygous genotypes compared with selfed diploids of genotype $Aa$.

**Solution:**

(*a*) Random assortment of 8 chromatids of the autotetraploid during meiotic prophase requires a gametic checkerboard with 64 squares.

| | $A$ | $A$ | $A$ | $A$ | $a$ | $a$ | $a$ | $a$ |
|---|---|---|---|---|---|---|---|---|
| $A$ | — | $AA$ | $AA$ | $AA$ | $Aa$ | $Aa$ | $Aa$ | $Aa$ |
| $A$ | | — | $AA$ | $AA$ | $Aa$ | $Aa$ | $Aa$ | $Aa$ |
| $A$ | | | — | $AA$ | $Aa$ | $Aa$ | $Aa$ | $Aa$ |
| $A$ | | | | — | $Aa$ | $Aa$ | $Aa$ | $Aa$ |
| $a$ | | | | | — | $aa$ | $aa$ | $aa$ |
| $a$ | | | | | | — | $aa$ | $aa$ |
| $a$ | | | | | | | — | $aa$ |
| $a$ | | | | | | | | — |

Summary of gametic types: $6\,AA : 16\,Aa : 6\,aa$ or $3:8:3$. Using these gametic expectations, we can now construct a zygotic checkerboard to generate the expected progeny.

| | $3\,AA$ | $8\,Aa$ | $3\,aa$ |
|---|---|---|---|
| $3\,AA$ | $9\,AAAA$ | $24\,AAAa$ | $9\,AAaa$ |
| $8\,Aa$ | $24\,AAAa$ | $64\,AAaa$ | $24\,Aaaa$ |
| $3\,aa$ | $9\,AAaa$ | $24\,Aaaa$ | $9\,aaaa$ |

Summary of progeny genotypes: $9\,AAAA$ (quadruplex) : $48\,AAAa$ (triplex) : $82\,AAaa$ (duplex) : $48\,Aaaa$ (simplex) : $9\,aaaa$ (nulliplex).

(*b*) Selfing a diploid of genotype $Aa$ produces 50% heterozygous progeny. Selfing the autotetraploid produces $178/196 = 91\%$ heterozygous progeny. The increase from 50% to 91% is $41/50 = 82\%$.

**9.3.** Pericarp is the outermost layer of the corn kernel and is maternal in origin. A dominant gene ($B$) produces brown pericarp, and its recessive allele ($b$) produces colorless pericarp. Tissue adjacent to the pericarp is aleurone (triploid). Purple pigment is deposited in the aleurone when the dominant gene $C$ is present; its recessive allele $c$ results in colorless aleurone. Aleurone is actually a single specialized layer of cells of the endosperm. The color of endosperm itself is modified by a pair of alleles. Yellow is governed by the dominant allele $Y$ and white by the recessive allele $y$. Both $C$ and $Y$ show xenia to their respective alleles. A plant which is $bbCcYy$ is pollinated by a plant of genotype $BbCcYy$. (*a*) What phenotypic ratio is expected among the progeny kernels? (*b*) If the $F_1$ is pollinated by plants of genotype $bbccyy$, in what color ratio will the resulting $F_2$ kernels be expected to appear?

**Solution:**

(*a*) If pericarp is colorless, then the color of the aleurone shows through. If aleurone is also colorless, then the color of the endosperm becomes visible. Since the maternal parent is *bb*, the pericarp on all $F_1$ seeds will be colorless. Any seeds with *C* will have purple aleurone. Only if the aleurone is colorless (*ccc*) can the color of the endosperm be seen.

Parents: $bbCcYy$ ♀ × $BbCcYy$ ♂

$F_1$: $\frac{3}{4}$ *C--* = $\frac{3}{4}$ purple

$\frac{1}{4}$ *ccc* <
- $\frac{3}{4}$ *Y--* = $\frac{3}{16}$ yellow
- $\frac{1}{4}$ *yyy* = $\frac{1}{16}$ white (colorless)

(*b*) Half of the $F_1$ embryos is expected to be *Bb* and will thus lay down a brown pericarp around their seeds ($F_2$); the other half is expected to be *bb* and will envelop its seeds with a colorless pericarp. Thus half of the seeds on the $F_1$ plants will be brown. Of the remaining half which has colorless pericarp, we need show only as much of the genotype as is necessary to establish the phenotype.

| *bb* × $F_1$ | | Diploid Fusion Nucleus | | Sperm Nucleus | | Triploid Tissue |
|---|---|---|---|---|---|---|
| $\frac{1}{2} \times \frac{1}{4}$ *CC* | = | $\frac{1}{8}$ *CC* | + | *c* | = | $\frac{1}{8}$ *CCc* purple |
| $\frac{1}{2} \times \frac{1}{2}$ *Cc* | = | $\frac{1}{4}$ { $\frac{1}{8}$ *CC* | + | *c* | = | $\frac{1}{8}$ *CCc* purple |
| | | $\frac{1}{8}$ *cc* { $\frac{1}{16}$ *YY* | + | *cy* | = | $\frac{1}{16}$ *cccYYy* yellow |
| | | $\frac{1}{16}$ *yy* | + | *cy* | = | $\frac{1}{16}$ *cccyyy* white |
| $\frac{1}{2} \times \frac{1}{4}$ *cc* | = | $\frac{1}{8}$ { $\frac{1}{16}$ *ccYY* | + | *cy* | = | $\frac{1}{16}$ *cccYYy* yellow |
| | | $\frac{1}{16}$ *ccyy* | + | *cy* | = | $\frac{1}{16}$ *cccyyy* white |

Summary of $F_2$ seed colors: $\frac{1}{2}$ brown : $\frac{1}{4}$ purple : $\frac{1}{8}$ yellow : $\frac{1}{8}$ white.

**9.4.** Eyeless is a recessive gene (*ey*) on the tiny fourth chromosome of *Drosophila*. A male trisomic for chromosome 4 with the genotype + + *ey* is crossed to a disomic eyeless female of genotype *ey ey*. Determine the genotypic and phenotypic ratios expected among the progeny by random assortment of the chromosomes to the gametes.

**Solution:**

Three types of segregation are possible in the formation of gametes in the triploid.

(1) + (+) *ey* → (+ (+)) (*ey*)  (2) + (+) *ey* → (+) ((+) *ey*)  (3) (+) + *ey* → ((+)) (+ *ey*)

Summary of sperm genotypes: 1 + + : 2 + *ey* : 2 + : 1 *ey*.

The union of these sperms with eggs of genotype *ey* results in the following progeny:

1 + + *ey*
2 + *ey ey*  } = 5 wild type
2 + *ey*
1 *ey ey* = 1 eyeless

## VARIATION IN THE ARRANGEMENT OF CHROMOSOME SEGMENTS

**9.5.** In 1931 Stern found two different translocations in *Drosophila* from which he developed females possessing heteromorphic X chromosomes. One X chromosome had a piece of the Y chromosome attached to it; the other X was shorter and had a piece of chromosome IV attached to it. Two sex-linked genes were used as markers for detecting crossovers, the recessive trait carnation eye color (*car*) and the dominant trait bar eye ($B$). Dihybrid bar females with heteromorphic chromosomes (both mutant alleles on the X portion of the X-IV chromosome) were crossed to hemizygous carnation males with normal chromosomes. The results of this experiment provided cytological proof that genetic crossing over involves an actual physical exchange between homologous chromosome segments. Diagram the expected cytogenetic results of this cross showing all genotypes and phenotypes.

**Solution:**

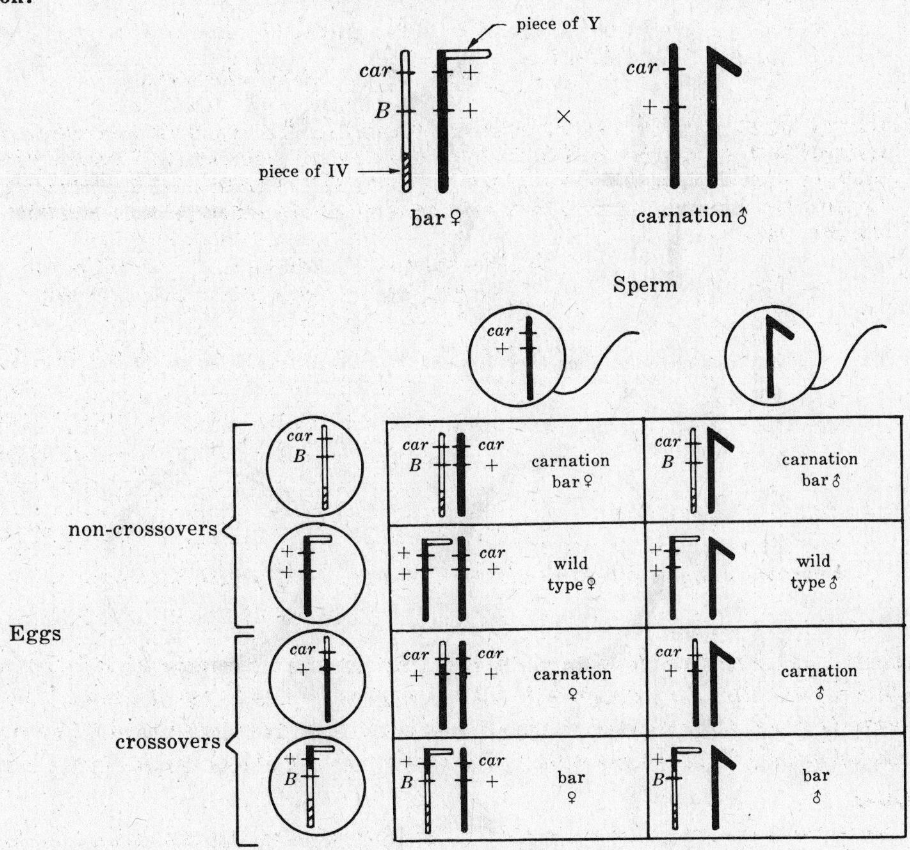


The existence of a morphologically normal X chromosome in recombinant male progeny with carnation eyes provides cytological proof that genetic crossing over is correlated with physical exchange between homologous chromosomes in the parents. Similarly, all other phenotypes correlate with the cytological picture.

**9.6.** Consider an organism with four pairs of chromosomes in standard order, the ends of which we shall label 1-2, 3-4, 5-6, 7-8. Strain A crossed to the standard strain gives a ring of four plus two bivalents during meiotic prophase. Strain B crossed to the standard strain also gives a ring of four plus two bivalents. In each of the four situations which follow, explain how a cross of strain A × strain B could produce (*a*) four bivalents, (*b*) ring of four plus two bivalents, (*c*) two rings of four, (*d*) ring of six plus one bivalent.

**Solution:**

A ring of four indicates a reciprocal translocation involving two non-homologous chromosomes. As a starting point, let us assume that strain A has experienced a single reciprocal translocation so that the order is 1-3, 2-4, 5-6, 7-8. Strain B also shows a ring of four with the standard, but we do not know whether the translocation involves the same chromosome as strain A or different chromosomes. The results of crossing A × B will indicate which of the B chromosomes have undergone translocations.

(*a*) Formation of only bivalents indicates that complete homology exists between the chromosomes in strains A and B. Therefore strain B has the same translocation as that in strain A.

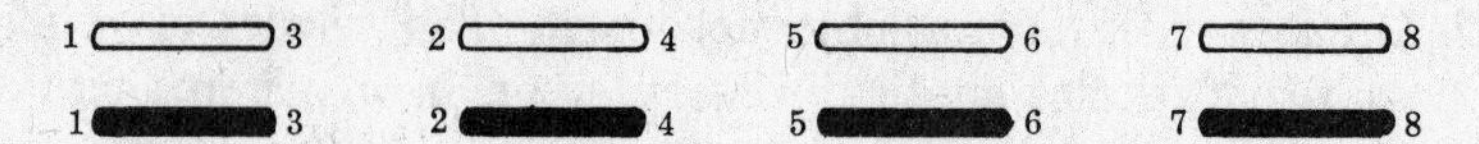


(*b*) A ring of four indicates that the same chromosomes which were interchanged in strain A are also involved in strain B, but with different end arrangements.

A: 1-3, 2-4, 5-6, 7-8

B: 1-4, 2-3, 5-6, 7-8

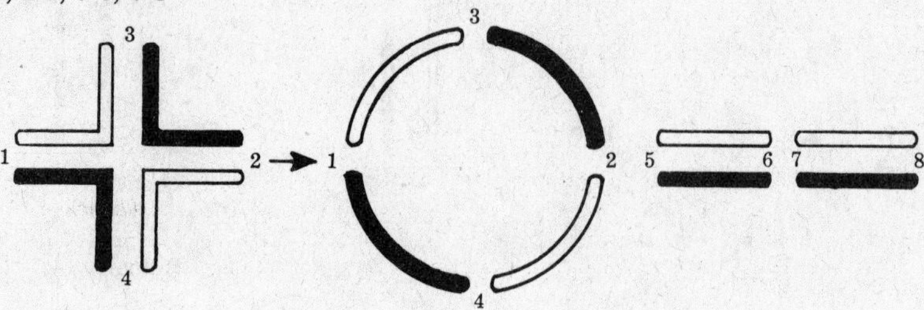


(*c*) Two rings of four chromosomes each indicate that B differs from A by two translocations.

A: 1-3, 2-4, 5-6, 7-8

B: 1-4, 2-3, 5-7, 6-8

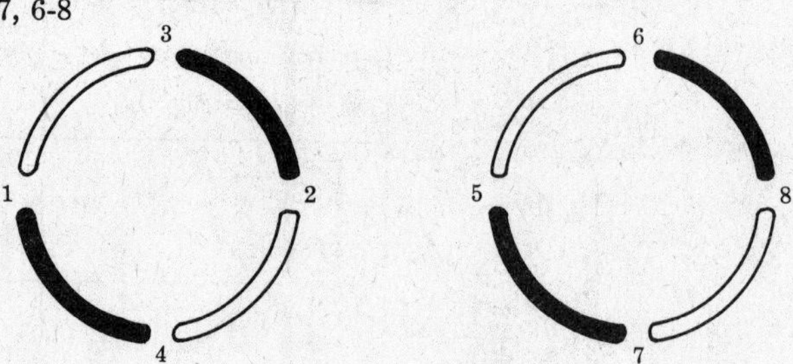


(*d*) A ring of six indicates that B differs from A by three translocations.

A: 1-3, 2-4, 5-6, 7-8

B: 1-2, 3-5, 4-6, 7-8

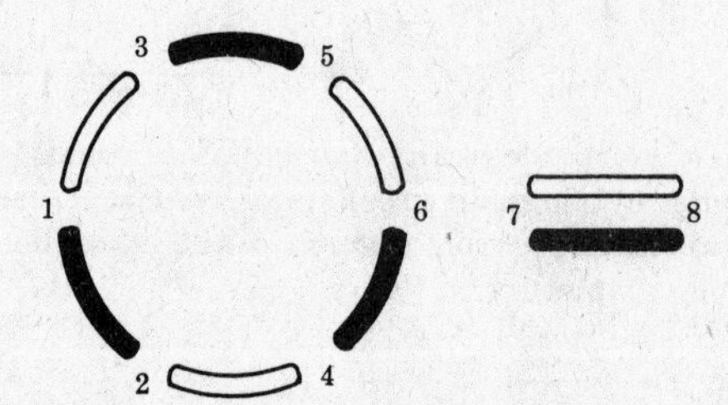


**9.7.** The centromere of chromosome V in corn is about 7 map units from the end. The gene for light yellow (virescent) seedling ($v$) is 10 map units from this end, and a gene which shortens internode length called brevis ($bv$) is 12 map units from this end. The break point of a translocation ($T$) is 20 map units from this end. A translocation heterozygote involving chromosomes V and VIII of genotype $+\ bv\ t/v\ +\ T$ is pollinated by a normal (non-translocated, $t$) plant of genotype $v\ bv\ t/v\ bv\ t$. If gametes are formed exclusively by alternate segregation from the ring of chromosomes formed by the translocation heterozygote, predict the ratio of progeny genotypes and phenotypes from this cross (considering multiple crossovers to be negligible).

**Solution:**

First let us diagram the effect which crossing over will have between the centromere and the point of translocation. We will label the ends of chromosome V with 1-2, and of chromosome VIII with 3-4. A cross-shaped pairing figure is formed during meiosis.

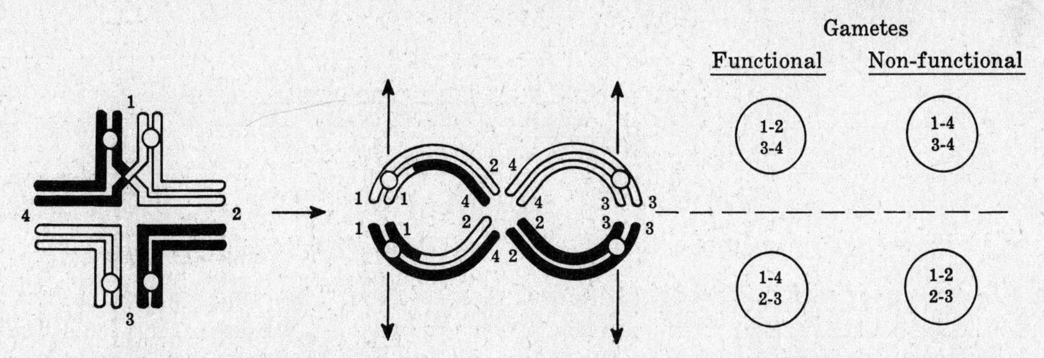


Alternate segregation produces half functional and half non-functional (duplication-deficiency) gametes. Note that the non-functional gametes derive only from the crossover chromatids. Thus recovery of chromatids which experience a crossover between the centromere and the point of translocation is prevented. The combination of genes in this region of the chromosome is prevented from being broken up by crossing over and are thus transmitted as a unit. This situation is analogous to the block of genes within an inversion which are similarly held together as a genetic unit. Non-crossover chromatids will form two types of functional gametes with equal frequency: $+\,bv\,t$ and $v+T$. Expected zygotes are: $\frac{1}{2}+bv\,t/v\,bv\,t$ = brevis, homozygous for the normal chromosome order and $\frac{1}{2}v+T/v\,bv\,t$ = virescent, heterozygous for the translocation.

**9.8.** Shrunken endosperm of corn is governed by a recessive gene *sh* and waxy endosperm by another recessive *wx*. Both of these loci are linked on chromosome 9. A plant which is heterozygous for a translocation involving chromosomes 8 and 9 and which developed from a plump, starchy kernel is pollinated by a plant from a shrunken, waxy kernel with normal chromosomes. The progeny are

171 shrunken, starchy, normal ear
205 plump, waxy, semisterile ear
82 plump, starchy, normal ear
49 shrunken, waxy, semisterile ear
17 shrunken, starchy, semisterile ear
40 plump, waxy, normal ear
6 plump, starchy, semisterile ear
3 shrunken, waxy, normal ear.

(*a*) How far is each locus from the point of translocation? (*b*) Diagram and label the pairing figure in the plump, starchy parent.

**Solution:**

(*a*) The point of translocation may be considered as a gene locus because it produces a phenotypic effect, namely semisterility. The conventional symbol for translocation is *T*, and *t* is used for the normal chromosome without a translocation. Gene order in the parents must be

$$\frac{+\ wx\ T}{sh\ +\ t} \times \frac{sh\ wx\ t}{sh\ wx\ t}$$

in order for double crossovers to produce the least frequent phenotypes

$$+ + T \quad = \quad \text{plump, starchy, semisterile ear}$$

$$sh\ wx\ t \quad = \quad \text{shrunken, waxy, normal ear}$$

The map distances are calculated in the usual way for a three point testcross.

Distance *sh*-*wx* $= (82+49+6+3)/573 = 24.4$ map units.

Distance *wx*-*T* $= (17+40+6+3)/573 = 11.5$ map units.

Distance *sh*-*T* $= 24.4 + 11.5 = 35.9$ map units.

(*b*)

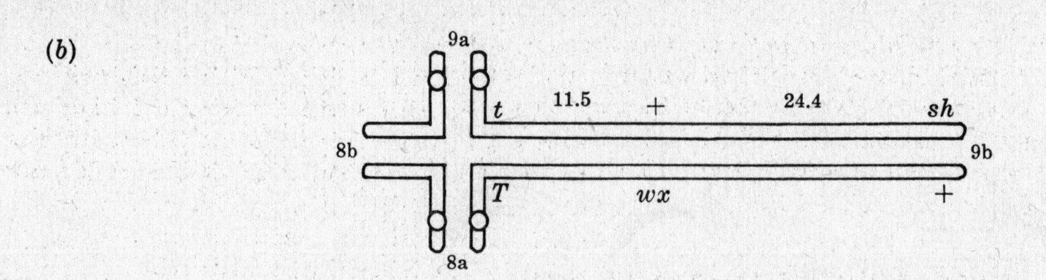


**9.9.** An inversion heterozygote possesses one chromosome in the normal order ● *a b c d e f g h* and one in the inverted order ● *a b f e d c g h*. A four strand double crossover occurs in the areas *f-e* and *d-c*. Diagram and label the first anaphase figures.

**Solution:**

A somewhat easier way to diagram the synapsing chromosomes when crossing over is only within the inversion is shown below. This is obviously not representative of the actual pairing figure. Let the crossover in the *c-d* region involve strands 2 and 3, and the crossover in the *e-f* region involve strands 1 and 4.

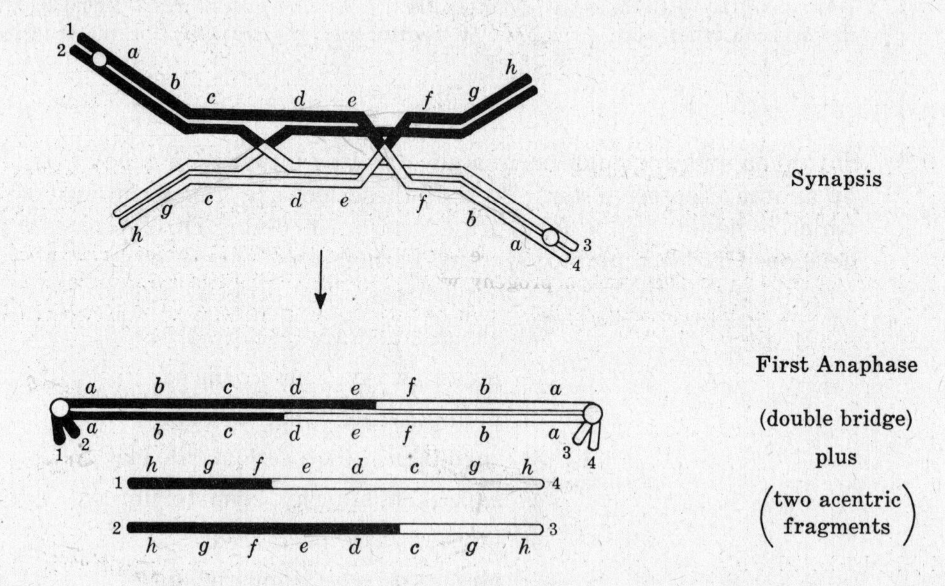


**9.10.** Eight regions of a dipteran chromosome are easily recognized cytologically and labeled *a* through *h*. Four different races within this species have the chromosomal orders as listed:

(1) *ahbdcfeg*, (2) *aedcfbhg*, (3) *ahbdgefc*, (4) *aefcdbhg*

Assuming that each race evolved by a single inversion from another race, show how the four races could have originated.

**Solution:**

An inversion in (1) involving *cfeg* produces the order for (3).

(1) *a h b d c f e g* ⟶ (3) *a h b d g e f c*

No single inversion in (3) can produce any of the other chromosomal orders. However, a different inversion in (1) can produce the order for (4).

(1) *a h b d c f e g* ⟶ (4) *a e f c d b h g*

Race 4 in turn can give rise to (2) by a single inversion.

(4) *a e f c d b h g* ⟶ (2) *a e d c f b h g*

If (1) was the original ancestor, the evolutionary pattern would be $2 \leftarrow 4 \leftarrow 1 \rightarrow 3$. If (2) was the original ancestor, the evolutionary pattern would be $2 \rightarrow 4 \rightarrow 1 \rightarrow 3$. If (3) was the original ancestor, the evolutionary pattern would be $3 \rightarrow 1 \rightarrow 4 \rightarrow 2$. If (4) was the original ancestor, the evolutionary pattern would be $2 \leftarrow 4 \rightarrow 1 \rightarrow 3$. Since we do not know which of the four was the original ancestor, we can briefly indicate all of these possibilities by using double-headed arrows: $2 \leftrightarrow 4 \leftrightarrow 1 \leftrightarrow 3$.

## VARIATION IN CHROMOSOME MORPHOLOGY

**9.11.** Yellow body color in *Drosophila* is produced by a recessive gene $y$ at the end of the X chromosome. A yellow male is mated to an attached-X female ($\widehat{XX}$) heterozygous for the $y$ allele. Progeny are of two types, yellow females and wild type females. What insight does this experiment offer concerning the stage (2 strand or 4 strand) at which crossing over occurs?

**Solution:**

Let us assume that crossing over occurs in the two strand stage, i.e. before the chromosome replicates into two chromatids.

The yellow male produces gametes with either a $y$-bearing X chromosome or one with the Y chromosome which is devoid of genetic markers. Trisomic X ($\widehat{XX}X$) flies seldom survive (superfemales). Those with $\widehat{XX}Y$ will be viable heterozygous wild type attached-X females. Crossing over fails to produce yellow progeny when it occurs in the two strand stage.

Let us assume that crossing over occurs after replication of the chromosome, i.e. in the four strand stage:

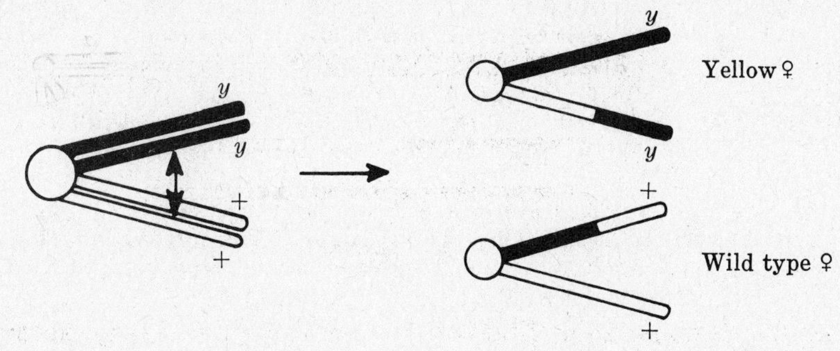


The appearance of yellow females in the progeny is proof that crossing over occurs in the four strand stage.

**9.12.** Data from *Drosophila* studies indicate that non-crossover (NCO) rings are recovered in equal frequencies with NCO rods from ring-rod heterozygotes. What light does this information shed on the occurrence of sister strand crossing over?

**Solution:**

Let us diagram the results of a sister strand crossover in a rod and in a ring chromosome.

(*a*) Rod Chromosome

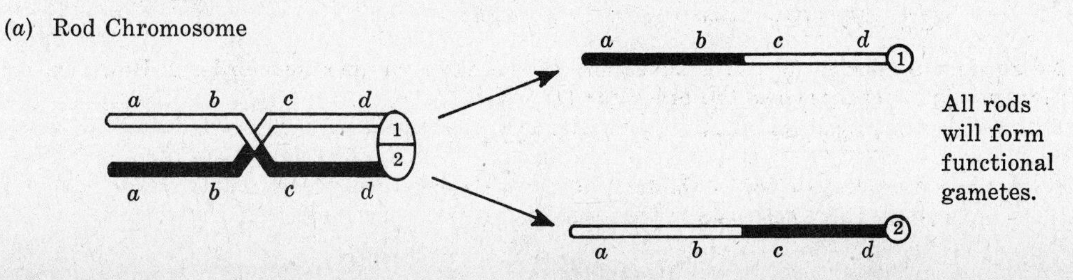

(*b*) Ring Chromosome

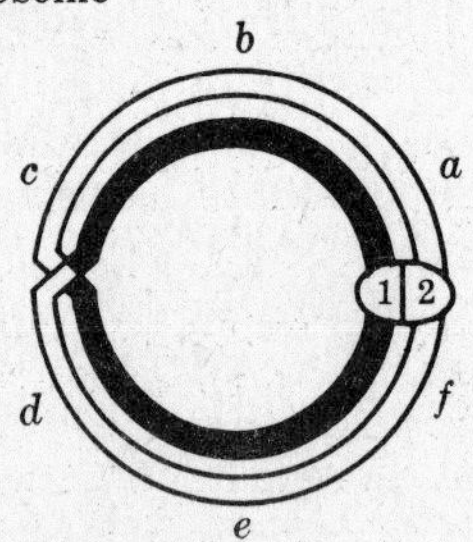


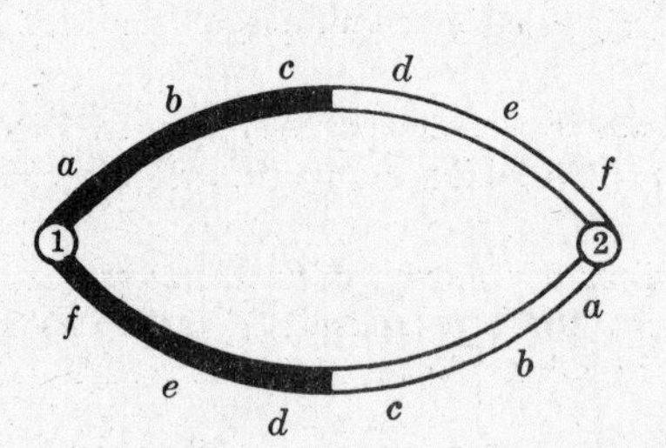


The double bridge at anaphase will rupture and produce non-functional gametes with duplications or deficiencies. These would fail to be recovered in viable offspring. The fact that both rings and rods are recovered with equal frequency argues against the occurrence of sister-strand crossing over.

Modern techniques (involving autoradiography with labeled thymidine or fluorescence microscopy of cultured cells that have incorporated 5-bromodeoxyuridine in place of thymine) reveal that some sister-strand exchanges occur by a repair mechanism when DNA is damaged. One of the initiating steps that transforms a normal cell to a cancer cell is DNA damage. Hence, screening chemicals for their ability to induce sister-strand exchanges is one method for detecting potential cancer-inducing agents (carcinogens).

## Supplementary Problems

### VARIATION IN CHROMOSOME NUMBER

**9.13.** Abyssinian oat (*Avena abyssinica*) appears to be a tetraploid with 28 chromosomes. The common cultivated oat (*Avena sativa*) appears to be a hexaploid in this same series. How many chromosomes does the common oat possess?

**9.14.** If two alleles, $A$ and $a$, exist at a locus, five genotypic combinations can be formed in an autotetraploid: quadruplex ($AAAA$), triplex ($AAAa$), duplex ($AAaa$), simplex ($Aaaa$), nulliplex ($aaaa$). Assume $A$ exhibits xenia over $a$. For each of these five genotypes determine the expected phenotypic ratio ($A : a$) when (*a*) the locus is tightly linked to its centromere (chromosomal assortment) and the genotype is selfed, (*b*) the locus is assorting chromosomally and the genotype is testcrossed, (*c*) the locus is far from its centromere so that chromatids assort independently and the genotype is selfed, (*d*) the locus assorts by chromatids and the genotype is testcrossed.

**9.15.** The loci of genes $A$ and $B$ are on different chromosomes. A dihybrid autotetraploid plant of genotype $AAaaBBbb$ is self pollinated. Assume that only diploid gametes are formed and that the loci of $A$ and $B$ are very close to their respective centromeres (chromosomal segregation). Find the phenotypic expectations of the progeny.

**9.16.** The European raspberry (*Rubus idaeus*) has 14 chromosomes. The dewberry (*Rubus caesius*) is a tetraploid with 28 chromosomes. Hybrids between these two species are sterile $F_1$ individuals. Some unreduced gametes of the $F_1$ are functional in backcrosses. Determine the chromosome number and level of ploidy for each of the following: (*a*) $F_1$, (*b*) $F_1$ backcrossed to *R. idaeus*, (*c*) $F_1$ backcrossed to *R. caesius*, (*d*) chromosome doubling of $F_1$ (*R. maximus*).

**9.17.** There are 13 pairs of chromosomes in Asiatic cotton (*Gossypium arboreum*) and also 13 pairs in an American species *G. thurberi*. Interspecific crosses between *arboreum* and *thurberi* are sterile because of highly irregular chromosome pairing during meiosis. The American cultivated cotton (*Gossypium hirsutum*) has 26 pairs of chromosomes. Crosses of *arboreum* × *hirsutum* or *thurberi* × *hirsutum* produce triploids with 13 bivalents (pairs of chromosomes) and 13 univalents (single unpaired chromosomes). How can this cytological information be used to interpret the evolution of *hirsutum*?

**9.18.** The flinty endosperm character in maize is produced whenever two or all three of the alleles in this triploid tissue are $F$. In the presence of its alternative allele $F'$ in double or triple dose, a floury endosperm is produced. White endosperm color is produced by a triple dose of a recessive allele $y$, its dominant allele $Y$ exhibiting xenia and producing yellow endosperm. The loci of $F$ and $Y$ assort independently. (*a*) In crosses between parents of genotype $FF'Yy$, what phenotypic ratio is expected in the progeny seed? (*b*) Pollen from a plant of genotype $FF'Yy$ is crossed onto a plant of genotype $FFyy$. Compare the phenotypic ratios produced by this cross with its reciprocal cross.

**9.19.** The diploid number of the garden pea is $2n = 14$. (*a*) How many different trisomics could be formed? (*b*) How many different double trisomics could be formed?

**9.20.** The diploid number of an organism is 12. How many chromosomes would be expected in (*a*) a monosomic, (*b*) a trisomic, (*c*) a tetrasomic, (*d*) a double trisomic, (*e*) a nullisomic, (*f*) a monoploid, (*g*) a triploid, (*h*) an autotetraploid?

**9.21.** Sugary endosperm of corn is regulated by a recessive gene *s* on chromosome IV and starchy endosperm by its dominant allele *S*. Assuming $n + 1$ pollen grains are non-functional, predict the genotypic and phenotypic ratios of endosperm expected in the progeny from the cross of (*a*) diploid *ss* pollinated by trisomic-IV of genotype *SSs*, (*b*) diploid *Ss* pollinated by trisomic-IV of genotype *SSs*.

**9.22.** A dominant gene $w^+$ produces yellow flowers in a certain plant species and its recessive allele *w* produces white flowers. Plants trisomic for the chromosome bearing the color locus will produce $n$ and $n+1$ functional female gametes, but viable pollen has only the $n$ number. Find the phenotypic ratio expected from each of the following crosses:

| | Seed Parent | | Pollen Parent |
|---|---|---|---|
| (*a*) | $++w$ | × | $++w$ |
| (*b*) | $+ww$ | × | $++w$ |
| (*c*) | $++w$ | × | $+w$ |
| (*d*) | $+ww$ | × | $+w$ |

**9.23.** Shrunken endosperm is the product of a recessive gene *sh* on chromosome III of corn; its dominant allele *Sh* produces full, plump kernels. Another recessive gene *pr* on chromosome V gives red color to the aleurone, and its dominant allele *Pr* gives purple. A diploid plant of genotype *Sh/sh, Pr/pr* was pollinated by a plant trisomic for chromosome III of genotype *Sh/Sh/sh, Pr/pr*. If $n+1$ pollen grains are non-functional, determine the phenotypic ratio expected in the progeny endosperms.

**9.24.** Normal women possess two sex chromosomes (XX) and normal men have a single X chromosome plus a Y chromosome which carries male determiners. Rarely a woman is found with marked abnormalities of primary and secondary sexual characteristics, having only one X chromosome (XO). The phenotypic expressions of this monosomic-X state is called Turner's syndrome. Likewise, men are occasionally discovered with an XXY constitution exhibiting corresponding abnormalities called Klinefelter's syndrome. Color blindness is a sex-linked recessive trait. (*a*) A husband and wife both had normal vision, but one of their children was a color blind Turner girl. Diagram this cross, including the gametes which produced this child. (*b*) In another family the mother is color blind and the father has normal vision. Their child is a Klinefelter with normal vision. What gametes produced this child? (*c*) Suppose the same parents in part (*b*) produced a color blind Klinefelter. What gametes produced this child? (*d*) The normal diploid number for humans is 46. A trisomic condition for chromosome 21 results in Down's syndrome. At least one case of Down-Klinefelter has been recorded. How many chromosomes would this individual be expected to possess?

## VARIATION IN ARRANGEMENT OF CHROMOSOME SEGMENTS

**9.25.** Colorless aleurone of corn kernels is a trait governed by a recessive gene *c* and is in the same linkage group (IX) with another recessive gene *wx* governing waxy endosperm. In 1931 Creighton and McClintock found a plant with one normal IX chromosome, but its homologue had a knob on one end and a translocated piece from another chromosome on the other end. A dihybrid colored, starchy plant with the heteromorphic IX chromosome shown below was testcrossed to a colorless, waxy plant with normal chromosomes. The results of this experiment provided cytological proof that genetic crossing over involves an actual physical exchange between homologous chromosome segments. Diagram the results of this cross, showing all genotypes and phenotypes.

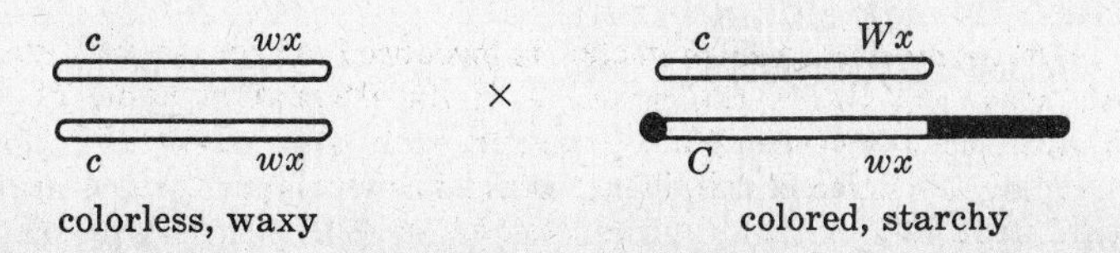

9.26. Nipple shaped tips on tomato fruit is the phenotypic expression of a recessive gene *nt* on chromosome V. A heterozygous plant (*Nt*/*nt*) which is also heterozygous for a reciprocal translocation involving chromosomes V and VIII is testcrossed to a plant with normal chromosomes. The progeny were: 48 normal fruit, fertile : 19 nipple fruit, fertile : 11 normal fruit, semisterile : 37 nipple fruit, semisterile. What is the genetic position of the locus of gene *Nt* with respect to the point of translocation?

9.27. Given a pericentric inversion heterozygote with one chromosome in normal order (1 2 3 4 . 5 6 7 8) and the other in the inverted order (1 5 . 4 3 2 6 7 8), diagram the first anaphase figure when a 4-strand double crossover occurs involving the regions between 4 and the centromere (.) and between the centromere and 5.

9.28. A four-strand double crossover occurs in an inversion heterozygote. The normal chromosome order is (. 1 2 3 4 5 6 7 8); the inverted chromosome order is (. 1 2 7 6 5 4 3 8). One crossover is between 1 and 2 and the other is between 5 and 6. Diagram and label the first anaphase figures.

9.29. Diagram and label the first anaphase figure produced by an inversion heterozygote whose normal chromosome is (*. a b c d e f g h*) and with the inverted order (*. a b f e d c g h*). Assume that a two-strand double crossover occurs in the regions *c-d* and *e-f*.

9.30. A chromosome with segments in the normal order is (*. a b c d e f g h*). An inversion heterozygote has the abnormal order (*. a b f e d c g h*). A three-strand double crossover occurs involving the regions between *a* and *b* and between *d* and *e*. Diagram and label the first and second anaphase figures.

9.31. Given the pairing figure for an inversion heterozygote with 3 crossovers as indicated on the right, diagram the first anaphase.

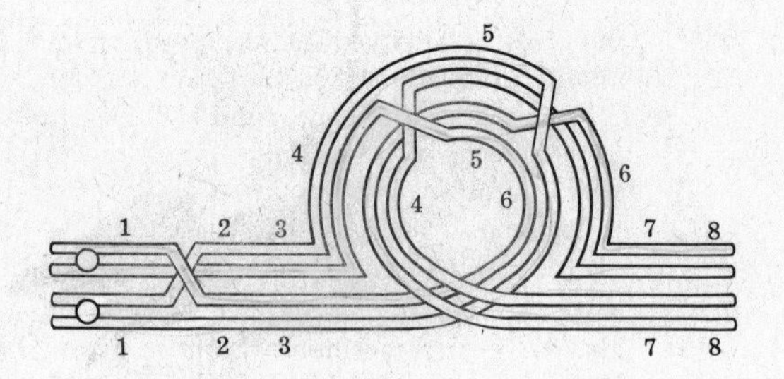


9.32. Four races of a species are characterized by variation in the segmental order (*a* through *h*) of a certain chromosome.

Race 1: *a b c d e f g h* Race 3: *g f e a c d b h*
Race 2: *g f e b d c a h* Race 4: *a c b d e f g h*

A fifth race, with still a different chromosomal order, is postulated to have existed in the past but is now extinct. Explain the evolutionary origin of these races in terms of single inversion differences. *Hint:* See Problem 9.10.

9.33. A species of the fruit fly is differentiated into five races on the basis of differences in the banding patterns of one of its giant chromosomes. Eight regions of the chromosome are designated *a* through *h*. If each of these races is separated by a single overlapping inversion, devise a scheme to account for the evolution of the five races: (1) *a d g h f c b e*, (2) *f h g d a c b e*, (3) *f h c a d g b e*, (4) *f h g b c a d e*, (5) *f a d g h c b e*

## VARIATION IN THE NUMBER OF CHROMOSOME SEGMENTS

9.34. In higher animals, even very small deficiencies, when homozygous, are usually lethal. A recessive gene *w* in mice results in an abnormal gait called "waltzing". A waltzing male was crossed to several homozygous normal females. Among several hundred offspring one was found to be a waltzer female. Presumably, a deficiency in the chromosome carrying the $w^+$ allele caused the waltzing trait to appear as pseudodominant. The pseudodominant waltzer female was then crossed to a homozygous normal male and produced only normal offspring. (*a*) List two possible genotypes for the normal progeny from the above cross. (*b*) Suppose that two males, one of each genotype produced in part (*a*), were backcrossed to their pseudodominant mother and each produced 12 zygotes. Assuming that homozygosity for the deletion is lethal, calculate the expected combined number of waltzer and normal progeny.

## VARIATION IN CHROMOSOME MORPHOLOGY

9.35. Vermilion eye color in *Drosophila* is a sex-linked recessive condition; bar eye is a sex-linked dominant condition. An attached-X female with vermilion eyes, also having a Y chromosome ($\widehat{XX}Y$), is mated to a bar eyed male. (*a*) Predict the phenotypic ratio which is expected in the $F_1$ flies. (*b*) How much death loss is anticipated in the $F_1$ generation? (*c*) What phenotypic ratio is expected in the $F_2$?

**9.36.** Two recessive sex-linked traits in *Drosophila* are garnet eye ($g$) and forked bristle ($f$). The attached-X chromosomes of females heterozygous for these genes are diagrammed on the right.

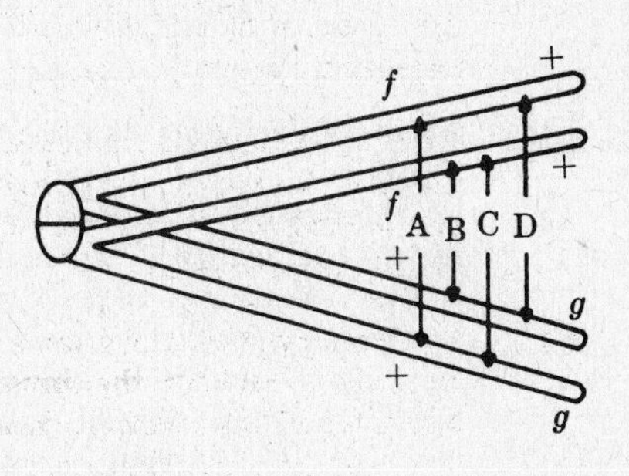


A crossover between two chromatids attached to the same centromere is called a reciprocal exchange; a crossover between two chromatids attached to different centromeres is a non-reciprocal exchange. Approximately 7% of the daughters from these attached-X females were $++/fg$, 7% were $f+/++$, 7% were $fg/+g$ and the remainder were $f+/+g$. (*a*) Which of the single exchanges (A, B, C, or D in the diagram) could produce the daughters (1) $++/fg$ and $f+/+g$, (2) $f+/++$ and $fg/+g$? (*b*) Are chromatids attached to the same centromere more likely to be involved in an exchange than chromatids attached to different centromeres? (*c*) Does the fact that neither homozygous wild type nor garnet-forked progeny were found shed any light on the number of chromatids which undergo exchange at any one locus?

**9.37.** Given the ring homozygote at the left (below), diagram the first anaphase figure when crossovers occur at positions (*a*) A and B, (*b*) A and C, (*c*) A and D.

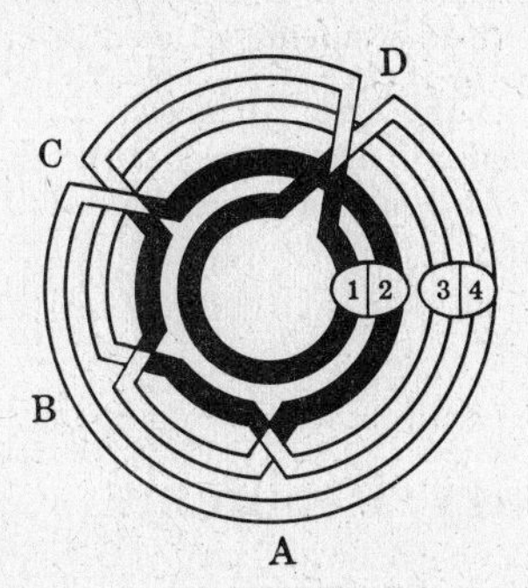


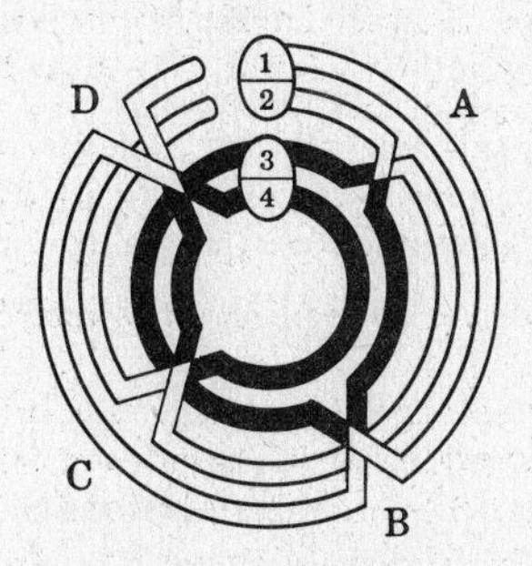


**9.38.** Given the ring-rod heterozygote at the right (above), diagram the first anaphase figure when crossovers occur at positions (*a*) A and B, (*b*) A and D.

## HUMAN CYTOGENETICS

**9.39.** Meiotic non-disjunction of the sex chromosomes in either parent can produce a child with Klinefelter's syndrome (XXY) or Turner's syndrome (XO). Color blindness is due to a sex-linked recessive gene. (*a*) If a color blind woman and man with normal vision produce a color blind Klinefelter child, in which parent did the non-disjunctional event occur? (*b*) If a heterozygous woman with normal vision and a man with normal vision produce a color blind Klinefelter child, how can this be explained?

**9.40.** Explain what type of abnormal sperm unites with a normal egg to produce an XYY offspring. Specifically, how does such an abnormal gamete arise?

**9.41.** In mosaics of XX and XO cell lines, the phenotype may vary from complete Turner's syndrome to a completely normal appearing female. Likewise, in XO/XY mosaics, the phenotypic variation ranges from complete Turner's syndrome to a normal appearing (but infertile) male. How can these variations be explained?

**9.42.** *Mosaicism* is the presence in an individual of two or more cell lines of different chromosomal constitution, each cell line being derived from the same zygote. In contrast, fusion of cell lines from different zygotes produces a *chimera.* Mosaicism results from abnormal postzygotic (mitotic) divisions of three kinds: (1) non-disjunction during the first cleavage division of the zygote, (2) non-disjunction during later mitotic divisions and (3) anaphase lag, in which one member of a chromosome pair fails to segregate chomatids from the metaphase plate, and that chromatid fails to be included in the daughter cell nuclei (the entire chromosome is thus lost). Assuming that non-disjunction of chromatids affects only one member of a pair of chromosomes of the diploid set, (*a*) specify the mosaic karyotypes expected from non-disjunction during the first cleavage division of a zygote. (*b*) If the first cleavage division is normal, but the second cleavage division involves a non-disjunctional event, what kind of mosaic is expected? (*c*) What kind of mosaic results from anaphase lag of the sex chromosomes in females? (*d*) What kind of mosaic results from anaphase lag of the sex chromosomes in males?

**9.43.** Suppose that part of the short arm of one chromosome 5 becomes non-reciprocally attached to the long arm end of one chromosome 13 in the diploid set. This is considered to be a "balanced translocation" because essentially all of the genetic material is present and the phenotype is normal. One copy of the short arm of chromosome 5 produces *cri du chat* syndrome; three copies lead to early postnatal death. If such a translocation individual has children by a chromosomally normal partner, predict the (*a*) chromosomal and (*b*) phenotypic expectations.

**9.44.** About 2% of patients with Down's syndrome have a normal chromosome number of 46. The extra chromosome 21 has been non-reciprocally translocated onto another autosome of the D or G group. These individuals are referred to as *translocation mongols*, and because this condition tends to be hereditary, it is also called *familial mongolism*. (*a*) Suppose that one phenotypically normal parent has 45 chromosomes, one of which is a translocation of the centromere and long arm of a D-group chromosome (either 14 or 15) and the long arm minus the centromere of a G-group chromosome (21). The short arms of each chromosome (presumably carrying no vital genes) are lost in previous cell divisions. If gametes from this translocated parent unite with those from a normal diploid individual, predict the chromosomal and phenotypic expectations in their progeny. (*b*) Assuming that in one parent the translocation is between chromosomes 21 and 22, that the centromere of the translocation is that of chromosome 22 (*like* centromeres go to opposite poles), and that the other parent is a normal diploid, predict the chromosomal and phenotypic expectations in their children. (*c*) Make the same analysis as in part (*b*), assuming that the centromere of the 21/22 translocation chromosome is that of chromosome 21. (*d*) Assuming that in one parent the translocation involves 21/21 and the other parent is a normal diploid, predict the chromosomal and phenotypic expectations in their children. (*e*) Among the live offspring of parts (*c*) and (*d*), what are the risks of having a Down's child?

**9.45.** The photograph accompanying this problem is at the back of the book. It shows the chromosomes from a human cell. Cut out the chromosomes and construct an idiogram. Do not look at the answer until you have solutions to the following questions. (*a*) Is the specimen from a male or a female? (*b*) What possible kinds of chromosomal abnormalities may be present in this patient?

## Answers to Supplementary Problems

**9.13.** 42

**9.14.**

| Genotype | (*a*) | (*b*) | (*c*) | (*d*) |
|---|---|---|---|---|
| quadruplex | all $A$ | all $A$ | all $A$ | all $A$ |
| triplex | all $A$ | all $A$ | 783 $A$ : 1 $a$ | 27 $A$ : 1 $a$ |
| duplex | 35 $A$ : 1 $a$ | 5 $A$ : 1 $a$ | 20.8 $A$ : 1 $a$ | 3.7 $A$ : 1 $a$ |
| simplex | 3 $A$ : 1 $a$ | 1 $A$ : 1 $a$ | 2.48 $A$ : 1 $a$ | 0.87 $A$ : 1 $a$ |
| nulliplex | all $a$ | all $a$ | all $a$ | all $a$ |

**9.15.** 1225 $AB$ : 35 $Ab$ : 35 $aB$ : 1 $ab$

**9.16.** (*a*) 21, triploid (*b*) 28, tetraploid (*c*) 35, pentaploid (*d*) 42, hexaploid

**9.17.** Half of the chromosomes of *hirsutum* have homology with *arboreum*, and the other half with *thurberi*. Doubling the chromosome number of the sterile hybrid (*thurberi* × *arboreum*) could produce an amphidiploid with the cytological characteristics of *hirsutum*.

**9.18.** (*a*) $\frac{3}{8}$ flinty, yellow : $\frac{1}{8}$ flinty, white : $\frac{3}{8}$ floury, yellow : $\frac{1}{8}$ floury, white

(*b*)

| | Original Cross | Reciprocal Cross |
|---|---|---|
| flinty, white | $\frac{1}{2}$ | $\frac{1}{4}$ |
| flinty, yellow | $\frac{1}{2}$ | $\frac{1}{4}$ |
| floury, white | | $\frac{1}{4}$ |
| floury, yellow | | $\frac{1}{4}$ |

**9.19.** (*a*) 7 (*b*) 21

**9.20.** (*a*) 11 (*b*) 13 (*c*) 14 (*d*) 14 (*e*) 10 (*f*) 6 (*g*) 18 (*h*) 24

**9.21.** (*a*) 2 *Sss* (starchy) : 1 *sss* (sugary) (*b*) $\frac{1}{3}$ *SSS* : $\frac{1}{6}$ *SSs* : $\frac{1}{3}$ *Sss* : $\frac{1}{6}$ *sss* : $\frac{5}{6}$ starchy : $\frac{1}{6}$ sugary

**9.22.** (*a*) 17 yellow : 1 white (*b*) 5 yellow : 1 white (*c*) 11 yellow : 1 white (*d*) 3 yellow : 1 white

**9.23.** $\frac{15}{24}$ plump, purple : $\frac{5}{24}$ plump, red : $\frac{3}{24}$ shrunken, purple : $\frac{1}{24}$ shrunken, red

**9.24.** (*a*) P: $X^CX^c \times X^CY$; gametes: ($X^c$) (O); $F_1$: $X^cO$ (*b*) ($X^c$) ($X^CY$) (*c*) ($X^cX^c$) (Y) (*d*) 48

**9.25.**

| | Non-crossovers | | Crossovers | |
|---|---|---|---|---|
| | *c* *Wx* | ● *C* *wx* | *c* *wx* | ● *C* *Wx* |
| *c* *wx* | *c* *Wx* / *c* *wx*<br>colorless, starchy | ● *C* *wx* / *c* *wx*<br>colored, waxy | *c* *wx* / *c* *wx*<br>colorless, waxy | ● *C* *Wx* / *c* *wx*<br>colored, starchy |

**9.26.** 26.1 map units from the point of translocation.

**9.27.** **9.28.** **9.29.**

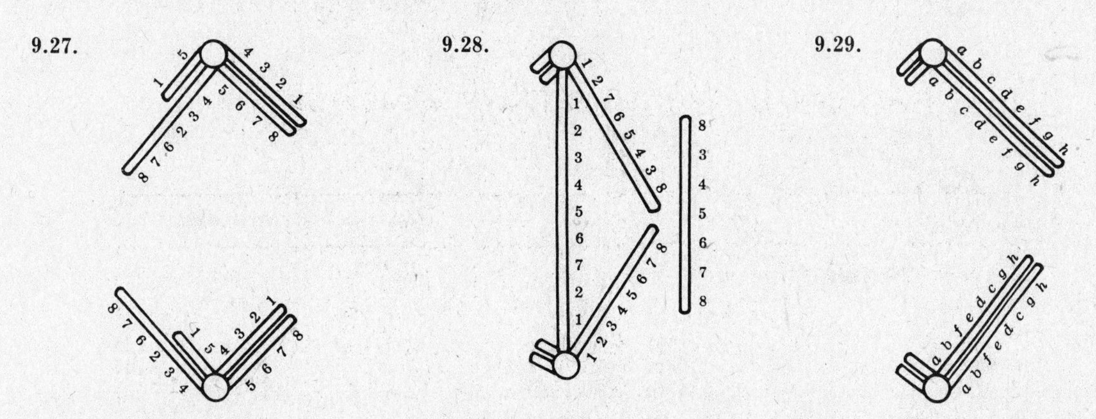

**9.30.** First anaphase: a diad, a loop chromatid and an acentric fragment; second anaphase: the diad splits into two monads, the loop forms a bridge and the acentric fragment becomes lost.

**9.31.** Two loop chromatids and two acentric fragments.

**9.32.** If the order of the extinct race (5) was *g f e d b c a h* or *a c d b e f g h*, then: $1 \leftrightarrow 4 \leftrightarrow 5 \leftrightarrow 2 \leftrightarrow 3$.

**9.33.**

```
4       1
  ↘   ↙
    2
  ↙   ↘
5       3
```

**9.34.** (*a*) +/*w* and +/(−) (heterozygous deficiency) (*b*) 9 waltzers : 12 normals

**9.35.** (*a*) All daughters have vermilion eyes ($\widehat{XX}Y$); all sons have bar eyes (XY). (*b*) 50% death loss; nullo-X is lethal (YY); superfemales ($\widehat{XX}X$) usually die. (*c*) Same as part (*a*).

**9.36.** (*a*) (1) D or C (2) B or A (*b*) No. Reciprocal vs. non-reciprocal exchanges are occurring in a 1 : 1 ratio, indicating that chromatids attached to the same centromere are involved in an exchange with the same frequency as chromatids attached to different centromeres. Daughters with genotype $f+/+g$ resulting from single crossovers of type C cannot be distinguished from non-exchange chromatids. (*c*) Two exchanges in the garnet-forked region involving all four strands, as well as one non-reciprocal exchange between $f$ and the centromere, are required to give homozygous wild type and garnet-forked daughters. Their absence is support for the assumption that only two of the four chromatids undergo exchange at any one locus.

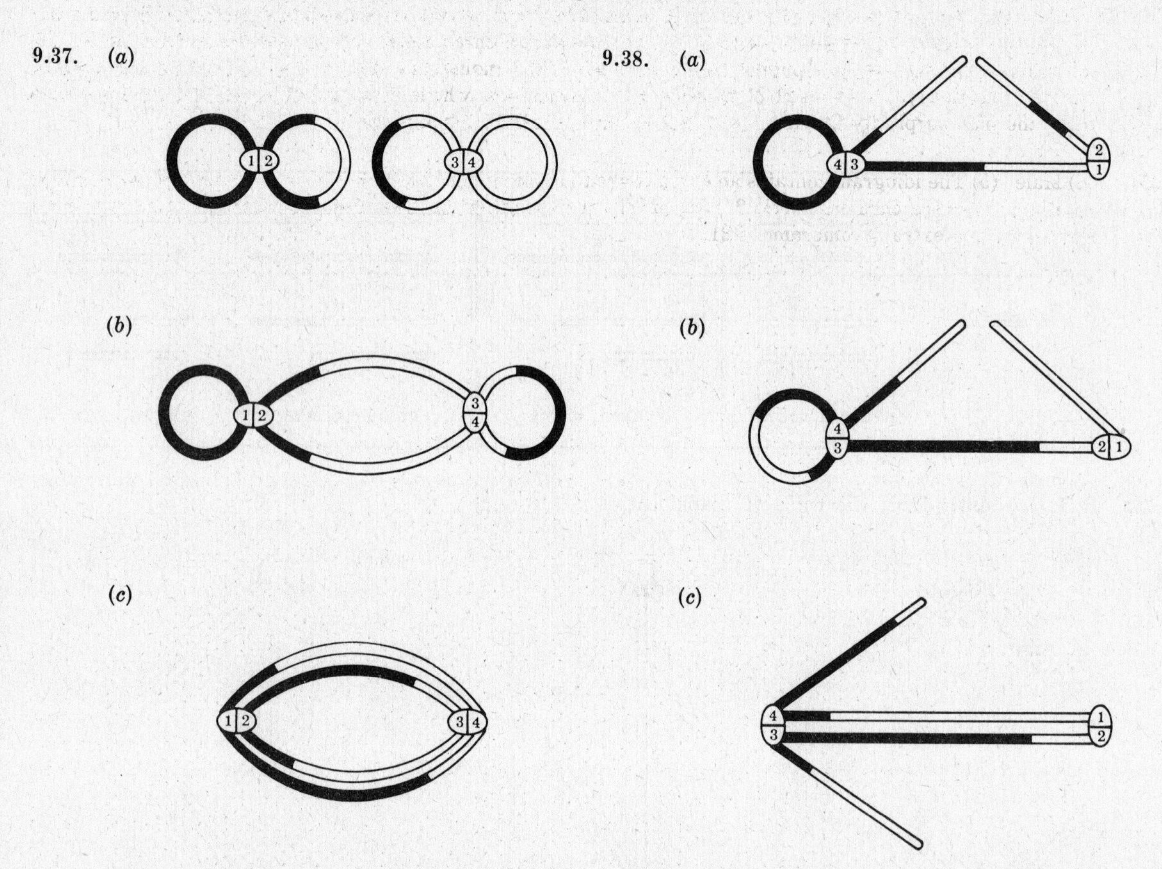


**9.39.** (*a*) Either non-disjunction of the two X chromosomes occurred in the mother in the first meiotic division or non-disjunction of the two sister chromatids occurred in the second meiotic division. (*b*) Non-disjunction during the second meiotic division of the sister chromatids of the X chromosome bearing the recessive color blind gene would produce an egg with two X chromosomes bearing only the color blind alleles. Alternatively, if crossing over occurs between the centromere and the color blind locus and is followed by non-disjunction of the X chromosomes at the first meiotic division, one of the four meiotic products would be expected to contain two recessive color blind alleles.

**9.40.** A sperm bearing two Y chromosomes is produced by non-disjunction of the Y sister chromatids during the second meiotic division. The other product of that same non-disjunctional second meiotic division would contain no sex chromosome; when united with a normal egg, an XO Turner female would be expected.

**9.41.** If mitotic non-disjunction occurs early in embryogenesis, mosaicism is likely to be widespread throughout the body. If it occurs late in embryogenesis, mosaicism may be limited to only one organ or to one patch of tissue. If chromosomally abnormal cells are extensive in reproductive tissue or in endocrine tissues responsible for gamete and/or sex hormone production, the effects on sterility are likely to be more intensively expressed.

**9.42.** (*a*) Half of the individual's cells should be trisomic ($2n + 1 = 47$); the other half should be monosomic ($2n - 1 = 45$). (*b*) Three cell lines are established (45/46/47). Each line should "breed true", barring further mitotic abnormalities. (*c*) XX/XO; sex-chromatin positive, Turner's syndrome. (*d*) XY/XO; may resemble Turner's syndrome or be a hermaphrodite with physical characteristics of both sexes.

**9.43.** (*a*) 1 normal karyotype : 1 balanced translocation : 1 deficient for short arm of chromosome 5 : 1 with three copies of the short arm of 5. (*b*) 2 normal : 1 *cri-du-chat* syndrome : 1 early childhood death

**9.44.** (*a*) 1 chromosomally and phenotypically normal ($2n = 46$) : 1 translocation carrier, phenotypically normal ($2n - 1 = 45$) : 1 monosomic ($2n = 45$) for a G-group chromosome (incompatible with life; aborted early in pregnancy) : 1 translocation Down's trisomic for the long arm of chromosome 21 ($2n = 46$). Among the live born offspring we expect 1/3 chromosomally normal : 1/3 translocation carriers : 1/3 Down's syndrome. (*b*) 1 chromosomally and phenotypically normal ($2n = 46$) : 1 that is a 21/22 translocation carrier, phenotypically normal ($2n - 1 = 45$) : 1 monosomic for chromosome 21 and aborted early in pregnancy ($2n - 1 = 45$) : 1 with a 21/22 translocation who is essentially trisomic for the long arms of 21 ($2n = 46$) and phenotypically Down's syndrome. (*c*) 1 chromosomally and phenotypically normal ($2n = 46$) : 1 that is a 21/22 translocation carrier, phenotypically normal ($2n - 1 = 45$) : 1 monosomic for 22 and aborted early in pregnancy ($2n - 1 = 45$) : 1 with a 21/22 translocation chromosome who is essentially trisomic for the long arms of 22 ($2n = 46$), phenotype unspecified. (*d*) 1 monosomic ($2n - 1 = 45$) for 21 and aborted early in pregnancy : 1 with a 21/21 translocation chromosome who is essentially trisomic for the long arms of 21 and phenotypically Down's. (*e*) 1 in 3 for part (*c*); 100% for part (*d*)

**9.45.** (*a*) Male (*b*) The idiogram contains an extra G-group chromosome ($2n + 1 = 47$). It cannot be determined whether the extra chromosome is 21, 22, or Y. If the patient has the physical characteristics of Down's syndrome, the extra chromosome is 21.

# Chapter 10

## Cytoplasmic Factors

### MATERNAL EFFECTS

Some attributes of progeny are not expressions of their own genes, but rather those of the maternal parent. Such effects may be ephemeral or may persist throughout the lifespan of the individual. The substances which produce maternal effects are not self-perpetuating, but must be synthesized anew for each generation of progeny by the appropriate maternal genotype.

**Example 10.1.** A dominant gene $K$ in the meal moth *Ephestia* produces a hormone-like substance called *kynurenine* which is involved in pigment synthesis. The recessive genotype $kk$ is kynurenineless and cannot synthesize pigment. Females with $K$ produce eggs containing a small amount of kynurenine. For a short time during early development, a larva may use this supply of kynurenine to develop pigment even though its own genotype is $kk$. The color fades as the larvae grow older because the maternally supplied kynurenine becomes depleted.

**Example 10.2.** The direction in which the shell coils in the snail *Limnaea* can be *dextral* like a right hand screw or *sinistral* like a left hand screw. The maternal genotype organizes the cytoplasm of the egg in such a way that cleavage of the zygote will follow either of these two patterns regardless of the genotype of the zygote. If the mother has the dominant gene $s^+$, all her progeny will coil dextrally; if she is of genotype $ss$, all her progeny will coil sinistrally. This coiling pattern persists for the life of the individual.

### PLASMAGENES

The behavior of some genetic elements indicates that they are not located on chromosomes. The smallest heritable extrachromosomal unit is called a *plasmagene*. All of the plasmagenes of a cell constitute the *plasmon*. Cytoplasm contains several organelles which may have physical continuity from one generation to the next. It is assumed that they do not arise *de novo* (i.e. manufactured by the cell from elementary particles), but are produced by replication of a preexisting organelle. The centrioles, which organize the polar regions during division, probably reproduce themselves as do the basal granules (sometimes called blepharoplasts or kinetosomes) associated with cilia or flagella. Mitochondria and plastids probably develop from undifferentiated structures such as proplastids. The fact that DNA has been found in mitochondria and plastids is one line of evidence which indicates that these organelles could carry genetic information. Other organelles such as microsomes, lysosomes, Golgi bodies, etc., may also be the residence of plasmagenes, but evidence at the present does not delimit the site of extranuclear genetic factors to these structures.

Traits with an extranuclear basis are identified by the accumulated evidence from a number of diagnostic criteria, as exemplified in the following cases.

(1) Differences in reciprocal crosses which cannot be attributed to sex-linkage or some other chromosomal basis tend to implicate extranuclear factors.

(*a*) When progeny shows only the characteristics of the female parent, *maternal inheritance* is operative. If this can be attributed to unequal cytoplasmic contributions of male and female parents, then plasmagenic inheritance is suspect.

**Example 10.3.** In higher plants, pollen usually contributes very little, if any, cytoplasm to the zygote. Most of the cytoplasmic elements are transmitted through the maternal parent. In the plant called "four-o'clock" (*Mirabilis jalapa*), there may be normal green, pale green, and variegated branches due to two types of chloroplasts. Plants grown from seeds which developed on normal green branches (with all normal chloroplasts) will all be normal green; those which developed on pale green branches (with abnormal chloroplasts) will all be pale green; those on variegated branches (with both normal and abnormal chloroplasts) will segregate green, pale green, and variegated in irregular ratios. The type of pollen used has no effect in this system. The irregularity of transmission from variegated branches is understandable if plasmagenes exist in the chloroplasts, because there is no mechanism to ensure the regular distribution of chloroplasts to daughter cells as there is for chromosomes.

(*b*) The unidirectional inheritance of a trait which cannot be attributed to unequal cytoplasmic contributions from parental gametes may nonetheless involve cytoplasmic factors.

**Example 10.4.** The uniting gametes of the single celled alga *Chlamydomonas reinhardi* (Fig. 5-1) are morphologically indistinguishable. One strain of the alga which is streptomycin resistant (*sr*) and of the "plus" mating type ($mt^+$) is crossed to a cell of "negative" mating type ($mt^-$) which is streptomycin sensitive (*ss*). All progeny are resistant, but the nuclear genes for mating type segregate as expected: $\frac{1}{2}mt^+, \frac{1}{2}mt^-$. The reciprocal cross $ss\,mt^+ \times sr\,mt^-$, again shows the expected segregation for mating type, but all progeny are sensitive. Repeated backcrossings of $sr\,mt^+$ to $ss\,mt^-$ fail to show segregation for resistance. It appears as though the plasmagenes of the $mt^-$ strain become lost in a zygote of $mt^+$. The mechanism which inactivates the plasmagenes of $mt^-$ in the zygote is unknown.

(2) Extranuclear factors may be detected by either the absence of segregation at meiosis (Example 10.5) or by segregation which fails to follow Mendelian laws (Example 10.6).

**Example 10.5.** Slow-growing yeast cells called *petites* lack normal activity of the respiratory enzyme cytochrome oxidase associated with the mitochondria. Petites can be maintained indefinitely in vegetative cultures through budding, but can sporulate only if crossed to wild type. When a haploid neutral petite cell fuses with a haploid wild type cell of opposite mating type, a fertile wild type diploid cell is produced. Under appropriate conditions, the diploid cell reproduces sexually (sporulates). The four ascospores of the ascus (Fig. 6-2) germinate into cells with a 1:1 mating type ratio (as expected for nuclear genes), but they are all wild type. The petite trait never appears again, even after repeated backcrossings of both mating types to petite. The mitochondrial factors for petite are able to perpetuate themselves vegetatively, but are "swamped", lost or permanently altered in the presence of wild type factors. Neutral petite behaves the same in reciprocal crosses regardless of mating type, and in this respect is different from the streptomycin resistance factors in *Chlamydomonas* (Example 10.4).

**Example 10.6.** Another type of petite in yeast called *suppressive* may segregate, but in a manner different from chromosomal genes. When haploid suppressive petites are crossed to wild types and each zygote is grown vegetatively as a diploid strain, both petites and wild types may appear, but in frequencies which are hardly Mendelian, varying from 1% to 99% petites. Diploid wild type cells may sporulate producing only wild type ascospores. By special treatment, all diploid zygotes can be made to sporulate. The majority of the ascospores thus induced germinates into petite clones. Some asci have 4, 3, 2, 1 or 0 petite ascospores, suggesting that environmental factors may alter their segregation pattern. Nuclear genes, such as mating type, maintain a 1 : 1 ratio in all asci.

(3) Repeated backcrossing of progeny to one of the parental types for several generations causes their chromosomal genetic endowment to rapidly approach 100% that of the parental line. The persistence of a trait in the progeny, when the backcross parent exhibits an alternative character, may be considered evidence for cytoplasmic factors.

**Example 10.7.** The protoperithecial parent in *Neurospora* (Fig. 6-3) supplies the bulk of the extrachromosomal material of the sexually produced ascospores. Very slow spore germination characterizes one strain of this fungus. The trait exhibits differences in reciprocal crosses, maternal inheritance, and fails to segregate at meiosis. When the slow strain acts as protoperithecial parent and the conidial strain has normal spore germination, all the progeny are slow, but possess 50% of the nuclear genes of the conidial parent. Each generation is then backcrossed to the conidial parent, so that the $F_2$ contains 75%, $F_3$ contains 87.5%, etc., of nuclear genes of the conidial parent. After the fifth or sixth backcross, the nuclear genes are almost wholly those of the conidial parent, but the slow germination trait persists in all of the progeny.

(4) Many substances are known which cause nuclear genes to mutate, but these *mutagens* are non-specific, i.e. they increase the frequency of mutations at many loci, and only a minority of the survivors of the treatment will have mutated. Extranuclear genetic elements, however, may not exhibit these mutational characteristics.

**Example 10.8.** Petite yeasts arise spontaneously at low frequency in most cultures. If yeasts are grown in media containing a small amount of acriflavine, all of the cells survive, and up to 100% of them will show only the petite trait. The specific induction by this mutagen and the high rate of petite mutants suggests changes in cytoplasmic factors rather than in nuclear genes.

## SPECIFIC INDUCTION OF PHENOTYPIC CHANGE

A specifically induced change in the phenotype need not emanate from a permanent heritable change in either the nuclear or extranuclear genetic systems. All of these changes collectively represent a broad spectrum of stabilities. An induced phenotypic change may persist for the life of the individual (Example 10.9); it may persist in subsequent generations or it may be induced to change at any time during the life cycle (Example 10.10); or it may gradually disappear in subsequent generations of progeny (Example 10.11).

**Example 10.9.** An environmentally induced change which resembles the effects of a gene mutation is called a *phenocopy*. By injecting boric acid into chick eggs at the appropriate stage of development, birds with shortened limbs are produced resembling the "creeper" trait attributed to chromosomal genes. The effect of the treatment which produces any phenocopy does not extend beyond the life of the individual. All of the sexually produced offspring will be normal. The inducing agent presumably interferes with the function of genes and/or plasmagenes during critical stages of development, but does not induce permanent changes in the hereditary material.

**Example 10.10.** The antigens of *Paramecium* are under the control of nuclear genes. An individual of a certain strain may have several loci, each specifying a different antigen, but only one locus is active at a given time, and thus only one antigen can be detected. In strain 90, low temperatures activate the locus which specifies antigen S, all other loci being inactivated; at intermediate temperatures, only the locus for antigen G becomes active; at high temperatures only antigen D appears. If the environment remains constant, the induced antigenic change persists through subsequent generations of progeny. At any time during the life cycle (Fig. 10-1) however, a change in temperature could cause a transformation in the serotype. In the absence of cytoplasmic exchange, reciprocal crosses give different results. Even though the nuclei are genetically identical, the serotype of each exconjugant is the same as that prior to conjugation. It appears that temperature can predictably modify the organization of the cytoplasm. The state of the cytoplasm then determines which locus can become active.

**Example 10.11.** *Dauermodifications* are transient, environmentally induced changes in plasmagenes, decreasing in penetrance and expressivity in succeeding generations in the absence of the inducing stimulus. Treatment of *Phaseolus vulgaris* with a 0.75% solution of chloral hydrate results in abnormal leaf development. Only abnormal plants are used as parents in each generation. The first generation of sexual progeny reared in a normal environment may have about 75% abnormal plants. As late as the fourth generation nearly 50% of the plants may still be abnormal. By the seventh generation only normal plants may be found. The reversion of a dauermodification to normal phenotype might be due to backmutation of altered plasmagenes or to the cumulative selection of normal cytoplasmic components which have survived the treatment.

## SYMBIONTS

Organisms which live together are called *symbionts*. Some symbionts live at the expense of their hosts, as parasites. Some symbionts help each other as exemplified by the association of algae and fungi in lichens. An intracellular symbiont such as a bacterium or a virus may produce a novel phenotype in its host. The distinction between a living and a non-living organism becomes tenuous when we consider a virus and the various phases in which it may exist. A virus may exist in an autonomous and infective state or its genetic material may be so similar to a portion of its host that it becomes a part of the host chromosome. The virus may be induced to return to an autonomous state under the appropriate environmental stimulus. Other genetic elements are known to have the option of a chromosomal or extrachromosomal existence. All such elements are non-essential to the life of the host cell and are called *episomes*. Symbionts may exhibit an infection-like transmission with a hereditary continuity of their own. Letters of the Greek alphabet (*kappa, lambda, mu,* etc.) are commonly used to name intracellular symbionts.

**Example 10.12.** *Killer* paramecia harbor parasitic or symbiotic particles called *kappa* in their cytoplasm. These particles are about the size of a large virus or small bacterium and contain some DNA. Kappa produce a poison, *paramecin,* which kills individuals lacking kappa (*sensitives*). A dominant nuclear gene $K$ must be present for the maintenance and replication of kappa. Cells of genotype $kk$ that have acquired kappa are *unstable*, will loose these particles, and eventually become sensitives. Kappa can be regained by conjugation with a killer in which cytoplasmic exchange occurs. Kappa cannot be generated *de novo* in sensitives of genotypes $KK$ or $Kk$. Sensitive cells are resistant to the effects of paramecin during conjugation with killer cells.

### *Paramecium aurelia*

The microscopic ciliated protozoan *Paramecium aurelia* possesses a large vegetative or somatic *macronucleus* consisting of many sets of chromosomes (polyploid) which regulates the physiological processes of the cell. Two smaller diploid *micronuclei* function in fertilization. Paramecia reproduce asexually (mitotically) in a process called *binary fission* (Fig. 10-1(*a*)). All of the cells derived from one ancestral cell by fission are genetically identical and collectively referred to as a *clone*.

Sexual reproduction is accomplished by *conjugation* between cells of different mating types (Fig. 10-1(*b*)). During conjugation the macronucleus degenerates and reforms in the cell following mating (exconjugant). After pairing, each of the two micronuclei in both conjugating cells undergoes meiosis forming eight haploid nuclei per cell, seven of which degenerate. The one remaining nucleus in each cell doubles by mitosis, producing two identical gametic nuclei, one being stationary, the other being mobile and migrating into the opposite cell. A fusion nucleus is formed by union of the stationary nucleus and the migratory nucleus from the opposite cell. Each fusion nucleus divides twice by mitosis; two of the products become micronuclei, two become new macronuclei. At the first fission following conjugation, each daughter cell receives one of the two macronuclei, restoring the normal complement of one macronucleus and two micronuclei. Normally, during reciprocal exchange of haploid nuclei, little if any cytoplasm is exchanged. Under certain condi-

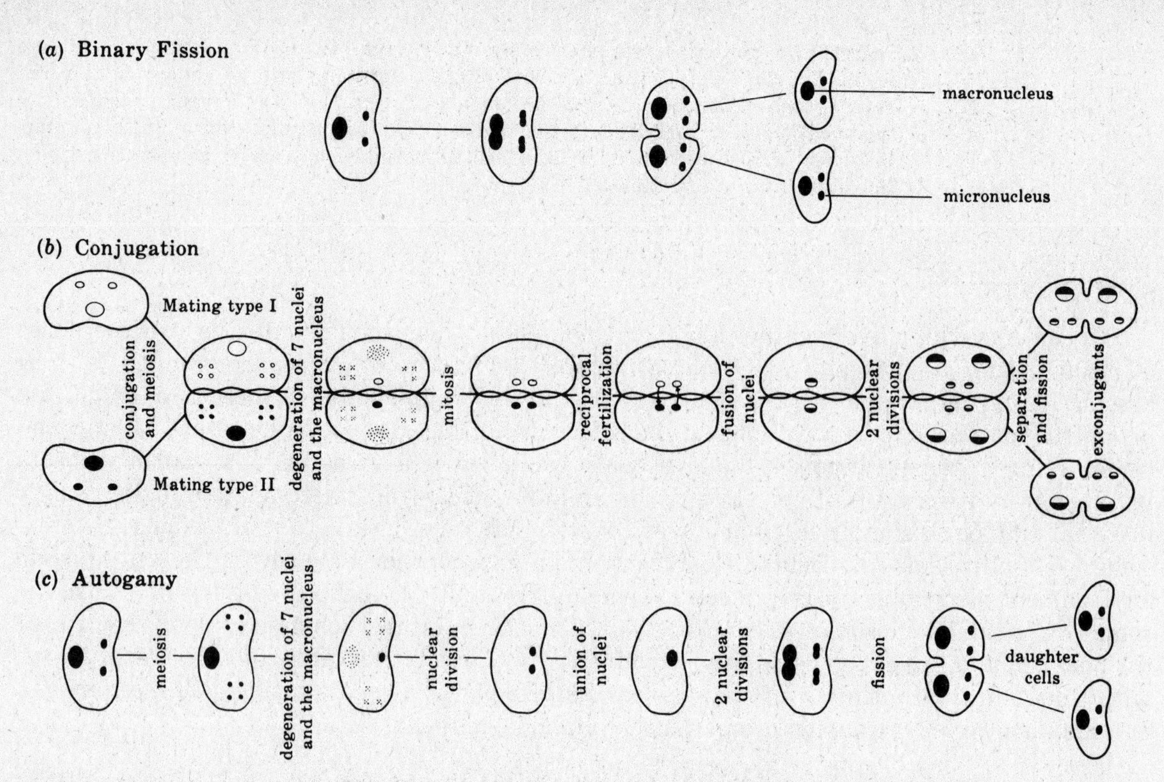


**Fig. 10-1.** Life cycle of *Paramecium aurelia*.

tions, however, the bridge connecting the conjugants persists after fertilization, allowing cytoplasmic exchange to occur.

Occasionally an individual *Paramecium* may reorganize its genetic endowment by a process of self-fertilization called *autogamy* (Fig. 10-1(*c*)). An isolated cell will follow the same sequence of events as in conjugation to the point where reciprocal fertilization would be possible. Having no pairing partner with which to exchange nuclei, the two haploid products unite forming a diploid fusion nucleus. The rest of the cycle is analogous to the corresponding steps in conjugation.

# Solved Problems

## MATERNAL EFFECTS

**10.1.** The hermaphroditic snail *Limnaea peregra* can reproduce either by crossing or by self-fertilization. The direction in which the shell coils is a maternal effect as explained in Example 10.2. A homozygous dextral snail is fertilized with sperm from a homozygous sinistral snail. The heterozygous $F_1$ undergoes two generations of self-fertilization. (*a*) What are the phenotypes of the parental individuals? (*b*) Diagram the parents, $F_1$, and two selfing generations, showing phenotypes and genotypes and their expected ratios.

**Solution:**

(*a*) Although we know the genotypes of the parents, we have no information concerning the genotype of the immediate maternal ancestor which was responsible for the organization of the egg cytoplasm from which our parental individuals developed. Therefore we are unable to determine what phenotypes these individuals exhibit. Let us assume for the purpose of diagramming part (*b*) that the maternal parent is dextral and the paternal parent is sinistral.

(b) Let D = dextrally organized cytoplasm, S = sinistrally organized cytoplasm.

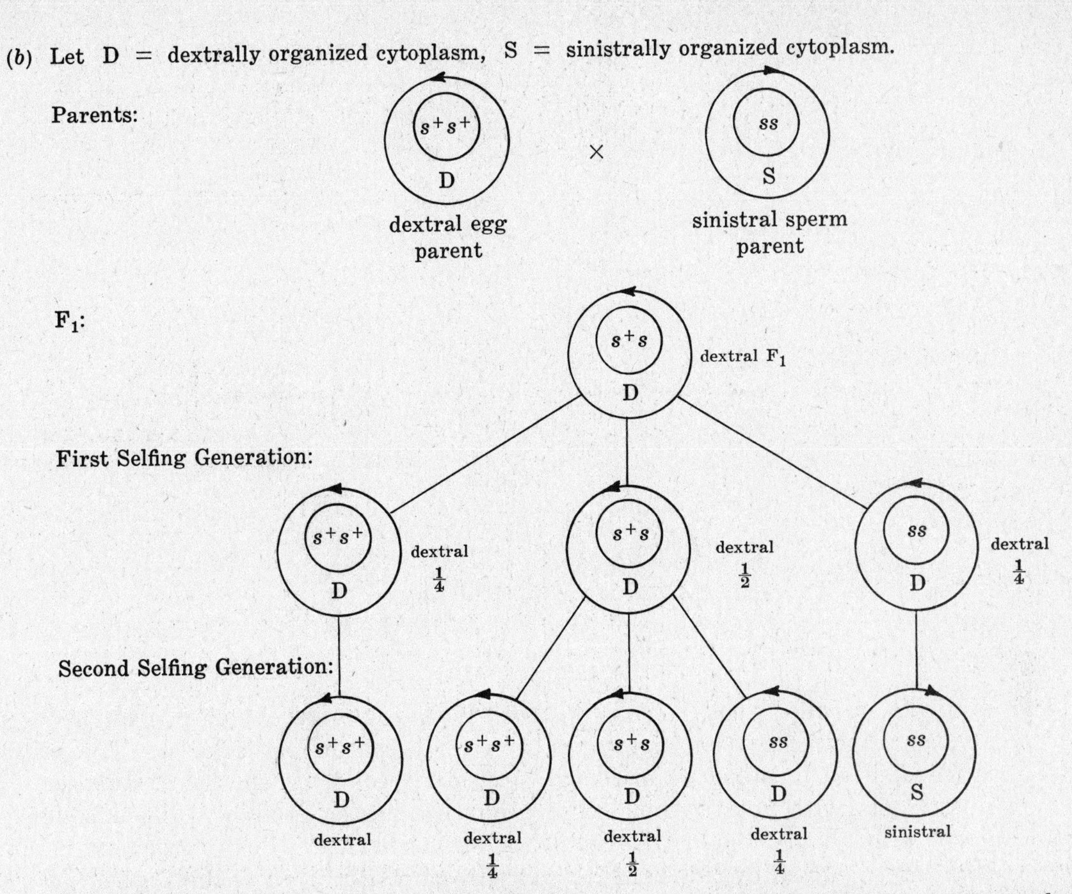


Notice that the $F_1$ is coiled dextrally, not because its own genotype is $s^+/s$, but because the maternal parent possessed the dominant dextral gene $s^+$. Likewise in the first selfing generation, all are phenotypically dextral regardless of their own genotype because the $F_1$ was $s^+/s$. In the second selfing generation, we expect the following:

| First Selfing Generation | | Second Selfing Generation | | | Summary Genotypes | | | | Phenotypes |
|---|---|---|---|---|---|---|---|---|---|
| $\frac{1}{4}\ s^+s^+$ | × | all $s^+s^+$ | = | $\frac{1}{4}\ s^+s^+$ | $s^+s^+$ | = | $\frac{1}{4}+\frac{1}{8}$ | = $\frac{3}{8}$ | $\frac{5}{8}$ dextral |
| | | $\frac{1}{4}\ s^+s^+$ | = | $\frac{1}{8}\ s^+s^+$ | $s^+s$ | = | $\frac{1}{4}$ | = $\frac{2}{8}$ | |
| $\frac{1}{2}\ s^+s$ | × | $\frac{1}{2}\ s^+s$ | = | $\frac{1}{4}\ s^+s$ | $ss$ | = | $\frac{1}{8}+\frac{1}{4}$ | = | $\frac{3}{8}$ sinistral |
| | | $\frac{1}{4}\ ss$ | = | $\frac{1}{8}\ ss$ | | | | | |
| $\frac{1}{4}\ ss$ | × | all $ss$ | = | $\frac{1}{4}\ ss$ | | | | | |

## PLASMAGENES

**10.2.** Assume that a neutral petite yeast (Example 10.5) has the chromosomal genes for normally functioning mitochondria, but has structurally defective mitochondria. Another kind of yeast is known, called *segregational petite*, which has structurally normal mitochondria which cannot function because of inhibition due to a recessive mutant chromosomal gene. What results would be expected among the sexual progeny when the neutral petite crosses with the segregational petite?

**Solution:**

The diploid zygote receives structurally normal mitochondria from the segregational petite parent which should be able to function normally in the presence of the dominant nuclear gene from the neutral petite parent. Sporulation would probably distribute at least some structurally normal mitochondria to each ascospore. The nuclear genes would segregate 1 normal : 1 segregational petite. Let shaded cytoplasm contain defective mitochondria. See figure below.

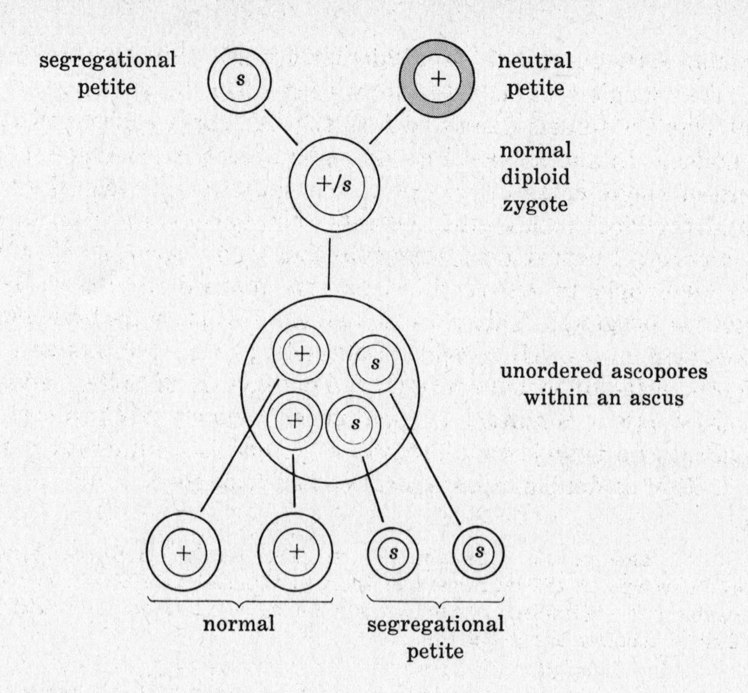


**10.3.** A condition called "poky" in *Neurospora* is characterized by slow growth due to an abnormal respiratory enzyme system similar to that of petite yeast. The poky trait is transmitted through the maternal (protoperithecial) parent. A chromosomal gene $F$ interacts with poky cytoplasm to produce a faster growing culture called "fast-poky" even though the enzyme system is still abnormal. Poky cytoplasm is not permanently modified by transient contact with an $F$ genotype in the zygote. It returns to the poky state when the genotype bears the alternative allele $F'$. Gene $F$ has no phenotypic expression in the presence of a normal cytoplasm. If the maternal parent is fast-poky and the paternal (conidial) parent is normal, predict the genotypes and phenotypes of the resulting ascospores.

**Solution:**

Let shaded cytoplasm contain poky mitochondria. The chromosomal alleles segregate in a 1 : 1 ratio, but poky cytoplasm follows the maternal (protoperithecial) line. The recovery of poky progeny indicates that poky cytoplasm has not been altered by its exposure to the $F$ gene in the diploid zygotic stage.

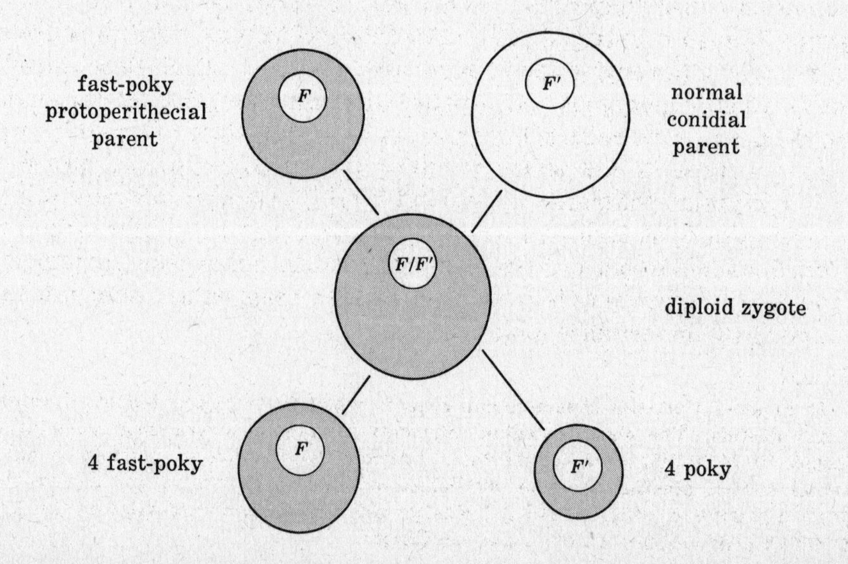

**10.4.** Commercial corn results from a "double-cross". Starting with four inbred lines (A, B, C, D), a single cross is made between A and B by growing the two lines together and removing the tassels from line A so that A cannot self-fertilize, and thus receives only B pollen. In another locality the same procedure is followed for lines C and D. The yield of single-cross hybrid seed is usually low because the inbred parent lacks vigor and produces small cobs. Plants that germinate from single-cross seed are usually vigorous hybrids with large cobs and many kernels. It is undesirable for the single-cross hybrid to self-fertilize, as this inbreeding process commonly produces less vigorous progeny. Therefore a double-cross is made by using only pollen from the CD hybrid on the AB hybrid. Detasseling is a laborious and expensive process. A cytoplasmic factor which prevents the production of pollen (male-sterile) is known. There also exists a dominant nuclear gene *R* which can restore fertility in a plant with male sterile cytoplasm. Propose a method for eliminating hand detasseling in the production of double-cross hybrid commercial seed.

**Solution:**

Let S = male sterile cytoplasm, F = male fertile cytoplasm. The ABCD double-cross hybrid seed develops on the large ears of the vigorous AB hybrid. When these seeds are planted they will grow into plants, half of which carry the gene for restoring fertility so that ample pollen will be shed to fertilize every plant.

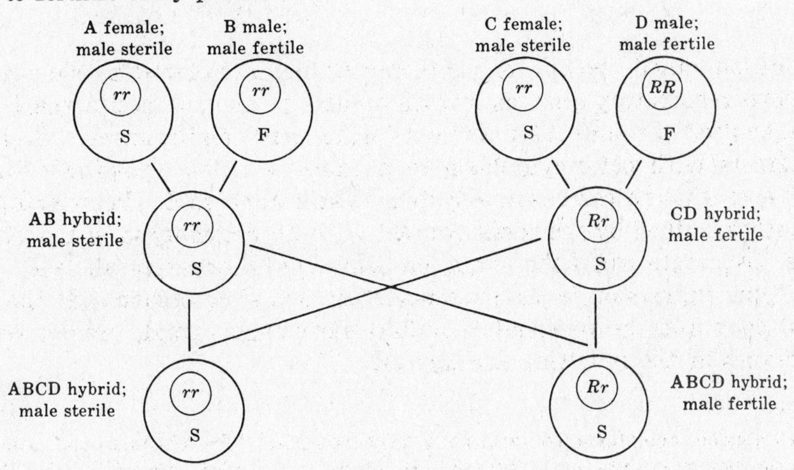


**10.5.** A recessive chromosomal gene produces green and white stripes in the leaves of maize, a condition called "japonica". This gene behaves normally in monohybrid crosses giving a 3 green : 1 striped ratio. Another striped phenotype was discovered in Iowa, named "iojap" (a contraction of Iowa and japonica), which is produced by a recessive gene *ij* when homozygous. If a plant with iojap striping serves as the seed parent, then the progeny will segregate green, striped and white in irregular ratios regardless of the genotype of the pollen parent. Backcrossing striped progeny of genotype *Ij/ij* to a green pollinator of genotype *Ij/Ij* produces progeny which continue to segregate green, striped and white in irregular ratios. White plants die due to lack of functional chloroplasts. Green plants produce only green progeny except when the genotype of the progeny is *ij/ij*; striping then reappears. Interpret this information to explain the inheritance of iojap.

**Solution:**

It appears that the chromosomal gene *ij* when homozygous induces irreversible changes in normal plastids. The plastids exhibit autonomy in subsequent generations, being insensitive to the presence of *Ij* in single or double dose. Random distribution of plastids to daughter cells could give all normal plastids to some, all defective plastids to others, and a mixture of normal and defective plastids to still others. All plastids are not rendered defective in the presence of *ij/ij* as this would produce only white (lethal) seedlings.

## SYMBIONTS

**10.6.** If a woman contracts German measles during the first trimester of pregnancy, the child may be seriously affected even though the mother herself suffers no permanent physical effects. Such anomalies as heart and liver defects, deafness, cataracts and blindness often occur in the affected children at birth. Can these phenotypic results be considered hereditary abnormalities?

**Solution:**

It is important to distinguish between *congenital* defects (recognizable at birth) which are acquired from the environment during embryonic development and genetic defects which are produced in response to the baby's own genotype. The former may be produced by infective agents such as the virus of German measles, which is not really a part of the baby's genotype but is acquired through agents external to the developing individual. An active case of this disease usually produces immunity so that subsequent children of this mother should not be susceptible to the crippling influences of this virus. A hereditary disease is one which is produced in response to instructions of an abnormal gene belonging to the diseased individual and which can be transmitted in Mendelian fashion from generation to generation.

**10.7.** The killer trait in *Paremecium* was explained in Example 10.12. A killer paramecium of genotype *Kk* conjugates with a sensitive paramecium of the same genotype without cytoplasmic transfer. One of the exconjugants is killer, the other is sensitive. Each of these conjugants undergoes autogamy. Predict the genotypes and phenotypes of the autogamous products.

**Solution:**

Reciprocal fertilization in paramecia produces identical genotypes in the exconjugants. In the micronuclear heterozygotes *Kk*, which of the four haploid nuclei produced by meiosis $(K, K, k, k)$ will survive depends on chance. Thus whether cytoplasmic mixing occurs or not, both exconjugants are expected to be *KK* 25% of the time, *Kk* 50% of the time, and *kk* 25% of the time. Autogamy of the exconjugants cannot produce any genotypic change in the homozygotes, but does produce homozygosity in heterozygotes. Half of the autogamous products of the monohybrid *Kk* are expected to be *KK* and half *kk*.

| Exconjugants with Kappa (Killers) | Autogamous Products | |
|---|---|---|
| $\frac{1}{4}$ *KK* | $= \frac{1}{4}$ *KK* | $\frac{1}{2}$ *KK* (killer) |
| $\frac{1}{2}$ *Kk* | $\frac{1}{2}$ *KK* $= \frac{1}{4}$ *KK* | |
| | $\frac{1}{2}$ *kk* $= \frac{1}{4}$ *kk* | $\frac{1}{2}$ *kk* (unstable) |
| $\frac{1}{4}$ *kk* | $= \frac{1}{4}$ *kk* | |

The same genotypic ratios would be expected in the autogamous products of exconjugants without kappa, but phenotypically they would all be sensitive.

**10.8.** The conjugation of paramecia occasionally results in the fusion of the two cells into a single animal with two gullets and other sets of organelles. Fission of doublets produces only more doublets. The genetic determinants for the doublet trait apparently reside neither in the nucleus nor the cytoplasm, but exclusively in the rigid outer envelope or cortex. A killer doublet of genotype *KK* conjugates with a sensitive singlet of genotype *kk*, in one case with and in another without cytoplasmic transfer. Predict the genotypic and phenotypic results of (*a*) fission in the exconjugants, (*b*) autogamy in the exconjugants.

**Solution:**

(*a*) Let us diagram this cross using shaded cytoplasm to represent the presence of kappa as shown below.

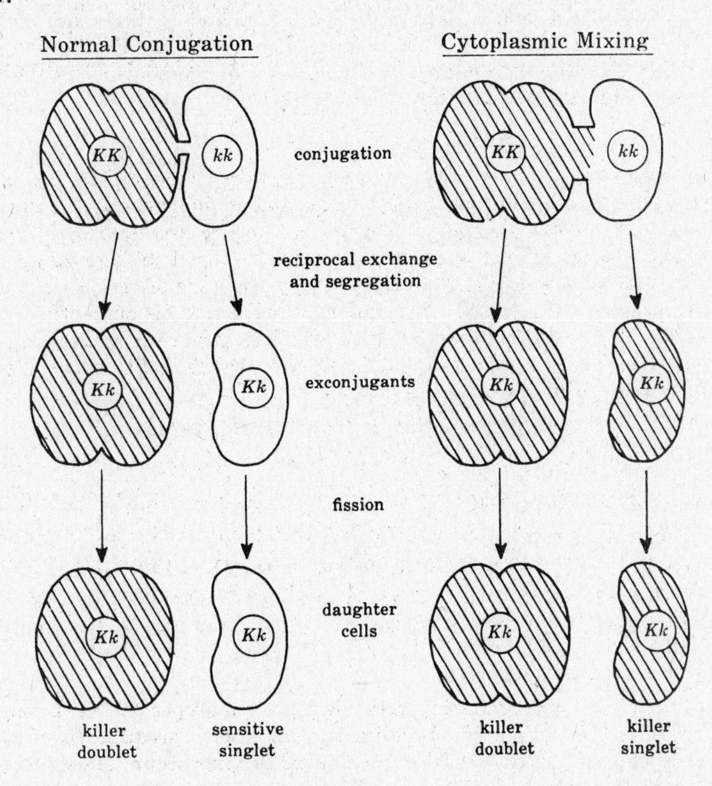


(*b*) Autogamy of the exconjugants produces homozygosity for its nuclear genes. There is a 50% chance that autogamy in a heterozygote (*Kk*) will produce homozygous *KK*, and a 50% chance for *kk*. Unstable individuals usually loose kappa within 8 or 10 fissions.

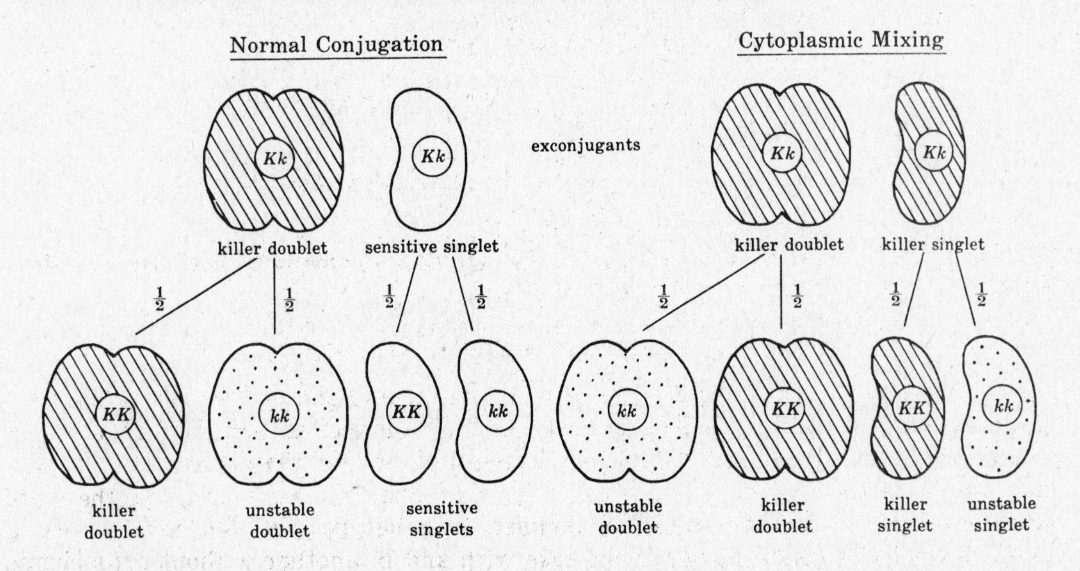


Notice in parts (*a*) and (*b*) that the predetermined structure of a doublet in its immobile ectoplasm does not segregate either with the nuclear gene *K* or the cytoplasmic marker *kappa*.

## Supplementary Problems

### MATERNAL EFFECTS

**10.9.** Continued use of certain drugs can lead to addiction. Addicted mothers by non-addicted fathers give birth to babies which are addicted. The children of addicted fathers by non-addicted mothers are normal. This difference in reciprocal crosses is not due to extranuclear genetic elements. Suggest a mechanism to account for the results.

**10.10.** *Erythroblastosis fetalis* or hemolytic disease of the newborn is caused by a blood group incompatibility between a mother and her child. The blood group antigen most frequently involved is D, governed by a dominant gene, *R*. Individuals who do not possess the D antigen are called Rh negative and may be considered homozygous recessive *rr*. When an Rh positive child develops in the uterus of an Rh negative woman, some of the baby's cells pass into the maternal circulation. The mother's body reacts to this foreign antigen (D) by producing Rh antibodies. The maternal antibodies pass across the placenta into the baby's circulation and react with the baby's blood cells, resulting in their destruction. A jaundiced condition and anemia are symptomatic of the disease. About 85% of the U.S. population is Rh positive. (*a*) Erythroblastotic children may appear from the cross $RR\,♂ \times rr\,♀$. Can the reciprocal cross produce Rh hemolytic disease? (*b*) What is the probability of a husband and wife having incompatible Rh blood types that could lead to erythroblastosis in some of their children?

**10.11** A snail produced by a cross between two individuals has a shell with right hand twist (dextral). This snail produces only left hand (sinistral) progeny by selfing. Determine the genotype of this snail and its parents. See Example 10.2 and Problem 10.1.

**10.12.** The genetic control of eye and skin color in larvae of the meal moth *Ephestia* is outlined in Example 10.1. (*a*) Determine the larval and adult phenotypes (colored or white) in the $F_1$ and $F_2$ from the mating $KK\,♀ \times kk\,♂$. (*b*) Do the same for the reciprocal cross.

**10.13.** A male and female *Ephestia* moth, both colored as larvae, were crossed. About half of the adult progeny were colored, half were white. What color did the male and female parents possess as adults?

### PLASMAGENES

**10.14.** Most strains of *Chlamydomonas* are sensitive to streptomycin. A strain is found which requires streptomycin in the culture medium for its survival. How could it be determined whether streptomycin-dependence is due to a chromosomal gene or to a cytoplasmic element?

**10.15.** Exposing a culture of white yeast to the mutagenic action of mustard gas produced some red individuals. When the red mutants were propagated vegetatively, some white cells frequently reappeared. How can these results be explained?

**10.16.** A yeast culture, when grown on medium containing acriflavine, produces numerous minute cells which grow very slowly. How could it be determined whether the slow growth was due to a cytoplasmic factor or to a nuclear gene?

**10.17.** The nuclear gene *F* in the presence of poky cytoplasm produces a fast-poky phenotype (Problem 10.3). In the presence of normal cytoplasm it has no phenotypic expression. How could a strain of *Neurospora* with normal growth be tested for the presence of *F* or its allele, *F'*?

**10.18.** Determine the genotypes and phenotypes of sexual progeny in *Neurospora* from the following crosses: (*a*) fast-poky male × normal female of genotype *F'*, (*b*) poky female × fast-poky male, (*c*) fast-poky female × poky male.

**10.19.** The cells of a *Neurospora* mycelium are usually multinucleate. Fusion of hyphae from different strains results in the exchange of nuclei. A mycelium which has genetically different nuclei in a common cytoplasm is called a *heterokaryon*. Moreover, the union results in a mixture of two different cytoplasmic systems called a *heteroplasmon* or a *heterocytosome*. The mycelia of two slow growing strains, each with an aberrant cytochrome spectrum, fuse to form a heteroplasmon that exhibits normal growth. Abnormal cytochromes *a* and *b* are still produced by the heteroplasmon. Offer an explanation for this phenomenon.

**10.20.** A striped phenotype in maize can be produced by either of two independently assorting recessive genes, japonica (*j*) or iojap (*ij*) (see Problem 10.5). Suppose that stocks of all the various genotypes for these two traits are available and you are given a striped plant which is either japonica or iojap. (*a*) What cross would you make to discover the genotype of this striped plant? (*b*) Give the genotype and phenotype expected in the $F_1$ if the plant was japonica. (*c*) List the phenotypes expected in the $F_1$ if the plant was iojap.

**10.21.** Male sterile plants in corn may be produced either by a chromosomal gene or by a cytoplasmic factor. (*a*) At least 20 different male sterile genes are known in maize, all of which are recessive. Why? Predict the $F_1$ and $F_2$ results of pollinating (*b*) a genetic male sterile by a normal, (*c*) a cytoplasmic male sterile by a normal.

**10.22.** Given seed from a male sterile line of corn, how would you determine if the sterility was genic or cytoplasmic?

**10.23.** Fireweed or the willow herb belongs to the genus *Epilobium*. Interspecific crosses of *E. luteum* ♀ × *E. hirsutum* ♂ gives a fertile $F_1$, but the reciprocal cross gives a pollen-sterile $F_1$. It has been suggested that *luteum* plastids function normally in combination with the hybrid genotype but *hirsutum* plastids function abnormally with adverse effects on fertility. To test this hypothesis it would be desirable to observe the effects of *hirsutum* plastids in *luteum* cytoplasm and of *luteum* plastids in *hirsutum* cytoplasm. Can you discern any biological impediment in the implementation of these experiments?

## SYMBIONTS

**10.24.** Determine which of the three paramecial phenotypes (killer, unstable, or sensitive) is produced by the following combinations of genotype and cytoplasmic state. *Hint:* See Example 10.12.

| | Genotype | Cytoplasm |
|---|---|---|
| (*a*) | *KK* | kappa |
| (*b*) | *Kk* | no kappa |
| (*c*) | *kk* | kappa |
| (*d*) | *KK* | no kappa |
| (*e*) | *Kk* | kappa |
| (*f*) | *kk* | no kappa |

**10.25.** When a killer strain of paramecia conjugates without cytoplasmic transfer with a sensitive strain, half of the exconjugants are sensitives and half are killers. Autogamy of the killer exconjugants produces only killers; autogamy of the sensitive exconjugants produces only sensitives. What are the genotypes of (*a*) the two original strains, (*b*) the exconjugants, (*c*) the autogamous products?

**10.26.** When two paramecia, each of genotype *Kk*, conjugate, with what frequencies are the following genotypes expected in the two exconjugants?

| | Exconjugant #1 | Exconjugant #2 |
|---|---|---|
| (*a*) | *KK* | *KK* |
| (*b*) | *KK* | *Kk* |
| (*c*) | *KK* | *kk* |
| (*d*) | *Kk* | *Kk* |
| (*e*) | *Kk* | *kk* |
| (*f*) | *kk* | *kk* |

**10.27.** When paramecia of genotype *Kk* possessing kappa conjugate with sensitive cells of genotype *kk*, and the exconjugants undergo autogamy, what percentage of the autogamous products are expected to possess and be able to maintain kappa through subsequent fissions if (*a*) conjugation is normal, (*b*) cytoplasmic mixing occurs?

**10.28.** From the following paramecial matings, predict the phenotypic ratios among the exconjugants under normal conditions and under conditions of cytoplasmic mixing.

| | Conjugating Paramecia | | | | | Exconjugant Phenotypes | |
|---|---|---|---|---|---|---|---|
| | Genotype | Cytoplasm | × | Genotype | Cytoplasm | Normal Conjugation | Cytoplasmic Mixing |
| (*a*) | *KK* | kappa | | *KK* | no kappa | | |
| (*b*) | *KK* | no kappa | | *KK* | no kappa | | |
| (*c*) | *KK* | kappa | | *kk* | no kappa | | |
| (*d*) | *Kk* | kappa | | *kk* | no kappa | | |

**10.29.** What percentage of exconjugants from the following paramecial matings will possess and be able to maintain kappa through subsequent asexual divisions?

| | Mating Type | | | Exconjugants Produced By: | |
|---|---|---|---|---|---|
| | Kappa Present | × | Kappa Absent | Normal Conjugation | Cytoplasmic Mixing |
| (*a*) | *KK* | | *KK* | | |
| (*b*) | *KK* | | *Kk* | | |
| (*c*) | *KK* | | *kk* | | |
| (*d*) | *Kk* | | *Kk* | | |
| (*e*) | *Kk* | | *kk* | | |
| (*f*) | *kk* | | *kk* | | |

**10.30.** With reference to the doublet trait described in Problem 10.8, where conjugations occur both with and without cytoplasmic mixing between sensitive doublets of genotype *kk* and killer singlets of genotype *KK*, predict the genotypic and phenotypic results of (*a*) fission in the exconjugants, (*b*) autogamy in the exconjugants.

**10.31.** Some *Drosophila* flies are known to be very sensitive to carbon dioxide gas, rapidly becoming anesthetized under its influence. Sensitive individuals possess a cytoplasmic particle called *sigma* which has many of the attributes of a virus. Resistant individuals do not have *sigma*. This trait shows strictly maternal inheritance. Predict the results of a cross between (*a*) sensitive female × resistant male, (*b*) sensitive male × resistant female.

**10.32.** Injection of body fluids or implantation of organs from a carbon dioxide sensitive fly (see Problem 10.31) into a normal resistant fly can render the recipient sensitive to $CO_2$. What kind of progeny would an inoculated female be likely to produce? Why?

**10.33.** A virus is suspected to be involved in the development of breast cancer in certain strains of mice. The virus is transmitted through the milk to the offspring. In crosses where the female carries the "milk factor" and the male is from a strain free of the factor, about 90% of the female progeny develop breast cancer prior to 18 months of age. The virus usually does not initiate cancer development in the infected mouse until she enters the nursing stage, and then only in conjunction with a hormone (estrone) from the ovaries. Males from a virus infected strain are crossed with females from a virus-free strain. (*a*) Predict the proportion of the offspring from this cross which, if individually isolated from weaning to 18 months of age, will probably exhibit breast cancer. (*b*) Predict the proportion of offspring from this cross which will probably exhibit breast cancer if housed in a group from weaning to 18 months of age. (*c*) Answer part (*a*) when the reciprocal cross is made. (*d*) Answer part (*b*) when the reciprocal cross is made.

**10.34.** Another case in which a disease is acquired through the milk (see Problem 10.33) is hemolytic anemia in newborn horses. A mare may produce two or three normal offspring by the same stallion and the next foal may develop severe jaundice within about 96 hours after birth and die. Subsequent matings to the same stallion often produces the same effect. Subsequent matings to another stallion could produce normal offspring. It has been found that if nursed for the first few days on a foster mother the foal will not become ill and develops normally. Evidently something is in the early milk (colostrum) which is responsible for this syndrome. If the foal should become ill, and subsequently recovers, the incompatibility is not transmitted to later generations. (*a*) How might this disease be generated? (*b*) How is the acquisition of this disease different from that of breast cancer in mice?

10.35. A bacterial spirochaete which is passed to the progeny only from the maternal parent has been found in *Drosophila willistoni.* This microorganism usually kills males during embryonic development but not females. The trait is called "sex ratio" (SR) for obvious reasons. Occasionally, a son of an SR female will survive. This allows reciprocal crosses to be made. The SR condition can be transferred between *D. equinoxialis* and *D. willistoni.* The spirochaete is sensitive to high temperatures which inactivates them, forming "cured" strains with a normal sex ratio. (*a*) What would you anticipate to be the consequence of repeated backcrossing of SR females to normal males? (*b*) A "cured" female is crossed to a rare male from an SR culture. Would the sex ratio be normal? Explain.

# Answers to Supplementary Problems

10.9. Small molecular weight molecules are continually diffusing across the placenta from the maternal circulation to that of the fetus and vice versa. Drugs taken by the mother enter the fetus in this manner.

10.10. (*a*) No. If the mother has the D factor she will not make antibodies against her own antigens. (*b*) $0.85 \times 0.15 = 0.13$

10.11. Parents: $s^+s$ ♀ × $s/?$ ♂; $F_1$: *ss*

10.12. (*a*) $F_1$: larvae and adults colored (*Kk*); $F_2$: $\frac{3}{4}$ colored larvae and adults (*K*-) : $\frac{1}{4}$ colored larvae, white adults (*kk*). (*b*) same as part (*a*)

10.13. Colored female (*Kk*); white male (*kk*)

10.14. Cross $ss\,mt^-$ (male) × $sd\,mt^+$ (female); if chromosomal, 25% of the sexual progeny should be $ss\,mt^-$, 25% $ss\,mt^+$, 25% $sd\,mt^-$, 25% $sd\,mt^+$; if cytoplasmic, almost all of the progeny should follow the maternal line (streptomycin dependent) as in Example 10.5, while mating type segregates 1 $mt^-$ : 1 $mt^+$.

10.15. Asexual reproduction cannot produce segregation of nuclear genes, but cytoplasmic constituents can be differentially distributed to fission products.

10.16. From a cross of minute × normal, a nuclear gene will segregate in the spores in a 1 : 1 ratio (e.g. segregational petite in Problem 10.2). If an extranuclear gene is involved, segregation will not be evident and all spores will be normal (e.g. neutral petite in Example 10.5).

10.17. Use the unknown as conidial parent on a standard poky strain. If the unknown carries *F*, half of the ascospores will be poky and half fast-poky. If the unknown carries *F′*, all of the sexual progeny will be poky.

10.18. (*a*) all phenotypically normal; $\frac{1}{2}F'$ (normal cytoplasm) : $\frac{1}{2}F$ (normal cytoplasm). (*b*) and (*c*) $\frac{1}{2}$ fast-poky; *F* (poky cytoplasm) : $\frac{1}{2}$ poky; *F′* (poky cytoplasm)

10.19. One strain may have an abnormal cytochrome *a* but a normal cytochrome *b*. The other strain might have an abnormal cytochrome *b* but a normal cytochrome *a*. The normal cytochromes in the heteroplasmon complement each other to produce rapid growth.

10.20. (*a*) Use the striped plant as seed parent; use pollen from a japonica plant of genotype *Ij/Ij, j/j*. (*b*) all japonica striped (*j/j*). (*c*) green, striped and white in unpredictable ratios.

10.21. (*a*) A plant in which a dominant genic male sterile gene arose by mutation of a normal gene would be unable to fertilize itself and would be lost unless cross-pollinated by a fertile plant. The gene would be rapidly eliminated from heterozygotes within a few generations by continuous back crossing to normal pollen parents. (*b*) $F_1$: $+/ms$, fertile; $F_2$: $\frac{1}{4}+/+$, $\frac{1}{2}+/ms$, $\frac{1}{4}ms/ms$; $\frac{3}{4}$ fertile, $\frac{1}{4}$ male sterile. (*c*) male sterile cytoplasm is transmitted to all $F_1$ progeny; a selfed $F_2$ cannot be produced because none of the $F_1$ plants can make fertile pollen.

**10.22.** Plant the seeds and pollinate the resulting plants with normal pollen from a strain devoid of male sterility. If the $F_1$ is sterile, then it is cytoplasmic; if the $F_1$ is fertile, it is genic.

**10.23.** Plastids are rarely transmitted through the pollen. For this reason it is not possible to get *hirsutum* plastids in *luteum* cytoplasm or vice versa.

**10.24.** Killer = (*a*), (*e*); unstable = (*c*); sensitive = (*b*), (*d*), (*f*)

**10.25.** (*a*) (*b*) (*c*) All killers are *KK* with kappa; all sensitives are *KK* without kappa.

**10.26.** (*b*) (*c*) (*e*) = 0%; (*a*) (*f*) = 25%; (*d*) = 50%

**10.27.** (*a*) 12.5% (*b*) 25%

**10.28.**

| | Normal Conjugation | Cytoplasmic Mixing |
|---|---|---|
| (*a*) (*c*) | 1 killer : 1 sensitive | both killers |
| (*b*) | both sensitive | both sensitive |
| (*d*) | 1 killer : 1 unstable : 2 sensitive | 1 killer : 1 unstable |

**10.29.**

| | Normal Conjugation | Cytoplasmic Mixing |
|---|---|---|
| (*a*) (*b*) (*c*) | 50% | 100% |
| (*d*) | 37.5% | 75% |
| (*e*) | 25% | 50% |
| (*f*) | 0% | 0% |

**10.30.**

| | Normal Conjugation | | Cytoplasmic Mixing | |
|---|---|---|---|---|
| | Doublet | Singlet | Doublet | Singlet |
| (*a*) | *Kk* sensitive | *Kk* killer | *Kk* killer | *Kk* killer |
| (*b*) | $\frac{1}{2}KK$ sensitive | $\frac{1}{2}KK$ killer | $\frac{1}{2}KK$ killer | $\frac{1}{2}KK$ killer |
| | $\frac{1}{2}kk$ sensitive | $\frac{1}{2}kk$ unstable | $\frac{1}{2}kk$ unstable | $\frac{1}{2}kk$ unstable |

**10.31.** (*a*) All progeny sensitive; (*b*) all progeny resistant

**10.32.** Only sensitive progeny would probably be produced because *sigma* is an infective agent which would invade the ooplasm.

**10.33.** (*a*) (*b*) None of the progeny is expected to develop breast cancer because non-infected females have nursed them. (*c*) None of the progeny is expected to develop breast cancer because in isolation the infected females could never produce a litter and subsequently enter a lactation period, a prerequisite for expression of the milk factor. (*d*) 50% females × 90% of females develop breast cancer = approximately 45%.

**10.34.** (*a*) This disease is similar to the Rh blood group system incompatibility between a human mother and her baby in Problem 10.10. In this case, antibodies are transferred to the offspring through the milk rather than across the placenta. (*b*) The particular stallion that is used has an immediate effect on the character. This is not true in the acquisition of breast cancer in mice. The incompatibility disease in horses cannot be transmitted to later generations, so there is no evidence of a specifically self-duplicating particle like the infective agent which causes mice cancer.

**10.35.** (*a*) If interspecific crosses can transmit the spirochaete, it is probably relatively insensitive to the chromosomal gene complement. Backcrossing would cause no change in the SR trait; indeed this is how the culture is maintained. (*b*) The sex ratio would probably be normal. It is unlikely that the spirochaete would be included in the minute amount of cytoplasm which surrounds the sperm nucleus.

# Chapter 11

## Quantitative Genetics and Breeding Principles

### QUALITATIVE VS. QUANTITATIVE TRAITS

The classical Mendelian traits encountered in the previous chapters have been qualitative in nature, i.e. traits which are easily classified into distinct phenotypic categories. These discrete phenotypes are under the genetic control of only one or a very few genes with little or no environmental modifications to obscure the gene effects. In contrast to this, the variability exhibited by many agriculturally important traits fails to fit into separate phenotypic classes (discontinuous variability), but instead forms a spectrum of phenotypes which blend imperceptively from one type to another (continuous variability). Economically important traits such as body weight gains, mature plant heights, egg or milk production records, yield of grain per acre, etc., are *quantitative* or *metric* traits with continuous variability. The basic difference between qualitative and quantitative traits involves the number of genes contributing to the phenotypic variability and the degree to which the phenotype can be modified by environmental factors. Quantitative traits may be governed by many genes (perhaps 10 to 100 or more), each contributing such a small amount to the phenotype that their individual effects cannot be detected by Mendelian methods. Genes of this nature are called *polygenes*. In many cases, most of the genetic variation of a quantitative trait may be attributed to segregation at relatively few loci with major effects plus a residue of minor pleiotropic effects from an undetermined number of other genes (the latter genes probably have major effects of their own, but not on the quantitative trait under consideration). In addition, the phenotypic variability expressed in most quantitative traits has a relatively large environmental component, and a correspondingly small genetic component. It is the task of the geneticist to determine the magnitude of the genetic and environmental components of the total phenotypic variability of each quantitative trait in a population. In order to accomplish this task, use is made of some rather sophisticated mathematics, especially of statistics. Only some of the more easily understood rudiments of this branch of genetics will be presented in this chapter. Below are summarized some of the major differences between quantitative and qualitative genetics.

| Qualitative Genetics | Quantitative Genetics |
|---|---|
| 1. Characters of *kind.* | 1. Characters of *degree.* |
| 2. *Discontinuous* variation; discrete phenotypic classes. | 2. *Continuous* variation; phenotypic measurements form a spectrum. |
| 3. *Single gene* effects discernible. | 3. *Polygenic* control; effects of single genes too slight to be detected. |
| 4. Concerned with *individual matings* and their progeny. | 4. Concerned with a *population* of organisms consisting of all possible kinds of matings. |
| 5. Analyzed by making *counts and ratios.* | 5. Statistical analyses give estimates of *population parameters* such as the mean and standard deviation. |

## QUASI-QUANTITATIVE TRAITS

In the early days of Mendelian genetics it was thought that there was a fundamental difference in the essence of qualitative and quantitative traits. One of the classical examples which helped bridge the gap between these two kinds of traits is the multiple gene model developed about 1910 by the Swedish geneticist Nilsson-Ehle to explain kernel color in wheat. When he crossed a certain red strain to a white strain he observed that the $F_1$ was all light red and that approximately $\frac{1}{16}$ of the $F_2$ was as extreme as the parents, i.e. $\frac{1}{16}$ were white and $\frac{1}{16}$ were red. He interpreted these results in terms of two genes, each with a pair of alleles exhibiting cumulative effects.

P: $R_1R_1R_2R_2$ (red) $\times$ $r_1r_1r_2r_2$ (white)

$F_1$: $R_1r_1R_2r_2$ (light red)

$F_2$:

$$\frac{1}{16} = R_1R_1R_2R_2 \; : \; \frac{4}{16} = \begin{cases} R_1R_1R_2r_2 \\ R_1r_1R_2R_2 \end{cases} : \; \frac{6}{16} = \begin{cases} R_1R_1r_2r_2 \\ R_1r_1R_2r_2 \\ r_1r_1R_2R_2 \end{cases} : \; \frac{4}{16} = \begin{cases} R_1r_1r_2r_2 \\ r_1r_1R_2r_2 \end{cases} : \; \frac{1}{16} = r_1r_1r_2r_2$$

red — medium red — light red — very light red — white

Each of the "active" alleles $R_1$ or $R_2$ adds some red to the phenotype, so that the genotype of whites contains neither of these alleles and a red genotype contains only $R_1$ and $R_2$ alleles. These results are plotted as histograms in Fig. 11-1. Note that the phenotype of the $F_1$ is intermediate between the two parental types and that the average phenotype of the $F_2$ is the same as that of the $F_1$ but is a much more variable population, i.e. the $F_2$ contains many more phenotypes (and genotypes) than in the $F_1$. The student should recognize the $F_2$ distribution as an expansion of the binomial $(a + b)^4$, where $a = b = \frac{1}{2}$.

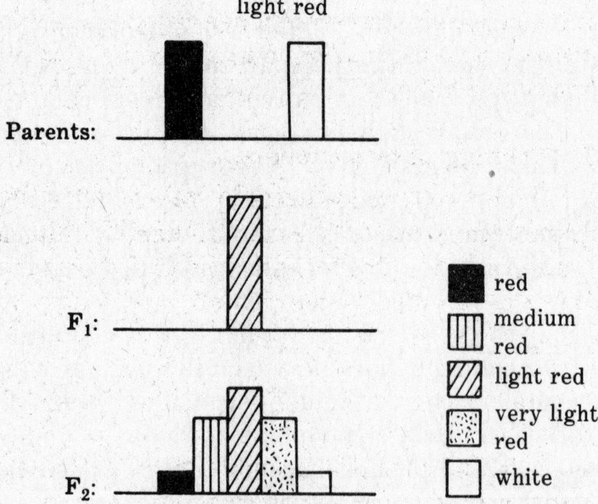


**Fig. 11-1.** Wheat color as an example of a quasi-quantitative trait.

Certain other strains of wheat with dark red kernels when crossed to whites exhibit an $F_1$ phenotype intermediate between the two parental types, but only $\frac{1}{64}$ of the $F_2$ are white. In this case the $F_1$ is probably segregating for three pairs of genes and only the genotype $r_1r_1r_2r_2r_3r_3$ produces white. Of course there would be more shades of red exhibited in the $F_2$ than in the previous case where only two genes are segregating. Even if the environment does not modify these color phenotypes (which it probably does to some extent), the ability of the eye to measure subtle differences in shading would probably be difficult with this many phenotypes and would become impossible if 4 to 5 genes were each contributing to kernel color.

Thus these multiple gene models, which are adequate to explain certain examples wherein discontinuous variation is still evident, may (by conceptual extension to include more genes plus environmental modifications) be useful in understanding the origin of continuous variation characterizing truly quantitative traits.

A rough estimate of the number of gene loci contributing to a quasi-quantitative trait can be obtained by determining the fraction of the $F_2$ (resulting from selfing the $F_1$ hybrid between two pure varieties) which is as extreme in its phenotype as that of one of the pure parental strains.

| Number of gene loci | 1 | 2 | 3 | ... | $n$ |
|---|---|---|---|---|---|
| Fraction of $F_2$ as extreme as one parent | 1/4 | 1/16 | 1/64 | ... | $(1/4)^n$ |

## THE NORMAL DISTRIBUTION

The study of a quantitative trait in a large population usually reveals that very few individuals possess the extreme phenotypes and that progressively more individuals are found nearer the average value for that population. This type of symmetrical distribution is characteristically bell-shaped as shown in Fig. 11-2 and is called a *normal distribution.* It is approximated by the binomial distribution $(p+q)^n$ introduced in Chapter 7 when the power of the binomial is very large and $p$ and $q$ are both $1/n$ or greater.

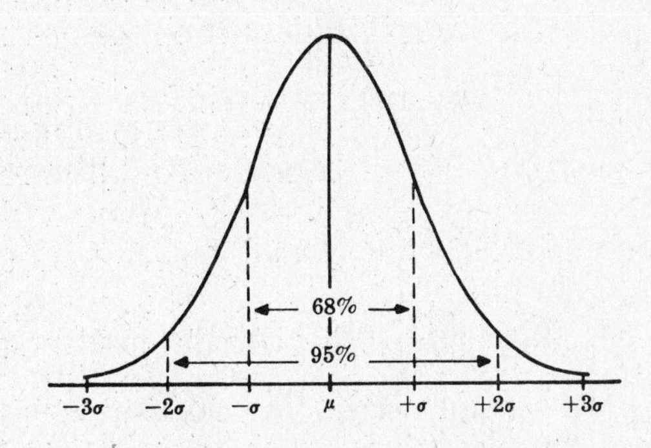


**Fig. 11-2.** A normal distribution.

### 1. Average Measurements.

The average phenotypic value for a normally distributed trait is expressed as the arithmetic mean ($\bar{X}$, read "X bar"). The arithmetic mean is the sum of the individual measurements ($\Sigma X$) divided by the number of individuals measured ($N$). The Greek letter "sigma" ($\Sigma$) directs the statistician to sum what follows.

$$\bar{X} = \frac{\sum_{i=1}^{N} X_i}{N} = \frac{X_1 + X_2 + X_3 + \cdots + X_N}{N} \tag{11.1}$$

It is usually not feasible to measure every individual in a population and therefore measurements are usually made on a sample from that population in order to estimate the population value (parameter). If the sample is truly representative of the larger population of which it is a part, then $\bar{X}$ will be an accurate estimate of the mean of the entire population ($\mu$). Note that letters from the English alphabet are used to represent *statistics,* i.e. measurements derived from a sample, whereas Greek letters are used to represent *parameters,* i.e. attributes of the population from which the sample was drawn. Parameters are seldom known and must be estimated from results gained by sampling. Obviously, the larger the sample size, the more accurately the statistic estimates the parameter.

### 2. Measurement of Variability.

Consider the three normally distributed populations shown in Fig. 11-3. Populations A and C have the same mean, but C is much more variable than A. A and B have different means, but otherwise appear to have the same shape (dispersion). Therefore, in order to adequately define a normal distribution, we must know not only its mean but also how much variability exists. One of the most useful measures of variability in a population for genetic purposes is the *standard deviation,* symbolized by the lower case Greek letter "sigma" ($\sigma$). A sample drawn from this population at random will have a sample standard deviation ($s$). To calculate $s$, the sample mean

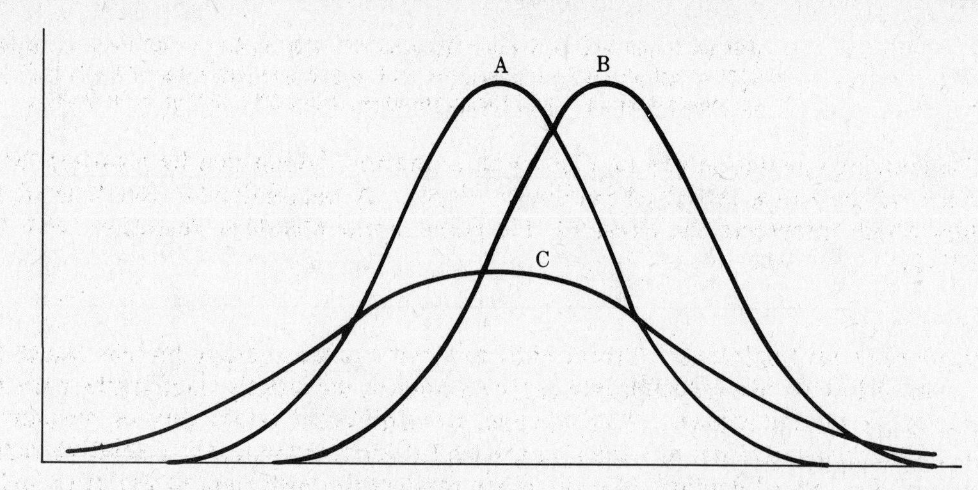


**Fig. 11-3.** Comparison of three populations (A, B, C) with respect to means and variances (see text; modeled after David T. Suzuki and A. J. Griffiths, *An Introduction to Genetic Analysis*, W. H. Freeman, 1976, p. 813.)

$(\overline{X})$ is subtracted from each individual measurement $(X_i)$ and the deviation $(X_i - \overline{X})$ is squared $(X_i - \overline{X})^2$, summed over all individuals in the sample $\left(\sum_{i=1}^{n}(X_i - \overline{X})^2\right)$, and divided by $n - 1$, where $n$ is the sample size. The calculation is completed by taking the square root of this value.

$$s = \sqrt{\frac{\sum_{i=1}^{n}(X_i - \overline{X})^2}{n - 1}} \tag{11.2}$$

To calculate $\sigma$, we substitute the total population size $(N)$ for $n$ in the above formula. For samples less than about 30, the appropriate correction factor for the denominator should be $n - 1$; for sample sizes greater than this, it makes little difference in the value of $s$ whether $n$ or $n - 1$ is used in the denominator. All other things being equal, the larger the sample size, the more accurately the statistic $s$ should estimate the parameter $\sigma$.

Relatively inexpensive electronic calculators are now available with the capacity to accumulate squared numbers. This usually makes it easier to calculate $s$ by the equivalent formula

$$s = \sqrt{\frac{\Sigma X^2 - [(\Sigma X)^2/n]}{n - 1}} \tag{11.3}$$

It is the property of every normal distribution that aproximately $\frac{2}{3}$ of the measurements (68%) will lie within plus or minus one standard deviation from the mean $(\mu \pm \sigma)$. Approximately $\frac{19}{20}$ of the measurements (95%) will lie within two standard deviations of the mean $(\mu \pm 2\sigma)$. More than 99% of the measurements will be found within plus or minus three standard deviations of the mean $(\mu \pm 3\sigma)$.

**Example 11.1.** The mean height of a sample from a plant population is 56 inches; the sample standard deviation is 6 inches. This indicates that approximately $\frac{2}{3}$ of the sample will be found between the values $56 \pm 6 = 50$ inches to 62 inches. Approximately $2\frac{1}{2}\%$ of all plants in this sample will measure smaller than $56 - (2 \times 6) = 56 - 12 = 44$ inches and $2\frac{1}{2}\%$ will measure larger than $56 + (2 \times 6) = 68$ inches.

The standard deviation can be plotted on a normal distribution by locating the point of inflection of the curve (point of maximum slope). A perpendicular constructed from the baseline which intersects the curve at this point is one standard deviation from the mean (Fig. 11-2).

***Coefficient of Variation.*** Traits with relatively large average metric values generally are expected to have correspondingly larger standard deviations than traits with relatively small average metric values. Furthermore, since different traits may be measured in different units, the *coefficients of variation* are useful for comparing their relative variabilities. Dividing the standard deviation by the mean renders the coefficient of variation independent of the units of measurement.

$$\begin{aligned}\text{Coefficient of Variation} &= \sigma/\mu \text{ for a population} \\ &= s/\bar{X} \text{ for a sample}\end{aligned} \qquad (11.4)$$

**3. Variance.**

The square of the standard deviation is called *variance* ($\sigma^2$). Unlike the standard deviation, however, variance cannot be plotted on the normal curve and can only be represented mathematically. Variance is widely used as an expression of variability because of the additive nature of its components. By a technique called "analysis of variance", the total phenotypic variance ($\sigma_P^2$) expressed by a given trait in a population can be statistically fragmented or partitioned into components of genetic variance ($\sigma_G^2$), non-genetic (or environmental) variance ($\sigma_E^2$), and variance due to genotype-environment interactions ($\sigma_{GE}^2$). Thus

$$\sigma_P^2 = \sigma_G^2 + \sigma_E^2 + \sigma_{GE}^2 \qquad (11.5)$$

It is beyond the scope of this text to present the analysis of variance, but a knowledge of variance components is essential to a discussion of breeding theory. Both the genetic variance and environmental variance can be further partitioned by this technique, so that the relative contributions of a number of factors influencing a metric trait can be ascertained. In order to simplify discussion, we shall ignore the interaction component.

**Example 11.2.** An analysis of variance performed on the birth weights of humans produced the following results:

| Variance Component | Percent of Total Phenotypic Variance |
|---|---|
| Offspring genotype | 16 |
| Sex | 2 |
| Maternal genotype | 20 |
| Maternal environment | 24 |
| Chronological order of child | 7 |
| Maternal age | 1 |
| Unaccountable variations (error) | 30 |
| | 100 |

***Variance Method of Estimating the Number of Genes.***

A population such as a line, a breed, a variety, a strain, a subspecies, etc., is composed of individuals which are more nearly alike in their genetic composition than those in the species as a whole. Phenotypic variability will usually be expressed even in a group of organisms which are genetically identical. All such variability within pure lines is obviously environmental in origin. Crosses between two pure lines produce a genetically uniform hybrid $F_1$. Phenotypic variability in the $F_1$ is likewise non-genetic in origin. In the formation of the $F_2$ generation, gene combinations are reshuffled and dealt out in new combinations to the $F_2$ individuals. It is a common observation that the $F_2$ generation is much more variable than the $F_1$ from which it was derived.

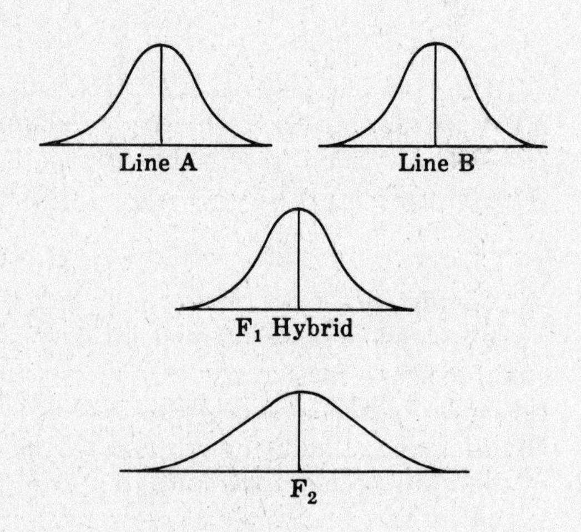


In a normally distributed trait, the means of the $F_1$ and $F_2$ populations tend to be intermediate between the means of the two parental lines. If there is no change in the environment from one generation to the next, then the environmental variation of the $F_2$ should be approximately the same as that of the $F_1$. An increase in phenotypic variance of the $F_2$ over that of the $F_1$ may then be attributed to genetic causes. Thus the genotypic variance of the $F_2$ ($\sigma^2_{GF2}$) is equal to the phenotypic variance of the $F_2$ ($\sigma^2_{PF2}$) minus the phenotypic variance of the $F_1$ ($\sigma^2_{PF1}$):

$$\sigma^2_{GF2} = \sigma^2_{PF2} - \sigma^2_{PF1}$$

The genetic variance of the $F_2$ is expressed by the formula $\sigma^2_{GF2} = (a^2N)/2$, where $a$ is the contribution of each active allele and $N$ is the number of *pairs* of genes involved in the metric trait. An estimate of $a$ is obtained from the formula $a = D/2N$, where $D$ is the numerical difference between the two parental means. Making substitutions and solving for $N$,

$$\sigma^2_{PF2} - \sigma^2_{PF1} = \sigma^2_{GF2} = a^2N/2 = D^2/8N$$

from which

$$N = \frac{D^2}{8(\sigma^2_{PF2} - \sigma^2_{PF1})} \tag{11.6}$$

This formula is an obvious over-simplification since it assumes all genes are contributing cumulatively the same amount to the phenotype, no dominance, no linkage and no interaction. Much more sophisticated formulas have been developed to take such factors into consideration, but these are beyond the scope of a first genetics course.

## TYPES OF GENE ACTION

Alleles may interact with one another in a number of ways to produce variability in their phenotypic expression. The following models may help us understand various modes of gene action.

(1) With dominance lacking, i.e. additive genes, each $A^1$ allele is assumed to contribute nothing to the phenotype (null allele), whereas each $A^2$ allele contributes one unit to the phenotype (active allele).

| Scale of phenotypic value: | 0 | 1 | 2 |
|---|---|---|---|
| Genotype: | $A^1A^1$ | $A^1A^2$ | $A^2A^2$ |

(2) With partial or incomplete dominance the heterozygote is almost as valuable as the $A^2A^2$ homozygote.

| | | | |
|---|---|---|---|
| Scale of phenotypic value: | 0 | 1.5 | 2 |
| Genotype: | $A^1A^1$ | $A^1A^2$ | $A^2A^2$ |

(3) In complete dominance identical phenotypes are produced by the heterozygote and $A^2A^2$ homozygote.

| | | |
|---|---|---|
| Scale of phenotypic value: | 0 | 2 |
| Genotype: | $A^1A^1$ | $A^1A^2$<br>$A^2A^2$ |

(4) In *overdominance* the heterozygote is more valuable than either homozygous genotype.

| | | | |
|---|---|---|---|
| Scale of phenotypic value: | 0 | 2 | 2.5 |
| Genotype: | $A^1A^1$ | $A^2A^2$ | $A^1A^2$ |

If allelic interaction is completely additive, a linear phenotypic effect is produced. In Fig. 11-4, a constant increment ($i$) is added to the phenotype for each $A^2$ allele in the genotype.

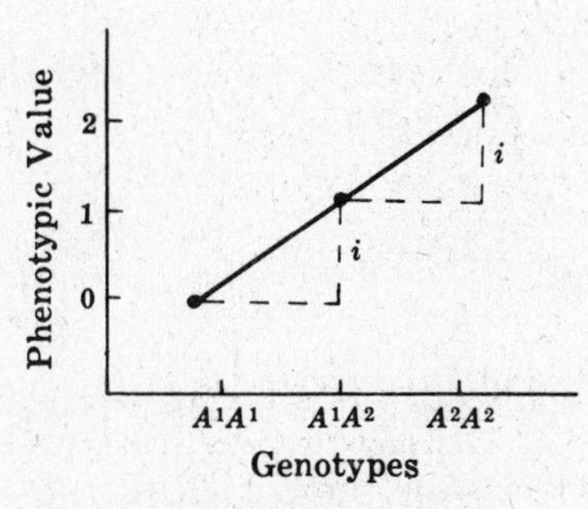


**Fig. 11-4.** Additive gene action.

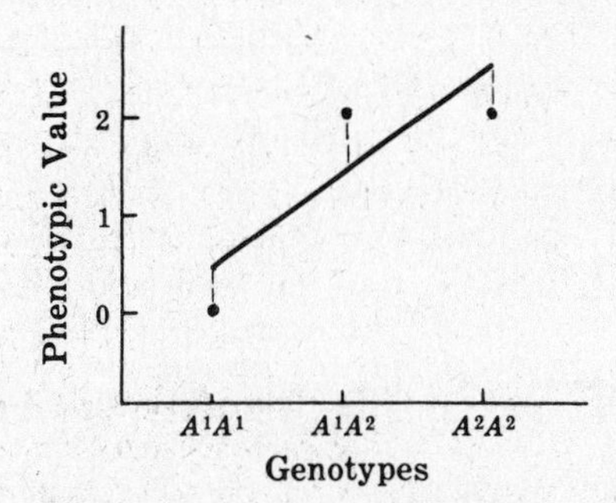


**Fig. 11-5.** Dominant gene action.

Even if complete dominance is operative, an underlying component of additivity (linearity) is still present (solid line in Fig. 11-5). The deviations from the additive scheme (dotted lines) due to many such genes with partial or complete dominance can be statistically estimated from appropriately designed experiments. The genetic contributions from such effects appear in the dominance component of variance ($\sigma_D^2$).

In a much more complicated way, deviations from an underlying additive scheme could be shown to exist for the interactions between genes at different loci (epistatic relationships). The contribution to the total genetic variance ($\sigma_G^2$) made by these genetic elements can be partitioned into a component called the epistatic or interaction variance ($\sigma_I^2$).

The sum of the additive gene effects produced by genes lacking dominance (additive genes) and by the additive contribution of genes with dominance or epistatic effects appears in the additive component of genetic variance ($\sigma_A^2$).

Thus the total genetic variance can be partitioned into three fractions:

$$\sigma_G^2 = \sigma_A^2 + \sigma_D^2 + \sigma_I^2 \qquad (11.7)$$

**Additive vs. Multiplicative Gene Action.**

Additive gene action produces an arithmetic series of phenotypic values such as 2, 4, 6, 8, ... representing the contributions of 1, 2, 3, 4, ... active alleles respectively. Additive gene action tends to produce a normal phenotypic distribution with the mean of the $F_1$ intermediate between means of the two parental populations. However, not all genes act additively. Some exhibit multiplicative gene action forming a geometric series such as 2, 4, 8, 16, ... representing the contributions of 1, 2, 3, 4, ... active alleles respectively. Traits governed by multiplicative gene action tend to be skewed into an asymmetrical curve such as that shown for the $F_2$ in Fig. 11-6. The means of the $F_1$ and $F_2$ are nearer to one of the parental means because the geometric mean of two numbers is the square root of their product.

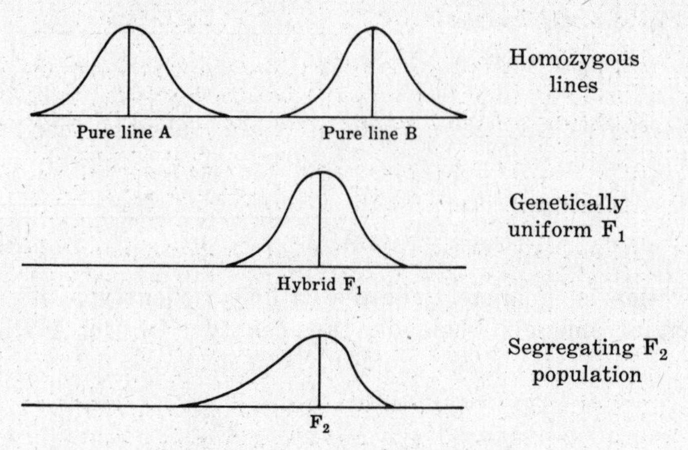


**Fig. 11-6.** Multiplicative gene action.

**Example 11.3.** (*a*) The geometric mean of 2 and 8 in the geometric series 2, 4, 8, which is increasing by a multiplicative increment of 2, is $\sqrt{2 \times 8} = 4$. The arithmetic mean of 2 and 8 is $(2+8)/2 = 5$.

(*b*) The geometric mean between 1.2 and 2.7 is $\sqrt{1.2 \times 2.7} = 1.8$, forming the geometric series 1.2, 1.8, 2.7, ... which is increasing by a multiplicative increment of 0.5. The arithmetic mean of 1.2 and 2.7 is $(1.2 + 2.7)/2 = 1.95$.

If a skewed distribution can be converted to a normal distribution by merely transforming the data to a logarithmic scale, this is evidence for multiplicative gene action.

**Example 11.4.** If the geometric series 1, 10, 100, 1000, ... (increasing by a multiplicative increment of 10) is converted to logarithms, we have the arithmetic series 0, 1, 2, 3, ... (increasing by an additive increment of one unit).

The variance and the mean are independent parameters in a normal distribution. That is, if the population mean is increased we cannot predict in advance to what degree the variance will be increased. In the case of multiplicative gene action, however, the variance is dependent upon the mean so that as the mean increases the variance increases proportionately. The coefficients of variation in segregating populations thereby remain constant.

The concepts of heritability and selection theory discussed in the following sections will deal only with normal distributions.

## HERITABILITY

One of the most important factors in the formulation of effective breeding plans for improving the genetic quality of crops and livestock is a knowledge of the relative contribution made by genes to the variability of a trait under consideration. The variability of phenotypic values for a quantitative trait can, at least in theory, be partitioned into genetic and nongenetic (environmental) components.

$$\sigma_P^2 = \sigma_G^2 + \sigma_E^2$$

*Heritability* (symbolized $h^2$) is the proportion of the total phenotypic variance due to gene effects.

$$h^2 = \sigma_G^2/\sigma_P^2 \tag{11.8}$$

The heritability of a given trait may be any number from 0 to 1.

**Example 11.5.** If all of the phenotypic variability of a trait is genetic in nature (as is true for most classical Mendelian traits, such as blood types), then environmental effects are absent and heritability equals one; i.e. if $\sigma_G^2 = \sigma_P^2$, the $h^2 = 1$.

**Example 11.6.** If all of the phenotypic variability is environmental in nature (as is true for any trait within a genetically homozygous line), then heritability of the trait is zero; i.e. if $\sigma_E^2 = \sigma_P^2$, then $\sigma_G^2 = 0$ and $h^2 = 0/\sigma_P^2 = 0$.

**Example 11.7.** If half of the phenotypic variability is due to gene effects, then heritability is 50%, i.e. if $\sigma_G^2 = \frac{1}{2}\sigma_P^2$, then $2\sigma_G^2 = \sigma_P^2$ and so $h^2 = \frac{1}{2} = 50\%$.

**Example 11.8.** If the environmental component of variance is three times as large as the genetic component, heritability is 25%, i.e. if $\sigma_E^2 = 3\sigma_G^2$, then

$$h^2 = \frac{\sigma_G^2}{\sigma_G^2 + \sigma_E^2} = \frac{\sigma_G^2}{\sigma_G^2 + 3\sigma_G^2} = \frac{1}{4} = 25\%$$

The parameter of heritability involves all types of gene action and thus forms a broad estimate of heritability. In the case of complete dominance, when a gamete bearing the active dominant allele $A^2$ unites with a gamete bearing the null allele $A^1$ the resulting phenotype might be two units. When two $A^2$ gametes unite, the phenotypic result would still be two units. On the other hand if genes lacking dominance (additive genes) are involved, then the $A^2$ gamete will add one unit to the phenotype of the resulting zygote regardless of the allelic contribution of the gamete with which it unites. Thus only the additive genetic component of variance has the quality of predictability necessary in the formulation of breeding plans. Heritability in this narrower sense is the ratio of the additive genetic variance to the phenotypic variance:

$$h^2 = \sigma_A^2/\sigma_P^2 \tag{11.9}$$

Unless otherwise specified in the problems of this book, heritability in the narrow sense is to be employed. It must be emphasized that the heritability of a trait only applies to a given population living in a particular environment. A genetically different population (perhaps a different variety, breed, race, or subspecies of the same species) living in an identical environment is likely to have a different heritability for the same trait. Likewise, the same population is likely to exhibit different heritabilities for the same trait when measured in different environments because a given genotype does not always respond to different environments in the same way. There is no one genotype that is adaptively superior in all possible environments. That is why natural selection tends to create genetically different populations within a species, each population being specifically adapted to local conditions rather than generally adapted to all environments in which the species is found.

Several methods can be used to estimate heritabilities of metric traits.

**1. Variance Components.**

Consider the simple, single-locus model (below) with alleles $b^1$ and $b^2$.

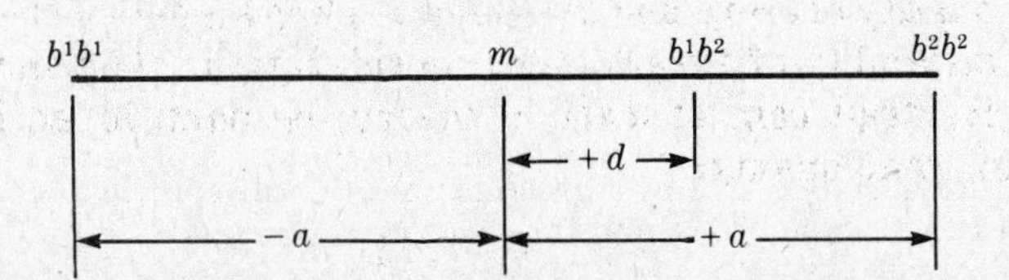

The midparent value, $m = \frac{1}{2}(b^1b^1 + b^2b^2)$. If the heterozygote does not have a phenotypic value equal to $m$, some degree of dominance ($d$) exists. If no dominance exists, then the alleles are completely additive. However, quantitative traits are governed by many loci and it might be possible that genotype $b^1b^2$ is dominant in a positive direction whereas genotype $c^1c^2$ is dominant in a negative direction, so that they cancel each other, giving the illusion of additivity. Dominance of all types can be estimated from the variances of $F_2$ and backcross generations. All of the phenotypic variance within pure lines $b^1b^1$ and $b^2b^2$, as well as in their genetically uniform $F_1$ ($b^1b^2$), is environmental. Hence, the phenotypic variances of each pure parental line ($V_{P1}$ and $V_{P2}$) as well as that of the $F_1$ ($V_{F1}$) serve to estimate the environmental variance ($V_E$). The $F_2$ segregates $\frac{1}{4}b^1b^1 : \frac{1}{2}b^1b^2 : \frac{1}{4}b^2b^2$. If each genotype departs from the midparent value as shown in the above model, then the average phenotypic value of $F_2$ should be $\frac{1}{4}(-a) + \frac{1}{2}(+d) + \frac{1}{4}(+a) = \frac{1}{2}d$. The contribution that each genotype makes to the total is its squared deviation from the mean ($m$) multiplied by its frequency $[f(X - \overline{X})^2]$. Therefore, the total $F_2$ variance (all genetic in this model) is

$$\begin{aligned} \tfrac{1}{4}(-a - \tfrac{1}{2}d)^2 + \tfrac{1}{2}(d - \tfrac{1}{2}d)^2 + \tfrac{1}{4}(a - \tfrac{1}{2}d)^2 &= \tfrac{1}{4}(a^2 + ad + \tfrac{1}{4}d^2) + \tfrac{1}{2}(\tfrac{1}{4}d^2) + \tfrac{1}{4}(a^2 - ad + \tfrac{1}{4}d^2) \\ &= \tfrac{1}{2}a^2 + \tfrac{1}{4}d^2 \end{aligned}$$

If we let $a^2 = A$, $d^2 = D$ and $E$ = environmental component, then the total $F_2$ phenotypic variance $(V_{F2}) = \frac{1}{2}A + \frac{1}{4}D + E$, representing the additive genetic variance ($V_A$) + the dominance genetic variance ($V_D$) + the environmental variance ($V_E$), respectively. Likewise it can be shown that $V_{B1}$ (the variance of backcross progeny $F_1 \times P_1$) or $V_{B2}$ (the variance of backcross progeny $F_1 \times P_2$) $= \frac{1}{4}A + \frac{1}{4}D + E$, and $V_{B1} + V_{B2} = \frac{1}{2}A + \frac{1}{2}D + 2E$. The degree of dominance is expressed as

$$\frac{d^2}{a^2} = \sqrt{\frac{D}{A}} \tag{11.10}$$

Heritability can be easily calculated from these variance components. The same is true of variance components derived from studies of identical (monozygotic) vs. non-identical (fraternal, dizygotic) twins. If twins reared together tend to be treated more alike than unrelated individuals, the heritabilities will be overestimated. This problem, and the fact that the environmental variance of fraternal twins tends to be greater than for identical twins, can be largely circumvented by studying twins that have been reared apart.

### 2. Genetic Similarity of Relatives.

If offspring phenotypes were always exactly intermediate between the parental values regardless of the environment, then such traits would have a narrow heritability of 1.0. On the other hand, if parental phenotypes (or phenotypes of other close relatives) could not be used to predict (with any degree of accuracy) the phenotypes of offspring (or other relatives), then such traits must have very low (or zero) heritabilities.

#### (a) *Regression Analysis.*

The regression coefficient ($b$) is an expression of how much (on the average) one variable ($Y$) may be expected to change per unit change in some other variable ($X$).

$$b = \frac{\sum_{i=1}^{n}(X_i - \overline{X})(Y_i - \overline{Y})}{\sum_{i=1}^{n}(X_i - \overline{X})^2} = \frac{\Sigma XY - (\Sigma X\,\Sigma Y)/n}{\Sigma X^2 - (\Sigma X)^2/n} \tag{11.11}$$

**Example 11.9.** If for every egg laid by a group of hens ($X$) the average production by their respective female progeny ($Y$) is 0.2, then the regression line of $Y$ on $X$ would have a slope ($b$) of 0.2.

0.2 unit $Y$

1 unit $X$

$$b = \frac{\Delta Y}{\Delta X} = \frac{0.2}{1.0} = 0.2$$

The regression line of $Y$ on $X$ has the formula

$$a = \overline{Y} - b\overline{X} \tag{11.12}$$

where $a$ is the "$Y$ intercept" (the point where the regression line intersects the $Y$ axis), $\overline{X}$ and $\overline{Y}$ are the respective mean values. The regression line also goes through the point $(\overline{X}, \overline{Y})$; establishing these two points allows the regression line to be drawn. Any $X$ value can then be used to predict the corresponding $Y$ value. Let $\hat{Y}$ = estimate of $Y$ from $X$; then

$$\hat{Y} = a + bX \qquad \text{(formula for a straight line)}$$

Since daughters receive only a sample half of their genes from each parent, the daughter-dam regression only estimates one-half of the narrow heritability of a trait (e.g. egg production in chickens). If the variances in the two populations are equal ($s_x = s_y$), then

$$h^2 = 2b_{\text{(daughter-dam)}} \tag{11.13}$$

Similarly, the regression of offspring on the average of their parents (midparent) is also an estimate of heritability

$$h^2 = b_{\text{(offspring-midparent)}} \tag{11.14}$$

Full sibs (having the same parents) are expected to share 50% of their genes in common; half sibs share 25% of their genes. Therefore,

$$h^2 = 2b_{\text{(full sibs)}} \tag{11.15}$$

$$h^2 = 4b_{\text{(half sibs)}} \tag{11.16}$$

If the variances of the two populations are unequal, the data can be converted to standardized variables (as discussed later in this chapter) and the resulting regression coefficients equated to heritabilities as described above.

(*b*) ***Correlation Analysis.***

The statistical correlation coefficient ($r$) measures how closely two sets of data are associated, is dimensionless and has the limits $\pm 1$. If all of the data points fall on the regression line, there is complete correlation. The regression coefficient ($b$) and the correlation coefficient ($r$) always have the same sign.

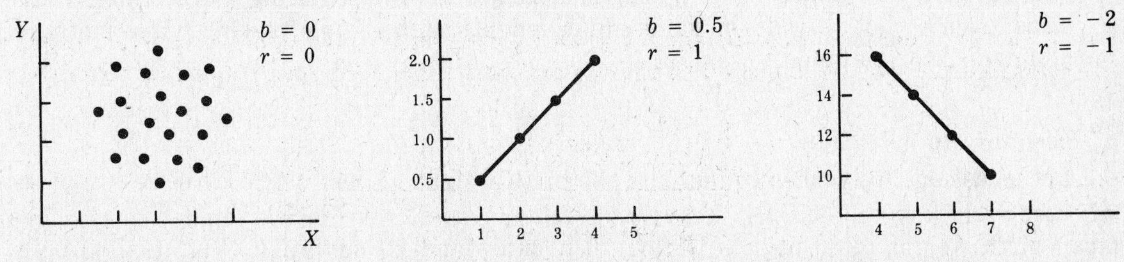


The correlation coefficient ($r$) of $Y$ on $X$ is defined as the linear change of $Y$, in standard deviations, for each increase of one standard deviation in $X$. The *covariance* (cov) of $X$ and $Y$ can be calculated from the following formula

$$\operatorname{cov}(X, Y) = \frac{\sum_{i=1}^{n} (X_i - \bar{X})(Y_i - \bar{Y})}{n - 1} \tag{11.17}$$

The covariance becomes the numerator in the formula for the correlation coefficient.

$$\begin{aligned} r &= \frac{\text{covariance}}{\text{geometric mean of variances}} \\ &= \frac{\operatorname{cov}(X, Y)}{s_x s_y} \\ &= \frac{\Sigma[(X - \bar{X})(Y - \bar{Y})]/(n - 1)}{\sqrt{[\Sigma(X - \bar{X})^2/(n - 1)][\Sigma(Y - \bar{Y})^2/(n - 1)]}} \\ &= \frac{\Sigma XY - [(\Sigma X \Sigma Y)/n]}{\sqrt{(\Sigma X^2 - [(\Sigma X)^2/n])(\Sigma Y^2 - [(\Sigma Y)^2/n])}} \end{aligned} \tag{11.18}$$

Notice that the numerators in the formulas for $r$ and $b$ are equivalent. Regression and correlation coefficients are related by

$$b = r\left(\frac{s_y}{s_x}\right) \tag{11.19}$$

so that if the variances of $X$ and $Y$ are identical, $b = r$. If the data are first converted to standardized variables, then the sample has a mean of zero and a standard deviation of one. Using standardized variables, regression and correlation coefficients become identical. Heritabilities can be estimated from $r$ just as they can from $b$.

**Example 11.10.** The correlation coefficient of Y offspring and midparent (X) is equivalent to narrow heritability; $h^2 = r$.

**Example 11.11.** If all the variation between offspring and one parent (e.g. their sires) is genetic, then $r$ should equal 0.5; if $r = 0.2$, then $h^2 = 2(0.2) = 0.4$.

**Example 11.12.** If litter mates were phenotypically correlated for a trait by $r = 0.15$, then $h^2 = 2(0.15) = 0.3$.

**Example 11.13.** If the correlation coefficient for half sibs is 0.08, then $h^2 = 4(0.08) = 0.32$.

All unbiased estimates of heritability based on correlations between relatives depend upon the assumption that there are no environmental correlations between relatives. Experimentally this can be fostered by randomly assigning all individuals in the study to their respective environments (field plots, pens, etc.), but this obviously is not possible for humans. Relatives such as full sibs usually share the same maternal and family environment and are likely to show a greater correlation between each other in phenotype than should rightly be attributed to common heredity. For this reason, the phenotypic correlation between sire and offspring is more useful for calculating heritabilities because sires often do not stay in the same environments with their offspring while mothers or siblings are prone to do so.

### 3. Response to Selection.

Let us assume we wished to increase the birth weight of beef cattle by selecting parents who themselves were relatively heavy at birth. Assume our initial population ($\bar{P}_1$) has a mean birth weight of 80 pounds with a 10 pound standard deviation (Fig. 11-7($a$) below). Further suppose that we will save all animals for breeding purposes which weigh over 95 pounds at birth. The mean of these animals which have been selected to be parents of the next generation ($\bar{P}_p$) is 100 pounds. The difference $\bar{P}_p - \bar{P}_1$ is called the *selection differential* symbolized $\Delta P$ (read "delta P") and sometimes referred to as "reach". Some individuals with an inferior genotype are expected to have high birth weights largely because of a favorable intrauterine environment. Others with a superior genotype may possess a low birth weight because of an unfavorable environment. In a large, normally distributed population, however, the plus and minus effects produced by good and poor environments are assumed to cancel each other so that the average phenotype ($\bar{P}_1$) reflects the effects of the average genotype ($\bar{G}_1$). Random mating among the selected group produces an offspring generation (Fig. 11-7(*b*) with its phenotypic mean ($\bar{P}_2$) also reflecting its average genotypic mean ($\bar{G}_2$). Furthermore, the mean genotype of the parents ($\bar{G}_p$) will be indicated in the mean phenotype of their offspring ($\bar{P}_2$) because only genes are transmitted from one generation to the next. Assuming the environmental effects remain constant from one generation to the next, we can attribute the difference $\bar{G}_2 - \bar{G}_1$ to the selection of genes for high birth weight in the individuals that we chose to use as parents for the next generation. This difference ($\bar{G}_2 - \bar{G}_1$) is called "genetic gain", symbolized $\Delta G$. If all of the variability in birth weight exhibited by a population was due solely to additive gene effects, and the environment was contributing nothing at all, then by selecting individuals on the basis of their birth weight records we would actually be selecting the genes which are responsible for high birth weight. That is, we will not be confused by the effects which a favorable environment can produce with a mediocre genotype or by the favorable interaction ("nick") of a certain combination of genes which will be broken up in subsequent generations. Realized heritability is defined as the ratio of the genetic gain to the selection differential:

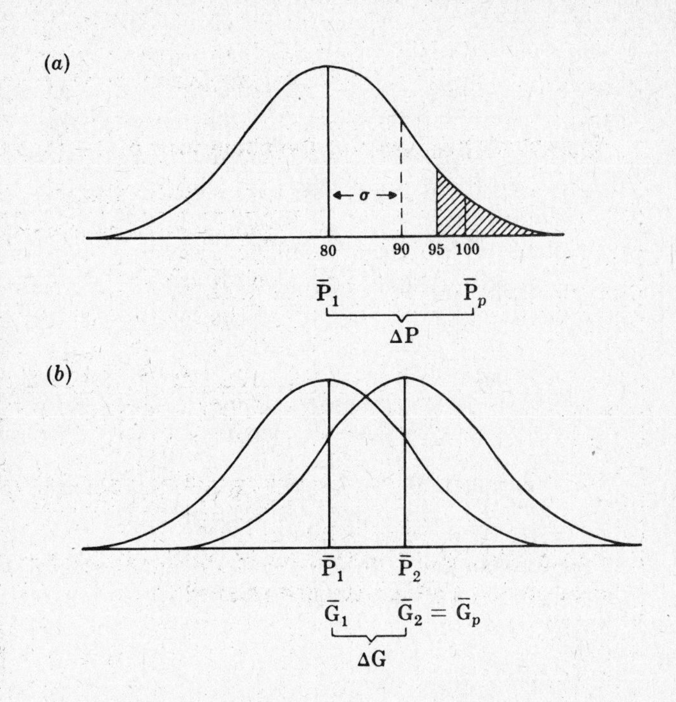


**Fig. 11-7.** Selection for birth weight in beef cattle.

$$h^2 = \Delta G/\Delta P \tag{11.20}$$

**Example 11.14.** If we gained in the offspring all that we "reached" for in the parents, then heritability is unity, i.e. if $\bar{P}_2 - \bar{P}_1 = 100 - 80 = 20$, and $\Delta P = \bar{P}_p - \bar{P}_1 = 100 - 80 = 20$, then $h^2 = \Delta G/\Delta P = 20/20 = 1$.

**Example 11.15.** If selection of parents with high birth weights fails to increase the mean birth weight of their offspring over that of the mean in the previous generation, then heritability is zero, i.e. if $\bar{P}_2$ and $\bar{P}_1 = 80$, then $\Delta G = \bar{P}_2 - \bar{P}_1 = 0$ and $h^2 = \Delta G/\Delta P = 0/20 = 0$.

**Example 11.16.** If the mean weight of the offspring is increased by half the selection differential, then heritability of birth weight is 50%, i.e. if $\Delta G = \frac{1}{2}\Delta P$, $\Delta P = 2\Delta G$, $h^2 = 0.5 = 50\%$. This is approximately the heritability estimate actually found for birth weight in beef cattle (Fig. 11-7(*b*)).

Most metric traits are not highly heritable. What is meant by high or low heritability is not rigidly defined, but the following values are generally accepted.

| | |
|---|---|
| High heritability | > 0.5 |
| Medium heritability | = 0.2-0.5 |
| Low heritability | < 0.2 |

Notice that when heritability is less than one the mean of the offspring in relationship to the mean of the parents tends to move back or "regress" toward the mean of the previous generation. The amount of this regression is directly related to the heritability of the trait. When heritability is 0.5, the mean of the offspring regresses 50% toward the mean of the previous generation. When heritability is 0.25, the mean of the offspring regresses 75% toward the mean of the previous generation. Thus *Heritability* = 100% − *Regression percentage.* The foregoing is not to be confused with the "statistical regression coefficient" (symbolized $b$) which indicates the amount one variable can be expected to change per unit change in some other variable. When one variable is the phenotype of the offspring and the other variable is the average phenotype of the two parents (midparent) then $b = h^2$.

**Example 11.17.** If $b = h^2 = 1$, then the offspring should have the same phenotypic value as the midparent value. That is, for each unit of increase in the phenotype of midparent, the offspring are expected to increase by the same amount.

**Example 11.18.** If $b = h^2 = 0.5$, for each unit of phenotypic increase in the midparent only $\frac{1}{2}$ unit increase is expected to appear in the offspring.

**Example 11.19.** If $b = h^2 = 0$, then the offspring are not expected to produce any better than the average of the population regardless of the midparent value.

## SELECTION METHODS

*Artificial selection* is operative when humans determine which individuals will be allowed to leave offspring (and/or the number of such offspring). Likewise, *natural selection* allows only those individuals to reproduce that possess traits adaptive to the environments in which they live. There are several methods by which artificial selection can be practiced.

### 1. Mass Selection.

If heritability of a trait is high, most of the phenotypic variability is due to genetic variation. Thus, a breeder should be able to make good progress by selecting from the masses those that excel phenotypically because the offspring-parent correlation should be high. This is called *mass selection,* but it is actually based on the individual's own performance record or phenotype. As the heritability of a trait declines, so does the prospect of making progress in improving the genetic quality of the selected line. In practice, selection is seldom made on the basis of one characteristic alone. Breeders usually desire to practice selection on several criteria simultaneously. However, the more traits selected for, the less selection "pressure" can be exerted on each trait. Selection should thus be limited to the two or three traits which the breeder considers to be the most important economically. It is probable that individuals scoring high in trait A will be mediocre or even poor in trait B (unless the two traits have a positive genetic correlation, i.e. some of the genes increasing trait A are also contributing positively to trait B). The breeder therefore must make compromises, selecting some individuals on a "total merit" basis which would probably not be saved for breeding if selection was being practiced on the basis of only a single trait.

The model used to illustrate the concept of genetic gain, wherein only individuals which score above a certain minimum value for a single trait would be saved for breeding, must now be modified to represent the more probable situation in which selection is based on the total merit of two or more traits (Fig. 11-8).

In selecting breeding animals on a "total merit" basis, it is desirable to reduce the records of performance on the important traits to a single score called the *selection index.* The index number has no meaning by itself, but is valuable in comparing several individuals on a

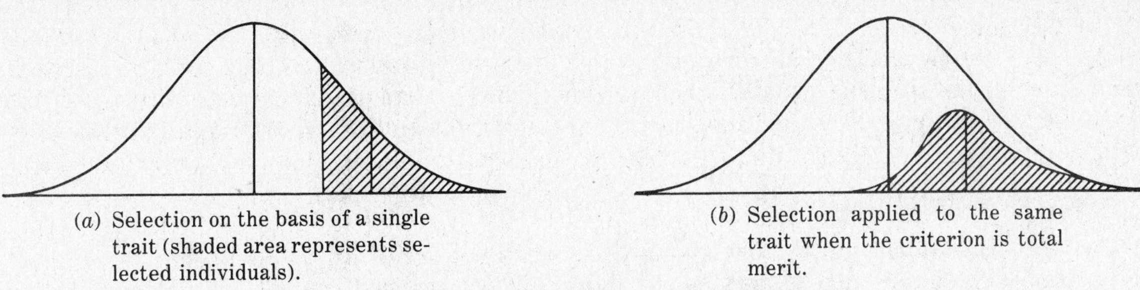

(*a*) Selection on the basis of a single trait (shaded area represents selected individuals).

(*b*) Selection applied to the same trait when the criterion is total merit.

**Fig. 11-8.** Selection on single trait vs. total merit.

*relative* basis. The methods used in constructing an index may be quite diverse, but they usually take into consideration the heritability and the relative economic importance of each trait in addition to the genetic and phenotypic correlations between the traits. An index (I) for three traits may have the general form

$$\text{I} = a\text{A}' + b\text{B}' + c\text{C}'$$

where $a$, $b$ and $c$ are coefficients correcting for the relative heritability and the relative economic importance for traits A, B and C respectively and where A′, B′ and C′ are the numerical values of traits A, B and C expressed in "standardized form". A *standardized variable* ($X'$) is computed by the formula

$$X' = \frac{X - \bar{X}}{\sigma_X} \tag{11.21}$$

where $X$ is the record of performance made by an individual, $\bar{X}$ is the average performance of the population, and $\sigma_X$ is the standard deviation of the trait. In comparing different traits, one is confronted by the fact that the mean and variability of each trait is different and often the traits are not even expressed in the same units.

**Example 11.20.** An index for poultry might use egg production (expressed in numbers of eggs per laying season), egg quality (expressed in terms of grades such as AA, A, B, etc.), and egg size (expressed in ounces per dozen).

**Example 11.21.** An index for swine might consider backfat thickness (in inches), feed conversion (pounds of feed per pound of gain), and conformation score (expressing the appearance of the individual in terms of points from a standard grading system).

The standardized variable, however, is a pure number (i.e. independent of the units used) based on the mean and standard deviation. Therefore any production record or score of a quantitative nature can be added to any other such trait if they are expressed in standardized form.

### 2. Family Selection.

When both broad and narrow heritabilities of a trait are low, environmental variance is high compared to genetic variance. Family selection is most useful when heritabilities of traits are low and family members resemble one another only because of their genetic relationship. It is usually more practical to first reduce environmental variance by changing the farming or husbandry practices before initiating selective breeding programs. Another way to minimize the effects of an inflated environmental variance is to save for breeding purposes all members of families that have the highest average performance even though some members of such families have relatively poor phenotypes. In practice, it is not uncommon to jointly use more than one selection method, e.g. choosing only the top 50% of individuals in only the families with the highest averages.

Family selection is most beneficial when members of a family have a high average genetic relationship to one another but the observed resemblance is low. If inbreeding increases the average genetic relationship within a family more than the increases in phenotypic resemblance, the gain from giving at least some weight to family averages may become relatively large.

### 3. Pedigree Selection.

In this method, consideration is given to the merits of ancestors. Rarely should pedigree selection be given as much weight as the individual's own merit unless the selected traits have low inheritabilities and the merits of the parents and grandparents are much better known than those of the individual in question. It may be useful for characteristics that can only be seen in the opposite sex or for traits that will not be manifested until later in life, perhaps even after slaughter or harvest. The value of pedigree selection depends upon how closely related the ancestor is to the individual in the pedigree, upon how many ancestors' or colateral ancestors' records exist, upon how completely the merits of such ancestors are known and upon the degree of heritability of the selected traits.

### 4. Progeny Test.

A *progeny test* is a method of estimating the breeding value of an animal by the performance or phenotype of its offspring. It has its greatest utility for those traits which (1) can only be expressed in one sex (e.g. estimating the genes for milk production possessed by a bull), (2) cannot be measured until after slaughter (e.g. carcass characteristics), (3) have low heritabilities so that individual selection is apt to be highly inaccurate.

Progeny testing cannot be practiced until after the animal reaches sexual maturity. In order to progeny test a male, he must be mated to several females. If the sex ratio is 1 : 1, then obviously every male in a flock or herd cannot be tested. Therefore males which have been saved for a progeny test have already been selected by some other criteria earlier in life. The more progeny each male is allowed to produce the more accurate the estimate of his "transmitting ability" (breeding value), but in so doing, fewer males can be progeny tested. If more animals could be tested, the breeder would be assured that he or she was saving only the very best for widespread use in the herd or flock. Thus a compromise must be made, in that the breeder fails to test as many animals as he or she would like because of the increased accuracy that can be gained by allotting more females to each male under test.

The information from a progeny test can be used in the calculation of the "equal parent index" (sometimes referred to as the "midparent index"). If the progeny receive a sample half of each of their parents' genotypes and the plus and minus effects of Mendelian errors and errors of appraisal tend to cancel each other in averages of the progeny and dams, then Average of progeny = sire/2 + (average of dams)/2 or

$$\text{Sire} = 2(\text{average of progeny}) - (\text{average of dams}) \qquad (11.22)$$

## MATING METHODS

Once the selected individuals have been chosen, they may be mated in various ways. The process known as "breeding" includes the judicious selection and mating of individuals for particular purposes.

### 1. Random Mating (Panmixis).

If the breeder places no mating restraints upon the selected individuals, their gametes are likely to randomly unite by chance alone. This is commonly the case with outcrossing (non-self-fertilizing) plants. Wind or insects carry pollen from one plant to another in essentially a random manner. Even livestock such as sheep and range cattle are usually bred panmicticly. The males locate females as they come into heat, copulate with ("cover") and inseminate them without any artificial restrictions as they forage for food over large tracts of grazing land. Most of the food that reaches our table is produced by random mating because it is the most economical mating method; relatively little manual labor is expended by the shepherd or herdsman other than keeping the flock or herd together, warding off predators, etc. This mating method is most likely to generate the greatest genetic diversity among the progeny.

## 2. Positive Assortative Mating.

This method involves mating individuals that are more alike, either phenotypically or genotypically, than the average of the selected group.

### (a) *Based on Genetic Relatedness.*

*Inbreeding* is the mating of individuals more closely related than the average of the population to which they belong. Fig. 11-9(*a*) shows a pedigree in which no inbreeding is evident because there is no common ancestral pathway from B to C (D, E, F and G all being unrelated). In the inbred pedigree of Fig. 11-9(*b*), B and C have the same parents and thus are full sibs (brothers and/or sisters). In the standard pedigree form shown in Fig. 11-9(*b*), sires appear on the upper lines and dams on the lower lines. Thus B and D are males; C and E are females. It is desirable to convert a standard pedigree into an arrow diagram for analysis (Fig. 11-9(*c*)). The *coefficient of relationship* (R) estimates the percentage of genes held in common by two individuals because of their common ancestry. Since each individual transmits only a sample half of his or her genotype to the offspring, each arrow in the diagram represents a probability of $\frac{1}{2}$. The sum ($\Sigma$) of all pathways between two individuals through common ancestors is the coefficient of relationship.

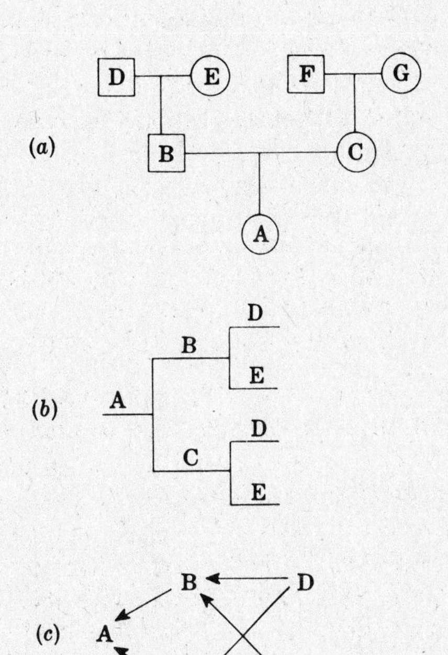


**Fig. 11-9.** Pedigree diagrams.

**Example 11.22.** In the arrow diagram of Fig. 11-9(*c*), there are two pathways connecting B and C. $R_{BC} = \Sigma(\frac{1}{2})^s$, where $s$ is the number of steps (arrows) from B to the common ancestor and back to C.

B and C probably contain $(\frac{1}{2})(\frac{1}{2}) = \frac{1}{4}$ of their genes in common through ancestor D.

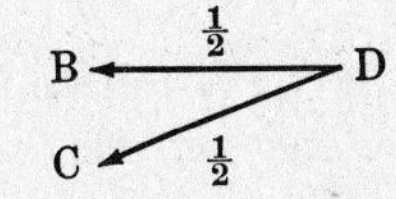


Similarly, B and C probably contain $\frac{1}{4}$ of their genes in common through ancestor E.

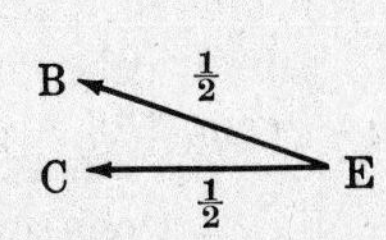


The sum of these two pathways is the coefficient of relationship between the full sibs B and C; $R_{BC} = \frac{1}{4} + \frac{1}{4} = \frac{1}{2}$ or 50%.

When matings occur only between closely related individuals (inbreeding) the genetic effect is an increase in homozygosity. The most intense form of inbreeding is self fertilization. If we start with a population containing 100 heterozygous individuals (*Aa*) as shown in Table 11.1, the expected number of homozygous genotypes is increased by 50% due to selfing in each generation.

**Table 11.1. Expected Increase in Homozygosity Due to Selfing**

| Generation | Genotypes *AA* | | *Aa* | | *aa* | Percent Heterozygosity | Percent Homozygosity |
|---|---|---|---|---|---|---|---|
| 0 | | | 100 | | | 100 | 0 |
| 1 | 25 | | 50 | | 25 | 50 | 50 |
| 2 | 25 | 12.5 | 25 | 12.5 | 25 | 25 | 75 |
| 3 | 37.5 | 6.25 | 12.5 | 6.25 | 37.5 | 12.5 | 87.5 |
| 4 | 43.75 | 3.125 | 6.25 | 3.125 | 43.75 | 6.25 | 93.75 |

Other less intense forms of inbreeding produce a less rapid approach to homozygosity, shown graphically in Fig. 11-10. As homozygosity increases in a population, either due to inbreeding or selection, the genetic variability of the population decreases. Since heritability depends upon the relative amount of genetic variability, it also decreases so that in the limiting case (pure line) heritability becomes zero.

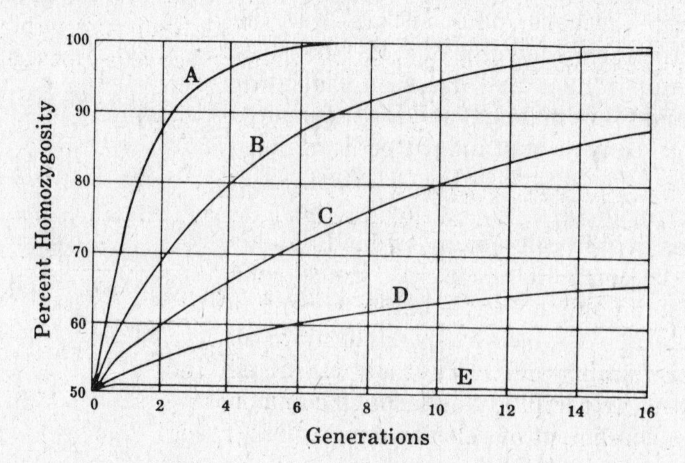


**Fig. 11-10.** Increase in percentage of homozygosity under various systems of inbreeding. (A) self fertilization, (B) full sibs, (C) double first cousins, (D) single first cousins, (E) second cousins.

When population size is reduced to a small isolated unit containing less than about 50 individuals, inbreeding very likely will result in a detectable increase in genetic uniformity. The *coefficient of inbreeding* (symbolized by F) is a useful indicator of the probable effect which inbreeding has had at two levels.

(1) On an *individual basis*, the coefficient of inbreeding indicates the probability that the two alleles at any locus are identical by descent, i.e. they are both replication products of a gene present in a common ancestor.

(2) On a *population basis*, the coefficient of inbreeding indicates the percentage of all loci which were heterozygous in the base population that now have probably become homozygous due to the effects of inbreeding. The base population is that point in the history of the population from which we desire to begin a calculation of the effects of inbreeding. Many loci are probably homozygous at the time we establish our base population. The inbreeding coefficient then measures the

*additional* increase in homozygosity due to matings between closely related individuals.

The coefficient of inbreeding (F) can be determined for an individual in a pedigree by several similar methods.

(1) If the common ancestor is not inbred, the inbreeding coefficient of an individual ($F_x$) is half the coefficient of relationship between the sire and dam ($R_{SD}$):

$$F_x = \tfrac{1}{2}R_{SD} \tag{11.23}$$

(2) If the common ancestors are not inbred, the inbreeding coefficient is given by

$$F_x = \sum (\tfrac{1}{2})^{p_1+p_2+1} \tag{11.24}$$

where $p_1$ is the number of generations (arrows) from one parent back to the common ancestor and $p_2$ is the number of generations from the other parent back to the same ancestor.

(3) If the common ancestors are inbred ($F_A$), the inbreeding coefficient of the individual must be corrected for this factor:

$$F_x = \sum [(\tfrac{1}{2})^{p_1+p_2+1}(1+F_A)] \tag{11.25}$$

(4) The coefficient of inbreeding of an individual may be calculated by counting the number of arrows ($n$) which connect the individual through one parent back to the common ancestor and back again to his or her other parent, and applying the formula

$$F_x = \sum (\tfrac{1}{2})^n (1+F_A) \tag{11.26}$$

The following table will be helpful in calculating F.

| $n$ | 1 | 2 | 3 | 4 | 5 | 6 | 7 | 8 | 9 |
|---|---|---|---|---|---|---|---|---|---|
| $(\frac{1}{2})^n$ | 0.5000 | 0.2500 | 0.1250 | 0.0625 | 0.0312 | 0.0156 | 0.0078 | 0.0039 | 0.0019 |

*Linebreeding* is a special form of inbreeding utilized for the purpose of maintaining a high genetic relationship to a desirable ancestor. Fig. 11-11 shows a pedigree in which close linebreeding to B has been practiced so that A possesses more than 50% of B's genes. D possesses 50% of B's genes and transmits 25% to C. B also contributes 50% of his genes to C. Hence C contains 50% + 25% = 75% B genes and transmits half of them (37.5%) to A. B also contributes 50% of his genes to A. Therefore A has 50% + 37.5% = 87.5% of B's genes.

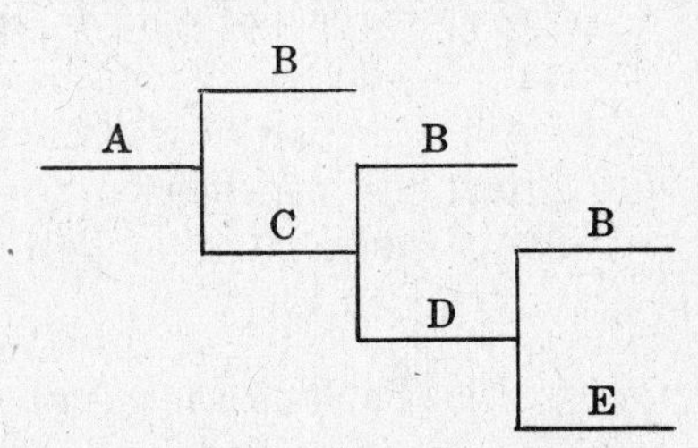

**Fig. 11-11.** Pedigree exemplifying close linebreeding.

(*b*) ***Based on Phenotypic Similarity.***

Positive phenotypic assortative mating is seldom practiced in its purest form among the selected individuals, i.e. mating only "look-alikes" or those with nearly the same selection indices. However, it can be used in conjunction with random mating; a few of the best among the selected group are "hand coupled", artificially cross-pollinated, or otherwise forced to breed.

**Example 11.23.** A beef cattle rancher may maintain a small "show string" in addition to a commercial herd. The few show animals would be closest to the ideal breed type (conformation of body parts, size for age, color markings, shape of horns, etc.) and would be mated like-to-like in hopes of generating more of the same for displaying at fairs and livestock expositions. The rest of the herd would be randomly mated to produce slaughter beef. Some of the cows from the commercial herd might eventually be selected for the show string; some of the young bulls or cows of the show string might not prove to be good enough to save for show and yet perform adequately as members of the commercial herd.

Both inbreeding and positive phenotypic assortative mating tend to reduce genetic heterozygosity, but the theoretical end results are quite different.

**Example 11.24.** As a model consider a metric trait governed by two loci, each with a pair of alleles both additive and equal in effect. Inbreeding among the five phenotypes would ultimately fix four homozygous lines (*AABB*, *AAbb*, *aaBB*, *aabb*). Positive phenotypic assortative mating would only fix two lines (*AABB* and *aabb*).

The rate at which heterozygous loci can be fixed (brought to homozygosity) in a population can be greatly accelerated by combining a system of close inbreeding with the additional restriction of positive phenotypic assortative mating; in other words, they must also "look" alike.

### 3. Negative Assortative Mating.

#### (*a*) *Based on Genetic Relatedness.*

When a mating involves individuals that are more distantly related than the average of the selected group it is classified as a negative genetic assortative mating. This may involve crossing individuals belonging to different families or crossing different inbred varieties of plants or crossing different breeds of livestock. It may occasionally involve crossing closely related species such as the horse and ass (donkey, burro) to produce the hybrid mule. The usual purpose of these "outcrosses" is an attempt to produce offspring of superior phenotypic quality (but not necessarily in breeding value) to that normally found in the parental populations.

Many recessives remain hidden in heterozygous conditions in non-inbred populations, but as homozygosity increases in an inbred population there is a greater probability that recessive traits, many of which are deleterious, will begin to appear. One of the consequences of inbreeding is a loss in vigor (i.e. less productive vegetatively and reproductively) which commonly accompanies an increase in homozygosity (*inbreeding depression*). Crosses between inbred lines usually produce a vigorous hybrid $F_1$ generation. This increased "fitness" of heterozygous individuals has been termed *heterosis.* The genetic basis of heterosis is still a subject of controversy, largely centered about two theories.

(1) The *dominance theory* of heterosis. Hybrid vigor is presumed to result from the action and interaction of dominant growth or fitness factors.

**Example 11.25.** Assume that four loci are contributing to a quantitative trait. Each recessive genotype contributes one unit to the phenotype and each dominant genotype contributes two units to the phenotype. A cross between two inbred lines could produce a more highly productive (heterotic) $F_1$ than either parental line.

| | | | |
|---|---|---|---|
| P: | *AA bb CC dd* | × | *aa BB cc DD* |
| Phenotypic Value: | $2+1+2+1 = 6$ | | $1+2+1+2 = 6$ |
| $F_1$: | | *Aa Bb Cc Dd* | |
| | | $2+2+2+2 = 8$ | |

(2) The *overdominance theory* of heterosis. Heterozygosity *per se* is assumed to produce hybrid vigor.

**Example 11.26.** Assume that four loci are contributing to a quantitative trait, recessive genotypes contribute 1 unit to the phenotype, heterozygous genotypes contribute 2 units, and homozygous dominant genotypes contribute $1\frac{1}{2}$ units.

| | | | |
|---|---|---|---|
| P: | $aa\ bb\ CC\ DD$ | $\times$ | $AA\ BB\ cc\ dd$ |
| Phenotypic Value: | $1+1+1\frac{1}{2}+1\frac{1}{2} = 5$ | | $1\frac{1}{2}+1\frac{1}{2}+1+1 = 5$ |
| $F_1$: | | $Aa\ Bb\ Cc\ Dd$ | |
| | | $2+2+2+2 = 8$ | |

Phenotypic variability in the hybrid generation is generally much less than that exhibited by the inbred parental lines (Fig. 11-12). This indicates that the heterozygotes are less subject to environmental influences than the homozygotes. Geneticists use the term "buffering" to indicate that the organism's development is highly regulated genetically ("canalized"). Another term often used in this connection is *homeostasis* which signifies the maintenance of a "steady state" in the development and physiology of the organism within the normal range of environmental fluctuations.

A rough guide to the estimation of heterotic effects (H) is obtained by noting the average excess in vigor which $F_1$ hybrids exhibit over the midpoint between the means of the inbred parental lines (Fig. 11-12).

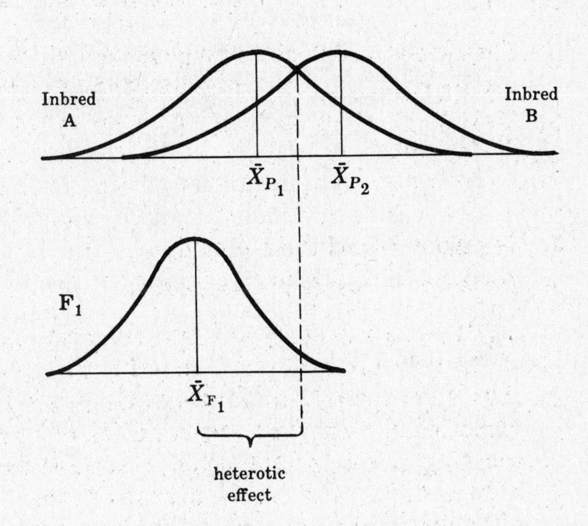


**Fig. 11-12.** Heterosis in the progeny from crossing inbred lines.

$$H_{F_1} = \bar{X}_{F_1} - \tfrac{1}{2}(\bar{X}_{P_1} + \bar{X}_{P_2}) \qquad (11.27)$$

The heterosis exhibited by an $F_2$ population is commonly observed to be half of that manifested by the $F_1$ hybrids.

(*b*) ***Based on Phenotypic Dissimilarity.***

When intermediate phenotypes are preferred, they are more likely to be produced by mating opposite phenotypes. For example, general purpose cattle can be produced by crossing a beef type with a dairy type. The offspring commonly produce an intermediate yield of milk and hang up a fair carcass when slaughtered (although generally not as good in either respect as the parental types). The same is true of the offspring from crossing an egg type (such as the Leghorn breed of chicken) with a meat type (such as the Cornish). Crossing phenotypic opposites may also be made to correct specific defects.

**Example 11.27.** Brahman cattle have more heat tolerance and resistance to certain insects than European cattle breeds. Brahmans are often crossed to these other breeds in order to create hybrids with the desirable qualities of both parental populations.

**Example 11.28.** Sometimes "weedy" relatives of agriculturally important crops may carry genes for resistance to specific diseases. Hybrids from such crosses may acquire disease resistance, and successive rounds of selection combined with backcrossing to the crop variety can eventually fix the gene or genes for disease resistance on a background that is essentially totally that of the cultivated species.

# Solved Problems

## QUASI-QUANTITATIVE TRAITS

**11.1.** Two homozygous varieties of *Nicotiana longiflora* have mean corolla lengths of 40.5 mm and 93.3 mm. The average of the $F_1$ hybrids from these two varieties was of intermediate length. Among 444 $F_2$ plants, none was found to have flowers either as long as or as short as the average of the parental varieties. Estimate the minimal number of pairs of alleles segregating from the $F_1$.

**Solution:**

If four pairs of alleles were segregating from the $F_1$, we expect $(\frac{1}{4})^4 = 1/256$ of the $F_2$ to be as extreme as one or the other parental average. Likewise, if five pairs of alleles were segregating, we expect $(\frac{1}{4})^5 = 1/1024$ of the $F_2$ to be as extreme as one parent or the other. Since none of the 444 $F_2$ plants had flowers this extreme, more than four loci (minimum of five loci) are probably segregating from the $F_1$.

**11.2.** The mean internode length of the Abed Binder variety of barley was found to be 3.20 mm. The mean length in the Asplund variety was 2.10 mm. Crossing these two varieties produced an $F_1$ and $F_2$ with average internode lengths of 2.65 mm. About 6% of the $F_2$ had an internode length of 3.2 mm and another 6% had a length of 2.1 mm. Determine the most probable number of gene pairs involved in internode length and the approximate contribution each gene makes to the phenotype.

**Solution:**

With one pair of genes we expect about $\frac{1}{4}$ or 25% of the $F_2$ to be as extreme as one of the parents. With two pairs of genes we expect approximately $\frac{1}{16}$ or 6.25% as extreme as one parent. Thus we may postulate two pairs of genes. Let $A$ and $B$ represent growth factors and $a$ and $b$ represent null genes.

P: *AA BB* (Abed Binder, 4 growth genes = 3.2 mm) × *aa bb* (Asplund, No growth genes = 2.1 mm)

$F_1$: *Aa Bb* (2 growth genes = 2.65 mm)

The difference $2.65 - 2.10 = 0.55$ mm is the result of two growth genes. Therefore each growth gene contributes 0.275 mm to the phenotype.

$F_2$:

| No. of Growth Genes | Mean Internode Length (mm) | Frequency | Genotypes |
|---|---|---|---|
| 4 | 3.200 | 1/16 | *AABB* |
| 3 | 2.925 | 1/4 | *AaBB*, *AABb* |
| 2 | 2.650 | 3/8 | *AAbb*, *AaBb*, *aaBB* |
| 1 | 2.375 | 1/4 | *aaBb*, *Aabb* |
| 0 | 2.100 | 1/16 | *aabb* (physiological minimum due to residual genotype) |

**11.3.** A large breed of chicken, the Golden Hamburg, was crossed to small Sebright Bantams. The $F_1$ was intermediate in size. The mean size of the $F_2$ was about the same as that of the $F_1$, but the variability of the $F_2$ was so great that a few individuals were found to exceed the size of either parental type (*transgressive variation*). If all of the alleles contributing to size act with equal and cumulative effects, and if the parents are considered to be homozygous, how can these results be explained?

**Solution:**

Let capital letters stand for growth genes (active alleles) and small letters stand for alleles which do not contribute to growth (null alleles). For simplicity we will consider only four loci.

P: $aa\ BB\ CC\ DD$ × $AA\ bb\ cc\ dd$

large Golden Hamburg, 6 active alleles — small Sebright Bantam, 2 active alleles

$F_1$: $Aa\ Bb\ Cc\ Dd$

intermediate sized hybrid
4 active alleles

$F_2$: Some genotypes could segregate out in the $F_2$ with phenotypic values which exceed that of the parents. For example:

|  |  | |
|---|---|---|
| $AA\ BB\ CC\ DD$ | 8 active alleles | larger than Golden Hamburg |
| $Aa\ BB\ CC\ DD$ | 7 active alleles | |
| $Aa\ bb\ cc\ dd$ | 1 active allele | smaller than Sebright Bantams |
| $aa\ bb\ cc\ dd$ | no active alleles (physiological minimum) | |

## THE NORMAL DISTRIBUTION

**11.4.** A representative sample of lamb weaning weights is shown below. Determine the weight limits within which 95% of all lambs from this population are expected to be found at weaning time.

| | | | | |
|---|---|---|---|---|
| 81 | 81 | 83 | 101 | 86 |
| 65 | 68 | 77 | 66 | 92 |
| 94 | 85 | 105 | 60 | 90 |
| 94 | 90 | 81 | 63 | 58 |

**Solution:**

The standard deviation is calculated as follows.

$$\bar{X} = \frac{\Sigma X}{N} = \frac{1620}{20} = 81.0$$

| $X$ | $X - \bar{X}$ | $(X - \bar{X})^2$ |
|---|---|---|
| 81 | 0 | 0 |
| 65 | −16 | 256 |
| 94 | +13 | 169 |
| 94 | +13 | 169 |
| 81 | 0 | 0 |
| 68 | −13 | 169 |
| 85 | +4 | 16 |
| 90 | +9 | 81 |
| 83 | +2 | 4 |
| 77 | −4 | 16 |
| 105 | +24 | 576 |
| 81 | 0 | 0 |
| 101 | +20 | 400 |
| 66 | −15 | 225 |
| 60 | −21 | 441 |
| 63 | −18 | 324 |
| 86 | +5 | 25 |
| 92 | +11 | 121 |
| 90 | +9 | 81 |
| 58 | −23 | 529 |
| $\Sigma X = 1620$ | $\Sigma (X - \bar{X}) = 0$ | $\Sigma (X - \bar{X})^2 = 3602$ |

$$s = \sqrt{\frac{\Sigma (X - \bar{X})^2}{N - 1}} = \sqrt{\frac{3602}{19}} = 13.77$$

95% of all weaning weights are expected to lie within $\pm 2s$ of the mean. Thus $\bar{X} \pm 2s = 81.0 \pm 2(13.77) = 81.0 \pm 27.54$. The upper limit is 108.54 and the lower limit is 53.46.

**11.5.** The average fleece weight in a large band of sheep together with its standard deviation was calculated to be $10.3 \pm 1.5$ lb. The statistics for fleece grade (on a scale from 0 to 10) was $5.1 \pm 0.7$ units. Which trait is relatively more variable?

**Solution:**

Relative variability may be determined from a comparison of their coefficients of variation.

$$\text{C.V. (fleece weight)} = s/\bar{X} = 1.5/10.3 = 0.146$$
$$\text{C.V. (fleece grade)} = 0.7/5.1 = 0.137$$

Fleece weight has a slightly higher coefficient of variation and thus is relatively more variable than fleece grade.

**11.6.** The Flemish breed of rabbits has an average body weight of 3600 grams. The Himalayan breed has a mean of 1875 grams. Matings between these two breeds produce an intermediate $F_1$ with a standard deviation of $\pm 162$ grams. The variability of the $F_2$ is greater as indicated by a standard deviation of $\pm 230$ grams. (*a*) Estimate the number of pairs of factors contributing to mature body weight in rabbits. (*b*) Estimate the average metric contribution of each active allele.

**Solution:**

(*a*) From equation (*11.6*),

$$N = \frac{D^2}{8(\sigma^2_{PF2} - \sigma^2_{PF1})} = \frac{(3600-1875)^2}{8(230^2 - 162^2)} = 13.95 \text{ or approximately 14 pairs}$$

(*b*) The difference $3600 - 1875 = 1725$ grams is attributed to 14 *pairs* of factors or 28 active alleles. The average contribution of each active allele is $1725/28 = 61.61$ grams.

**11.7.** In a population having a phenotypic mean of 55 units, a total genetic variance for the trait of 35 units², and an environmental variance of 14 units², between what two phenotypic values will approximately 68% of the population be found?

**Solution:**

$\sigma^2_P = \sigma^2_G + \sigma^2_E = 35 + 14 = 49, \quad \sigma_P = 7.$

68% of a normally distributed population is expected to be found within the limits $\mu \pm \sigma = 55 \pm 7$, or between 48 units and 62 units.

## TYPES OF GENE ACTION

**11.8.** The $F_1$ produced by crossing two varieties of tomatoes has a mean fruit weight of 50 grams and a phenotypic variance of 225 gram². The $F_1$ is backcrossed to one of the parental varieties having a mean of 150 grams. Assuming that tomato fruit weight is governed by multiplicative gene action, predict the variance of the backcross progeny.

**Solution:**

The expected mean of the backcross progeny is the geometric mean of 150 and 50, or $\sqrt{150 \times 50} = \sqrt{7500} = 86.6$. If multiplicative gene action is operative the variance is dependent upon the mean, thus producing a constant coefficient of variation in segregating populations.

$F_1$ coefficient of variation $= s/\bar{X} = \sqrt{225}/50 = 0.3$, and so the coefficient of variation in the backcross generation is also expected to be 0.3.

Standard deviation of backcross $= 0.3(86.6) = 25.98$.

Variance of backcross $= (25.98)^2 = 675$ gram².

## HERITABILITY

**11.9.** Identical twins are derived from a single fertilized egg (monozygotic). Fraternal twins develop from different fertilized eggs (dizygotic) and are expected to have about half of their genes in common. Left-hand middle finger length measurements were taken on the fifth birthday in samples (all of the same sex) of identical twins, fraternal twins, and unrelated individuals from a population of California Caucasians. Using only variances between twins and between unrelated members of the total population, devise a formula for estimating the heritability of left-hand middle finger length at five years of age in this population.

**Solution:**

Let $V_i$ = phenotypic variance between identical twins, $V_f$ = phenotypic variance between fraternal twins, and $V_t$ = phenotypic variance between randomly chosen pairs from the total population of which these twins are a part. All of the phenotypic variance between identical twins is nongenetic (environmental); thus $V_i = V_e$. The phenotypic variance between fraternal twins is partly genetic and partly environmental. Since fraternal twins are 50% related, their genetic variance is expected to be only half that of unrelated individuals; thus $V_f = \frac{1}{2}V_g + V_e$. The difference $(V_f - V_i)$ estimates half of the genetic variance.

$$V_f - V_i = \underbrace{\tfrac{1}{2}V_g + V_e}_{V_f} - \underbrace{V_e}_{V_i} = \tfrac{1}{2}V_g$$

Therefore, heritability is twice that difference divided by the total phenotypic variance.

$$h^2 = \frac{2(V_f - V_i)}{V_t}$$

**11.10.** Two homozygous varities of *Nicotiana longiflora* were crossed to produce $F_1$ hybrids. The average variance of corolla length for all three populations was 8.76. The variance of the $F_2$ was 40.96. Estimate the heritability of flower length in the $F_2$ population.

**Solution:**

Since the two parental varieties and the $F_1$ are all genetically uniform, their average phenotypic variance is an estimate of the environmental variance ($V_e$). The phenotypic variance of the $F_2$ ($V_t$) is partly genetic and partly environmental. The difference ($V_t - V_e$) is the genetic variance ($V_g$).

$$h^2 = \frac{V_g}{V_t} = \frac{V_t - V_e}{V_t} = \frac{40.96 - 8.76}{40.96} = 0.79$$

**11.11.** Given the phenotypic variances of a quantitative trait in two pure lines ($V_{P1}$ and $V_{P2}$), in their $F_1$ and $F_2$ progenies ($V_{F1}$ and $V_{F2}$), in the backcross progeny $F_1 \times P_1(V_{B1})$ and in the backcross progeny $F_1 \times P_2(V_{B2})$, show how estimates of additive genetic variance ($V_A$), dominance genetic variance ($V_D$), and environmental variance ($V_E$) can be derived.

**Solution:**

Since all of the phenotypic variance within pure lines and their genetically uniform $F_1$ progeny is environmental,

$$V_E = \frac{V_{P1} + V_{P2} + V_{F1}}{3}$$

Therefore,

$$V_{F2} - V_E = (\tfrac{1}{2}A + \tfrac{1}{4}D + E) - E = \tfrac{1}{2}A + \tfrac{1}{4}D \qquad (1)$$

Likewise,

$$(V_{B1} + V_{B2}) - V_E = \tfrac{1}{2}A + \tfrac{1}{2}D + 2E = \tfrac{1}{2}A + \tfrac{1}{2}D \qquad (2)$$

By multiplying equation (1) by two and subtracting equation (2) from the result, we can solve for $A$.

$$\begin{aligned} A + \tfrac{1}{2}D &= 2(V_{F2} - V_E) \\ \tfrac{1}{2}A + \tfrac{1}{2}D &= (V_{B1} + V_{B2}) - V_E \\ \hline \tfrac{1}{2}A \quad &= V_A \end{aligned}$$

By substituting $\frac{1}{2}A$ into equation (1), we can solve for $\frac{1}{4}D = V_D$.

**11.12.** The pounds of grease fleece weight was measured in a sample from a sheep population. The data listed below is for the average of both parents ($X$, midparent) and their offspring ($Y$).

| $X$ | 11.8 | 8.4 | 9.5 | 10.0 | 10.9 | 7.6 | 10.8 | 8.5 | 11.8 | 10.5 |
|---|---|---|---|---|---|---|---|---|---|---|
| $Y$ | 7.7 | 5.7 | 5.8 | 7.2 | 7.3 | 5.4 | 7.2 | 5.6 | 8.4 | 7.0 |

(*a*) Calculate the regression coefficient of offspring on midparent and estimate the heritability of grease fleece weight in this population. (*b*) Plot the data and draw the regression line. (*c*) Calculate the correlation coefficient and from that estimate the heritability.

**Solution:**

| $X$ | $Y$ | $X^2$ | $Y^2$ | $XY$ |
|---|---|---|---|---|
| 11.8 | 7.7 | 139.24 | 59.29 | 90.86 |
| 8.4 | 5.7 | 70.56 | 32.49 | 47.88 |
| 9.5 | 5.8 | 90.25 | 33.64 | 55.10 |
| 10.0 | 7.2 | 100.00 | 51.84 | 72.00 |
| 10.9 | 7.3 | 118.81 | 53.29 | 79.57 |
| 7.6 | 5.4 | 57.76 | 29.16 | 41.04 |
| 10.8 | 7.2 | 116.64 | 51.84 | 77.76 |
| 8.5 | 5.6 | 72.25 | 31.36 | 47.60 |
| 11.8 | 8.4 | 139.24 | 70.56 | 99.12 |
| 10.5 | 7.0 | 110.25 | 49.00 | 73.50 |
| $\Sigma X = 99.8$ | $\Sigma Y = 67.3$ | $\Sigma X^2 = 1015.00$ | $\Sigma Y^2 = 462.47$ | $\Sigma XY = 684.43$ |

$$n = 10$$

$$\frac{(\Sigma X)^2}{n} = \frac{(99.8)^2}{10} = \frac{9960.04}{10} = 996.0$$

$$\frac{\Sigma X \, \Sigma Y}{n} = \frac{(99.8)(67.3)}{10} = \frac{6716.54}{10} = 671.65$$

$$b = \frac{\Sigma XY - [(\Sigma X \, \Sigma Y)/n]}{\Sigma X^2 - [(\Sigma X)^2/n]} = \frac{684.43 - 671.65}{1015 - 996} = \frac{12.78}{19} = 0.6726$$

The regression of offspring on midparent is an estimate of heritability:

$$h^2 = 0.67$$

(*b*) Data plot and regression line:

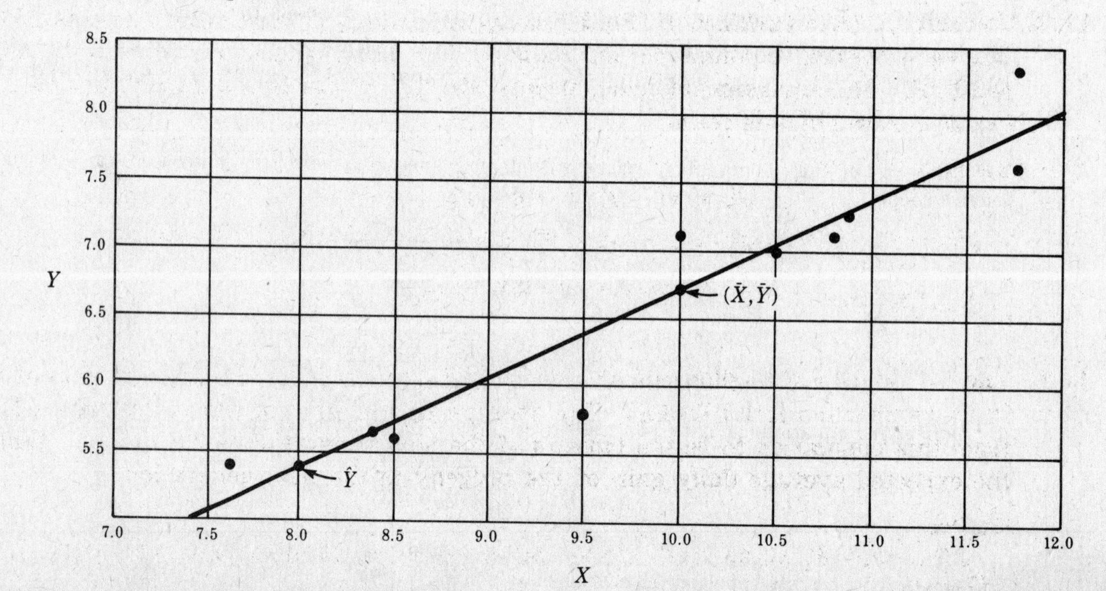

The regression line goes through the intersection of the two means $(\overline{X}, \overline{Y})$; $\overline{X} = 99.8/10 = 9.98$; $\overline{Y} = 67.3/10 = 6.73$. The regression line intersects the $Y$ axis at the $Y$ intercept ($a$).

$$a = \overline{Y} - b\overline{X}$$
$$= 6.73 - 0.673(9.98) = 0.01$$

Now let us choose a value of $X$ that is distant from $(\overline{X}, \overline{Y})$ but easily plotted on the graph (e.g. $X = 8.0$). The corresponding value of Y is estimated to be

$$\hat{Y} = a + bX = 0.01 + 0.673(8.0) = 5.39$$

These two points $((\overline{X}, \overline{Y})$ and $\hat{Y})$ establish the regression line with slope $b = 0.67$. For every one-pound increase in midparent values, offspring tend to produce 0.67 pound.

(c)

$$r = \frac{\overbrace{\Sigma XY - [(\Sigma X\, \Sigma Y)/n]}^{\text{same numerator as } b}}{\sqrt{\underbrace{\{\Sigma X^2 - [(\Sigma X)^2/n]\}}_{\text{same denominator as } b}\{\Sigma Y^2 - [(\Sigma Y)^2/n]\}}} \qquad \Sigma Y^2 = 462.47 \qquad \frac{(\Sigma Y)^2}{n} = \frac{4529.29}{10} = 452.93$$

$$r = \frac{12.78}{\sqrt{(19)(462.47 - 452.93)}} = \frac{12.78}{\sqrt{181.26}} = \frac{12.78}{13.46} = 0.95$$

Therefore, the $X$ and $Y$ values are very highly positively correlated. Note that two variables can be highly correlated without also being nearly equal. Two variables are perfectly correlated if for one unit change in one variable there is a constant change (either plus or minus) in the other. Negative correlations for heritability estimates are biologically meaningless. Different traits, however, may have negative genetic correlations (e.g. total milk production vs. butterfat percentage in dairy cattle); many of the same genes that contribute positively to milk yield also contribute negatively to butterfat content.

$$h^2 = b = r\left(\frac{s_Y}{s_X}\right)$$

$$s_Y = \sqrt{\frac{\Sigma Y^2 - [(\Sigma Y)^2/n]}{n-1}} = \sqrt{\frac{462.47 - 452.93}{9}} = 1.03$$

$$s_X = \sqrt{\frac{\Sigma X^2 - [(\Sigma X)^2/n]}{n-1}} = \sqrt{\frac{1015 - 996}{9}} = 1.45$$

$$h^2 = b = 0.95\left(\frac{1.03}{1.45}\right) = 0.95(0.71) = 0.67$$

**11.13.** The total genetic variance of 180 day body weight in a population of swine is 250 $lb^2$. The variance due to dominance effects is 50 $lb^2$. The variance due to epistatic effects is 20 $lb^2$. The environmental variance is 350 $lb^2$. What is the heritability estimate (narrow sense) of this trait?

**Solution:**

$$\sigma_P^2 = \sigma_G^2 + \sigma_E^2 = 250 + 350 = 600$$
$$\sigma_G^2 = \sigma_A^2 + \sigma_D^2 + \sigma_I^2, \quad 250 = \sigma_A^2 + 50 + 20, \quad \sigma_A^2 = 180$$
$$h^2 = \sigma_A^2/\sigma_P^2 = 180/600 = 0.3$$

**11.14.** The heritability of feedlot rate of gain in beef cattle is 0.6. The average rate of gain in the population is 1.7 lb/day. The average rate of gain of the individuals selected from this population to be the parents of the next generation is 2.8 lb/day. What is the expected average daily gain of the progeny in the next generation?

**Solution:**

$\Delta P = \overline{P}_p - \overline{P}_1 = 2.8 - 1.7 = 1.1$. $\Delta G = h^2(\Delta P) = 0.6(1.1) = 0.66$. $\overline{P}_2 = \overline{P}_1 + \Delta G = 2.36$ lb/day.

## SELECTION METHODS

**11.15.** Fifty gilts (female pigs) born each year in a given herd can be used for proving sires. Average litter size at birth is 10 with 10% mortality to maturity. Only the five boars (males) with the highest sire index will be saved for further use in the herd. If each test requires 18 mature progeny, how much culling can be practiced among the progeny tested boars, i.e. what proportion of those tested will not be saved?

**Solution:**

Each gilt will produce an average of $10 - (0.1)(10) = 9$ progeny raised to maturity. If 18 mature progeny are required to prove a sire, then each boar should be mated to two gilts. (50 gilts)/(2 gilts per boar) = 25 boars can be proved. $20/25 = 4/5 = 80\%$ of these boars will be culled.

**11.16.** Given the following pedigree with butterfat records on the cows and equal-parent indices on the bulls, estimate the index for young bull X (*a*) using information from A and B, (*b*) when the record made by B is only one lactation, and that made in another herd.

X

A (750 lb index)

B (604 lb record)

C (820 lb index)

D (492 lb record)

**Solution:**

(*a*) The midparent index (estimate of transmitting ability) for X is $(750 + 604)/2 = 677$.

(*b*) Since we cannot rely on B's record, we should use information from C and D, recalling that X is separated by two Mendelian segregations from the grandparents. Then $X = 750/2 + 820/4 + 492/4 = 703$.

## MATING METHODS

**11.17.** Calculate the inbreeding coefficient for A in the following pedigree.

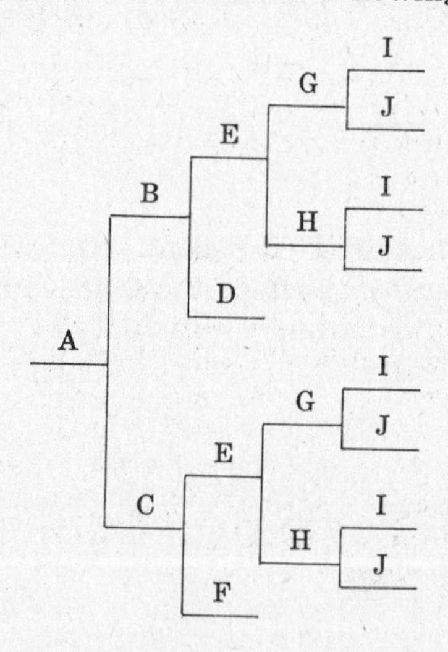


**Solution:**

First we must convert the pedigree to an arrow diagram.

D B G I A E C H J F

There is only one pathway from B to C and that goes through ancestor E. However, ancestor E is himself inbred. Note that the parents of E are full sibs, i.e. G and H are 50% related (see Example 11.22). By formula (*11.23*),

$$F_E = \tfrac{1}{2}R_{GH} = \tfrac{1}{2}(0.5) = 0.25$$

The inbreeding coefficient of A is given by equation (*11.26*),

$$F_A = \sum (\tfrac{1}{2})^n (1 + F_{\text{ancestor}}) = (\tfrac{1}{2})^3(1+0.25) = 0.156$$

where $n$ is the number of arrows connecting the individual (A) through one parent (B) back to the common ancestor (E) and back again to the other parent (C).

**11.18.** The average plant heights of two inbred tobacco varieties and their hybrids have been measured with the following results: Inbred parent ($P_1$) = 47.8 inches, Inbred parent ($P_2$) = 28.7 inches, $F_1$ hybrid ($P_1 \times P_2$) = 43.2 inches. (*a*) Calculate the amount of heterosis exhibited by the $F_1$. (*b*) Predict the average height of the $F_2$.

**Solution:**

(*a*) The amount of heterosis is expressed by the excess of the $F_1$ average over the midpoint between the two parental means.

$$\text{Heterosis of } F_1 = \bar{X}_{F_1} - \tfrac{1}{2}(\bar{X}_{P_1} + \bar{X}_{P_2}) = 43.2 - \tfrac{1}{2}(47.8 + 28.7) = 43.2 - 38.25 = 4.95 \text{ inches.}$$

(*b*) As a general rule the $F_2$ shows only about half the heterosis of the $F_1$: $\tfrac{1}{2}(4.95) = 2.48$. Hence the expected height of $F_2$ plants = 38.25 + 2.48 = 40.73 inches.

# Supplementary Problems

## QUASI-QUANTITATIVE TRAITS

**11.19.** Beginning at some arbitrary date, two varieties of wheat were scored for the length of time (in days) to heading, from which the following means were obtained: variety X = 13.0 days, variety Y = 27.6 days. From a survey of 5,504,000 $F_2$ progeny, 86 were found to head out in 13 days or less. How many pairs of factors are probably contributing to early flowering?

**11.20.** Suppose that the average skin color of one racial population is 0.43 (measured by the reflectance of skin to red light of 685 nm wavelength); the average skin color of a population racially distinct from the first is 0.23; and racial hybrids between these two populations average 0.33. If about 1/150 offspring from hybrid (racially mixed) parents have skin colors as extreme as the average of either race, estimate the number of segregating loci in the hybrid parents that contribute to skin color variability in their offspring.

**11.21.** Suppose that 5 pairs of genes with equal and cumulative effects are contributing to body weight in laboratory albino rats. Two highly inbred strains (homozygous) have the extreme high and low mature weights respectively. A hybrid $F_1$ generation is produced by crossing these two lines. The average cost of raising a rat to maturity is \$2.00. What will it probably cost the breeder to recover in the $F_2$ a rat which is as extreme as the high parental line?

## THE NORMAL DISTRIBUTION

**11.22.** From a sample of 10 pig body weights determine (*a*) mean body weight, (*b*) sample standard deviation ($s$), (*c*) the weight which will probably be exceeded by $2\tfrac{1}{2}\%$ of this population. Pig weights: 210, 215, 220, 225, 215, 205, 220, 210, 215, 225.

## VARIANCE

**11.23.** Suppose 6 pairs of genes were contributing to a metric trait in a cultivated crop. Two parental lines with averages of 13,000 lb/acre and 7,000 lb/acre produced an intermediate hybrid $F_1$ with a variance of 250,000 $lb^2$. Estimate the standard deviation of the $F_2$ by formula (*11.6*).

**11.24.** Two strains of mice were tested for susceptibility to a carcinogenic drug. The susceptible strain had an average of 75.4 tumorous lung nodules, whereas the resistant strain failed to develop nodules. The $F_1$ from crossing these two strains had an average of 12.5 nodules with a standard deviation of $\pm 5.3$; the $F_2$ had $10.0 \pm 14.1$ nodules. Estimate the number of gene pairs contributing to tumor susceptibility by use of formula (*11.6*).

## TYPES OF GENE ACTION

**11.25.** Calculate the metric values of the parents and their $F_1$ hybrids in the cross $AA\,B'B'\,CC\,D'D' \times A'A'\,BB\,C'C'\,DD$ assuming (*a*) additive gene action where unprimed alleles contribute 3 units each to the phenotype and primed alleles contribute 6 units each, (*b*) primed alleles are fully dominant to unprimed alleles; at a given locus, genotypes with one or two primed alleles produce 12 units and the recessive genotype produces 6 units.

**11.26.** Several generations of selection and inbreeding in a laboratory strain of mice produced a giant strain with a mean of 40 grams at two months of age and a midget strain with an average of 12 grams. The phenotypic variance of the giant strain is 26.01 and of the midget strain is 2.92. (*a*) Calculate the coefficient of variation for the giant and the midget strains. (*b*) On the basis of your findings in part (*a*), calculate the expected mean of the $F_1$ produced by crossing these two lines.

**11.27.** Several examples of multiplicative gene action are known in crosses between tomato varieties. Calculate the arithmetic mean and geometric mean for each $F_1$ and compare each of their absolute deviations from the $F_1$ mean. The mean fruit weight in grams for each group is given in parentheses.

| | Variety #1 | Variety #2 | $F_1$ |
|---|---|---|---|
| (*a*) | Yellow Pear (12.4) | Honor Bright (150.0) | (47.5) |
| (*b*) | Yellow Pear (12.4) | Peach (42.6) | (23.1) |
| (*c*) | Dwarf Aristocrat (112.4) | Peach (42.6) | (67.1) |
| (*d*) | Tangerine (173.6) | Red Currant (1.1) | (8.3) |

## HERITABILITY

**11.28.** Let $V_i$ = phenotypic variance between identical twins, $V_f$ = phenotypic variance between fraternal twins, and heritability $(h^2) = (V_f - V_i)/V_f$. Given the following differences in intelligence quotients (IQ) of 20 pairs of twins (all females, reared together, and identically tested at the same age), estimate the heritability of IQ.

| Identical twins | 6 | 2 | 7 | 2 | 4 | 4 | 3 | 5 | 5 | 2 |
|---|---|---|---|---|---|---|---|---|---|---|
| Fraternal twins | 10 | 7 | 13 | 15 | 12 | 11 | 14 | 9 | 12 | 17 |

**11.29.** Suppose that population A has a mean IQ of 85 and that of population B is 100. Estimates of heritability of IQ in both populations are relatively high (0.4 to 0.8). Explain why each of the following statements is false.

(*a*) Heritability estimates measure the degree to which a trait is determined by genes.

(*b*) Since the heritability of IQ is relatively high, the average differences between the two populations must be largely due to genetic differences.

(*c*) Since population B has a higher average IQ than population A, population B is genetically superior to A.

**11.30.** Flower lengths were measured in two pure lines, their $F_1$ and $F_2$ and backcross progenies. To eliminate multiplicative effects, logarithms of the measurements were used. The phenotypic variances were $P_1 = 48$, $P_2 = 32$, $F_1 = 46$, $B_1(F_1 \times P_1) = 85.5$ and $B_2(F_1 \times P_2) = 98.5$. (*a*) Estimate the environmental variance ($V_E$), the additive genetic variance ($V_A$) and the dominance genetic variance ($V_D$). (*b*) Calculate the degree of dominance. (*c*) Estimate the narrow heritability of flower length in the $F_2$.

**11.31.** Let $r_1$ = phenotypic correlation of full sibs, $r_2$ = phenotypic correlation of half sibs, $r_3$ = correlation of offspring with one parent, $r_4$ = correlation of monozygotic twins, and $r_5$ = correlation of dizygotic twins. In the following formulas, determine the values of $x$ and or $y$:

(*a*) $h^2 = x(r_1 - r_2)$ (*b*) $h^2 = xr_1 - yr_3$ (*c*) $h^2 = x(r_4 - r_5)$

**11.32.** In the following table, $Y$ represents the average number of bristles on a specific thoracic segment of *Drosophila melanogaster* in four female offspring and $X$ represents the number of bristles in the mother (dam) of each set of four daughters.

| Family | 1 | 2 | 3 | 4 | 5 | 6 | 7 | 8 | 9 | 10 |
|---|---|---|---|---|---|---|---|---|---|---|
| $X$ | 9 | 6 | 9 | 6 | 7 | 8 | 7 | 7 | 8 | 9 |
| $Y$ | 8 | 6 | 7 | 8 | 8 | 7 | 7 | 9 | 9 | 8 |

(*a*) Calculate the daughter-dam regression. (*b*) Estimate the heritability of bristle number in this population assuming $s_X = s_Y$.

**11.33.** The regression coefficient (*b*) represents how much one variable is expected to change per unit change in some other variable. The correlation coefficient (*r*) reflects how closely the data points are to the regression line (perfect correlation = $\pm 1$). Using only these definitions (no formulas), determine the regression of $X_1$ on $X_2$ and their correlation from the following pairs of measurements:

| $X_1$ | 11 | 12 | 13 | 14 | 15 |
|---|---|---|---|---|---|
| $X_2$ | 13 | 15 | 17 | 19 | 21 |

**11.34.** Around 1903, Pearson and Lee collected measurements of brother and sister heights from more than a thousand British families. A sample of eleven of such families is shown below.

| Family No. | 1 | 2 | 3 | 4 | 5 | 6 | 7 | 8 | 9 | 10 | 11 |
|---|---|---|---|---|---|---|---|---|---|---|---|
| Brother | 71 | 68 | 66 | 67 | 70 | 71 | 70 | 73 | 72 | 65 | 66 |
| Sister | 69 | 64 | 65 | 63 | 65 | 62 | 65 | 64 | 66 | 59 | 62 |

(*a*) Calculate the regression coefficient of sisters' height on brothers' height. (*b*) Calculate the regression coefficient of brothers' height on sisters' height. (*c*) The correlation coefficient (*r*) is the geometric mean of the above two regression coefficients; determine *r*. (*d*) If the variance of brothers' heights $s_B^2 = 74$ and that of sisters' $s_S^2 = 66$, calculate the heritability of body height from *r*. Does the answer make biological sense? Explain.

**11.35.** A flock of chickens has an average mature body weight of 6.6 lb. Individuals saved for breeding purposes have a mean of 7.2 lb. The offspring generation has a mean of 6.81 lb. Estimate the heritability of mature body weight in this flock.

**11.36.** Yearly wool records (in pounds) are taken from a sample of 10 sheep: 11.8, 8.4, 9.5, 10.0, 10.9, 7.8, 10.8, 8.5, 11.8, 10.5. (*a*) Calculate the range within which approximately 95% of the sheep in this population are expected to be found. (*b*) If the additive genetic variance is 0.60, what is the heritability estimate of wool production in this breed?

**11.37.** Determine (*a*) the dominance variance and (*b*) the environmental variance from the following information: heritability (formula (*11.9*) = 0.3, phenotypic variance = 200 lb$^2$, total genetic variance = 100 lb$^2$, and epistatic variance is absent.

**11.38.** Thickness of backfat in a certain breed of swine has been estimated to have a heritability of 80%. Suppose the average backfat thickness of this breed is 1.2 inches and the average of individuals selected from this population to be the parents of the next generation is 0.8 inch. What is the expected average of the next generation?

**11.39.** The average yearly milk production of a herd of cows is 18,000 lb. The average milk production of the individuals selected to be parents of the next generation is 20,000 lb. The average milk production of the offspring generation is 18,440 lb. (*a*) Estimate the heritability of milk production in this population. (*b*) If the phenotypic variance of this population is 4,000,000 lb$^2$, estimate the additive genetic variance. (*c*) Between what two values is the central 68% of the original (18,000 lb average) population expected to be found?

**11.40.** The average weight at 140 days in a swine population is 180 lb. The average weight of individuals selected from this population for breeding purposes is 195 lb. Heritability of 140 day weight in swine is 30%. Calculate (*a*) selection differential, (*b*) expected genetic gain in the progeny, (*c*) predicted average 140 day weight of the progeny.

**11.41.** About 1903 Johannsen, a Danish botanist, measured the weight of seeds in the Princess Variety of bean. Beans are self fertilizing and therefore this variety is a pure line. The weights in centigrams of a small but representative sample of beans are listed below.

| | | | | | | | | |
|---|---|---|---|---|---|---|---|---|
| 19 | 31 | 18 | 24 | 27 | 28 | 25 | 30 | 29 |
| 22 | 29 | 26 | 23 | 20 | 24 | 21 | 25 | 29 |

(*a*) Calculate the mean and standard deviation for bean weight in this sample. (*b*) Calculate the environmental variance. (*c*) Estimate the heritability of bean weight in this variety. (*d*) If the average bean weight of individuals selected to be parents from this population is 30 cg, predict the average bean weight of the next generation.

## SELECTION METHODS

**11.42.** The length of an individual beetle is 10.3 mm or 0.5 when expressed in "standardized" form. The average measurement for this trait in the beetle population is 10.0 mm. What is the variance of this trait?

**11.43.** Given the swine selection index I = 0.14W − 0.27S, where W is the pig's own 180 day weight and S is its market score. (*a*) Rank the following three animals according to index merit:

| Animal | Weight | Score |
|---|---|---|
| X | 220 | 48 |
| Y | 240 | 38 |
| Z | 200 | 30 |

(*b*) If differences in index score are 20% heritable, and parents score 3.55 points higher than the average of the population, how much increase in the average score of the progeny is expected?

**11.44.** A beef cattle index (I) for selecting replacement heifers takes the form I = 6 + 2WW′ + WG′, where WW′ is weaning weight in standardized form and WG′ is weaning grade in standardized form. The average weaning weight of the herd = 505 lb with a standard deviation of ± 34.5 lb. The average weaning grade (a numerical score) is 88.6 with a standard deviation of ± 2.1. Which of the following animals has the best overall merit?

| Animal | Actual Weaning Weight | Actual Weaning Grade |
|---|---|---|
| A | 519 | 88 |
| B | 486 | 91 |

**11.45.** Suppose 360 ewes (female sheep) are available for proving sires. All ewes lamb; 50% of ewes lambing have twins. The ten rams with the highest progeny test scores will be kept as flock sires. How much selection can be practiced among the progeny tested individuals, i.e. what proportion of those tested can be saved if a test requires (*a*) 18 progeny, (*b*) 12 progeny, (*c*) 6 progeny?

**11.46.** During the same year three dairy bulls were each mated to a random group of cows. The number of pounds of butterfat produced by the dams and their daughters (corrected to a 305 day lactation at maturity with twice daily milking) was recorded as shown below.

| Bull | Dam | Dam's Record | Daughter's Record |
|---|---|---|---|
| A | 1 | 600 | 605 |
| | 2 | 595 | 640 |
| | 3 | 615 | 625 |
| | 4 | 610 | 600 |
| B | 5 | 585 | 610 |
| | 6 | 590 | 620 |
| | 7 | 620 | 605 |
| | 8 | 605 | 595 |
| C | 9 | 590 | 590 |
| | 10 | 590 | 595 |
| | 11 | 610 | 600 |
| | 12 | 600 | 605 |

(*a*) Calculate the sire index for each of the three sires. (*b*) Which sire would you save for extensive use in your herd?

## MATING METHODS

**11.47.** A is linebred to B in the following pedigree. Calculate the inbreeding coefficient of A.

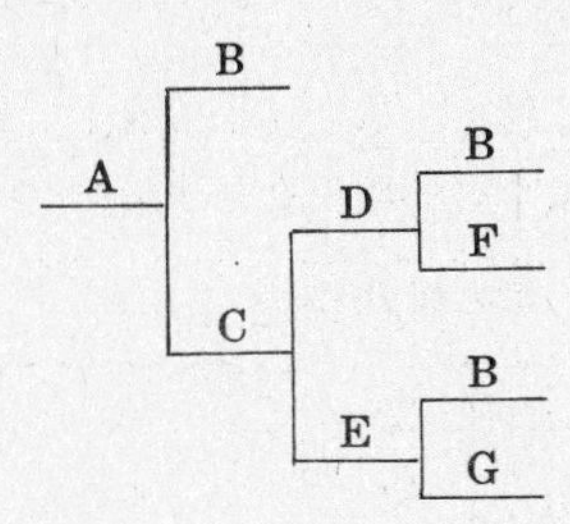


**11.48.** Given the following arrow diagram, calculate the inbreeding coefficient of A.

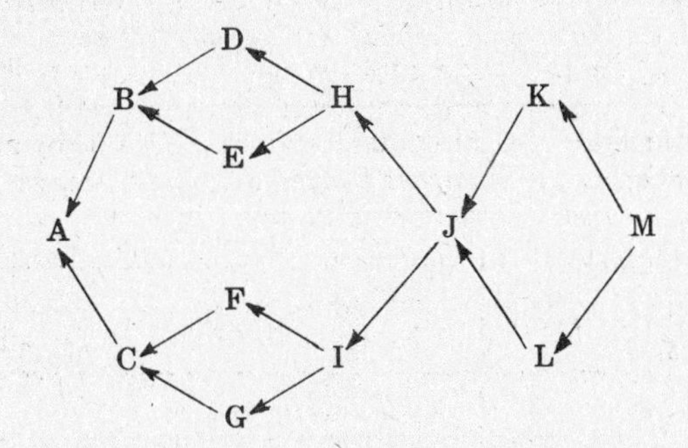


**11.49.** Calculate the inbreeding of A in the following. *Hint:* There are 9 pathways between B and C.

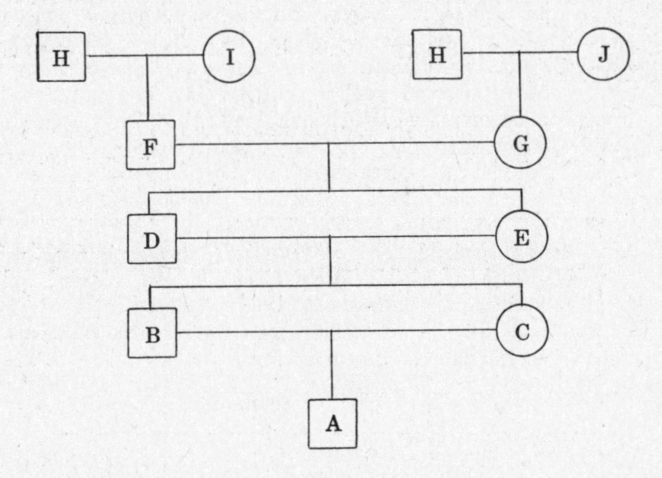


**11.50.** The yield of seed (in bushels per acre) and plant height (in centimeters) was measured on several generations of corn. Calculate by equation (*11.27*), (*a*) the amount of heterosis in the $F_1$ resulting from crossing the parental varieties with the inbreds, (*b*) the yield and height expectations of the $F_2$.

| | Seed Yield | Plant Height |
|---|---|---|
| Parental varieties | 73.3 | 265 |
| Inbreds | 25.0 | 193 |
| $F_1$ hybrids | 71.4 | 257 |

## Answers to Supplementary Problems

**11.19.** 8 pairs of factors

**11.20.** 3 or 4 loci (pairs of alleles)

**11.21.** $2048

**11.22.** (*a*) 216 lb (*b*) ±6.58 lb (*c*) 229.16 lb

**11.23.** ±1000 lb/acre

**11.24.** 4.16 or approximately 4 gene pairs

**11.25.** (*a*) Both parents, $F_1$ = 36 units (*b*) Both parents = 36 units, $F_1$ = 48 units

**11.26.** (*a*) Giant strain = 0.13, midget strain = 0.14 (*b*) Multiplicative gene action is indicated by the similarity of the two coefficients of variation; $\bar{X}_{F_1}$ = 21.9 grams.

**11.27.**

| | Arithmetic mean | Absolute deviation from $F_1$ mean | Geometric mean | Absolute deviation from $F_1$ mean |
|---|---|---|---|---|
| (*a*) | 81.2 | 33.7 | 43.1 | 4.4 |
| (*b*) | 27.5 | 4.4 | 23.0 | 0.1 |
| (*c*) | 77.5 | 10.4 | 69.2 | 2.1 |
| (*d*) | 87.3 | 79.0 | 13.8 | 5.5 |

**11.28.** 0.64

**11.29.** (*a*) Heritability estimates measure the proportion of the total phenotypic variation for a trait among individuals of a population that is due to genetic variation. There is no genetic variation in a pure line (heritability = 0), but blood groups (for example) would still be 100% determined by genes. (*b*) Suppose that a group of identical twins were divided, one member of each pair to the two populations A and B. Each population would then have the same genetic constitution. If population A is not given equal social, educational and vocational opportunities with population B, then A might be expected to show lower average IQ. In other words, the average IQs of these populations would be solely reflective of nongenetic (environmental) differences regardless of the heritability estimates made in each population. (*c*) The answer to part (*b*) demonstrates that the difference between phenotypic averages of two populations does not necessarily imply that one population is genetically superior to the other. Important environmental differences may be largely responsible for such deviations. We could imagine that a pure line (heritability = 0) would be very well suited to a particular environment, whereas a highly genetically heterogeneous population with a high heritability for the same trait might be relatively poorly adapted to that same environment. In other words, the high heritabilities for IQ within populations A and B reveal nothing about the causes of the average phenotypic differences between them.

**11.30.** (*a*) $V_E = 42.0$, $V_A = 77.0$, $V_D = 11.5$ (*b*) 0.55 (*c*) 0.59

**11.31.** (*a*) $x = 4$ (*b*) $x = 4$, $y = 2$ (*c*) $x = 2$

**11.32.** (*a*) $b = 0.22$ (*b*) $h^2 = 2b = 0.49$

**11.33.** $b = 0.5$, $r = 1.0$

**11.34.** (*a*) 0.527 (*b*) 0.591 (*c*) 0.558 (*d*) Since neither sister nor brother can be considered dependent variables, two solutions are possible; $h^2 = 2r(s_S/s_B) = 1.054$ or $h^2 = 2r(s_B/s_S) = 1.182$. Heritability cannot be greater than 1.0. Since human siblings are usually reared together, their common environments have probably made them more alike than they would have been had they been reared in randomly chosen environments. Heritability estimates made from regression or correlation calculations assume that the populations are normally distributed and there are no environmental correlations between relatives. If either or both of these assumptions are invalid, so is the corresponding heritability elements.

**11.35.** 0.35

**11.36.** (*a*) 7.17 to 12.83 lb (*b*) 0.3

**11.37.** (*a*) 40 $lb^2$ (*b*) 100 $lb^2$

**11.38.** 0.88 inch

**11.39.** (*a*) 0.22 (*b*) 880,000 $lb^2$ (*c*) 16,000 to 20,000 lb

**11.40.** (*a*) 15 lb (*b*) 4.5 lb (*c*) 184.5 lb

**11.41.** (*a*) $\bar{X}$ = 25 cg, $s = \pm 3.94$ cg (*b*) 15.53 $cg^2$; note that this is the square of the phenotypic standard deviation in part (*a*). In pure lines, all of the variance is environmentally induced. (*c*) $h^2 = 0$, since a pure line is homozygous; there is no genetic variability. (*d*) $\bar{X}$ = 25 cg; no genetic gain can be made by selecting in the absence of genetic variability.

**11.42.** 0.36 $mm^2$

**11.43.** (*a*) Y = 23.34, Z = 19.90, X = 17.84 (*b*) 0.71 point

**11.44.** $I_A = 6.526$, $I_B = 6.041$; A excels in overall merit.

**11.45.** (*a*) 1/3 (*b*) 22.2% (*c*) 1/9

**11.46.** (*a*) A = 630.0, B = 615.0, C = 597.5 (*b*) Sire A

**11.47.** 0.25

**11.48.** 0.0351

**11.49.** 0.4375

**11.50.** (*a*) Heterosis for seed yield = 22.2 bu/acre, for plant height = 28 cm (*b*) 60.3 bu/acre, 243 cm

# Chapter 12

## Population Genetics

### HARDY-WEINBERG EQUILIBRIUM

A *Mendelian population* may be considered to be a group of sexually reproducing organisms with a relatively close degree of genetic relationship (such as a species, subspecies, breed, variety, strain, etc.) residing within defined geographical boundaries wherein interbreeding occurs. If all the gametes produced by a Mendelian population are considered as a hypothetical mixture of genetic units from which the next generation will arise, we have the concept of a *gene pool.*

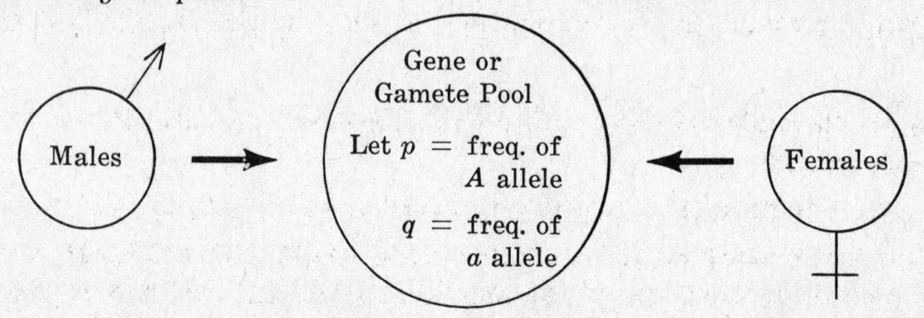


If we consider a pair of alleles ($A$ and $a$), we will find that the percentage of gametes in the gene pool bearing $A$ or $a$ will depend upon the genotypic frequencies of the parental generation whose gametes form the pool. For example, if most of the population was of the recessive genotype $aa$ then the frequency of the recessive allele in the gene pool would be relatively high, and the percentage of gametes bearing the dominant ($A$) allele would be correspondingly low.

When matings between members of a population are completely at random, i.e. when every male gamete in the gene pool has an equal opportunity of uniting with every female gamete, then the zygotic frequencies expected in the next generation may be predicted from a knowledge of the gene (allelic) frequencies in the gene pool of the parental population. That is, given the relative frequencies of $A$ and $a$ gametes in the gene pool, we can calculate (on the basis of the chance union of gametes) the expected frequencies of progeny genotypes and phenotypes. If $p$ = percentage of $A$ alleles in the gene pool and $q$ = percentage of $a$ alleles, then we can use the checkerboard method to produce all the possible chance combinations of these gametes.

| ♀ \ ♂ | $p$ (A) | $q$ (a) |
|---|---|---|
| $p$ (A) | $p^2$ $AA$ | $pq$ $Aa$ |
| $q$ (a) | $pq$ $Aa$ | $q^2$ $aa$ |

Note that $p + q = 1$, i.e. the percentage of $A$ and $a$ gametes must add to 100% in order to account for all of the gametes in the gene pool. The expected genotypic (zygotic) frequencies in the next generation then may be summarized as follows:

$$(p+q)^2 \;=\; \underset{AA}{p^2} + \underset{Aa}{2pq} + \underset{aa}{q^2} \;=\; 1.0$$

Thus $p^2$ is the fraction of the next generation expected to be homozygous dominant ($AA$), $2pq$ is the fraction expected to be heterozygous ($Aa$), and $q^2$ is the fraction expected to be recessive ($aa$). All of these genotypic fractions must add to unity to account for all genotypes in the progeny population.

This formula, expressing the genotypic expectations of progeny in terms of the gametic (allelic) frequencies of the parental gene pool, is called the *Hardy-Weinberg law*. If a population conforms to the conditions on which this formula is based, there should be no change in the gametic or the zygotic frequencies from generation to generation. Should a population initially be in disequilibrium, one generation of random mating is sufficient to bring it into genetic equilibrium and thereafter the population will remain in equilibrium (unchanging in gametic and zygotic frequencies) as long as the Hardy-Weinberg conditions persist.

Several assumptions underlie the attainment of genetic equilibrium as expressed in the Hardy-Weinberg equation.

(1) The population is infinitely large and mates at random (*panmictic*).

(2) No selection is operative, i.e. each genotype under consideration can survive just as well as any other (no differential mortality), and each genotype is equally efficient in the production of progeny (no differential reproduction).

(3) The population is closed, i.e. no immigration of individuals from another population into nor emigration from the population under consideration is allowed.

(4) There is no mutation from one allelic state to another. Mutation may be allowed if the forward and back mutation rates are equivalent, i.e. $A$ mutates to $a$ with the same frequency that $a$ mutates to $A$.

(5) Meiosis is normal so that chance is the only factor operative in gametogenesis.

If we define *evolution* as any change in a population from the equilibrium conditions, then a violation of one or more of the Hardy-Weinberg restrictions could cause the population to move away from the gametic and zygotic equilibrium frequencies. Changes in gene frequencies can be produced by a reduction in population size, by selection, migration or mutation pressures, or by *meiotic drive* (nonrandom assortment of chromosomes). No population is infinitely large, spontaneous mutations cannot be prevented, selection and migration pressures usually exist in most natural populations, etc., so it may be surprising to learn that despite these violations of Hardy-Weinberg restrictions many genes do conform, within statistically acceptable limits, to equilibrium conditions between two successive generations. Changes too small to be statistically significant deviations from equilibrium expectations between any two generations can nonetheless accumulate over many generations to produce considerable alterations in the genetic structure of a population. The mathematics of evolutionary problems is presented in Chapter 13.

A *race* is a genetically (and usually geographically) distinctive interbreeding population of a species. The number of races one wishes to recognize generally depends on the purpose of the investigation. Populations that differ significantly in gene frequencies at one or more loci may be considered as different races. Human races are defined on the basis of gene frequency differences in qualitative traits such as blood groups, hair texture, eye color, etc., as well as by mean differences in quantitative traits such as skin color, body build, shapes of noses, lips, eyes, etc. Races of a given species can freely interbreed with one another.

Members of different species, however, are reproductively isolated to a recognizable degree. *Subspecies* are races that have been given distinctive taxonomic names. Varieties, breeds, strains, etc. of cultivated plants or domesticated animals may also be equated with the racial concept. Geographic isolation is usually required for populations of a species to become distinctive races. Race formation is a prerequisite to the splitting of one species into two or more species (*speciation*). Differentiation at many loci over many generations is generally required to reproductively isolate these groups by time of breeding, behavioral differences, growth requirements, hybrid inviability, hybrid sterility and other such mechanisms.

Equilibrium at an autosomal genetic locus becomes fully established in a non-equilibrium population after one generation of random mating under Hardy-Weinberg conditions regardless of the number of alleles at that locus. However, when autosomal allelic frequencies are dissimilar in the sexes, they become equilibrated after one generation of random mating, but the genotypic frequencies do not become equilibrated until the second generation of random mating. If the frequencies of sex-linked alleles are unequal in the sexes, the equilibrium value is approached rapidly during successive generations of random mating in an oscillatory manner by the two sexes. This phenomenon derives from the fact that females (XX) carry twice as many sex-linked alleles as do males (XY). Females receive their sex-linked heredity equally from both parents, but males receive their sex-linked heredity only from their mothers. The difference between the allelic frequencies in males and females is halved in each generation under random mating. Within each sex, the deviation from equilibrium is halved in each generation, with sign reversed. The average frequency of one allele ($\bar{p}$) in the entire population is also the equilibrium approached by each sex during successive generations of random mating.

$$\bar{p} = \tfrac{2}{3}p_f + \tfrac{1}{3}p_m$$

Although alleles at a single autosomal locus reach equilibrium following one generation of random mating, gametic equilibrium involving two independently assorting genes is approached rapidly over a number of generations. At equilibrium, the product of coupling gametes equals the product of repulsion gametes.

**Example 12.1.** Consider one locus with alleles $A$ and $a$ at frequencies represented by $p$ and $q$ respectively. A second locus has alleles $B$ and $b$ at frequencies $r$ and $s$ respectively. The expected frequencies of coupling gametes $AB$ and $ab$ are $pr$ and $qs$ respectively. The expected frequencies of repulsion gametes $Ab$ and $aB$ are $ps$ and $qr$ respectively. At equilibrium, $(pr)(qs) = (ps)(qr)$. Also at equilibrium, the disequilibrium coefficient ($d$) is $d = (pr)(qs) - (ps)(qr) = 0$.

For independently assorting loci under random mating, the disequilibrium value of $d$ is halved in each generation during the approach to equilibrium because unlinked genes experience 50% recombination. The approach to equilibrium by linked genes, however, is slowed by comparison because they recombine less frequently than unlinked genes (i.e. less than 50% recombination). The closer the linkage, the longer it takes to reach equilibrium. The disequilibrium ($d_t$) that exists at any generation ($t$) is expressed as

$$d_t = (1 - r)d_{t-1}$$

where $r$ = frequency of recombination and $d_{t-1}$ = disequilibrium in the previous generation.

**Example 12.2.** If $d = 0.25$ initially and the two loci experience 20% recombination (i.e. the loci are 20 map units apart), the disequilibrium that would be expected after one generation of random mating is $d_t = (1 - 0.2)(0.25) = 0.2$. This represents $0.20/0.25 = 0.8$ or 80% of the maximum disequilibrium that could exist for a pair of linked loci.

## CALCULATING GENE FREQUENCIES

### 1. Autosomal Loci With Two Alleles.

#### (a) *Codominant Autosomal Alleles.*

When codominant alleles are present in a two-allele system, each genotype has a distinctive phenotype. The numbers of each allele in both homozygous and heterozygous conditions may be counted in a sample of individuals from the population and expressed as a percentage of the total number of alleles in the sample. If the sample is representative of the entire population (containing proportionately the same numbers of genotypes as found in the entire population) then we can obtain an estimate of the allelic frequencies in the gene pool. Given a sample of N individuals of which D are homozygous for one allele ($A^1A^1$), H are heterozygous ($A^1A^2$), and R are homozygous for the other allele ($A^2A^2$), then $N = D + H + R$. Since each of the N individuals are diploid at this locus, there are 2N alleles represented in the sample. Each $A^1A^1$ genotype has two $A^1$ alleles. Heterozygotes have only one $A^1$ allele. Letting $p$ represent the frequency of the $A^1$ allele and $q$ the frequency of the $A^2$ allele, we have

$$p = \frac{2D + H}{2N} = \frac{D + \frac{1}{2}H}{N}, \qquad q = \frac{H + 2R}{2N} = \frac{\frac{1}{2}H + R}{N}$$

#### (b) *Dominant and Recessive Autosomal Alleles.*

Determining the gene frequencies for alleles which exhibit dominance and recessive relationships requires a different approach from that used with codominant alleles. A dominant phenotype may have either of two genotypes, *AA* or *Aa*, but we have no way (other than by laboriously testcrossing each dominant phenotype) of distinguishing how many are homozygous or heterozygous in our sample. The only phenotype whose genotype is known for certain is the recessive (*aa*). If the population is in equilibrium, then we can obtain an estimate of $q$ (the frequency of the recessive allele) from $q^2$ (the frequency of the recessive genotype or phenotype).

**Example 12.3.** If 75% of a population was of the dominant phenotype (*A*-), then 25% would have the recessive phenotype (*aa*). If the population is in equilibrium with respect to this locus, we expect $q^2$ = frequency of *aa*.

Then $q^2 = 0.25$, $q = 0.5$, $p = 1 - q = 0.5$.

#### (c) *Sex-Influenced Traits.*

The expression of dominance and recessive relationships may be markedly changed in some genes when exposed to different environmental conditions, most notable of which are the sex hormones. In sex-influenced traits (Chapter 5), the heterozygous genotype usually will produce different phenotypes in the two sexes, making the dominance and recessive relationships of the alleles appear to reverse themselves. We shall consider only those sex-influenced traits whose controlling genes are on autosomes. Determination of allelic frequencies must be indirectly made in one sex by taking the square root of the frequency of the recessive phenotype ($q = \sqrt{q^2}$). A similar approach in the opposite sex should give an estimate of $p$. Corroboration of sex-influence is obtained if these estimates of $p$ and $q$ made in different sexes add close to unity.

### 2. Autosomal Loci With Multiple Alleles.

If we consider three alleles, $A$, $a'$ and $a$, with the dominance hierarchy $A > a' > a$, occurring in the gene pool with respective frequencies $p$, $q$ and $r$, then random mating will generate zygotes with the following frequencies:

$$(p+q+r)^2 = p^2 + 2pq + 2pr + q^2 + 2qr + r^2 = 1$$

$$\begin{array}{ll} \text{Genotypes:} & \underbrace{AA \quad Aa' \quad Aa}_{A} \quad \underbrace{a'a' \quad a'a}_{a'} \quad \underbrace{aa}_{a} \\ \text{Phenotypes:} & \end{array}$$

For ease in calculation of a given allelic frequency, it may be possible to group the phenotypes of the population into just two types.

**Example 12.4.** In a multiple allelic system where $A > a' > a$, we could calculate the frequency of the top dominant allele $A$ by considering the dominant phenotype ($A$) in contrast to all other phenotypes produced by alleles at this locus. The latter group may be considered to be produced by an allele $a^x$, which is recessive to $A$.

Let $p$ = frequency of allele $A$, $q$ = frequency of allele $a^x$.

$q^2$ = frequency of phenotypes other than $A$

$q = \sqrt{q^2}$

$p = 1 - q$ = frequency of gene $A$.

Many multiple allelic series involve codominant relationships such as $(A^1 = A^2) > a$, with respective frequencies $p$, $q$ and $r$. More genotypes can be phenotypically recognized in codominant systems than in systems without codominance.

$$(p+q+r)^2 = p^2 + 2pr + 2pq + q^2 + 2qr + r^2 = 1$$

$$\begin{array}{ll} \text{Genotypes:} & \underbrace{A^1A^1 \quad A^1a}_{A^1} \quad \underbrace{A^1A^2}_{A^1A^2} \quad \underbrace{A^2A^2 \quad A^2a}_{A^2} \quad \underbrace{aa}_{a} \\ \text{Phenotypes:} & \end{array}$$

The use of this formula in calculating multiple allelic frequencies is presented in Problems 12.9 and 12.10. Similar methods may be utilized to derive other formulas for calculating gene frequencies in multiple allelic systems with more than three alleles, but their computation becomes too involved for our purposes at the introductory level. Therefore multiple allelic problems in this chapter will be mainly concerned with three alleles.

### 3. Sex-Linked Loci.

#### (a) *Codominant Sex-Linked Alleles.*

Data from both males and females can be used in the direct computation of sex-linked codominant allelic frequencies. Bear in mind that in organisms with an X-Y mechanism of sex determination, the heterozygous condition can only appear in females. Males are hemizygous for sex-linked genes.

**Example 12.5.** In domestic cats, black melanin pigment is deposited in the hair by a sex-linked gene; its alternative allele inhibits melanin production, resulting in yellow hair. Random inactivation of one of the X chromosomes occurs in female embryos producing a genetic mosaic. Consequently, heterozygous females have patches of black and yelow hair called calico or tortoiseshell. The calico trait had formerly been erroneously attributed to the action of sex-linked codominant alleles $C^b$ and $C^y$.

| | Phenotypes | | |
|---|---|---|---|
| | Black | Tortoise-shell | Yellow |
| Females | $C^bC^b$ | $C^bC^y$ | $C^yC^y$ |
| Males | $C^b$Y | – | $C^y$Y |

Let $p$ = frequency of $C^b$, $q$ = frequency of $C^y$.

$$p = \frac{2\begin{pmatrix}\text{no. of black}\\\text{females}\end{pmatrix} + \begin{pmatrix}\text{no. of tortoise-}\\\text{shell females}\end{pmatrix} + \begin{pmatrix}\text{no. of black}\\\text{males}\end{pmatrix}}{2(\text{no. females}) + \text{no. males}}$$

$$q = \frac{2\begin{pmatrix}\text{no. of yellow}\\ \text{females}\end{pmatrix} + \begin{pmatrix}\text{no. of tortoise-}\\ \text{shell females}\end{pmatrix} + \begin{pmatrix}\text{no. of yellow}\\ \text{males}\end{pmatrix}}{2(\text{no. females}) + \text{no. males}}$$

(*b*) ***Dominant and Recessive Sex-Linked Alleles.***

Since each male possesses only one sex-linked allele, the frequency of a sex-linked trait among males is a direct measure of the allelic frequency in the population, assuming of course that the allelic frequencies thus determined are representative of the allelic frequencies among females as well.

## TESTING A LOCUS FOR EQUILIBRIUM

In cases where dominance is involved, the heterozygous class is indistinguishable phenotypically from the homozygous dominant class. Hence there is no way of checking the Hardy-Weinberg expectations against observed sample data unless the dominant phenotypes have been genetically analyzed by observation of their progeny from test-crosses. Only when codominant alleles are involved can we easily check our observations against the expected equilibrium values through the chi-square test (Chapter 7).

### Degrees of Freedom.

The number of variables in chi-square tests of Hardy-Weinberg equilibrium is not simply the number of phenotypes minus one (as in chi-square tests of classical Mendelian ratios). The number of observed variables (number of phenotypes $= k$) is further restricted by testing their conformity to an expected Hardy-Weinberg frequency ratio generated by a number of additional variables (number of alleles, or allelic frequencies $= r$). We have $k - 1$ degrees of freedom in the number of phenotypes, $(r - 1)$ degrees of freedom in establishing the frequencies for the $r$ alleles. The combined number of degrees of freedom is $(k - 1) - (r - 1) = k - r$. Even in most chi-square tests for equilibrium involving multiple alleles, the number of degrees of freedom is the number of phenotypes minus the number of alleles.

# Solved Problems

## HARDY-WEINBERG EQUILIBRIUM

**12.1.** In a population gene pool, the alleles $A$ and $a$ are at initial frequencies $p$ and $q$ respectively. Prove that the gene frequencies and the zygotic frequencies do not change from generation to generation as long as the Hardy-Weinberg conditions are maintained.

**Solution:**

Zygotic frequencies generated by random mating are

$$p^2(AA) + 2pq(Aa) + q^2(aa) = 1$$

All of the gametes of $AA$ individuals and half of the gametes of heterozygotes will bear the dominant allele ($A$). Then the frequency of $A$ in the gene pool of the next generation is

$$p^2 + pq = p^2 + p(1-p) = p^2 + p - p^2 = p$$

Thus each generation of random mating under Hardy-Weinberg conditions fails to change either the allelic or zygotic frequencies.

**12.2.** Prove the Hardy-Weinberg law by finding the frequencies of all possible kinds of matings and from these generating the frequencies of genotypes among the progeny using the symbols shown below.

| | Alleles | | Genotypes | | |
|---|---|---|---|---|---|
| | $A$ | $a$ | $AA$ | $Aa$ | $aa$ |
| Frequency: | $p$ | $q$ | $p^2$ | $2pq$ | $q^2$ |

**Solution:**

There are six kinds of matings (ignoring male-female differences) which are easily generated in a mating table.

| | | Male Parent | | |
|---|---|---|---|---|
| | | $AA$ $p^2$ | $Aa$ $2pq$ | $aa$ $q^2$ |
| Female Parent | $AA$ $p^2$ | $p^4$ | $2p^3q$ | $p^2q^2$ |
| | $Aa$ $2pq$ | $2p^3q$ | $4p^2q^2$ | $2pq^3$ |
| | $aa$ $q^2$ | $p^2q^2$ | $2pq^3$ | $q^4$ |

The matings $AA \times Aa$ occur with the frequency $4p^3q$. Half the offspring from this mating are expected to be $AA$ $[\frac{1}{2}(4p^3q) = 2p^3q]$, and half are expected to be $Aa$ (again with the frequency $2p^3q$). Similar reasoning generates the frequencies of genotypes among the progeny shown in the following table.

| | Mating | Frequency | Genotypic Frequencies Among Progeny | | |
|---|---|---|---|---|---|
| | | | $AA$ | $Aa$ | $aa$ |
| (1) | $AA \times AA$ | $p^4$ | $p^4$ | – | – |
| (2) | $AA \times Aa$ | $4p^3q$ | $2p^3q$ | $2p^3q$ | – |
| (3) | $AA \times aa$ | $2p^2q^2$ | – | $2p^2q^2$ | – |
| (4) | $Aa \times Aa$ | $4p^2q^2$ | $p^2q^2$ | $2p^2q^2$ | $p^2q^2$ |
| (5) | $Aa \times aa$ | $4pq^3$ | – | $2pq^3$ | $2pq^3$ |
| (6) | $aa \times aa$ | $q^4$ | – | – | $q^4$ |

$$\text{Sums:}\quad \begin{aligned} (AA) &= p^4 + 2p^3q + p^2q^2 = p^2(p^2 + 2pq + q^2) = p^2 \\ (Aa) &= 2p^3q + 4p^2q^2 + 2pq^3 = 2pq(p^2 + 2pq + q^2) = 2pq \\ (aa) &= p^2q^2 + 2pq^3 + q^4 = q^2(p^2 + 2pq + q^2) = q^2 \\ & \qquad\qquad\qquad\qquad \text{Total} = 1.00 \end{aligned}$$

**12.3.** At what allelic frequency does the homozygous recessive genotype ($aa$) become twice as frequent as the heterozygous genotype ($Aa$) in a Hardy-Weinberg population?

**Solution:**

Let $q$ = frequency of recessive allele, $p$ = frequency of dominant allele.
The frequency of homozygous recessives ($q^2$) is twice as frequent at heterozygotes ($2pq$) when

$$\begin{aligned} q^2 &= 2(2pq) \\ &= 4pq \\ &= 4q(1-q) \\ &= 4q - 4q^2 \\ 0 &= 4q - 5q^2 \\ 0 &= q(4 - 5q) \end{aligned}$$

Therefore, either $q = 0$ (which is obviously an incorrect solution), or

$$\begin{aligned} 4 - 5q &= 0 \\ 5q &= 4 \\ q &= 4/5 \text{ or } 0.8 \end{aligned}$$

*Proof:*

$$q^2 = 2(2pq)$$
$$(0.8)^2 = 4(0.2)(0.8)$$
$$0.64 = 0.64$$

## CALCULATING GENE FREQUENCIES

### Autosomal Loci With Two Alleles.

### *Codominant Autosomal Alleles.*

**12.4.** In Shorthorn cattle, the genotype $C^RC^R$ is phenotypically red, $C^RC^W$ is roan (a mixture of red and white) and $C^WC^W$ is white. (*a*) If 108 red, 48 white and 144 roan animals were found in a sample of Shorthorns from the central valley of California, calculate the estimated frequencies of the $C^R$ allele and the $C^W$ allele in the gene pool of the population. (*b*) If this population is completely panmictic, what zygotic frequencies would be expected in the next generation? (*c*) How does the sample data in part (*a*) compare with the expectations for the next generation in part (*b*)? Is the population represented in part (*a*) in equilibrium?

**Solution:**

(*a*)

| Numbers | Phenotypes | Genotypes |
|---|---|---|
| 108 | red | $C^RC^R$ |
| 144 | roan | $C^RC^W$ |
| 48 | white | $C^WC^W$ |
| 300 | | |

First, let us calculate the frequency of the $C^R$ allele. There are 108 red individuals each carrying two $C^R$ alleles; $2 \times 108 = 216$ $C^R$ alleles. There are 144 roan individuals each carrying only one $C^R$ allele; $1 \times 144 = 144$ $C^R$ alleles. Thus the total number of $C^R$ alleles in our sample is $216 + 144 = 360$. Since each individual is a diploid (possessing two sets of chromosomes, each bearing one of the alleles at the locus under consideration), the total number of alleles represented in this sample is $300 \times 2 = 600$. The fraction of all alleles in our sample of type $C^R$ becomes $360/600 = 0.6$ or 60%. The other 40% of the alleles in the gene pool must be of type $C^W$. We can arrive at this estimate for $C^W$ by following the same procedure as above. There are $48 \times 2 = 96$ $C^W$ alleles represented in the homozygotes and 144 in the heterozygotes; $96 + 144 = 240$; $240/600 = 0.4$ or 40% $C^W$ alleles.

(*b*) Recall that panmixis is synonymous with random mating. We will let the frequency of the $C^R$ allele be represented by $p = 0.6$, and the frequency of the $C^W$ allele be represented by $q = 0.4$. Then according to the Hardy-Weinberg law, we would expect as genotypic frequencies in the next generation

$$p^2 = (0.6)^2 = 0.36\ C^RC^R \ :\ 2pq = 2(0.6)(0.4) = 0.48\ C^RC^W \ :\ q^2 = (0.4)^2 = 0.16\ C^WC^W$$

(*c*) In a sample of size 300 we would expect $0.36(300) = 108$ $C^RC^R$ (red), $0.48(300) = 144$ $C^RC^W$ (roan), and $0.16(300) = 48$ $C^WC^W$ (white). Note that these figures correspond exactly to those of our sample. Since the genotypic and gametic frequencies are not expected to change in the next generation, the original population must already be in equilibrium.

### *Dominant and Recessive Autosomal Alleles.*

**12.5.** White wool is dependent upon a dominant allele $B$ and black wool upon its recessive allele $b$. Suppose that a sample of 900 sheep of the Rambouillet breed in Idaho gave the following data: 891 white and 9 black. Estimate the allelic frequencies.

**Solution:**

$$p^2(BB) + 2pq(Bb) + q^2(bb) = 1.0$$

If we assume the population is in equilibrium, we can take the square root of that percentage of the population which is of the recessive genotype (phenotype) as our estimator for the frequency of the recessive allele.

$q = \sqrt{q^2} = \sqrt{9/900} = 0.1 =$ frequency of allele $b$.

Since $p + q = 1$, the frequency of the allele $B$ is 0.9.

**12.6.** It is suspected that the excretion of the strongly odorous substance methanethiol is controlled by a recessive gene $m$ in humans; nonexcretion is governed by the dominant allele $M$. If the frequency of $m$ is 0.4 in Iceland, what is the probability of finding two nonexcretor boys and one excretor girl in Icelandic families of size three where both parents are nonexcretors?

**Solution:**

In order that two nonexcretor parents produce an excretor child, they must both be heterozygous *Mm*, in which case $\frac{1}{4}$ of their children would be expected to be excretors (*mm*). Girls are expected with a frequency of 0.5. Therefore the probability of $Mm \times Mm$ parents producing an excretor girl is $(\frac{1}{4})(\frac{1}{2}) = \frac{1}{8}$. The probability of having a nonexcretor boy $= (\frac{3}{4})(\frac{1}{2}) = \frac{3}{8}$. The probability of a nonexcretor individual in this population being heterozygous can be estimated from the equilibrium expectations. Let $q = 0.4$, then $p = 0.6$.

$$\begin{array}{ccccccc} p^2\,MM & + & 2pq\,Mm & + & q^2\,mm & = & 1.0 \\ (0.6)^2 & + & 2(0.6)(0.4) & + & (0.4)^2 & = & 1.0 \\ 0.36 & + & 0.48 & + & 0.16 & = & 1.0 \end{array}$$

$$\underbrace{0.36 + 0.48}_{\text{nonexcretor}} \quad \underbrace{0.16}_{\text{excretor}}$$

The probability of a nonexcretor individual being heterozygous is $48/(36+48) = 0.57$. The probability of both parents being heterozygous $= (0.57)^2 = 0.325$.

Let $a$ = probability of heterozygous parents producing a nonexcretor boy $= \frac{3}{8}$,

$b$ = probability of heterozygous parents producing an excretor girl $= \frac{1}{8}$.

The probability of heterozygous parents producing two nonexcretor boys and one excretor girl is found in the second term of the expansion $(a+b)^3 = a^3 + 3a^2b + \cdots$; thus $3(\frac{3}{8})^2(\frac{1}{8}) = 0.053$. The probability that both nonexcretor parents are heterozygous *and* produce two nonexcretor boys and one excretor girl is $(0.325)(0.053) = 0.017$ or 1.7%.

**12.7.** Two independently assorting recessive genes govern production of salmon silks (*sm*) and shrunken endosperm (*sh*) of maize. A sample from a population which is mating at random yielded the following data: 6 shrunken : 10 salmon, shrunken : 30 wild type : 54 salmon. Determine the frequencies of the salmon allele $q$ and shrunken allele $t$.

**Solution:**

| | | |
|---|---|---|
| 6 | *Sm - sh sh* | shrunken |
| 10 | *sm sm sh sh* | salmon, shrunken |
| 30 | *Sm - Sh -* | wild type |
| 54 | *sm sm Sh -* | salmon |
| 100 | | |

Let $q^2$ = frequency of the recessive trait, salmon silks $= (10 + 54)/100 = 0.64$; $q = 0.08$.

Let $t^2$ = frequency of the recessive trait, shrunken endosperm $= (6 + 10)/100 = 0.16$; $t = 0.4$.

### *Sex-Influenced Traits.*

**12.8.** In the human population, an index finger shorter than the ring finger is governed by a sex-influenced gene which appears to be dominant in males and recessive in females. A sample of the males in this population was found to contain 120 short and 210 long index fingers. Calculate the expected frequencies of long and short index fingers in females of this population.

**Solution:**

Since the dominance relationships are reversed in the two sexes, let us use all lower case letters with superscripts to avoid confusion with either dominance or codominance symbolism.

| Genotype | Phenotypes | |
|---|---|---|
| | Males | Females |
| $s^1s^1$ | short | short |
| $s^1s^2$ | short | long |
| $s^2s^2$ | long | long |

Let $p$ = frequency of $s^1$ allele, $q$ = frequency of $s^2$ allele. $p^2(s^1s^1) + 2pq(s^1s^2) + q^2(s^2s^2) = 1.0$.

In males, the allele for long finger $s^2$ is recessive. Then $q = \sqrt{q^2} = \sqrt{210/(120+210)} = \sqrt{0.64} = 0.8$; $p = 1 - 0.8 = 0.2$.

In females, short index finger is recessive. Then $p^2 = (0.2)^2 = 0.04$ or 4% of the females of this population will probably be short fingered. The other 96% should possess long index fingers.

## Autosomal Loci With Multiple Alleles.

**12.9.** A multiple allelic system governs the coat colors of rabbits; $C$ = full color, $c^h$ = Himalayan, $c$ = albino, with dominance expressed as $C > c^h > c$ and occurring with the frequencies $p$, $q$ and $r$ respectively. (*a*) If a population of rabbits containing full colored individuals, Himalayans and albinos, is mating at random, what is the expected genotypic ratio in the next generation in terms of $p$, $q$ and $r$? (*b*) Derive a formula for the calculation of allelic frequencies from the expected phenotypic frequencies. (*c*) A sample from a rabbit population contains 168 full color, 30 Himalayan and 2 albino. Calculate the allelic frequencies $p$, $q$ and $r$. (*d*) Given the allelic frequencies $p = 0.5$, $q = 0.1$ and $r = 0.4$, calculate the expected genotypic ratios among the full colored rabbits.

**Solution:**

(*a*) The zygotic expectations from a population mating at random with allelic frequencies $p$, $q$ and $r$ can be found by expanding $(p + q + r)^2$.

| | $p\,(C)$ | $q\,(c^h)$ | $r\,(c)$ |
|---|---|---|---|
| $p\,(C)$ | $p^2\,CC$ | $pq\,Cc^h$ | $pr\,Cc$ |
| $q\,(c^h)$ | $pq\,Cc^h$ | $q^2\,c^hc^h$ | $qr\,c^hc$ |
| $r\,(c)$ | $pr\,Cc$ | $qr\,c^hc$ | $r^2\,cc$ |

Summary:

| Genotypic Frequencies | Genotypes | Phenotypes |
|---|---|---|
| $p^2$ | $CC$ | full color |
| $2pq$ | $Cc^h$ | full color |
| $2pr$ | $Cc$ | full color |
| $q^2$ | $c^hc^h$ | Himalayan |
| $2qr$ | $c^hc$ | Himalayan |
| $r^2$ | $cc$ | albino |

(*b*) $r$ = frequency of allele $c$ = $\sqrt{\text{frequency of albinos}} = \sqrt{r^2}$

q = frequency of allele $c^h$; let H = frequency of the Himalayan phenotype.

$$q^2 + 2qr = H$$

Completing the square,

$$q^2 + 2qr + r^2 = H + r^2, \quad (q+r)^2 = H + r^2, \quad q + r = \sqrt{H + r^2}, \quad q = \sqrt{H + r^2} - r$$

$p$ = frequency of allele $C = 1 - q - r$.

(*c*) The frequency of allele $c = r = \sqrt{2/(168 + 30 + 2)} = \sqrt{2/200} = 0.1$. To calculate the frequency of the allele $c^h$, let H represent the frequency of the Himalayan phenotype in the population; then $q = \sqrt{H + r^2} - r = \sqrt{30/200 + 0.01} - 0.1 = 0.3$. The frequency of the allele $C = p = 1 - q - r = 1 - 0.1 - 0.3 = 0.6$.

(*d*)

$$\left.\begin{array}{lll} p^2\,CC & = (0.5)^2 & = 0.25 \\ 2pq\,Cc^h & = 2(0.5)(0.1) & = 0.10 \\ 2pr\,Cc & = 2(0.5)(0.4) & = 0.40 \end{array}\right\} \text{full color}$$

$$\overline{0.75}$$

Therefore considering only the full colored rabbits, we expect

$$25/75\ CC = 33\tfrac{1}{3}\%$$
$$10/75\ Cc^h = 13\tfrac{1}{3}\%$$
$$40/75\ Cc = 53\tfrac{1}{3}\%$$

100% of the full colored rabbits

**12.10.** The ABO blood group system is governed by a multiple allelic system in which some codominant relationships exist. Three alleles, $I^A$, $I^B$ and $i$ form the dominance hierarchy $(I^A = I^B) > i$. (*a*) Determine the genotypic and phenotypic expectations for this blood group locus from a population in genetic equilibrium. (*b*) Derive a formula for use in finding the allelic frequencies at the ABO blood group locus. (*c*) Among New York Caucasians, the frequencies of the ABO blood groups were found to be approximately 49% type O, 36% type A, 12% type B and 3% type AB. What are the allelic frequencies in this population? (*d*) Given the population in part (*c*) above, what percentage of type A individuals is probably homozygous?

**Solution:**

(*a*) Let $p$ = frequency of $I^A$ allele, $q$ = frequency of $I^B$ allele, $r$ = frequency of $i$ allele. The expansion of $(p + q + r)^2$ yields the zygotic ratio expected under random mating.

| Genotypic Frequencies | Genotypes | Phenotypes (Blood Groups) |
|---|---|---|
| $p^2$ | $I^AI^A$ | A |
| $2pr$ | $I^Ai$ | A |
| $q^2$ | $I^BI^B$ | B |
| $2qr$ | $I^Bi$ | B |
| $2pq$ | $I^AI^B$ | AB |
| $r^2$ | $ii$ | O |

(*b*) Let $\bar{A}$, $\bar{B}$ and $\bar{O}$ represent the *phenotypic* frequencies of blood groups A, B and O respectively. Solving for the frequency of the recessive allele $i$, $r = \sqrt{r^2} = \sqrt{\bar{O}}$.

Solving for the frequency of the $I^A$ allele,

$$p^2 + 2pr + r^2 = \bar{A} + \bar{O}, \quad (p+r)^2 = \bar{A} + \bar{O}, \quad p = \sqrt{\bar{A} + \bar{O}} - r = \sqrt{\bar{A} + \bar{O}} - \sqrt{\bar{O}}$$

Solving for the frequency of the $I^B$ allele, $q = 1 - p - r$. Or, following the method for obtaining the frequency of the $I^A$ allele, $q = \sqrt{\bar{B} + \bar{O}} - \sqrt{\bar{O}}$.

Presenting the solutions in a slightly different form,

$$\underbrace{\sqrt{\bar{A}+\bar{O}}-\sqrt{\bar{O}}}_{p}+\underbrace{\sqrt{\bar{B}+\bar{O}}-\sqrt{\bar{O}}}_{q}+\underbrace{\sqrt{\bar{O}}}_{r} = 1.0$$

$$p = 1-\sqrt{\bar{B}+\bar{O}}, \quad q = 1-\sqrt{\bar{A}+\bar{O}}, \quad r = \sqrt{\bar{O}}$$

(*c*) Frequency of allele $i = \sqrt{\bar{O}} = \sqrt{0.49} = 0.70 = r$

Frequency of $I^B$ allele $= 1-\sqrt{\bar{A}+\bar{O}} = 1-\sqrt{0.36+0.49} = 0.08 = q$

Frequency of allele $I^A = 1-\sqrt{\bar{B}+\bar{O}} = 1-\sqrt{0.12+0.49} = 0.22 = p$

**Check:** $p+q+r = 0.22+0.08+0.70 = 1.00$

(*d*) $p^2 = I^AI^A = (0.22)^2 = 0.048$

$2pr = I^Ai = 2(0.22)(0.7) = 0.308$

$0.356$ = total group A individuals

Thus $48/356 = 0.135$ or $13\frac{1}{2}\%$ of all group A individuals in this population is expected to be homozygous.

## Sex-Linked Loci.

### *Codominant Sex-Linked Alleles.*

**12.11.** The genetics of coat colors in cats was presented in Example 12.5: $C^BC^B$ ♀♀ or $C^BY$♂♂ are black, $C^YC^Y$♀♀ or $C^YY$♂♂ are yellow, $C^BC^Y$♀♀ are tortoise-shell (blotches of yellow and black). A population of cats in London was found to consist of the following phenotypes:

| | Black | Yellow | Tortoise-Shell | Totals |
|---|---|---|---|---|
| Males | 311 | 42 | 0 | 353 |
| Females | 277 | 7 | 54 | 338 |

Determine the allelic frequencies using all of the available information.

**Solution:**

The total number of $C^B$ alleles in this sample is $311 + 2(277) + 54 = 919$. The total number of alleles (X chromosomes) in this sample is $353 + 2(388) = 1029$. Therefore the frequency of the $C^B$ allele is $919/1029 = 0.893$. The frequency of the $C^Y$ allele would then be $1 - 0.893 = 0.107$.

### *Dominant and Recessive Sex-Linked Alleles.*

**12.12.** White eye color in *Drosophila* is due to a sex-linked recessive gene $w$, and wild type (red) eye color to its dominant allele $w^+$. A laboratory population of *Drosophila* was found to contain 170 red eyed males and 30 white eyed males. (*a*) Estimate the frequency of the $w^+$ allele and the $w$ allele in the gene pool. (*b*) What percentage of the females in this population would be expected to be white eyed?

**Solution:**

(*a*)

| Observed No. of Males | Genotypes of Males | Phenotypes of Males |
|---|---|---|
| 170 | $w^+$Y | wild type (red eye) |
| 30 | $w$ Y | white eye |
| 200 | | |

Thus 30 of the 200 X chromosomes in this sample carry the recessive allele $w$.

$q = 30/200 = 0.15$ or 15% $w$ alleles.

$p = 1-q = 1-0.15 = 0.85$ or 85% $w^+$ alleles.

(*b*) Since females possess two X chromosomes (hence two alleles), their expectations may be calculated in the same manner as that used for autosomal genes.

$$p^2(w^+w^+) + 2pq(w^+w) + q^2(ww) = 1.0 \text{ or } 100\% \text{ of the females.}$$

$q^2 = (0.15)^2 = 0.0225$ or 2.25% of all females in the population is expected to be white eyed.

## TESTING A LOCUS FOR EQUILIBRIUM

**12.13.** A human serum protein called haptoglobin has two major electrophoretic variants produced by a pair of codominant alleles $Hp^1$ and $Hp^2$. A sample of 100 individuals has 10 $Hp^1/Hp^1$, 35 $Hp^1/Hp^2$ and 55 $Hp^2/Hp^2$. Are the genotypes in this sample conforming to the frequencies expected for a Hardy-Weinberg population within statistically acceptable limits?

**Solution:**

First we must calculate the allelic frequencies.

$$\text{Let } p = \text{frequency of } Hp^1 \text{ allele} = \frac{2(10) + 35}{2(100)} = \frac{55}{200} = 0.275$$

$$q = \text{frequency of } Hp^2 \text{ allele} = 1 - 0.275 = 0.725$$

From these gene (allelic) frequencies we can determine the genotypic frequencies expected according to the Hardy-Weinberg equation.

$$Hp^1/Hp^1 = p^2 = (0.275)^2 = 0.075625$$

$$Hp^1/Hp^2 = 2pq = 2(0.275)(0.725) = 0.39875$$

$$Hp^2/Hp^2 = q^2 = (0.725)^2 = 0.525625$$

Converting these genotypic frequencies to numbers based on a total sample size of 100, we can do a chi-square test.

| Genotypes | Observed | Expected | Deviation (o − e) | (o − e)$^2$ | (o − e)$^2$/e |
|---|---|---|---|---|---|
| $Hp^1/Hp^1$ | 10 | 7.56 | 2.44 | 5.95 | 0.79 |
| $Hp^1/Hp^2$ | 35 | 39.88 | −4.88 | 23.81 | 0.60 |
| $Hp^2/Hp^2$ | 55 | 52.56 | 2.44 | 5.95 | 0.11 |
| Totals | 100 | 100.00 | 0 | | $\chi^2 = 1.50$ |

df = *k* phenotypes − *r* alleles = 3 − 2 = 1; P = 0.2–0.3 (Table 7.2)

This is not a significant $\chi^2$ value, and we may accept the hypothesis that this sample (and hence presumably the population from which it was drawn) is conforming to the equilibrium distribution of genotypes.

**12.14.** One of the "breeds" of poultry has been largely built on a single gene locus, that for "frizzled" feathers. The frizzled phenotype is produced by the heterozygous genotype $M^NM^F$. One homozygote $M^FM^F$ produces extremely frizzled birds called "woolies". The other homozygous genotype $M^NM^N$ has normal plumage. A sample of 1000 individuals of this "breed" in the United States contained 800 frizzled, 150 normal and 50 wooly birds. Is this population in equilibrium?

**Solution:**

$$\text{Let } p = \text{frequency of the } M^F \text{ allele} = \frac{2(50) + 800}{2(1000)} = 0.45$$

$$q = \text{frequency of } M^N \text{ allele} = 1 - 0.45 = 0.55$$

| Genotypes | Equilibrium Frequencies | Calculations | | | | Expected Numbers |
|---|---|---|---|---|---|---|
| $M^FM^F$ | $p^2$ | $(0.45)^2$ | = | 0.2025(1000) | = | 202.5 |
| $M^FM^N$ | $2pq$ | 2(0.45)(0.55) | = | 0.4950(1000) | = | 495.0 |
| $M^NM^N$ | $q^2$ | $(0.55)^2$ | = | 0.3025(1000) | = | 302.5 |

Chi-square test for conformity to equilibrium expectations.

| Phenotypes | o | e | (o − e) | $(o-e)^2$ | $(o-e)^2/e$ |
|---|---|---|---|---|---|
| wooly | 50 | 202.5 | −152.5 | 23,256 | 114.8 |
| frizzle | 800 | 495.0 | +305.0 | 93,025 | 187.9 |
| normal | 150 | 302.5 | −152.5 | 23,256 | 76.9 |
| | | | | | $\chi^2 = 379.6$ |

df = 1, $p < 0.01$ (Table 7.2)

This highly significant chi-square value will not allow us to accept the hypothesis of conformity with equilibrium expectations. The explanation for the large deviation from the equilibrium expectations is two-fold. Much *artificial selection* (by people) is being practiced. The frizzled heterozygotes represent the "breed" type and are kept for show purposes as well as for breeding by bird fanciers. They dispose of (cull) many normal and wooly types. *Natural selection* is also operative on the wooly types because they tend to lose their feathers (loss of insulation) and eat more feed just to maintain themselves, are slower to reach sexual maturity, and lay fewer eggs than do the normal birds.

# Supplementary Problems

## HARDY-WEINBERG EQUILIBRIUM

**12.15.** At what allelic frequency is the heterozygous genotype ($Aa$) twice as frequent as the homozygous genotype ($aa$) in a Hardy-Weinberg population?

**12.16.** There is a singular exception to the rule that genetic equilibrium at two independently assorting autosomal loci is attained in a non-equilibrium population only after a number of generations of random mating. Specify the conditions of a population that should reach genotypic equilibrium after a single generation of random mating.

**12.17.** Let the frequencies of a pair of autosomal alleles ($A$ and $a$) be represented by $p_m$ and $q_m$ in males and by $p_f$ and $q_f$ in females respectively. Given $q_f = 0.6$ and $q_m = 0.2$, (*a*) determine the equilibrium gene frequencies in both sexes after one generation of random mating, and (*b*) give the genotypic frequencies expected in the second generation of random mating.

**12.18.** The autosomal gametic disequilibrium in a population is expressed as $d = 0.12$. The two loci under consideration recombine with a frequency of 16%. Calculate the disequilibrium ($d$ value) which existed in the gamete pool of (*a*) the previous generation, and (*b*) the next generation.

**12.19.** The frequency of a sex-linked allele is 0.4 in males and 0.8 in females of a population (XY sex determination) not in genetic equilibrium. Find the equilibrium frequency of this allele in the entire population.

**12.20.** A laboratory population of flies contains all females homozygous for a sex-linked dominant allele and all males hemizygous for the recessive allele. Calculate the frequencies expected in each sex for the dominant allele in the first three generations of random mating.

**12.21.** For two independently assorting loci under Hardy-Weinberg conditions, (*a*) what is the maximum value of the disequilibrium coefficient ($d$)? (*b*) Specify the two conditions in which a population must be in order to maximize $d$.

**12.22.** Given gene $A$ at frequency 0.2 and gene $B$ at frequency 0.6, find the equilibrium frequencies of the gametes $AB$, $Ab$, $aB$ and $ab$.

## CALCULATING GENE FREQUENCIES

### Autosomal Loci With Two Alleles

#### *Codominant Autosomal Alleles*

**12.23.** A population of soybeans is segregating for the colors golden, light green, and dark green produced by the codominant genotypes $C^GC^G$, $C^GC^D$, $C^DC^D$ respectively. A sample from this population contained 2 golden, 36 light green and 162 dark green. Determine the frequencies of the alleles $C^G$ and $C^D$.

**12.24** The M-N blood group system in humans is governed by a pair of codominant alleles ($L^M$ and $L^N$). A sample of 208 Bedouins in the Syrian Desert was tested for the presence of the M and N antigens and found to contain 119 group M ($L^ML^M$), 76 group MN ($L^ML^N$) and 13 group N ($L^NL^N$). (*a*) Calculate the gene frequencies of $L^M$ and $L^N$. (*b*) If the frequency of $L^M = 0.3$, how many individuals in a sample of size 500 would be expected to belong to group MN?

#### *Dominant and Recessive Autosomal Alleles*

**12.25.** The ability of certain people to taste a chemical called PTC is governed by a dominant allele $T$, and the inability to taste PTC by its recessive allele $t$. If 24% of a population is homozygous taster and 40% is heterozygous taster, what is the frequency of $t$? *Hint:* Use the same method as that employed for codominant alleles for greatest accuracy.

**12.26.** Gene $A$ governs purple stem and its recessive allele $a$ produces green stem in tomatoes; $C$ governs cut-leaf and $c$ produces potato-leaf. If the observations of phenotypes in a sample from a tomato population were 204 purple, cut : 194 purple, potato : 102 green, cut : 100 green, potato, determine the frequency of (*a*) the cut allele, (*b*) the allele for green stem.

**12.27.** An isolated field of corn was found to be segregating for yellow and white endosperm. Yellow is governed by a dominant allele and white by its recessive allele. A random sample of 1000 kernels revealed that 910 were yellow. Find the allelic frequency estimates for this population.

**12.28.** The $R$ locus controls the production of one system of antigens on the red blood cells of humans. The dominant allele results in Rh-positive individuals, whereas the homozygous recessive condition results in Rh-negative individuals. Consider a population in which 85% of the people are Rh-positive. Assuming the population to be at equilibrium, what is the gene frequency of alleles at this locus?

**12.29.** What is the highest frequency possible for a recessive lethal which kills 100% of its bearers when homozygous? What is the genetic constitution of the population when the lethal allele reaches its maximum?

**12.30.** Dwarf corn is homozygous recessive for gene $d$ which constitutes 20% of the gene pool of a population. If two tall corn plants are crossed in this population, what is the probability of a dwarf offspring being produced?

**12.31.** A dominant gene in rabbits allows the breakdown of the yellowish xanthophyll pigments found in plants so that white fat is produced. The recessive genotype $yy$ is unable to make this conversion, thus producing yellow fat. If a heterozygous buck (male) is mated to a group of white fat does (females) from a population in which the frequency of $Y$ is $\frac{2}{3}$, how many offspring with yellow fat would be expected among 32 progeny?

**12.32.** A metabolic disease of humans called "phenylketonuria" is the result of a recessive gene. If the frequency of phenylketonurics is 1/10,000 what is the probability that marriages between normal individuals will produce a diseased child?

**12.33.** Two recessive genes in *Drosophila*, $h$ and $b$, producing hairy and black phenotypes respectively, assort independently of one another. Data from a large population mating at random follows: 9.69% wild type, 9.31% hairy, 41.31% black, 39.69% hairy and black. Calculate the frequencies for the hairy and the black alleles.

#### *Sex-Influenced Traits*

**12.34.** Baldness is governed by a sex-influenced trait which is dominant in men and recessive in women. In a sample of 10,000 men, 7225 were found to be non-bald. In a sample of women of equivalent size, how many non-bald women are expected?

**12.35.** The presence of horns in some breeds of sheep is governed by a sex-influenced gene which is dominant in males and recessive in females. If a sample of 300 female sheep is found to contain 75 horned individuals, (*a*) what percentage of the females is expected to be heterozygous, (*b*) what percentage of the males is expected to be horned?

### Autosomal Loci With Multiple Alleles

**12.36.** The genetics of the ABO human blood groups is presented in Problem 12.10. (*a*) A sample of a human population was blood grouped and found to contain 23 group AB, 441 group O, 371 group B and 65 group A. Calculate the allelic frequencies of $I^A$, $I^B$ and $i$. (*b*) Given the gene frequencies $I^A = 0.36$, $I^B = 0.20$, and $i = 0.44$, calculate the percentage of the population expected to be of groups A, B, AB and O.

**12.37.** The color of screech owls is under the control of a multiple allelic series: $G^r$ (red) $>$ $g^i$ (intermediate) $>$ $g$ (gray). A sample from a population was analyzed and found to contain 38 red, 144 intermediate and 18 gray owls. Calculate the allelic frequencies.

**12.38.** Several genes of the horse are known to control coat colors. The $A$ locus apparently governs the distribution of pigment in the coat. If the dominant alleles of the other color genes are present, the multiple alleles of the $A$ locus produce the following results: $A^+$ = wild type (Prejvalski) horse (bay with zebra markings), $A$ = dark or mealy bay (black mane and tail), $a^t$ = seal brown (almost black with lighter areas), $a$ = recessive black (solid color). The order of dominance is $A^+ > A > a^t > a$. If the frequency of $A^+ = 0.4$, $A = 0.2$, $a^t = 0.1$ and $a = 0.3$, calculate the equilibrium phenotypic expectations.

### Sex-Linked Loci

**12.39** A genetic disease of humans called hemophilia (excessive bleeding) is governed by a sex-linked recessive gene which constitutes 1% of the gametes in the gene pool of a certain population. (*a*) What is the expected frequency of hemophilia among men of this population? (*b*) What is the expected frequency of hemophilia among women?

**12.40.** Color blindness in humans is due to a sex-linked recessive gene. A survey of 500 men from a local population revealed that 20 were color blind. (*a*) What is the gene frequency of the normal allele in the population? (*b*) What percentage of the females in this population would be expected to be normal?

**12.41.** The white eyes of *Drosophila* are due to a sex-linked recessive gene and wild type (red eyes) to its dominant allele. In a *Drosophila* population the following data were collected: 15 white eyed females, 52 white eyed males, 208 wild type males, 365 wild type females (112 of which carried the white allele). Using all the data, calculate the frequency of the white allele.

## TESTING A LOCUS FOR EQUILIBRIUM

**12.42.** A pair of codominant alleles governs coat colors in Shorthorn cattle: $C^RC^R$ is red, $C^RC^W$ is roan and $C^WC^W$ is white. A sample of a cattle population revealed the following phenotypes: 180 red, 240 roan and 80 white. (*a*) What is the frequency of the $C^R$ allele? (*b*) What is the frequency of the $C^W$ allele? (*c*) Does the sample indicate that the population is in equilibrium? (*d*) What is the chi-square value? (*e*) How many degrees of freedom exist? (*f*) What is the probability that the deviation of the observed from the expected values is due to chance?

**12.43.** A blood group system in sheep, known as the X-Z system, is governed by a pair of codominant alleles ($X$ and $X^z$). A large flock of Rambouillet sheep was blood grouped and found to contain 113 $X/X$, 68 $X/X^z$ and 14 $X^z/X^z$. (*a*) What are the allelic frequencies? (*b*) Is this population conforming to the equilibrium expectations? (*c*) What is the chi-square value? (*d*) How many degrees of freedom exist? (*e*) What is the probability of the observed deviation being due to chance?

**12.44.** The frequency of the $T$ allele in a human population = 0.8, and a sample of 200 yields 90% tasters ($T$-) and 10% non-tasters ($tt$). (*a*) Does the sample conform to the equilibrium expectations? (*b*) What is the chi-square value? (*c*) How many degrees of freedom exist? (*d*) What is the probability that the observed deviation is due to chance?

**12.45.** In poultry, the autosomal gene $F^B$ produces black feather color and its codominant allele $F^W$ produces splashed-white. The heterozygous condition produces Blue Andalusian. A splashed-white hen is mated to a black rooster and the $F_2$ was found to contain 95 black, 220 blue and 85 splashed-white. (*a*) What $F_2$ ratio is expected? (*b*) What is the chi-square value? (*c*) How many degrees of freedom exist? (*d*) What is the probability that the observed deviation is due to chance? (*e*) May the observations be considered to conform to the equilibrium expectations?

## Answers to Supplementary Problems

**12.15.** 0.5

**12.16.** All individuals are *AaBb*.

**12.17.** (*A*) $p_m = p_f = 0.6$, $q_m = q_f = 0.4$ (*b*) $AA = 0.36$, $Aa = 0.48$, $aa = 0.16$

**12.18** (*a*) $\frac{1}{7} = 0.143$ (*b*) 0.1008

**12.19.** $\frac{2}{3} = 0.67$

**12.20.** Males: (1) = 1.0, (2) = 0.5, (3) = 0.75; Females: (1) = 0.5, (2) = 0.75, (3) = 0.625

**12.21.** (*a*) 0.25 (*b*) $\frac{1}{2}AABB : \frac{1}{2}aabb$ or $\frac{1}{2}aaBB : \frac{1}{2}AAbb$

**12.22.** $AB = 0.12$, $Ab = 0.08$, $aB = 0.48$, $ab = 0.32$

**12.23.** $C^D = 0.9$, $C^G = 0.1$

**12.24.** (*a*) $L^M = 75.5\%$, $L^N = 24.5\%$ (*b*) 210

**12.25.** $t = 0.56$

**12.26.** (*a*) $C = 0.30$ (*b*) $a = 0.58$

**12.27.** $Y = 0.7$, $y = 0.3$

**12.28.** $R = 0.613$, $r = 0.387$

**12.29.** 0.5; all individuals are heterozygous carriers of the lethal allele.

**12.30.** $\left(\frac{2pq}{1-q^2}\right)^2\left(\frac{1}{4}\right) = \frac{p^2q^2}{(1-q^2)^2} = \frac{1}{36} = 0.0277778$

**12.31.** 4

**12.32.** $\left(\frac{2(0.99)(0.01)}{(0.99)^2 + 2(0.99)(0.01)}\right)^2 (0.25) = 0.01\%$

**12.33.** $h = 0.7$, $b = 0.9$

**12.34.** 9775

**12.35.** (*a*) 50% (*b*) 75%

**12.36.** (*a*) $I^A = 0.05$, $I^B = 0.25$, $i = 0.70$ (*b*) A = 44.6%, B = 21.6%, AB = 14.4%, O = 19.4%

**12.37.** $G^r = 0.1$, $g^i = 0.6$, $g = 0.3$

**12.38.** 64% wild type, 20% dark bay, 7% seal brown, 9% black

**12.39.** (*a*) 1/100 (*b*) 1/10,000

**12.40.** (*a*) 0.96 (*b*) 99.84%

**12.41.** $w = 0.19$

**12.42.** (*a*) 0.6 (*b*) 0.4 (*c*) Yes (*d*) 0 (*e*) 1 (*f*) 1

**12.43.** (*a*) $X = 0.75$, $X^z = 0.25$ (*b*) Yes (*c*) 0.7 (*d*) 1 (*e*) 0.3-0.5

**12.44.** (*a*) No (*b*) 18.75 (*c*) 1 (*d*) $< 0.001$

**12.45.** (*a*) $\frac{1}{4}$ black : $\frac{1}{2}$ blue : $\frac{1}{4}$ splashed-white (*b*) 4.50 (*c*) 1 (*d*) 0.01-0.05 (*e*) No

# Chapter 13

## Principles of Evolution

Evolution is the *process* whereby living organisms change from one form into another. Most of these changes are considered to be produced gradually over long periods of geological time. As explained in the previous chapter, when Hardy-Weinberg conditions exist, the population remains in equilibrium and no change is possible. Several factors are known to have the potential for profoundly changing the genetic structure of the population. These factors are violations of the Hardy-Weinberg conditions and include migration, mutation, selection and genetic drift. The first three factors are *systematic processes* which tend to change the gene frequency in a predictable manner both in amount and direction. The latter factor (genetic drift) is a *dispersive process* and is predictable in amount but not direction. The impact of each of these factors will be mathematically defined in the following sections. Derivation of the more complex formulas can be found in advanced treatises on population genetics and will not be presented here.

### MIGRATION

Equilibrium conditions assume that the population is closed, i.e. it is not subject to outside influences by immigration. Gene frequencies can be altered whenever a population is exposed to immigrants from another population. The change in gene frequency ($\Delta q$) produced by immigration depends upon the proportion of immigrants (i) in each generation and the difference in gene frequencies between the natives and the immigrants ($q_n - q_i$).

$$\Delta q = i(q_n - q_i) \qquad (13.1)$$

In order for two populations to diverge genetically on different evolutionary paths, which can eventually lead to speciation (the formation of new species), they must first become geographically isolated from each other. If a species is fragmented into two or more local populations (*demes*) that are geographically isolated from one another, the independent units will follow different evolutionary paths because

(1) Different mutations appear in separated groups of organisms.

(2) The mutations and gene combinations which appear in separate populations will have different adaptive values in the new environments.

(3) Organisms which originally colonize a certain geographical area and form an isolated population may not be representative of the group from which they came so that different gene frequencies exist from the beginning (*founder principle*).

(4) The size of the new population may become quite small at various times so that a genetic "bottleneck" is formed from which all subsequent organisms will arise. Gene frequencies will fluctuate in unpredictable directions during periods of small population size (*genetic drift*).

When migration is eliminated between two populations in different geographical areas, their gene pools become the isolated entities which are prerequisite to the accumulation of genetic differences. If the two gene pools of what was originally one population can remain isolated for a sufficient length of time, these differences can become so great that members of one population will no longer mate (interbreed) with or can no longer produce viable or fertile offspring by a member of the other population, even when they are no longer geographically isolated from one another. The two populations have become *reproductively isolated* from each other and can then be said to represent different *biological species*. Much evolution (genetic change) can occur in a population without its division into different species. Speciation, being a special aspect of evolution, is beyond the scope of the fundamentals considered in this chapter.

## MUTATION

Mutation itself is one of the major elements in evolution. As a species becomes exposed to different environments through long periods of time, its ability to survive becomes dependent upon its store of genetic diversity to generate new genotypes with new ranges of tolerance which can allow certain members of the population to survive and reproduce their kind. The sexual mechanism can produce a large but finite amount of genetic combinations. Even the best adapted genotypes under present conditions may not be able to survive at some time in the future under a different set of environmental circumstances. Unless new genetic material is introduced into the gene pool by mutation, evolution is limited to the range of tolerance of genotypes which already exist in the population. Spontaneous mutations are occurring continuously without regard to their immediate need or usefulness. Most mutations are either of no value or are deleterious under present environmental conditions. Harmful mutations tend to be eliminated from a population or kept at low frequency by natural selection. Occasionally, when a beneficial mutation occurs, selective forces act to increase its frequency in the population at the expense of its less favored allele(s). Thus, mutations may be considered as the raw materials and natural selection as the driving force in evolution.

Most gene loci have a mutation rate of $10^{-5}$ to $10^{-6}$ or one mutation in 100,000 to one mutation in a million. Thus, mutation pressure by itself is not usually a factor of major importance in changing gene frequency. In fact, a very small selective advantage or disadvantage of a particular allele can entirely offset the effects on gene frequency change produced by mutation pressure. However, if the only pressure acting on a given locus is that of mutation, given enough time, gene frequencies will gradually change and come to rest at their equilibrium values dependent solely upon the rates of forward and back mutation.

Each allele has a characteristic mutation rate in the population. Forward mutation and back mutation rates are usually not equivalent. Suppose that the initial frequency of the $A_1$ allele is $p_0$ and that of the $A_2$ allele is $q_0$; the mutation rate of $A_1$ to $A_2$ is $u$, and the back mutation rate of $A_2$ to $A_1$ is $v$. The change in frequency of the $A_2$ allele ($\Delta q$) due to one generation of differential mutation pressures is

$$\Delta q = up_0 - vq_0 \tag{13.2}$$

The initial disequilibrium in gene frequencies is progressively diminished in successive generations until at equilibrium $\Delta q = 0$, $up = vq$, and

$$\hat{q} = \frac{u}{u+v} \qquad \hat{p} = \frac{v}{u+v} \tag{13.3}$$

## SELECTION

Charles Darwin laid the foundation of modern evolutionary principles in his book, *On the Origin of Species by Means of Natural Selection or the Preservation of Favored Races in the Struggle for Life*, published in 1859. Darwin observed that living organisms have a reproductive potential which far exceeds the number of breeding individuals observed in the population. Moreover, the size of most populations tends to remain relatively stable from generation to generation. For example, a single female cod fish lays millions of eggs, but no increase is observed in the size of the natural population in successive generations. Competition for such basic necessities as food, shelter and space severely limits the number of organisms which can occupy the same ecological niche (i.e. pursue the same way of living). Darwin reasoned that the difference between the reproductive potential and the actual population size represents the force of *natural selection*. Only those individuals which are the best adapted or the most fit will survive to reproduce their kind.

Selection may be briefly defined as the non-random *differential reproduction* of genotypes. The laws governing heredity were unknown when Darwin wrote his thesis. With the knowledge of modern genetics, evolutionary theory has gained the status of a science and is now able to define many of its principles in mathematical terms.

One of the Hardy-Weinberg restrictions is an absence of selection. Gene frequencies will change if one genotype produces more offspring on the average than another. The *adaptive value* or *fitness* of a genotype is measured by its proportionate contribution of progeny to the next generation. Selection occurs when one genotype is better adapted or more fit than another. The intensity of selection is expressed as the *coefficient of selection* ($s$), which is defined as the proportionate reduction in the gametic contribution of one genotype compared with some other genotype (usually the most fit).

**Example 13.1.** Let us consider three situations in which fitness varies among the genotypes of a locus with two alleles, $A_1$ and $A_2$.

**Case 1.** Complete dominance with respect to fitness (a consideration entirely apart from dominance with respect to phenotype).

| Genotype | Fitness |
|---|---|
| $A_1A_1$ | 1 |
| $A_1A_2$ | 1 |
| $A_2A_2$ | $1 - s$ |

When the fitness of the most favored type is set at 1, then the unfavored genotype against which selection is acting with a force $s$ has an average fitness in the population of $1 - s$.

If $s = 0.2$, for every 100 zygotes produced by the favored genotypes ($A_1A_1$ or $A_1A_2$), the unfavored genotype ($A_2A_2$) produces 20% fewer offspring or $1 - 0.2 = 0.8$ or 80 zygotes.

**Case 2.** Partial dominance with respect to fitness occurs when the heterozygote is less adapted than one homozygote, but more fit than the other homozygote.

| Genotype | Fitness |
|---|---|
| $A_1A_1$ | 1 |
| $A_1A_2$ | $1 - s_1$ |
| $A_2A_2$ | $1 - s_2$ |

If $s_1 = 0.05$ and $s_2 = 0.35$, then on the average for every 100 zygotes produced by $A_1A_1$, only 95 and 65 zygotes would be produced by $A_1A_2$ and $A_2A_2$ respectively. If the $A_2A_2$ genotype is lethal, $s_2 = 1$ and fitness $= 0$.

**Case 3.** Overdominance with respect to fitness exists when the heterozygote produces more offspring on the average than either homozygous genotype.

| Genotype | Fitness |
|---|---|
| $A_1A_1$ | $1 - s_1$ |
| $A_1A_2$ | 1 |
| $A_2A_2$ | $1 - s_2$ |

Whenever heterozygotes are superior to either homozyous class, both alleles tend to remain in the population and, as a consequence, both homozygous types are commonly found in the population. A situation wherein two or more phenotypes (depending on the number of alleles and dominance conditions at the locus) are found in relatively high frequencies in a population is called *polymorphism* (many forms). One of the explanations for polymorphism is overdominance with respect to fitness (*balanced* polymorphism).

Longevity and vegetative vigor should not be confused with adaptive value or fitness. A genotype may have an average life span of 100 years or more and exhibit a healthy and vigorous phenotype, but unless it produces some offspring its fitness in the evolutionary scheme is zero. Phenotypic changes which occur within an individual during its life span have no evolutionary implications. Evolution operates only between successive generations in a population of organisms.

The rapidity with which gene frequencies change at any given time depends upon the intensity of selection ($s$) on the gene frequency, and on the conditions of dominance with respect to fitness.

### 1. Selection Against a Recessive Lethal.

A genotype which does not allow the organism to live to breeding age, or one which renders the organism sterile, has a fitness of zero and selection against it is complete ($s = 1$). In each generation, selection will eliminate all of the recessive lethal genotypes; the recessive gene survives only in heterozygous dominant individuals. If $q$ is the frequency of the recessive gene, then selection is most effective in changing the frequency of the deleterious allele when $q$ is relatively high because a greater proportion ($q^2$) of the population will be phenotypically recessive and inviable. As $q$ becomes smaller, a greater proportion of the recessive alleles will be "hidden" from selection in heterozygotes, and the change in gene frequency from one generation to the next ($\Delta q$) will become increasingly smaller.

$$\Delta q = -\frac{q^2}{1+q} \qquad (13.4)$$

The number of generations ($n$) required to change the frequency of the lethal allele from its initial frequency ($q_0$) to any desired frequency ($q_n$) is given by

$$n = \frac{1}{q_n} - \frac{1}{q_0} \qquad (13.5)$$

All formulas which predict the number of generations required to accomplish a given change in gene frequency assume that the environmental conditions remain constant throughout the period of change so that the selection coefficient is also a constant. This assumption obviously cannot be applied to long periods of evolutionary time because the environment is continually changing, sometimes rapidly, sometimes quite slowly. A genotype which is at a selective disadvantage in one environment may become a well-adapted genotype in a new environment. As a consequence, gene frequencies may fluctuate up and down as the environment changes the fitness values of the various genotypes in the population.

### 2. Partial Selection Against Recessives.

If the fitness of the two dominant genotypes is 1, then the fitness of the less favored recessive genotype is $1 - s$. The change in gene frequency of the recessive allele per generation ($\Delta q$) is small when $q$ itself is either very large or very small. The largest values of $\Delta q$ are obtained when $q$ is intermediate.

$$\Delta q = \frac{-sq^2(1-q)}{1-sq^2} \qquad (13.6)$$

**Example 13.2.** When $s = 0.2$ and $q = 0.9$ then

$$\Delta q = \frac{-0.2(0.9)^2(0.1)}{1-(0.2)(0.9)^2} = -0.01933$$

Similarly, with the same coefficient of selection and $q = 0.5$, $\Delta q = -0.02631$; with $q = 0.1$, $\Delta q = -0.001803$.

When the coefficient of selection ($s$) against the recessive is small, the change in gene frequencies will be very slow. The quantity $sq^2$ in the denominator of equation (13.6) becomes so small in comparison with unity that for all practical purposes the denominator may be considered to be unity. Under these conditions, the number of generations ($n$) required to change the initial gene frequency ($q_0$) to any desired frequency ($q_n$) is estimated by the following formula:

$$n = \frac{(q_0 - q_n)/q_0 q_n + \log_e \{q_0(1 - q_n)/[q_n(1 - q_0)]\}}{s} \tag{13.7}$$

### 3. Selection in the Absence of Dominance.

Consider a case where the fitness of heterozygotes is exactly intermediate between the two homozygotes.

| Genotypes: | $A_1A_1$ | $A_1A_2$ | $A_2A_2$ |
|---|---|---|---|
| Fitness: | 1 | $1 - \frac{1}{2}s$ | $1 - s$ |

$$\Delta q = \frac{-\frac{1}{2}qs(1 - q)}{1 - sq} \tag{13.8}$$

Selection against zygotes without dominance is equivalent to selection directly against the gametes. For a constant intensity of selection, $\Delta q$ changes more rapidly under gametic selection than under zygotic selection.

The number of generations ($n$) required to effect a change in the initial gene frequency ($q_0$) to any desired value ($q_n$) is

$$n = \frac{\log_e\{q_0(1 - q_n)/[q_n(1 - q_0)]\}}{s} \tag{13.9}$$

### 4. Selection Favoring Heterozygotes.

Gene frequencies tend to reach a stable equilibrium whenever the heterozygote is superior in fitness to either homozygote. Let the fitness of $A_1A_1 = 1 - s_1$, $A_1A_2 = 1$ and $A_2A_2 = 1 - s_2$, then the change in gene frequency per generation is

$$\Delta q = \frac{pq(s_1p - s_2q)}{1 - s_1p^2 - s_2q^2} \tag{13.10}$$

The equilibrium values of the $A_1$ allele ($\hat{p}$) and the $A_2$ allele ($\hat{q}$) are independent of the initial gene frequencies and are determined entirely by the selection coefficients against the homozygotes as expressed in the following equations:

$$\hat{p} = \frac{s_2}{s_1 + s_2} \quad \text{and} \quad \hat{q} = \frac{s_1}{s_1 + s_2} \tag{13.11}$$

## GENETIC DRIFT

One of the important restrictions on the validity of the equilibrium law is the assumption of a large (actually infinitely large) population size. The dispersive process of genetic drift becomes an important factor in changing gene frequency only in small populations. Inbreeding is unavoidable in very small populations. Recall from Chapter 11 that inbreeding reduces heterozygosity in the population and increases the proportion of homozygotes. Thus genetic drift can change gene frequencies through inbreeding. As explained in Chapter 7 for the chi-square analysis, relatively large deviations from the expected values are commonly encountered in small samples. The larger the sample size, the smaller the relative deviation will be from the expected values. Drift can also be considered as a direct result of sampling of gametes from a small gene pool. Gene frequencies in small populations fluctuate randomly from generation to generation if unopposed by any of the systematic processes.

**Example 13.3.** Consider a small population with a constant size of 10 individuals. Beginning with all heterozygous individuals ($A_1A_2$), sampling of gametes from this small gene pool could produce the random changes in gene frequencies shown in the following table.

| | No. of Genotypes | | | Gene Frequencies | |
|---|---|---|---|---|---|
| Generation | $A_1A_1$ | $A_1A_2$ | $A_2A_2$ | $p = A_1$ | $q = A_2$ |
| 0 | 0 | 10 | 0 | 0.50 | 0.50 |
| 1 | 4 | 5 | 1 | 0.65 | 0.35 |
| 2 | 7 | 3 | 0 | 0.85 | 0.15 |
| 3 | 7 | 2 | 1 | 0.80 | 0.20 |

Eventually, gene frequencies may shift so far that one allele becomes *lost* from the population (frequency = 0) and the other allele becomes *fixed* in the population (frequency = 1).

**Example 13.4.** Consider a very small population of four diploid individuals, three of which are $A_1A_1$ and the fourth is $A_1A_2$. Eight alleles are represented by these four individuals. The frequency of the $A_2$ allele ($q$) in this population is 1 out of 8 or 0.125. If the heterozygote does not reproduce, the $A_2$ allele becomes lost ($q = 0$) from the population and the $A_1$ allele becomes fixed in the population ($p = 1$).

The *breeding size* of a population is the number of individuals which contributes gametes to the next generation. The *effective breeding size* ($N_e$) is equivalent to the breeding size only in a hypothetical "ideal" population. The assumptions underlying the concept of an idealized population are

(1) Within each line (subgroup of a population) mating is at random, including self-fertilization in random amount.

(2) Each generation is distinct and does not overlap with another generation.

(3) The number of breeding individuals is the same in all lines and all generations.

(4) Systematic factors are inoperative, i.e. no migration, mutation or selection.

Effective breeding size indicates the size of the "ideal" population whose genetic behavior would be the same as that of the population under consideration. This concept allows comparisons to be made between populations varying in size and sex composition. When the numbers of males ($N_m$) and females ($N_f$) are unequal, the effective breeding size ($N_e$) is approximated by

$$N_e = \frac{4\,N_m\,N_f}{N_m + N_f} \tag{13.12}$$

**Example 13.5.** In a population consisting of 40 breeding males and 280 breeding females, the effective breeding size is

$$N_e = \frac{4(40)(280)}{40 + 280} = 140$$

Thus, when sexes are unequal in numbers, the effective breeding size is always smaller than the actual breeding population, which in turn is always smaller than the total number of individuals in the population.

Most natural populations experience considerable fluctuations in size which may be seasonal, annual or longer in period. An estimate of the effective population size $N_e$ over any number of generations ($n$) where the first generation has an effective population size $N_1$ to the last generation with an effective population size $N_n$ is given by

$$\frac{1}{N_e} = \frac{1}{n}\left(\frac{1}{N_1} + \frac{1}{N_2} + \cdots + \frac{1}{N_n}\right) \tag{13.13}$$

**Example 13.6.** Given four generations with effective sizes 100, 800, 20, 5000, their average effective size is

$$\frac{1}{N_e} = \frac{1}{4}\left(\frac{1}{100} + \frac{1}{800} + \frac{1}{20} + \frac{1}{5000}\right)$$
$$= 0.25(0.01 + 0.00125 + 0.05 + 0.0002) = 0.25(0.06145)$$
$$= 0.0153625$$
$$N_e = 65$$

Thus, the average effective size is much closer to the minimum number than to the maximum number over any range of generations. Obviously, the generations with the smallest numbers have the greatest effect on the genetic structure of the population.

The effective size of an inbred population is approximated by

$$N_e = \frac{N}{1 + F} \tag{13.14}$$

where N is the actual number of breeding individuals in the line and F is Wright's coefficient of inbreeding as explained in Chapter 11.

The rate of inbreeding ($\Delta F$) is the increase of F in one generation relative to the proportion of loci that is still heterozygous and is expressed by

$$\Delta F = \tfrac{1}{2}N = \frac{N_f + N_m}{8N_m N_f} \tag{13.15}$$

where $N_f$ is the number of females and $N_m$ is the number of males.

**Example 13.7.** With 10 males and 50 females in each generation, the rate of inbreeding is

$$\Delta F = \frac{50 + 10}{8(10)(50)} = \frac{60}{4000} = 0.015 \text{ or } 1\tfrac{1}{2}\% \text{ per generation}$$

With five males and five females in each generation, the rate of inbreeding is

$$\Delta F = \frac{5 + 5}{8(5)(5)} = \frac{10}{200} = 0.05 \text{ or } 5\% \text{ per generation}$$

Since the inbreeding coefficient (F) expresses the proportionate increase in homozygotes at the expense of heterozygotes, we can make the following comparisons:

| Genotypes | Original Frequencies | Change Due to Inbreeding |
|---|---|---|
| $A_1A_1$ | $p^2$ | $+pq$F |
| $A_1A_2$ | $2pq$ | $-2pq$F |
| $A_2A_2$ | $q^2$ | $+pq$F |

Thus, one of the major consequences of drift, when it is considered as an inbreeding process, is the reduction of heterozygosity or the increase of genetic uniformity within a line. It is equally obvious that an increase in the frequency of homozygous genotypes at all loci will be produced in different lines. Because of the random nature of genetic drift, different lines will tend to become homozygous for different allelic combinations so that the isolated subpopulations (lines) become genetically differentiated from each other.

If a population could be initially subdivided into a very large number of different lines, the frequency of $q$ might average about 0.5 over all lines and might vary among the different lines from 0.2 to 0.8, forming a distribution something like that shown in Fig. 13-1($a$).

Gene frequencies experience different amounts of drift in the various lines and begin to spread out across the entire range from 0 to 1. After these subpopulations have been exposed for several generations to the sampling process, an equal proportion of lines would be expected to have reached fixation or loss and might appear as in Fig. 13-1(*b*) (lines which have undergone fixation or loss are excluded). Finally, after many generations, a steady state would be expected to be reached, where all gene frequencies of $q$ except the two limits, would be equally probable among the lines (Fig. 13-1(*c*)).

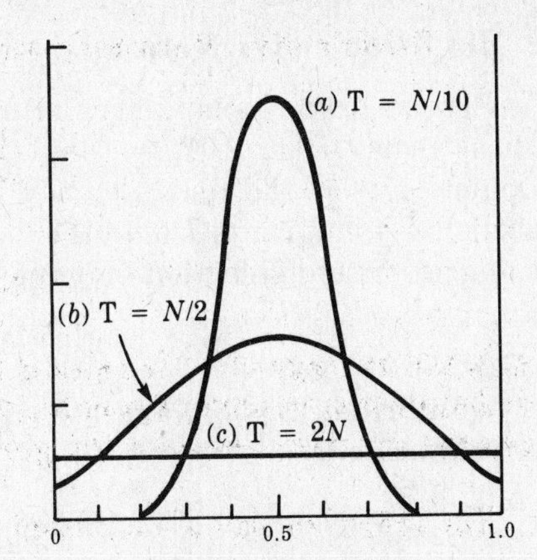


**Fig. 13-1.** Theoretical distributions of gene frequencies among idealized populations (lines) initially at 0.5 produced by genetic drift (exclusive of populations that have become fixed). T = time in generations; all lines are of uniform size $N$. (Idea from Douglas S. Falconer, *Introduction to Quantitative Genetics*, John Wiley & Sons, New York, 1960, p. 55.)

An analogy could be made with a pile of sand in a long trough. Jiggling the trough back and forth would tend to spread the sand out randomly over its entire length. When sand grains reach the ends of the trough, they fall off, representing fixation and loss of gene frequencies. After a period of time, the sand would be evenly spread over the entire length of the trough (steady state, where all gene frequencies are equally probable), but many grains (representing genetic lines) would by this time have already dropped from the ends of the trough. If the pile of sand is initially placed nearer one end of the trough than the other, we would expect a greater amount of sand to drop off the nearer end. Thus, the proportion of lines in which different alleles at a locus are fixed is equal to the initial frequencies of the alleles.

**Example 13.8.** A population is subdivided into 400 isolated subgroups. The initial gene frequency in the entire population is $\bar{q} = 0.8$. When all groups have become homozygous, 80% of them should have reached fixation for the gene in question and in 20% of the lines the gene should have been lost. Thus, $0.8(400) = 320$ lines should have $q = 1$ and 80 lines should have $q = 0$.

The changes wrought by drift may not be adaptive in nature. It is possible, however, for some adaptive types to become fixed in the population by chance along with non-adaptive types. Small populations in which every locus has become fixed have no genetic variability left. When the environment changes, these populations will be unable to generate new adaptive types through genetic recombinations. Thus, no matter how well adapted a homozygous population may be under present conditions, its ultimate fate is likely to be extinction in the long process of evolution as long as it remains isolated.

## JOINT PRESSURES

Contents of the preceding sections have dealt with one evolutionary pressure at a time. This, of course, is not the situation encountered in natural populations. Usually all systematic processes are acting jointly on a population. In addition, natural populations are usually fragmented into local subgroups of various sizes and in the smaller ones the dispersive process can cause appreciable changes in gene frequencies. Migration in variable amounts between the local subgroups of a species often becomes of major importance in these cases. It is not within the scope of this text to cover any of the evolutionary forces in detail. When two or more pressures are considered simultaneously, the situation can become quite complex. A few simple examples of the effects which concurrent forces have on a population will be offered as an introduction to the subject.

**1. Migration and Selection.**

For the sake of simplicity, let us consider a case where heterozygotes are of intermediate fitness ($A_1A_1 = 1$, $A_1A_2 = 1 - s$, $A_2A_2 = 1 - 2s$). The frequency of the $A_2$ allele in the population is $\bar{q}$, and its frequency within a subgroup of this population is $q$. This subgroup receives i immigrants from other semi-isolated groups in each generation. The joint effect of migration and selection on gene frequency is

$$\Delta q = sq^2 - q(\mathrm{i} + s) + \mathrm{i}\bar{q} \tag{13.16}$$

At equilibrium, $\Delta q = 0$ and the equilibrium value of the $A_2$ allele ($\hat{q}$) in the subgroup under consideration is

$$\hat{q} = \frac{(\mathrm{i} + s) \pm \sqrt{(\mathrm{i} + s)^2 - 4\mathrm{i}s\bar{q}}}{2s} \tag{13.17}$$

**2. Migration and Genetic Drift.**

Small populations which are completely isolated from other groups experience more matings between closely related individuals than in a large population. Thus, inbreeding is expected to drive the small population toward homozygosity. As it does so, the coefficient of inbreeding (F) approaches unity. Obviously, if immigration is allowed the larger the proportion of immigrants, the smaller the change in F. At equilibrium, the increased value of F (due to inbreeding in a small population) would be offset by the effect of immigrants from other partially isolated subgroups of the total population. Thus, the values of F in two successive generations would be equal, as expressed in the following:

$$\mathrm{F} = (1 - \mathrm{i})^2\left(\frac{1 + 2\mathrm{FN} - \mathrm{F}}{2\mathrm{N}}\right) \tag{13.18}$$

where the population has been subdivided into many small groups of size N.

Drift generates genetic diversification among the subgroups of a population. Some of these subgroups may be very well adapted to present conditions even though they are highly homozygous. Without some migration between the subgroups, they are unlikely to survive when the environment changes because of their inability to generate new adaptive combinations. Thus, the greatest opportunity for evolutionary success is offered to a population which is subdivided into a number of lines when some migrations occur between the lines.

**3. Mutation and Selection.**

When both mutation and selection are exerting pressure in the same direction, the change in gene frequency will be faster than by either pressure alone. But if they oppose each other, their effects can cancel each other, resulting in a stable equilibrium.

For a fully recessive gene with frequency $q$, with mutation rate to it $u$, and from it $v$, and selection coefficient against it $s$, the equilibrium frequency ($\hat{q}$) is approximately

$$\hat{q} = \sqrt{\frac{u}{s}} \tag{13.19}$$

For a gene lacking dominance (codominant allele) an estimate of the equilibrium frequency is

$$\hat{q} = \frac{u}{s} \tag{13.20}$$

When selection acts against a completely dominant gene, the equilibrium frequency of the recessive allele is approximated by

$$\hat{q} = \frac{v}{s - sq} \tag{13.21}$$

The mutation rates of most genes are of the magnitude of $10^{-5}$ to $10^{-8}$. Only a small selection against the mutant gene is required to hold the gene frequency at a very low equilibrium.

**Example 13.9.** If a recessive gene appears in a population by mutation with a rate of $10^{-5}$, a selection coefficient of 0.001 is sufficient to hold the frequency of the recessive homozygote at 1%; when $s = 0.1$ the frequency of $\hat{q}^2 = 0.0001$, or 1 in 10,000.

# Solved Problems

## MIGRATION

**13.1.** Suppose that a proportion (i) of a large population consists of new immigrants each generation. The frequency of a recessive gene ($a$) among the immigrants is represented by $q_i$ and the frequency among the natives is $q_n$.

(*a*) Derive a formula which will predict the gene frequency of $a$ in the next generation ($q_1$).

(*b*) Derive formula (*13.1*) for the change in gene frequency ($\Delta q$) which occurs between each generation due to migration pressure.

**Solution:**

(*a*) If i is the proportion of the population that is immigrants, then (1 − i) represents the proportion of natives. Gene frequency after immigration has occurred has been changed from $q_n$ to

$$\begin{aligned} q_1 &= \mathrm{i}q_i + (1 - \mathrm{i})q_n \\ &= \mathrm{i}q_i + q_n - \mathrm{i}q_n \\ &= \mathrm{i}(q_i - q_n) + q_n \end{aligned}$$

(*b*) The change in gene frequency ($\Delta q$) which accrues as a result of one generation of immigration is the difference between the frequencies before and after immigration:

$$\begin{aligned} \Delta q &= q_1 - q_n \\ &= [\mathrm{i}(q_i - q_n) + q_n] - q_n \\ &= \mathrm{i}(q_i - q_n) \end{aligned}$$

The degree to which migration can change gene frequency in a population is thus dependent upon the proportion of immigrants (i) and the difference in gene frequencies between immigrants and natives ($q_i - q_n$).

## MUTATION

**13.2.** Suppose that the only factor operative in changing the frequencies of a pair of alleles, $A$ and $a$, is their mutation rate. Let $u$ = mutation rate of $A$ to $a$, and $v$ = mutation rate of $a$ to $A$. The initial frequencies of $A$ and $a$ are $p_0$ and $q_0$ respectively. Derive equation (*13.3*) which expresses the equilibrium frequencies of the recessive allele, $q$, and the dominant allele, $p$.

**Solution:**

The frequency of $A$ will be decreased in the next generation by the amount $up_0$; the frequency of $a$ will be decreased by the amount $vq_0$. The change in gene frequency of the $a$ allele in one generation due to mutation pressure is expressed by

$$\Delta q = up_0 - vq_0$$

Once equilibrium is reached, there will be no change in gene frequency, i.e. $\Delta q = 0$ or $u\hat{p} - v\hat{q} = 0$.

$$u\hat{p} = v\hat{q}$$

$$\hat{q} = \frac{u\hat{p}}{v} = \frac{u(1-\hat{q})}{v}$$

$$\hat{q}v = u - u\hat{q}$$

$$\hat{q}v + u\hat{q} = u$$

$$\hat{q}(v+u) = u$$

$$\hat{q} = \frac{u}{u+v}$$

In like manner, the equilibrium value for the $A$ allele can be shown to be

$$\hat{p} = \frac{v}{u+v}$$

**13.3.** The allele $A_2$ has the initial frequency $q_0$ and mutates to $A_1$ with a rate $u$. Back mutation of $A_1$ to $A_2$ is negligible. Derive a formula which will predict the number of generations required to reduce the frequency of the $A_2$ allele from $q_0$ to any desired value of $q_n$.

**Solution:**

Because $A_2$ mutates to $A_1$ with a rate $u$, the frequency of $A_1$ gametes will be increased by an amount $uq_0$ in the next generation. Similarly, $A_2$ gametes will be reduced by an equal amount. Thus, after one generation of mutation, gametes no longer are in the ratio $p_0(A_1) : q_0(A_2)$, but are changed to the ratio

$$[p_0 + uq_0](A_1) : [q_0 - uq_0](A_2)$$

The frequency of the $A_2$ allele in the next generation is expressed as

$$q_1 = q_0 - uq_0 = q_0(1-u)$$

In each successive generation, the value of $q$ will be reduced by a fraction $u$, forming a geometrical series:

$$\begin{aligned} q_1 &= q_0(1-u) \\ q_2 &= q_1(1-u) = q_0(1-u)^2 \\ q_3 &= q_2(1-u) = q_0(1-u)^3 \\ &\vdots \\ q_n &= q_0(1-u)^n \end{aligned}$$

This formula can be solved for the number of generations ($n$) by converting it to logarithmic form:

$$\log q_n = n\log(1-u) + \log q_0$$

$$n\log(1-u) = \log q_n - \log q_0$$

$$n = \frac{\log q_n - \log q_0}{\log(1-u)}$$

## SELECTION

**13.4.** Let $p_0$ = initial frequency of the dominant allele ($A$) and $q_0$ = initial frequency of the recessive allele ($a$) in a population. Recessive genotypes ($aa$) are lethal. For a population mating at random, determine the gene frequency of the recessive allele in the next generation ($q_1$) in terms of the initial frequency ($q_0$).

**Solution:**

A population mating at random should conform to the Hardy-Weinberg expectations:

$$\underbrace{p_0^2(AA) + 2p_0q_0(Aa)}_{\text{This fraction gives rise to the next generation.}} + \underbrace{q_0^2(aa)}_{\text{This fraction is eliminated.}} = 1$$

The homozygous genotype $AA$ produces only $A$-bearing gametes. The heterozygous genotype $Aa$ yields $\frac{1}{2}A$- and $\frac{1}{2}a$-bearing gametes. The effective breeding population will produce gametes in the ratio

$$\begin{aligned}[p_0^2 + \tfrac{1}{2}(2p_0q_0)](A) : \tfrac{1}{2}(2p_0q_0)(a) &= (p_0^2 + p_0q_0)(A) : p_0q_0(a) \\ &= p_0(p_0 + q_0)(A) : p_0q_0(a) \\ &= p_0(A) : p_0q_0(a) \qquad [\text{since } (p_0 + q_0) = 1] \\ &= 1(A) : q_0(a) \qquad [\text{both sides divided by } p_0]\end{aligned}$$

These gametic proportions do not add up to unity but to $(1 + q_0)$. In order to compare them with those of the previous generation we must express each as a fraction of the new total $(1 + q_0)$:

$$\frac{1}{1 + q_0}(A) : \frac{q_0}{1 + q_0}(a)$$

The frequency of the recessive allele in this generation, designated $q_1$, is $q_0/(1 + q_0)$.

**13.5.** To determine the gene frequency ($q_n$) of an autosomal recessive gene after $n$ generations of complete selection against the recessive homozygote, we employ the formula

$$q_n = \frac{q_0}{1 + nq_0}$$

where $q_0$ is the initial frequency of the recessive gene. Show how this formula is derived.

**Solution:**

As was shown in the preceding problem, gene frequencies in the first generation can be expressed as

$$p_1 = \frac{1}{1 + q_0} \qquad \text{and} \qquad q_1 = \frac{q_0}{1 + q_0}$$

Let us make a ratio of $p_1$ to $q_1$:

$$\frac{p_1}{q_1} = \frac{1}{1 + q_0} \div \frac{q_0}{1 + q_0} = \frac{1}{q_0}$$

One generation of complete selection against the recessive genotype produces a change in the initial ratio of $p/q$ expressed by

$$\frac{p_1}{q_1} - \frac{p_0}{q_0} = \frac{1}{q_0} - \frac{p_0}{q_0} = \frac{1 - p_0}{q_0} = 1$$

The Greek letter delta ($\Delta$) signifies an increment of change. The change in the ratio of the two gene frequencies $= \Delta(p/q) = 1$. This means that in a single generation, the ratio of $A$ to $a$ alleles has increased numerically by one and will continue to increase by 1 in each of the subsequent generations, so that by the $n$th generation the ratio will have increased by $n$. Hence, in the $n$th generation, the ratio will be

$$\frac{p_n}{q_n} = \frac{p_0}{q_0} + n$$

Substituting $(1 - q_n)$ for $p_n$ and $(1 - q_0)$ for $p_0$, we have

$$\frac{1 - q_n}{q_n} = \frac{1 - q_0}{q_0} + n$$

$$\frac{1}{q_n} - 1 = \frac{1}{q_0} - 1 + n$$

$$\frac{1}{q_n} = \frac{1}{q_0} + n = \frac{1 + nq_0}{q_0}$$

Therefore,

$$q_n = \frac{q_0}{1 + nq_0}$$

**13.6.** Suppose that some fraction ($s$) of the recessive genotype ($aa$) is eliminated in each generation. Beginning with the initial gene frequencies $q_0$ for $a$ and $p_0$ for $A$ in a population mating at random, derive a formula for expressing the frequency of $a$ in the next generation ($q_1$) in terms of $q_0$ and $s$.

**Solution:**

The initial distribution of genotypes should conform to the equilibrium law if mating is at random:

$$p_0^2(AA) + 2p_0q_0(Aa) + q_0^2(aa) = 1$$

If a fraction ($s$) of the $aa$ zygotes is eliminated from the effective breeding population, this leaves $AA$, $Aa$ and $aa$ individuals in the ratio

$$p_0^2(AA) + 2p_0q_0(Aa) + q_0^2(1 - s)(aa) = 1$$

The $p_0^2$ homozygous dominant individuals ($AA$) produce only $A$ gametes. Half of the gametes of heterozygotes ($Aa$) bear the $A$ allele $= \frac{1}{2}(2p_0q_0) = p_0q_0$; the other half of the gametes or $p_0q_0$ bear the $a$ allele. Homozygous recessives produce only $a$ gametes in the proportion of $(1 - s)q_0^2$. One generation of partial selection of magnitude $s$ against the recessive genotype produces gametes in the ratio $(p_0^2 + p_0q_0)(A) : [p_0q_0 + (1 - s)q_0^2](a)$ or $p_0(p_0 + q_0)(A) : (p_0q_0 + q_0^2 - sq_0^2)(a)$.

Since $p_0 + q_0 = 1$, we have the ratio

$$p_0(\text{A}):(p_0q_0 + q_0^2 - sq_0^2)(a) = p_0(\text{A}):[q_0(p_0 + q_0) - sq_0^2](a)$$
$$= p_0(\text{A}):(q_0 - sq_0^2)(a)$$

In order for these gametic proportions to add up to unity, they must be expressed in terms of their sum, $p_0 + q_0 - sq_0^2$ or $1 - sq_0^2$. Allelic frequencies in the next generation then become

$$\frac{p_0}{1 - sq_0^2}(A) : \frac{q_0 - sq_0^2}{1 - sq_0^2}(a)$$

The frequency of the recessive allele ($q_1$) after one generation of selection is given by the formula

$$q_1 = \frac{q_0 - sq_0^2}{1 - sq_0^2}$$

**13.7.** A selection coefficient of 0.001 is acting against a recessive genotype. How many generations would be required to reduce the incidence of recessives from 50% to 1%?

**Solution:**

If $q_0^2 = 0.5$, then $q_0 = \sqrt{0.5} = 0.7071$; $q_n^2 = 0.01$, $q_n = \sqrt{0.01} = 0.1$.

By equation (*13.7*)

$$n = \frac{[(q_0 - q_n)/q_0q_n] + \log_e[q_0(1 - q_n)/q_n(1 - q_0)]}{s}$$

Natural logarithms (to the base $e$) can be converted to common logarithms (base 10) by multiplying by 2.303.

$$n = \frac{[(0.7071 - 0.1)/(0.7071)(0.1)] + 2.303 \log_{10}[0.707(1 - 0.1)/0.1(1 - 0.707)]}{0.001}$$
$$= \left[\frac{0.6071}{0.0707} + 2.303 \log_{10}\left(\frac{0.707 \times 0.90}{0.10 \times 0.293}\right)\right] \div 0.001 = \left[8.587 + 2.303 \log_{10}\frac{0.6363}{0.0293}\right] \div 0.001$$
$$= [8.587 + 2.303 \log_{10} 21.7167] \div 0.001 = [8.587 + (2.303)(1.33686)] \div 0.001$$
$$= (8.587 + 3.079) \div 0.001 = 11.666 \div 0.001$$
$$= 11{,}666 \text{ generations}$$

**13.8** Consider a pair of alleles ($A_1$ and $A_2$) in which the heterozygote is superior to either homozygote (overdomimance). Different selection coefficients, $s_1$ and $s_2$, are active against the $A_1A_1$ and $A_2A_2$ homozygotes respectively. The initial frequencies and fitness of the genotypes are shown below.

| Genotypes | Frequency | Fitness |
|---|---|---|
| $A_1A_1$ | $p^2$ | $1 - s_1$ |
| $A_1A_2$ | $2pq$ | $1$ |
| $A_2A_2$ | $q^2$ | $1 - s_2$ |

Derive formula (*13.10*) which expresses the change in frequency of the $A_2$ allele ($\Delta q$) as a result of one generation of selection.

**Solution:**

$$p^2(1 - s_1)(A_1A_1) + 2pq(A_1A_2) + q^2(1 - s_2)(A_2A_2) = 1$$
$$p^2 - p^2s_1 + 2pq + q^2 - q^2s_2 = 1$$
$$(p^2 + 2pq + q^2) - p^2s_1 - q^2s_2 = 1$$
$$1 - p^2s_1 - q^2s_2 = 1$$

$$q_1 = \frac{pq + q^2 - q^2s_2}{1 - p^2s_1 - q^2s_2} = \frac{q(1-q) + q^2 - q^2s_2}{1 - p^2s_1 - q^2s_2}$$
$$= \frac{q - q^2 + q^2 - q^2s_2}{1 - p^2s_1 - q^2s_2} = \frac{q - q^2s_2}{1 - p^2s_1 - q^2s_2}$$

The change in $q$ after one generation of selection is

$$\Delta q = q_1 - q$$
$$= \frac{q - q^2s_2}{1 - p^2s_1 - q^2s_2} - q = \frac{q - q^2s_2 - q + qp^2s_1 + q^2s_2(1-p)}{1 - p^2s_1 - q^2s_2}$$
$$= \frac{-q^2s_2 + qp^2s_1 + q^2s_2 - pq^2s_2}{1 - p^2s_1 - q^2s_2} = \frac{pq(ps_1 - qs_2)}{1 - p^2s_1 - q^2s_2}$$

**13.9.** In some parts of Africa the frequency of sickle-cell anemia is as high as 40% of the population. Many homozygotes ($S^1S^1$) suffer severe anemia and often die. Heterozygotes ($S^1S^2$) incur a much less debilitating malady called "sickling trait". An abnormal hemoglobin is found in both $S^1S^1$ homozygotes and heterozygotes. Approximately 3% of abnormals are homozygotes. The fitness of $S^1S^1$ homozygotes is only about $\frac{1}{4}$ that of heterozygotes. Where malaria is prevalent, heterozygotes exhibit overdominance with respect to fitness. It can be shown that the equilibrium frequency ($\hat{q}$) of the $S^1$ gene is expressed by

$$\hat{q} = \frac{s_1}{s_1 + s_2}$$

where $s_1$ is the selection coefficient of normal homozygotes ($S^2S^2$) and $s_2$ is that of the sickling homozygotes. How much of a selective advantage do heterozygotes have over normal individuals?

**Solution:**

When 40% of the population is abnormal

$$q^2\ S^1S^1 + 2pq\ S^1S^2 = 0.4$$

Because 3% of abnormals are homozygous, $q^2 = 3\%$ of 0.4 or 0.012.

$$2pq = 0.4 - 0.012 = 0.388$$

$q$ = frequency of $S^1$ allele = frequency of $S^1S^1$ homozygotes + $\frac{1}{2}$ frequency of heterozygotes
$= 0.012 + \frac{1}{2}(0.388) = 0.012 + 0.194$
$= 0.206$

When heterozygotes are superior to either homozygous class the fitness values of genotypes are

$$S^1S^1 = 1 - s_2$$
$$S^1S^2 = 1$$
$$S^2S^2 = 1 - s_1$$

If the fitness of $S^1S^1$ is $\frac{1}{4}$ that of $S^1S^2$, then

$$1 - s_2 = 0.25$$
$$s_2 = 1 - 0.25 = 0.75 \qquad (13.11)$$
$$\hat{q} = \frac{s_1}{s_1 + s_2}$$

Assuming that this locus has reached its equilibrium values, we have

$$0.206 = \frac{s_1}{s_1 + 0.75}$$
$$0.206s_1 + 0.1545 = s_1$$
$$0.794s_1 = 0.1545$$
$$s_1 = 0.195$$

We can now express the selective advantage of heterozygotes relative to normal homozygotes as

$$\frac{1}{1 - s_1} = \frac{1}{1 - 0.195} = \frac{1}{0.805} = 1.24$$

Thus, the heterozygotes have a selective advantage which is 24% greater than normal homozygotes due to their resistance against malaria.

**13.10.** Hemolytic disease of the newborn (*erythroblastosis fetalis*) is a condition commonly encountered when an Rh-positive child is carried by an Rh-negative mother. If blood from the child gets into the maternal circulation, the mother produces antibodies against the Rh antigens on the blood cells of her baby. In severe cases, the child may die. The Rh antigen is produced in response to a dominant gene ($R$) and the absence of the Rh antigen by its recessive allele ($r$). If the father is Rh-positive ($RR$ or $Rr$) and the mother is Rh-negative ($rr$), there is a risk that the heterozygous Rh-positive child ($Rr$) will be erythroblastotic. Thus, selection is not directed at all heterozygotes, but only those which are born to Rh-negative mothers. Derive a formula which expresses the change in gene frequency ($\Delta q$) of the $r$ allele between each generation due to this type of selection.

**Solution:**

Let $p$ = frequency of $R$ allele and $q$ = frequency of $r$ allele; fitness of heterozygotes born to recessive mothers is $1 - s$; all other heterozygotes and other genotypes are normal.

When the mother is $RR$, the frequency of matings is as follows:

(1) $RR \times RR = p^2 \times p^2 = p^4$
(2) $RR \times Rr = p^2 \times 2pq = 2p^3q$
(3) $RR \times rr = p^2 \times q^2 = p^2q^2$

All the children from the first mating will be $RR$ $= p^4$

Half of the children from the second mating will be $RR$ $= p^3q$

Total $= p^4 + p^3q$

$= p^3(p + q)$

$= p^3(1) = p^3$

Half of the children from the second mating will be $Rr$ $= p^3q$

All of the children from the third mating will be $Rr$ $= p^2q^2$

Total $= p^3q + p^2q^2$

$= p^2q(p + q)$

$= p^2q(1) = p^2q$

Similar reasoning for all other mating types produces the following table of expectations:

| Mother × Father | Frequency of Mating | Children | | | Total |
|---|---|---|---|---|---|
| | | $RR$ | $Rr$ | $rr$ | |
| $RR \times -$ | $p^2$ | $p^3$ | $p^2q$ | 0 | $p^2$ |
| $Rr \times -$ | $2pq$ | $p^2q$ | $pq$ | $pq^2$ | $2pq$ |
| $rr \times RR$ | $p^2q^2$ | 0 | $p^2q^2(1-s)$ | 0 | $q^2 - spq^2$ |
| $rr \times Rr$ | $2pq^3$ | 0 | $pq^3(1-s)$ | $pq^3$ | |
| $rr \times rr$ | $q^4$ | 0 | 0 | $q^4$ | |
| Total | 1.00 | $p^2$ | $2pq - spq^2$ | $q^2$ | $1 - spq^2$ |

The totals at the bottom of the table are derived as follows.

Total of the first column: $p^2 + 2pq + p^2q^2 + 2pq^3 + q^4 = \underbrace{p^2 + 2pq}_{1} + q^2\underbrace{(p^2 + 2pq + q^2)}_{1} = 1$

Total of $RR$ children: $p^3 + p^2q = p^3 + p^2(1-p) = p^3 + p^2 - p^3 = p^2$

Total of $Rr$ children:

$$
\begin{aligned}
&p^2q + pq + p^2q^2(1-s) + pq^3(1-s)\\
&= p^2q + pq + p^2q^2 - p^2q^2s + pq^3 - pq^3s\\
&= p^2q + pq + p^2q^2 + pq^3 - pq^2s(p+q)\\
&= p^2(1-p) + p(1-p) + p^2(1-p)^2 + p(1-p)^3 - pq^2s\\
&= p^2 - p^3 + p - p^2 + p^2 - 2p^3 + p^4 + p - 3p^2 + 3p^3 - p^4 - pq^2s\\
&= 2p - 2p^2 - pq^2s = 2p(1-p) - pq^2s = 2pq - pq^2s
\end{aligned}
$$

Total of $rr$ children:

$$
\begin{aligned}
&pq^2 + pq^3 + q^4 = q^2(1-q) + q^3(1-q) + q^4\\
&= q^2 - q^3 + q^3 - q^4 + q^4 = q^2
\end{aligned}
$$

Total of last column:

$$
\begin{aligned}
&(p^2 + 2pq + q^2) - spq^2\\
&1 - spq^2
\end{aligned}
$$

The frequency of the recessive allele ($r$) in the next generation is the total of $rr$ homozygotes plus one-half of the total of heterozygotes.

$$
\begin{aligned}
q_1 &= \frac{pq - \frac{1}{2}spq^2 + q^2}{1 - spq^2} = \frac{q(1-q) - \frac{1}{2}spq^2 + q^2}{1 - spq^2} = \frac{q - q^2 - \frac{1}{2}spq^2 + q^2}{1 - spq^2}\\
&= \frac{q - \frac{1}{2}spq^2}{1 - spq^2}
\end{aligned}
$$

The change in gene frequency due to one generation of this peculiar kind of selection is

$$
\begin{aligned}
\Delta q &= q_1 - q\\
&= \frac{q - \frac{1}{2}spq^2}{1 - spq^2} - q = \frac{q - \frac{1}{2}spq^2 - q(1 - spq^2)}{1 - spq^2}\\
&= \frac{q - \frac{1}{2}spq^2 - q + spq^3}{1 - spq^2} = \frac{spq^2(q - \frac{1}{2})}{1 - spq^2}
\end{aligned}
$$

When the selection coefficient is small, the denominator is approximately one and the formula reduces to

$$\Delta q = spq^2(q - \tfrac{1}{2})$$

Notice that an unstable equilibrium exists only when $q = \frac{1}{2}$. The $r$ allele will increase if $q$ exceeds $\frac{1}{2}$, and will decrease if $q$ is less than $\frac{1}{2}$.

## GENETIC DRIFT

**13.11.** Only when the number of males ($N_m$) is equal to the number of females ($N_f$) is their sum ($N_m + N_f$) equal to the effective population size (N). Under these conditions, prove that equation (*13.15*) is equivalent to $\Delta F = 1/2N$.

**Solution:**

$$\Delta F = \frac{N_f + N_m}{8N_mN_f} \qquad (13.15)$$

Since $N_f + N_m = N$, then $N_f = N_m = N/2$

$$\Delta F = \frac{N}{8(N/2)(N/2)} = \frac{N}{8N^2/4} = \frac{N}{2N^2} = \frac{1}{2N}$$

**13.12.** Approximately 20% inbreeding has occurred in a small population. Determine the amount of genetic change which has occurred as a result of the reduction in population size when the initial gene frequencies of $A_1$ and $A_2$ were respectively (*a*) $p = 0.5$, $q = 0.5$, (*b*) $= 0.9$, $q = 0.1$.

**Solution:**

(*a*) Originally the population was in equilibrium with

$$p^2A_1A_1 + 2pqA_1A_2 + q^2A_2A_2 = 1.0$$
$$(0.5)^2 + 2(0.5)(0.5) + (0.5)^2 = 1.0$$
$$25\% + 50\% + 25\% = 1.0$$

Now, because of 20% inbreeding (F = 0.2), the genotypic frequencies are expected to be

$$\begin{aligned} A_1A_1 &= p^2 + pqF = 0.25 + (0.25)(0.2) = 0.300 \\ A_1A_2 &= 2pq - 2pqF = 0.50 - (0.5)(0.2) = 0.400 \\ A_2A_2 &= q^2 + pqF = 0.25 + (0.25)(0.2) = \underline{0.300} \\ & \qquad\qquad\qquad\qquad\qquad\qquad\quad 1.000 \end{aligned}$$

(*b*) Originally the population was in equilibrium with

$$p^2A_1A_1 + 2pqA_1A_2 + q^2A_2A_2 = 1.0$$
$$(0.9)^2 + 2(0.9)(0.1) + (0.1)^2 = 1.0$$
$$81\% + 18\% + 1\% = 1.0$$

Now, because of inbreeding, the frequencies are expected to be

$$\begin{aligned} A_1A_1 &= p^2 + pqF = 0.81 + (0.09)(0.2) = 0.828 \\ A_1A_2 &= 2pq - 2pqF = 0.18 - 0.18(0.2) = 0.144 \\ A_2A_2 &= q^2 + pqF = 0.01 + (0.09)(0.2) = \underline{0.028} \\ & \qquad\qquad\qquad\qquad\qquad\qquad\quad 1.000 \end{aligned}$$

A 20% reduction in the proportion of heterozygotes occurs in both parts (*a*) and (*b*), but in the former where gene frequencies are intermediate, a much larger numerical reduction is expected (50% − 40% = 10% vs. 18% − 14.4% = 3.6%). Thus, as is true in selection, greater changes in gene frequency ($\Delta q$) can be made with those genes at intermediate values.

## JOINT PRESSURES

**13.13.** Let the percentage of immigrants (i) be equivalent to the coefficient of selection. Taking $\bar{q} = 0.40$ for the total population, determine the equilibrium value ($\hat{q}$) when (*a*) selection favors the gene; (*b*) selection acts against the gene.

**Solution:**

(*a*) If selection favors the gene, $s$ will have a *negative* value. Thus, if $|i| = -s$ then the combined forces of migration and selection have the value $i + s = 0$. Substituting into equation (*13.17*), we obtain

$$\hat{q} = \frac{(i + s) \pm \sqrt{[(i + s)^2 - 4is\bar{q}]}}{2s} = \frac{0 \pm \sqrt{0 - 4is\bar{q}}}{2s}$$

If $i = -s$, then we have

$$\hat{q} = \frac{\sqrt{-4(-s)s\bar{q}}}{2s} = \frac{\sqrt{4s^2\bar{q}}}{2s} = \frac{2s\sqrt{\bar{q}}}{2s}$$
$$= \sqrt{\bar{q}} = \sqrt{0.4} = 0.63$$

(*b*) If selection acts against the gene, then $s$ will have a positive value. Thus, when i = $s$, i + $s$ = $2s$. Again substituting into equation (*13.17*), we obtain

$$\hat{q} = \frac{2s \pm \sqrt{(2s)^2 - 4s^2\bar{q}}}{2s} = \frac{2s \pm \sqrt{4s^2 - 4s^2\bar{q}}}{2s}$$

$$= \frac{2s \pm \sqrt{4s^2(1-\bar{q})}}{2s} = \frac{2s \pm 2s\sqrt{1-\bar{q}}}{2s} = 1 \pm \sqrt{1-\bar{q}}$$

Obviously the equilibrium frequency of the gene cannot be more than unity, so the negative root must be the correct solution.

$$\hat{q} = 1 - \sqrt{1-\bar{q}} = 1 - \sqrt{1-0.4} = 1 - \sqrt{0.6} = 1 - 0.77 = 0.23$$

**13.14.** Consider a fully recessive gene ($a$) with frequency $q$, mutation rate $u$ representing mutation of $A$ to $a$ and $v$ representing back mutation of $a$ to $A$, and coefficient of selection $s$ against the recessive genotype. Develop formula (*13.19*) which approximately represents the equilibrium value of $a$ ($\hat{q}$) generated by the opposing forces of selection and mutation.

**Solution:**

The change in $q$ produced by one generation of opposing mutation pressures was developed in Problem 13.2.

$$\Delta q_m = up_0 - vq_0$$

Usually we are interested in genes at low equilibrium frequencies. If $q$ is small, then the value $vq_0$ is negligible and we may ignore it to simplify computations. The change in $q$ produced by one generation of selection of intensity $s$ is given by formula (*13.6*):

$$\Delta q_s = \frac{-sq^2(1-q)}{1-sq^2}$$

If $q$ is small, then in the denominator $sq^2$ is also negligible. At equilibrium

$$\Delta q_m = \Delta q_s$$
$$u(1-q) = -sq^2(1-q)$$
$$u = -sq^2$$
$$q^2 = -\frac{u}{s}$$
$$\hat{q} = \sqrt{\frac{u}{s}} \qquad (13.19)$$

**13.15.** Derive a formula which will predict the equilibrium value of $q^2$, the frequency of recessive homozygotes ($aa$), when recurrent mutation of $A$ to $a$ is opposed by complete selection against $aa$ zygotes.

**Solution:**

It was shown in Problem 13.4 that one generation of complete selection against the homozygous recessive genotype changes the gene frequencies from $p_0$ ($A$) : $q_0$ ($a$) to

$$\frac{1}{1+q_0}(A) : \frac{q_0}{1+q_0}(a)$$

If $u$ is the mutation rate of $A$ to $a$ and we assume that the back mutation rate ($a$ to $A$) is negligible, then the gene frequencies in the next generation are corrected to

$$\left(\frac{1}{1+q_0} - \frac{u}{1+q_0}\right)(A) : \left(\frac{q_0}{1+q_0} + \frac{u}{1+q_0}\right)(a)$$

The frequency of the recessive allele in the next generation is

$$q_1 = \frac{q_0 + u}{1 + q_0}$$

The change in $q$ resulting from a single generation of the opposing forces of selection against the recessive and mutation favoring the recessive is

$$q_1 - q_0 = \frac{q_0 + u}{1 + q_0} - q_0 = \frac{q_0 + u - q_0(1 + q_0)}{1 + q_0}$$
$$= \frac{q_0 + u - q_0 - q_0^2}{1 + q_0} = \frac{u - q_0^2}{1 + q_0}$$

The change in $q$ in each generation is generally expressed as $\Delta q = (u - q^2)/(1 + q)$. When equilibrium has been attained, there is no change in gene frequency from generation to generation and $\Delta q = 0$.

$$\frac{u - q^2}{1 + q} = 0$$

Therefore

$$u - q^2 = 0$$
$$q^2 = u$$

That is to say, equilibrium will have been reached between selection and mutation pressures when the frequency of the homozygous recessive genotype ($aa$) equals the rate of mutation of $A$ to $a$.

**13.16** Chondrodystrophy, a form of dwarfism in humans, is governed by a dominant gene. Fitness of these dwarfs is estimated to be 20% that of normal individuals. The frequency of dwarfs in the population is about 1 in 10,000. Assuming that this gene has already reached equilibrium, estimate the mutation rate to it by equation (*13.21*).

**Solution:**

$$\hat{q} = \frac{v}{s - sq} \qquad (13.21)$$
$$v = qs - q^2 s = qs(1 - q)$$
$$qp = \frac{v}{s}$$

The frequency of heterozygotes is $2pq$

$$2pq = \frac{2v}{s}$$

If the frequency of dwarfs is rare (in this case $1 \times 10^{-4}$), virtually all dwarfs will be heterozygotes. If fitness of dwarfs is $1 - s = 0.2$, then $s = 0.8$. Substituting into the above formula, we have

$$0.0001 = \frac{2v}{0.8}$$
$$0.00008 = 2v$$
$$v = 0.00004 \text{ or } 4 \times 10^{-5}$$

## Supplementary Problems

### MIGRATION

**13.17.** One generation of immigration into a population results in a change of gene frequency expressed as $\Delta q = \mathrm{i}(q_\mathrm{i} - q_n)$. Prove that $\Delta q$ may also be expressed as $\Delta q = \mathrm{i}p_n q_\mathrm{i} - \mathrm{i}p_\mathrm{i} q_\mathrm{n}$. *Hint:* Add and subtract the quantity $\mathrm{i}q_\mathrm{i}q_\mathrm{n}$ to right-hand side of the equation.

**13.18.** Suppose that 5% of a population consists of immigrants in each generation. The initial gene frequency, $q_\mathrm{n}$, of a certain gene among the natives before immigration is 0.3. The frequency of this gene among the immigrants, $q_\mathrm{i}$, is 0.7. (*a*) What will be the gene frequency, $q_1$, in the mixed population after immigration has occurred? (*b*) Suppose that the gene frequency among the immigrants ($q_\mathrm{i}$) was only 0.4. What immigration pressure, i, would be required to yield the same $q_1$ value as in part (*a*)?

## MUTATION

**13.19.** A close approximation to the formula developed in Problem 13.3 is given by

$$n = \frac{2.30}{u}(\log q_0 - \log q_n)$$

A recessive trait ($aa$) is found in 81% of the population. The recessive allele, $a$, mutates to $A$ with a frequency $u = 7 \times 10^{-5}$. Back mutation of $A$ to $a$ is negligible. How many generations ($n$) are required to decrease the incidence of the recessive *trait* to 25%?

**13.20.** The initial frequency of the $A_1$ allele is $p_0$ and that of the $A_2$ allele is $q_0$; the mutation rate of $A_1$ to $A_2$ is $u = 5 \times 10^{-5}$, and the back mutation rate of $A_2$ to $A_1$ is $v = 1 \times 10^{-5}$. Calculate the change in frequency of the $A_2$ allele ($\Delta q$) due to one generation of mutation by equation (*13.2*) when (*a*) $p = 0.9$, $q = 0.1$; (*b*) $p = 0.5$, $q = 0.5$; and (*c*) $p = 0.1$, $q = 0.9$.

## SELECTION

**13.21.** Approximately 1% of all zygotes formed by a population dies as a result of a recessive lethal gene. (*a*) What change in gene frequency ($\Delta q$) is effected by one generation of selection against the lethal allele? *Hint:* Use equation (*13.4*). (*b*) What will be the frequency ($q_1$) of the lethal allele in the next generation?

**13.22.** Beginning with a population where 16% of the individuals are recessive ($aa$) lethal homozygotes (*a*) determine the frequencies, $q_1$ and $q_2$, of the recessive allele in the next two generations. *Hint:* Use the formula developed in Problem 13.5. (*b*) Determine the percentage of lethal genotypes in each of the two generations of part (*a*).

**13.23.** The frequency of the recessive allele ($q_n$) in the $n$th generation of complete selection against the recessive genotype can be expressed in terms of the initial frequency ($q_0$) as follows:

$$n = \frac{1}{q_n} - \frac{1}{q_0} \qquad (13.5)$$

How many generations of complete selection against the recessive genotype ($aa$) will be required to reduce the frequency of $aa$ zygotes from 16% to 1%?

**13.24.** Albinism, governed by a single recessive gene, is not considered to be a very serious genetic defect. Approximately 1 out of 20,000 individuals in the population are now afflicted with albinism. Suppose that it is suggested that all albinos should henceforth be prevented from reproducing so that the incidence of albinos would be reduced in future generations. Assuming a generation to be 25 years, how many years would it take to reduce the frequency of albinism to half of its present value?

**13.25.** In Problem 13.6, it was shown that one generation of selection against a recessive genotype ($aa$) changes the gene frequency of the recessive allele from $q_0$ to $q_1 = (q_0 - sq_0^2)/(1 - sq_0^2)$. From this, derive formula (*13.6*) which represents the change in gene frequency ($\Delta q$) which occurs as a result of the selection pressure.

**13.26.** Selection against a recessive genotype produces in each generation a change ($\Delta q$) in the gene frequency of the recessive allele expressed as

$$\Delta q = \frac{-sq^2(1 - q)}{1 - sq^2} \qquad (13.6)$$

If the coefficient of selection against the recessive is small, then the denominator of the above equation approximates unity so that $\Delta q \cong -sq^2(1 - q)$. At what value of $q$ would a constant selection pressure produce the greatest change in gene frequency?

**13.27.** Consider the case of a pair of codominant alleles ($A_1$ and $A_2$). In a random mating population, the initial frequencies and fitnesses of the genotypes are as shown below:

| Genotypes | Frequency | Fitness |
|---|---|---|
| $A_1A_1$ | $p^2$ | 1 |
| $A_1A_2$ | $2pq$ | $1 - \frac{1}{2}s$ |
| $A_2A_2$ | $q^2$ | $1 - s$ |

Derive formula (*13.8*) which expresses the change in frequency of the $A_2$ allele ($\Delta q$) as a result of one generation of selection.

**13.28.** In a case where a pair of alleles lacking dominance is concerned, the fitness values of the genotypes are as follows: $A_1A_1 = 1$, $A_1A_2 = 0.99$, $A_2A_2 = 0.98$. How many generations are required to reduce the frequency of the $A_2$ allele ($q$) from its initial value of 0.40 to 0.04? *Hint:* Use formula (*13.9*) and see Problem 13.7.

**13.29.** When homozygotes are inferior to heterozygotes the change in gene frequency per generation is given by formula (*13.10*).

$$\Delta q = \frac{pq(s_1p - s_2q)}{1 - s_1p^2 - s_2q^2}$$

If $s_1p > s_2q$, then $\Delta q$ will be positive and the $A_2$ allele will become more frequent in the gene pool; if $s_1p < s_2q$, then $\Delta q$ will be negative and the $A_2$ allele will become less frequent. Equilibrium is reached when $\Delta q = 0$ and $s_1p = s_2q$. Using the symbol $\hat{p}$ and $\hat{q}$ for the equilibrium values of $A_1$ and $A_2$ respectively, we have $s_1\hat{p} = s_2\hat{q}$. Prove that the equilibrium gene frequencies are determined solely by the magnitude of the selection coefficients, i.e.

$$\hat{q} = \frac{s_1}{s_1 + s_2} \quad \text{and} \quad \hat{p} = \frac{s_2}{s_1 + s_2} \qquad (13.11)$$

**13.30.** When complete dominance with respect to fitness is operative with selection against the $A_1$ allele, we have

| Genotypes: | $A_1A_1$ | $A_1A_2$ | $A_2A_2$ |
|---|---|---|---|
| Fitness: | $1 - s$ | $1 - s$ | 1 |

Prove that the change in gene frequency per generation of the $A_2$ allele ($\Delta q$) can be expressed as $sq^2(1 - q)/[1 - s(1 - q^2)]$.

**13.31.** In a case where selection is against heterozygotes, the fitness values of genotypes are: $A_1A_1 = 1$, $A_1A_2 = 1 - s$, $A_2A_2 = 1$. (*a*) Prove that the change in frequency of the $A_2$ allele per generation ($\Delta q$) can be expressed as $\Delta q = 2spq(q - \frac{1}{2})/(1 - 2spq)$. (*b*) When $s$ is small, the denominator of the equation is approximately unity. Will $\Delta q$ be positive or negative when $q$ is greater than 0.5; when $q$ is less than 0.5?

## GENETIC DRIFT

**13.32.** Given a population consisting of 20 breeding males and 100 breeding females, calculate the effective breeding size by equation (*13.12*).

**13.33** A line is maintained with an indefinitely large number of females, but with only one male per generation. Using formula (*13.12*) calculate the approximate effective breeding size.

**13.34.** From the results of Problem 13.11, prove equation (*13.12*). *Hint:* Set equation (*13.15*) equal to 1/2N.

**13.35.** A certain population of insects experiences wide fluctuations in population size over 3 consecutive years: $N_1 = 20{,}000$, $N_2 = 1400$, $N_3 = 66{,}000$. Estimate the effective population size during this interval by equation (*13.13*).

**13.36.** Under what condition may the actual size of an inbred population ($n$) be considered equivalent to the *effective* size (N)? *Hint:* Use equation (*13.14*).

**13.37.** Using equation (*13.14*), determine the effective population size (N) of an isolated line which is 50% inbred and consists of 20 breeding pairs.

## JOINT PRESSURES

**13.38.** The average gene frequency of a certain allele in the entire population is $\bar{q} = 0.40$. When the proportion of immigrants (i) is equivalent to the selection coefficient ($s$) determine the equilibrium gene frequency ($\hat{q}$) when selection (*a*) favors the gene; (*b*) acts against the gene. *Hint:* See Problem 13.13.

**13.39.** Equation (*13.16*) expresses the change in frequency of the $A_2$ allele when the heterozygote lacks dominance with respect to fitness. Assume that the frequency of the $A_2$ allele in the entire population ($\bar{q}$) is 0.3, and in a partially isolated subgroup the frequency ($q$) is 0.5. Ten percent of the individuals in the subgroup are new immigrants each generation from other subgroups of the population. The coefficient of selection ($s$) against the $A_2$ allele is 0.05. Determine the expected gene frequency in the next generation.

**13.40.** Let us consider a case where the coefficient of selection ($s$) is much larger than the proportion of immigrants (i). Assume that $i = 0.01$ and the average frequency of the gene in the entire population is $\bar{q} = 0.40$. Determine the equilibrium gene frequency ($\hat{q}$) when (*a*) $s = -0.15$, (*b*) $s = +0.15$. *Hint:* Use equation (*13.17*).

**13.41.** Suppose that the proportion of immigrants (i) is a much larger value than the absolute value of the selection coefficient. When the average frequency of the gene in the entire population is $\bar{q} = 0.40$ and $i = 0.15$, calculate the equilibrium gene frequency ($\hat{q}$) for (*a*) $s = +0.01$, (*b*) $s = -0.01$. *Hint:* Use equation (*13.17*).

**13.42.** When a population is subdivided into a large number of groups of size 50 each, what percentage of immigrants must each group receive per generation so that inbreeding will not go above 10%. *Hint:* Use equation (*13.18*).

**13.43.** A very small selection coefficient ($s = 0.01$) is active against an allele ($A_2$) which occurs with frequency $q = 0.6$ in a population. The mutation rate to $A_2$ is $u = 8 \times 10^{-5}$ and from it is $v = 4 \times 10^{-5}$. Determine the equilibrium gene frequency ($\hat{q}$) of the $A_2$ allele when it is (*a*) fully recessive (equation *13.19*); (*b*) lacking dominance (equation *13.20*); (*c*) completely dominant (equation *13.21*).

**13.44.** One of the highest known mutation rates in humans is that from the normal recessive allele to the dominant mutant allele which causes neurofibromatosis (von Recklinghausen's disease). This syndrome is characterized by spots of abnormal pigmentation of the skin and by numerous tumors associated with the nervous system. If the average fitness of affected individuals is only 40% that of normals, and approximately 1 out of every 3000 births has an abnormality, estimate the mutation rate from the normal to the abnormal allele. *Hint:* See Problem 13.16.

## Answers to the Supplementary Problems

**13.18.** (*a*) 0.32 (*b*) 20%

**13.19.** Approximately 8387 generations

**13.20.** (*a*) $4.4 \times 10^{-5}$ (*b*) $2 \times 10^{-5}$ (*c*) $-4 \times 10^{-6}$

**13.21.** (*a*) $-0.0909$ (*b*) 0.0091

**13.22.** (*a*) $q_1 = 0.286$, $q_2 = 0.222$ (*b*) $q_1^2$ = approx. 8.2%, $q_2^2$ = approx. 4.9%

**13.23.** 7.5 theoretical generations or in reality 8 full generations

**13.24.** 1475 years

**13.26.** $\frac{2}{3}$

**13.28.** 277 generations

**13.31.** (*b*) If $q > \frac{1}{2}$, $\Delta q$ will be positive. If $q < \frac{1}{2}$, $\Delta q$ will be negative.

**13.32.** $66\frac{2}{3}$

**13.33.** 4

**13.35.** approx. 3855

**13.36.** When the line is completely inbred ($F = 1$).

**13.37.** $26\frac{2}{3}$ (or 27)

**13.38.** (*a*) $\hat{q} = 0.6325$ (*b*) $\hat{q} = 0.2254$

**13.39.** 0.4675

**13.40.** (*a*) 0.9611 (*b*) 0.0256

**13.41.** (*a*) 0.416 (*b*) 0.384

**13.42.** 4.2% immigrants per generation

**13.43.** (*a*) 0.0894 (*b*) 0.008 (*c*) 0.01

**13.44.** approx. $1 \times 10^{-4}$

# Chapter 14

## The Chemical Basis of Heredity

### NUCLEIC ACIDS

The chromosomes of higher organisms are composed of nucleoprotein, i.e. a conjugate of nucleic acids (organic acids found predominantly in the nucleus of the cell) and proteins such as histones and/or protamines. Only the nucleic acids carry genetic information. The nucleoprotein units (*nucleosomes*) each consist of a short, disk-shaped cylinder of histones wrapped by $1\frac{3}{4}$ turns of nucleic acid; they are spaced at approximately 100-Å intervals when chromatin is "condensed" into nucleosomes.

The nucleic acid which serves as the carrier of genetic information in all organisms other than some viruses is *deoxyribonucleic acid* (DNA). The double helical structure of this extremely long molecule is shown in Fig. 14-1. The backbone of the helix is composed of two chains with alternating sugar (S)-phosphate (P) units. The sugar is a pentose (5-carbon) called *deoxyribose,* differing from its close relative called *ribose* by one oxygen atom in the 2′ position (Fig. 14-2 below). The phosphate group ($PO_4$) links adjacent sugars through a 3′-5′ phosphodiester linkage; in one chain the linkages are polarized 3′-5′, in the other chain they are in the reverse order 5′-3′. The steps in the spiral staircase (i.e. the units connecting one strand of DNA to its polarized complement) consist of paired organic bases of four species (symbolized A, G, T, C) classified into two groups, the *purines* and the *pyrimidines.* Purines only pair with pyrimidines and vice versa, thus producing a symmetrical double helix. Furthermore, the base pairs are linked in such a manner that the number of bonds between them is always maximized. For this reason, adenine (A) pairs only with thymine (T) and guanine (G) only with cytosine (C); see Fig. 14-3. A base plus its sugar is termed a *nucleoside*; a nucleoside plus its phosphate is called a *nucleotide.* The DNA molecule is thus a long *polymer* (i.e. a macromolecule composed of a number of similar or identical subunits, *monomers,* covalently bonded) of thousands of nucleotide pairs, even in such simple organisms as viruses.

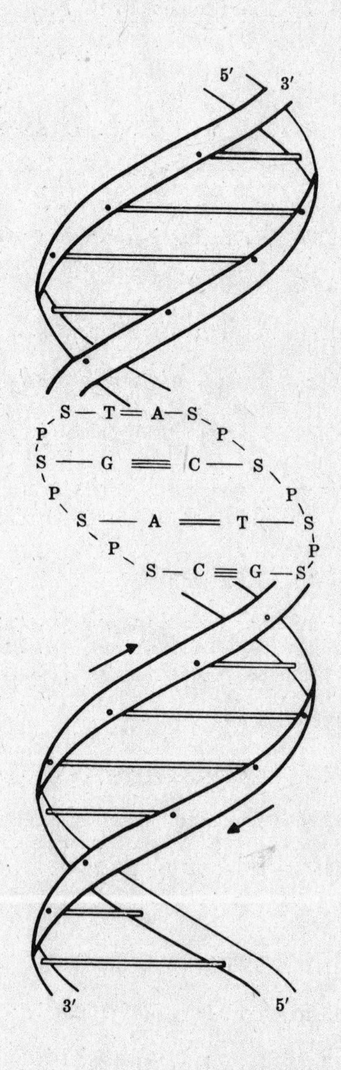


Fig. 14-1. Diagram of Watson-Crick model of DNA.

Another class of nucleic acids, called *ribonucleic acid* (RNA), is slightly different from DNA in the following respects.

(1) RNA is considered to be single stranded; DNA is double stranded.

(2) RNA contains ribose sugar; DNA contains deoxyribose sugar.

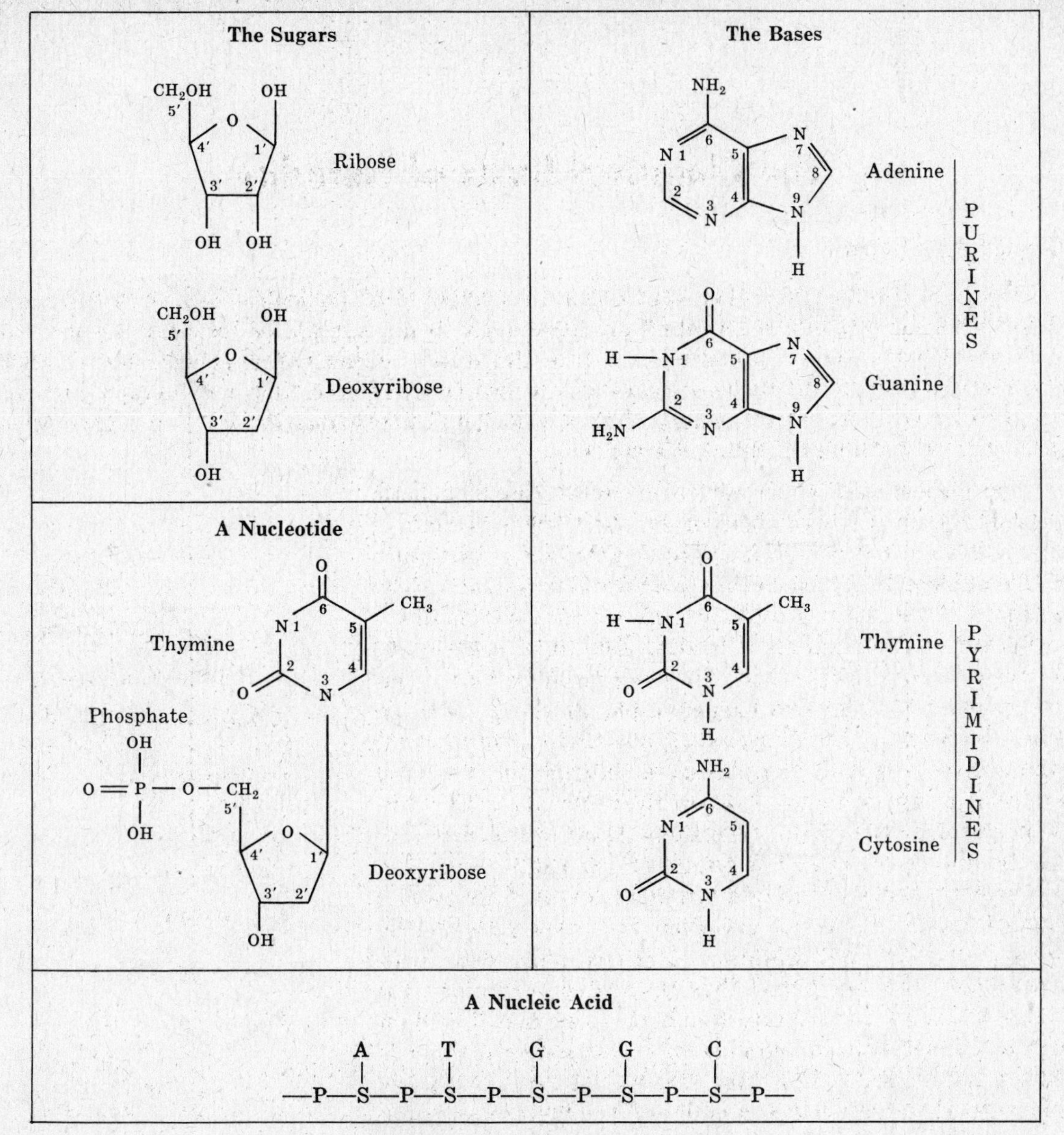


Fig. 14-2. Structure of nucleic acids.

(3) RNA contains the pyrimidine uracil (U) instead of thymine (U pairs with A).

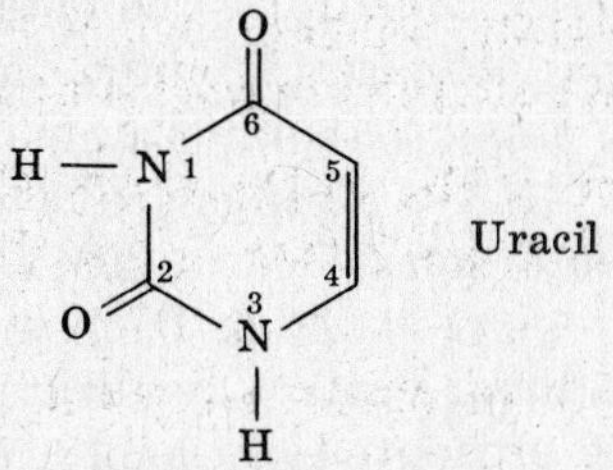

Uracil

(4) RNA molecules are generally much shorter than DNA molecules.

RNA functions primarily in protein synthesis, acting in one capacity as a messenger (mRNA) carrying information from the instructions coded into the DNA to the ribosomal sites of protein synthesis in the cell. Ribosomes contain a special form of RNA called

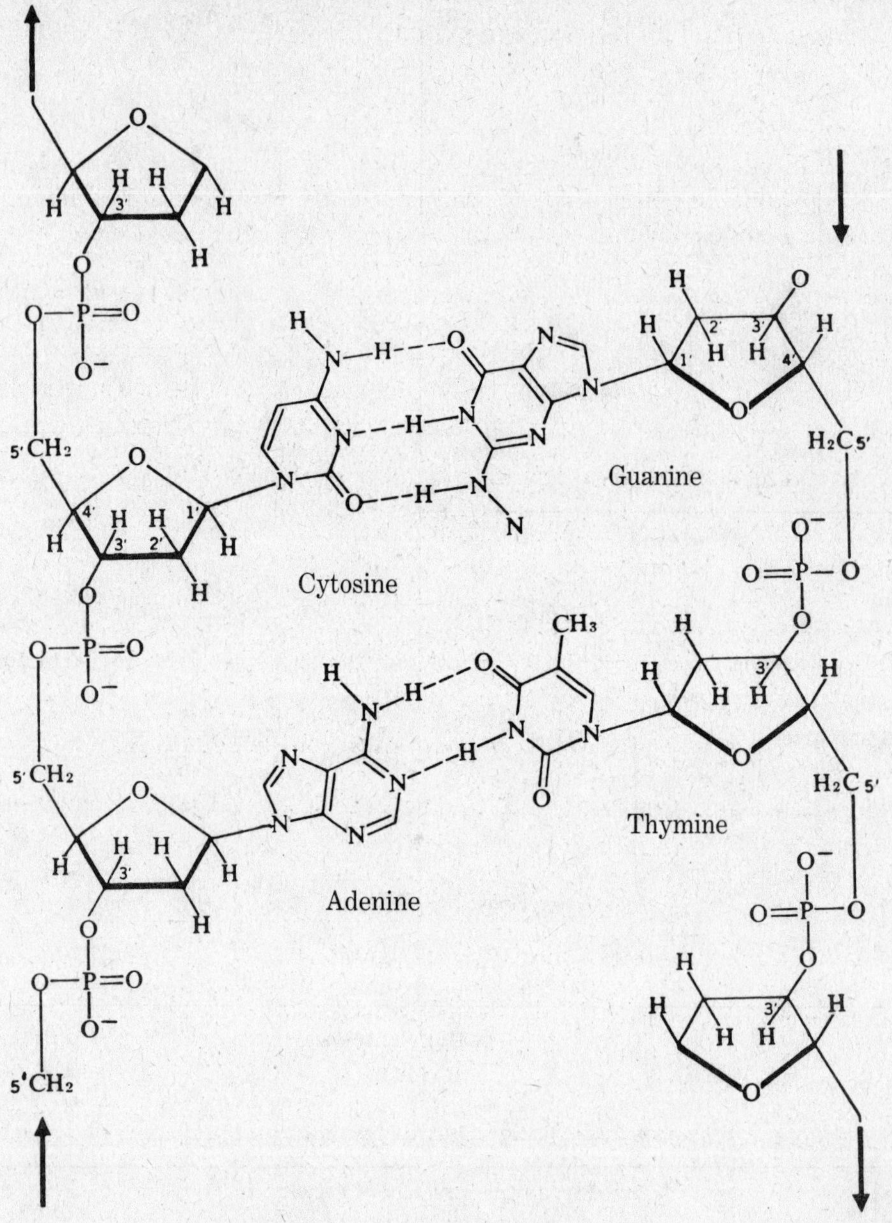

Fig. 14-3. Base pairing in DNA.

ribosomal RNA (rRNA), which constitutes the bulk of cellular RNA. A third kind of RNA, called transfer RNA (tRNA), attaches to amino acids and during protein synthesis brings them each into proper positioning with other amino acids using the mRNA-ribosome complex as a template. All cellular RNA molecules are made from a DNA template. In some of the plant viruses such as tobacco mosaic virus (TMV), DNA is absent and RNA functions as genetic material without the assistance of DNA at any stage of its "life cycle". However, some animal viruses (RNA tumor viruses) replicate via the intermediate of a DNA template. Details of viral genetics are presented in Chapter 15.

## CENTRAL DOGMA

The major functions of DNA are summarized in the so-called "central dogma" of molecular genetics. DNA has both an *autocatalytic* function (self-replication) and a *heterocatalytic* function (transcription into RNA's). Protein translation occurs on mRNA templates, never on DNA templates. Genes become phenotypically expressed by their ability to specify the structures of biologically active proteins.

## GENETIC CODE

Biochemical reactions are mediated by enzymes. All enzymes are proteins. Proteins are polymers of subunits (monomers) called amino acids, often spoken of as "residues". Each amino acid has an amino group ($NH_2$) at one end and a carboxyl group (COOH) at the other end. Twenty different kinds of amino acids occur naturally in proteins. Each enzyme consists of a certain number of amino acid residues in a precisely ordered sequence. The blueprint for making proteins is coded in the nucleotide sequence of DNA. The number of nucleotides which code for an amino acid is termed a *codon*. There are 20 common amino acids but only 4 different nucleotides. Obviously a singlet code (i.e. one nucleotide coding for one amino acid) could code for only 4 amino acids. A doublet code (i.e. two nucleotides coding for one amino acid) allows for only 16 combinations. Mathematically, therefore, a triplet code (i.e. three nucleotides coding for one amino acid) is the smallest coding unit (codon) that can accommodate the 20 amino acids.

The first experimental evidence supporting the triplet code concept was provided by the study of single base pair additions in the single linkage group of a bacterial virus (phage T4). If the transcription of a functional genetic unit (*cistron*) of DNA into mRNA is always read from a fixed position, then the first six codons in one chain of a DNA cistron might be as follows:

| 1 | 2 | 3 | 4 | 5 | 6 |
|---|---|---|---|---|---|
| TCA | GGC | TAA | AGT | CGG | TCG |

The addition of a single base (e.g. G) at the end of the second codon would shift all other codons one nucleotide out of register and prevent the correct reading of all codons to the right of the base addition.

| 1 | 2 | 3 | 4 | 5 | 6 | |
|---|---|---|---|---|---|---|
| TCA | GGC | [G]TA | AAG | TCG | GTC | G |

By successively adding bases in a nearby region, it should be possible to place the reading of the codons back into register. It was found that one or two base additions failed to produce a functionally normal protein. But three base additions apparently can place the reading of the codons back into register for all codons to the right of the third base addition.

| 1 | 2 | 3 | 4 | 5 | 6 | 7 |
|---|---|---|---|---|---|---|
| TCA | [A]GG | C[G]T | A[C]A | AGT | CGG | TCG |
| | Missense Codons (out of register) | | | Codons in correct register | | |

The same has also been found to be true for single nucleotide deletions. Three deletions or multiples thereof can correct the reading frame in the synthesis of an active protein. Several other lines of evidence indicate that the codon is a sequence of three nucleotides and the genetic code is generally referred to as a "triplet code".

High concentrations of ribonucleotides in the presence of the enzyme polynucleotide phosphorylase can generate synthetic mRNA molecules *in vitro* by forming an internucleotide 3′-5′ phosphodiester bond. In this way a number of uracil molecules can become linked together to form a synthetic poly-U with mRNA activity. The addition of poly-U to bacterial cell extracts results in the limited synthesis of polypeptides containing only the amino acid phenylalanine. Thus 3 uracils probably code for phenylalanine. Mixtures of different ribonucleotides can also form synthetic mRNA molecules with the nucleotides in random order.

**Example 14.1.** Poly-AU made from a mixture of adenine and uracil in concentrations of 2:1 respectively is expected to form AAA triplets most frequently, AAU (or AUA or UAA) triplets would be next in frequency, AUU (or UAU or UUA) triplets next, and UUU triplets would be least frequent. The frequencies with which various amino acids are incorporated into polypeptides under direction of the synthetic poly-AU can then be correlated with the expected frequencies of various triplets, allowing tentative assignment of specific triplets to specific amino acids.

A combination of organic chemical and enzymatic techniques can be used to prepare synthetic polyribonucleotides with known repeating sequences as, for example, AUAUAUAU... which alternately codes for the amino acids isoleucine and tyrosine, CUCUCUCU... which codes for leucine and serine alternately, etc.

Even in the absence of mRNA and protein synthesis, a trinucleotide will bind to a ribosome. Trinucleotides of known sequence can then be used to specifically bind one out of a mixture of 20 different amino acids to ribosomes *in vitro*. For example, UUG binds only leucine-loaded tRNA to ribosomes, UGU binds only cysteine-loaded tRNA, etc.

The genetic code is *degenerate* because more than one codon exists for most amino acids. The code is basically the same for all organisms. Apparently only three of the 64 possible three letter codons fail to code for any amino acid. Such codons are termed *nonsense* triplets. The mRNA codons for the twenty amino acids are listed in Table 14.1.

**Table 14.1. mRNA Codons**

| First Letter | Second Letter: U | | C | | A | | G | | Third Letter |
|---|---|---|---|---|---|---|---|---|---|
| U | UUU | Phe | UCU | Ser | UAU | Tyr | UGU | Cys | U |
| | UUC | | UCC | | UAC | | UGC | | C |
| | UUA | Leu | UCA | | UAA | Nonsense | UGA | Nonsense | A |
| | UUG | | UCG | | UAG | | UGG | Trp | G |
| C | CUU | Leu | CCU | Pro | CAU | His | CGU | Arg | U |
| | CUC | | CCC | | CAC | | CGC | | C |
| | CUA | | CCA | | CAA | Gln | CGA | | A |
| | CUG | | CCG | | CAG | | CGG | | G |
| A | AUU | Ile | ACU | Thr | AAU | Asn | AGU | Ser | U |
| | AUC | | ACC | | AAC | | AGC | | C |
| | AUA | | ACA | | AAA | Lys | AGA | Arg | A |
| | AUG | Met | ACG | | AAG | | AGG | | G |
| G | GUU | Val | GCU | Ala | GAU | Asp | GGU | Gly | U |
| | GUC | | GCC | | GAC | | GGC | | C |
| | GUA | | GCA | | GAA | Glu | GGA | | A |
| | GUG | | GCG | | GAG | | GGG | | G |

The abbreviated names of amino acids are as follows: ala = alanine, arg = arginine, asn = asparagine, asp = aspartic acid, cys = cysteine, gln = glutamine, glu = glutamic acid, gly = glycine, his = histidine, ile = isoleucine, leu = leucine, lys = lysine, met = methionine, phe = phenylalanine, pro = proline, ser = serine, thr = threonine, trp = tryptophan, tyr = tyrosine, val = valine.

## PROTEIN SYNTHESIS

The coded information in a deoxyribonucleotide sequence is *transcribed* into the ribonucleotide sequence of an RNA molecule by a specific enzyme (RNA polymerase). This enzyme recognizes certain A,T-rich paired sequences of the intact double helix as initiation sites (*promotor* regions) and begins transcription of one of the two strands in a nearby region; the promotor itself is not transcribed. Within a given region of the DNA, only one of the two strands "makes sense", i.e. is transcribed into RNA. In some other region of that same DNA molecule, the other strand may be the sense strand. However, the information for making any given RNA molecule or polypeptide chain resides totally on one of the two strands. That is to say, RNA polymerase does not jump from one strand to the other in the process of transcription of a given gene or group of adjacent genes into a single RNA molecule. Both rRNA and mRNA molecules become associated with proteins as precursors and then move from the nucleus to the cytoplasm. There the message is unidirectionally read beginning at the 5′-hydroxyl end by one or more ribosomes (*polysomes*). The 5′ end of the mRNA corresponds to the N-terminal amino acid in the completed protein. The ribosome serves as a binding site where mRNA and tRNA interact (Fig. 14-4). Messenger RNA is very unstable in bacterial systems and must be continually replaced by newly synthesized molecules.

If each of the 20 amino acids is recognized by its own species of tRNA, then at least 20 species of tRNA would exist in the cytoplasm. The attachment of an amino acid to its

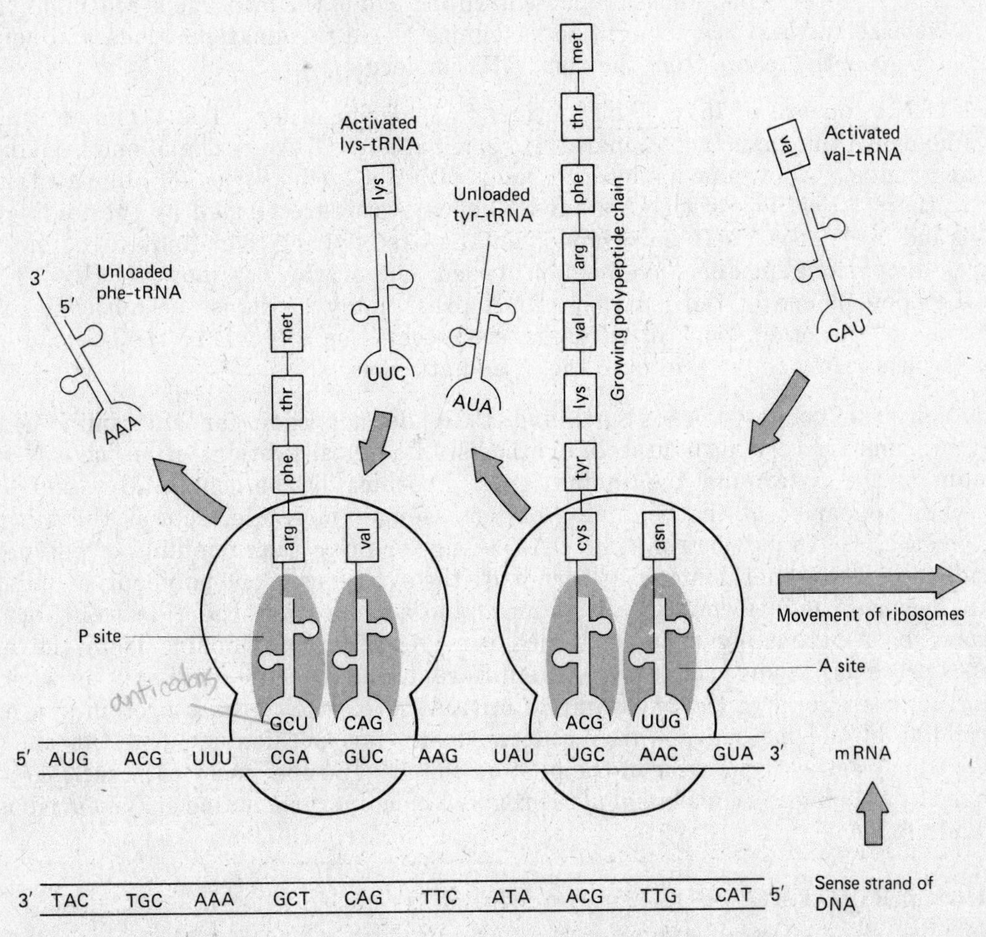


**Fig. 14-4.** Diagram of protein synthesis. (Reprinted, by permission, from William D. Stansfield, *The Science of Evolution*, © 1977 by Macmillan Publishing Co., Inc.)

tRNA molecule is mediated by a specific enzyme in a process called *activation* or *loading*. Somewhere on the tRNA is a sequence of three nucleotides (the *anticodon*) which is complementary to the mRNA codon. The affinity of the complementary triplets brings each amino acid into its proper relationship with other amino acids in the growing polypeptide chain. A ribosomal enzyme links the amino group and the carboxyl group of adjacent amino acids together by a *peptide bond*. The tRNA molecule is then released from its amino acid, from the mRNA, and from the ribosome. The tRNA is then free to acivate another free amino acid of the same species. When the ribosome reaches the end of the messenger, *translation* of the nucleotide code into an amino acid sequence is completed.

Bacterial ribosomes consist of two major subunits, a larger 50S subunit and a smaller 30S subunit (S = Svedberg units, a sedimentation coefficient for molecules in an ultracentrifuge). More than 30 different proteins are found associated with rRNA in ribosomes. The specific function which these constituents perform has yet to be elucidated. At least two sites seem to exist on a functional ribosome: (1) a "P" (peptidal) site and (2) an "A" (amino-acyl) site. During protein synthesis, mRNA is bound to the 30S subunit. The first loaded tRNA molecule enters the "P" site (perhaps by passing through the "A" site). The next loaded tRNA enters the "A" site and a peptide bond is enzymatically formed between the two adjacent amino acids. The unloaded tRNA in the "P" site now leaves the ribosome and may become enzymatically loaded again with another molecule of its own species of amino acid. The remaining tRNA-dipeptide complex moves from the "A" to the "P" site coincident with a movement of the ribosome along the messenger which displays the next mRNA codon in the vacant "A" site. This process repeats until the end of the message is reached. Specific proteins (release factors) are thought to recognize these terminator codons and sever the completed polypeptide chain from the last tRNA molecule.

Each tRNA consists of 75 to 80 nucleotides in a specific order. The 3′ end of all tRNA's which is bound to the amino acid apparently terminates in CCA$^{3'}$. The 5′ end terminates in a guanine residue. A few unusual bases occur in tRNA's, such as pseudoradilic acid, inosinic acid and others. Portions of tRNA's have a helical structure formed by the folding of the single strand back upon itself and complementary base pairing over limited regions. Several three-dimensional models have been proposed, one of which is shown in Fig. 14-4. An unpaired region in one of the loops of tRNA presumably contains the anticodon triplet. Perhaps one of the other loops in tRNA is the *recognition site* where the specific enzyme that loads the specific amino acid onto the 3′ end attaches.

The nonsense codons UAA, UAG and UGA do not code for any amino acid and thereby presumably terminate protein synthesis. Bacterial proteins often have N-formylmethionine at the N-terminal (beginning) end. Presumably the codon AUG (and perhaps GUG), when appearing at the beginning of a messenger molecule, acts as the initiator of protein synthesis. In polycistronic mRNA (coding for more than one kind of polypeptide), AUG coding for formylmethionine may terminate the synthesis of one polypeptide and initiate the next because the N-formyl group on an amino acid prevents peptide bond formation. There may be a "trimming enzyme" which removes formylmethionine from the ends of completed proteins. Some codons may have different meanings, depending upon where they occur in the message: if at the beginning of mRNA, AUG may initiate protein synthesis; if in the middle of the message, it may code normally for methionine; if at the end of the message, it may act as terminator of the peptide chain. The specificity of genetic coding can be changed by altering any component of the readout machinery including enzymes, ribosomes, tRNA and mRNA.

## PROTEIN STRUCTURE

Knowledge of protein structure and bonding forces is essential for a keen understanding of how various genetic factors (mutations) and environmental factors (e.g. pH, temperature, salt concentrations, chemical treatments) can modify proteins and either reduce or destroy their biological activities. Such knowledge is also important for developing techniques to extract functional proteins from genetically engineered cells.

All completely ionized biological amino acids (except proline) have the general structure shown in Fig. 14-5. The α-carbon is the central atom to which an amino ($NH^{3+}$) and a carboxyl ($COO^-$) group is attached. As pH increases above neutrality (pH 7), the more basic nature of the environment tends to neutralize the acidic carboxyl group. As pH decreases below neutrality, the more acidic nature of the environment tends to neutralize the basic amino group. *Polar molecules* are those with separate positive and negative charges, as exemplified by an amino acid at pH 7. Water is also a polar molecule because the two positive hydrogen atoms are near one end of the molecule and the oxygen (2−) atom is at the other end. *Nonpolar molecules* (such as methane, $CH_4$) have no noticeable charge (un-ionized).

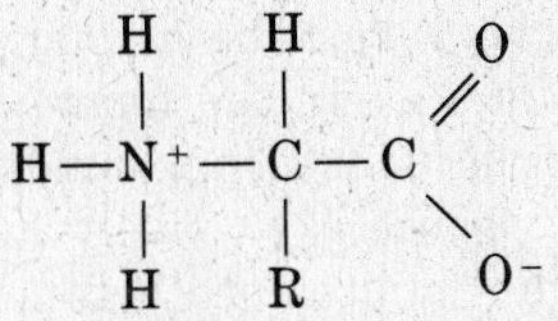

**Fig. 14-5.** General structure of an amino acid in completely ionized form. R represents a side chain or radical.

The peptide bond that joins adjacent amino acids during protein synthesis is a strong covalent bond (a bond in which atoms are coupled by sharing electrons). By the removal of water, the carboxyl group of one amino acid becomes joined to the amino group of the adjacent amino acid as shown in Fig. 14-6. This union is an example of dehydration synthesis. One of the proteins of the bacterial 50S ribosomal subunit, an enzyme called peptidyl transferase, is responsible for making the peptide bond. Each complete polypeptide chain has an uncomplexed ("free") amino group at one end and a free carboxyl group at its other end. The N- (amino) terminal end of the polypeptide corresponds to the 5′ end of its respective mRNA. The C- (carboxy) terminal end of the polypeptide corresponds to the 3′ end of the same mRNA.

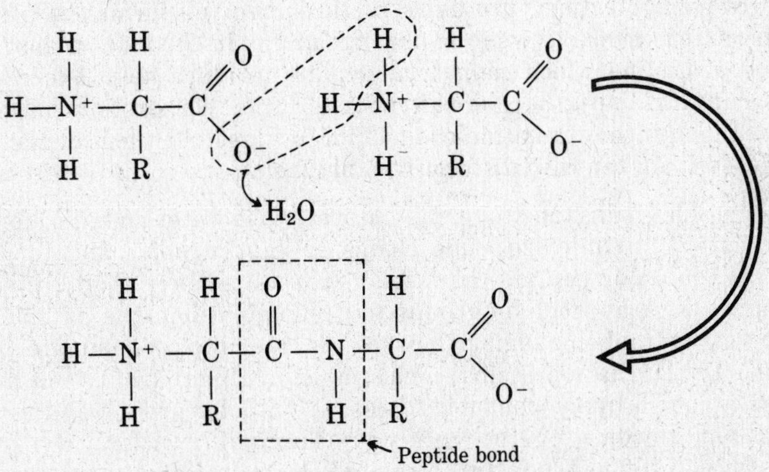

$H_2O$

Peptide bond

**Fig. 14-6.** Dehydration synthesis of a dipeptide and formation of a peptide bond.

One amino acid differs from another according to the nature of the side chain (R) attached to the α-carbon. Glycine has the simplest side chain consisting of a hydrogen atom. Other amino acids have side chains of hydrocarbons of various lengths; some of these chains are ionized positively (basic proteins such as lysine and arginine), others are negatively charged (acidic amino acids such as aspartic and glutamic acids), and still others are un-ionized (e.g. valine, leucine). Different proteins can be separated on the basis of their net electrical charges by a technique known as *electrophoresis*. Closely related proteins differing by a single amino acid can sometimes be resolved in this way. Some amino acids, such as phenylalanine and tyrosine, have aromatics (ring structures) in their side chains. The amino acid proline does not contain a free amino group because its nitrogen atom is involved in a ring structure with the side chain. Only two amino acids contain sulfur in their side chain (cysteine and methionine). The sulfurs of different cysteine residues can be covalently linked into a disulfide bond (S—S) that is responsible for helping to stabilize the tertiary and quaternary shapes of proteins containing them.

The linear sequence of amino acids forms the primary structure; at least some portion of many proteins have a secondary structure in the form of an alpha-helix. The protein chain may then fold back upon itself, forming internal bonds (including strong disulfide bonds) which stabilize its tertiary structure into a precisely and often intricately folded pattern. Two or more tertiary structures may unite into a functional quarternary structure. For example, hemoglobin consists of four polypeptide chains, two identical α-chains and two identical β-chains. A protein does not become an active enzyme until it has assumed its tertiary or quarternary pattern (Fig. 14-7 below). Any subsequent disturbance of its final configuration may inactivate the enzyme. For example, heating destroys enzymatic activity because bonds, which hold the protein in its higher structural forms, are ruptured. The shape of an active enzyme molecule is believed to fit its *substrate* (the substance which is catalyzed by the enzyme) in a way that is analogous to a lock and key situation (Fig. 14-8). An altered enzyme, either genetically or physically, may not fit the substrate and therefore would be incapable of catalyzing its conversion into the normal product.

Relatively weak bonds, such as hydrogen bonds (electronegative atoms that share a proton or hydrogen atom) and *ionic bonds* (attraction of positively and negatively charged ionic groups) are mainly responsible for the secondary and higher structural levels of protein organization. Enzymes are not involved in the formation of weak bonds. For example, the alpha-helix is formed by hydrogen bonding of the carbonyl group (CO) of one residue to the imino group (NH) four residues down the polypeptide chain. The extent to which a protein contains alpha-helical regions is dependent upon at least three factors. The most important factor governing tertiary protein structure involves formation of the energetically most favorable interactions between atomic groupings in the side chains. A second factor is the presence of proline which cannot participate in alpha-helical formation because it has no imino group. Proline is therefore often found at the "corners" or "hairpin turns" of polypeptide chains. Finally, the formation of disulfide bridges between cysteine residues on the same polypeptide chain tends to distort the alpha-helix.

Ionized side chains tend to readily interact with water and therefore are called *hydrophilic* ("loves water"). Un-ionized side chains are *hydrophobic* and tend to avoid contact with water. When a polypeptide chain folds into its tertiary shape, these forces tend to cause amino acids with hydrophilic groups to predominate on the outside of the protein; by the same token, hydrophobic segments of the chain tend to predominate in the interior of globular proteins. Quarternary proteins (consisting of aggregates of two or more polypeptide chains) are usually joined by hydrophobic forces. Non-polar groups on the surface of the different subunits tend to come together as a way of excluding water. Hydrogen bonds, ionic bonds and possibly disulfide bonds may also participate in forming quarternary structures. Some quarternary proteins consist of identical subunits (e.g. the enzyme β-galactosidase consists of four identical polypeptide chains). Such proteins are called *homopolymers*. Other

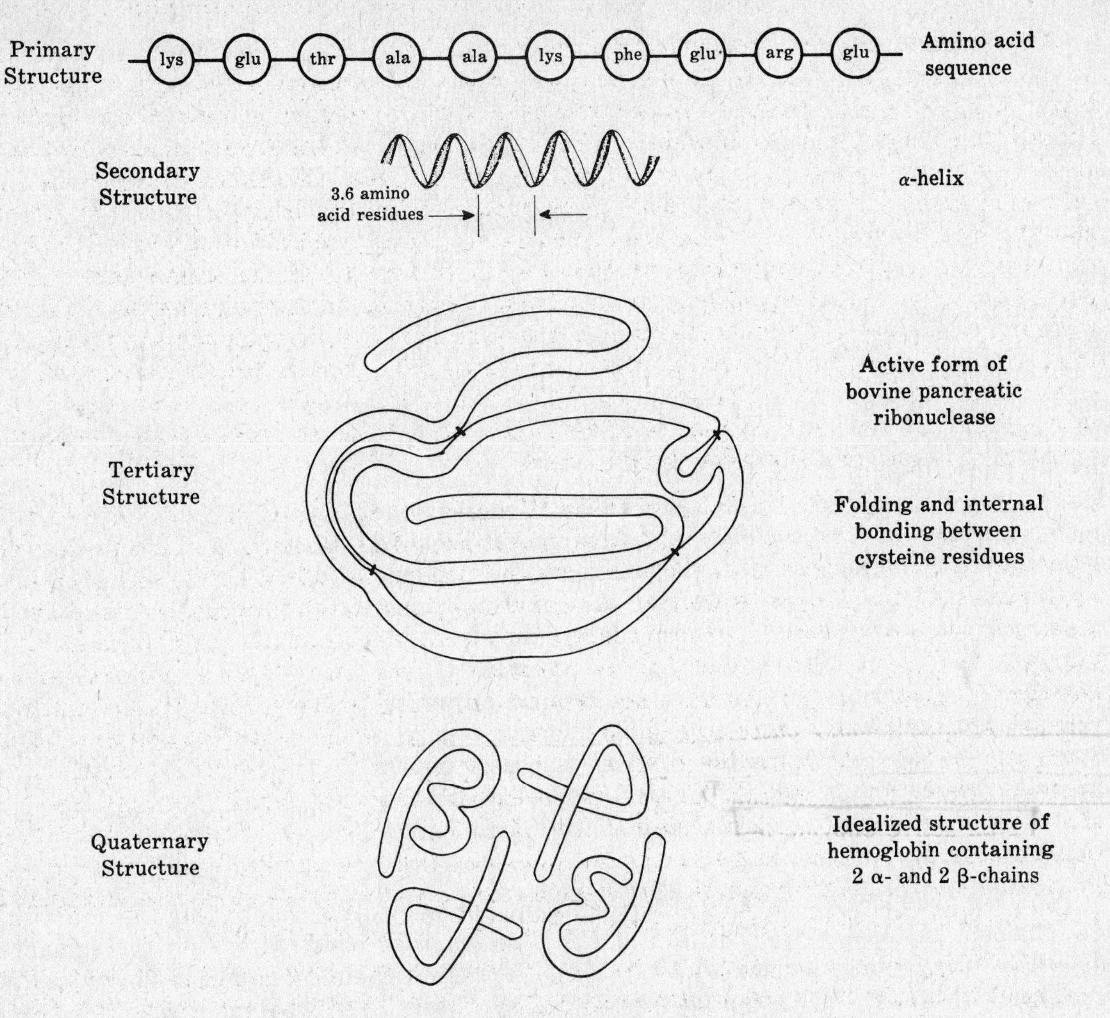


**Fig. 14-7.** Stages in the development of a functional protein.

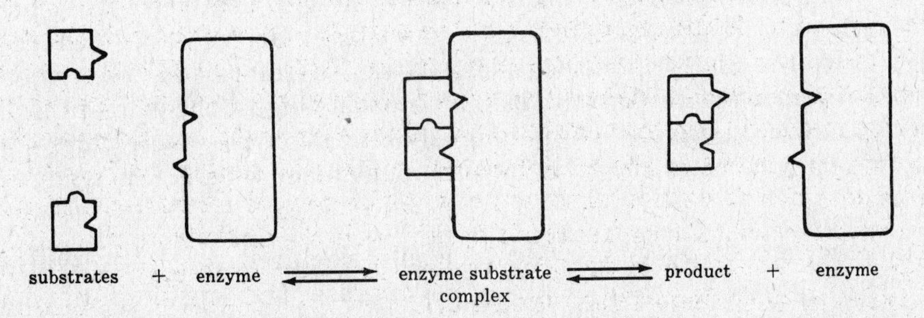


**Fig. 14-8.** Diagram of enzymatic action.

quarternary proteins (such as hemoglobin) consist of non-identical subunits and are called *heteropolymers*. Some polypeptide chains are subject to posttranslational modifications before they assume their biologically active forms. For example, chymotrypsinogen is cut at one position by an enzyme and the active split-product is chymotrypsin.

## NUCLEIC ACID ENZYMOLOGY

Numerous enzymes are required for the normal activity of nucleic acids, including DNA replication and repair, genetic recombination of linked loci, transcription of RNA's and translation of proteins.

## 1. DNA Replication.

The hydrogen bonds linking base pairs together are relatively weak bonds. During DNA replication, the two strands separate along this line of weakness in zipperlike fashion (Fig. 14-9). Each strand of the DNA molecule can serve as a template against which a complementary strand can form (according to the rules of specific base pairing) by the catalytic activity of the enzyme *DNA polymerase.* At least three forms of DNA polymerase have been identified in bacteria. All DNA molecules are synthesized by these polymerases adding deoxyribonucleotide-5′-triphosphates to the 3′-OH groups at the ends of growing chains (5′ to 3′ polymerization). All three kinds of DNA polymerase can also degrade DNA in the 3′ to 5′ direction. Enzymes that degrade nucleic acids are called *nucleases.* If the enzyme cleaves nucleotides from the ends of the polymer it is called an *exonuclease*; if it makes cuts in the interior of the molecule it is termed an *endonuclease.* As long as deoxyribonucleotide precursors are present in even moderate amounts, the synthetic activity of DNA polymerases is greatly favored over their degradation activity. During replication, incorrectly paired bases have a high probability of being removed by the exonuclease activity of the DNA polymerases before the next nucleotide is added. This is part of the so-called "proofreading system" that constantly monitors the DNA for errors or lesions.

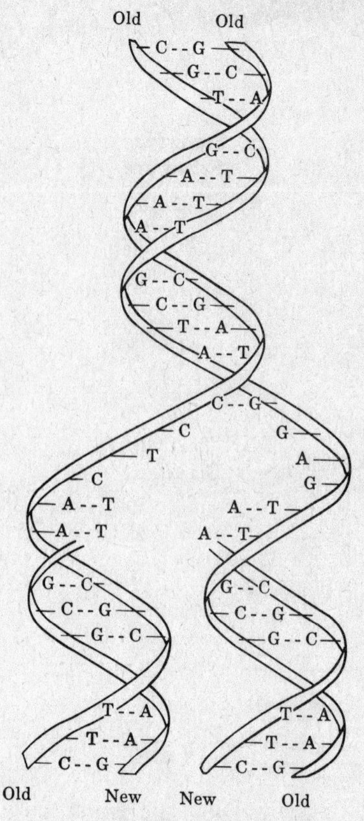


**Fig. 14-9.** Replication of DNA.

All DNA polymerases can only extend existing polynucleotide chains; they cannot initiate new chains. Instead, a special kind of *RNA polymerase* constructs a short segment of ribonucleotides by base pairing with the 5′ end of a DNA chain. The formation of an *RNA primer* constitutes the *initiation phase* of DNA synthesis. DNA polymerase III then adds the appropriate deoxyribonucleotides to extend the chain. DNA polymerase I removes the RNA sequences from RNA/DNA hybrids and closes the gaps between RNA-primed DNA fragments. Polymerase I also has 5′ to 3′ exonuclease activity and can remove both ribonucleotides as well as deoxyribonucleotides. It can therefore add deoxyribonucleotides to the 3′ ends of RNA primers behind it and simultaneously digest the primer that lies ahead. The only enzyme known to be capable of joining adjacent deoxyribonucleotides is *polynucleotide ligase.* This enzyme seals the terminal fragment made by DNA polymerase I to the rest of its chain.

At least three other clases of enzymes are also required for DNA synthesis. The *unwinding proteins* proceed ahead of DNA polymerase III, opening the double helix and producing single-stranded templates for replication. These single-stranded regions are stabilized when complexed with *helix-destabilizing* (HD) *proteins.* A third kind of enzyme, called *DNA gyrase,* unwinds certain circular DNA molecules (bacterial, mitochondrial) and removes superhelical twists that would otherwise be generated by unwinding the double helix. Several additional proteins are already known to be required for DNA replication and more will likely be discovered.

Following local separation of the DNA strands and formation of the initiation primer, DNA polymerase III extends the primer in the 5′ to 3′ direction, forming the *leading strand.* The opposite template strand ends with a 5′ nucleotide and DNA polymerase cannot replicate it from that end. Instead, a so-called *lagging strand* is replicated in short sections by the 5′ to 3′ activity of DNA polymerase. These segments that are synthesized in an opposite

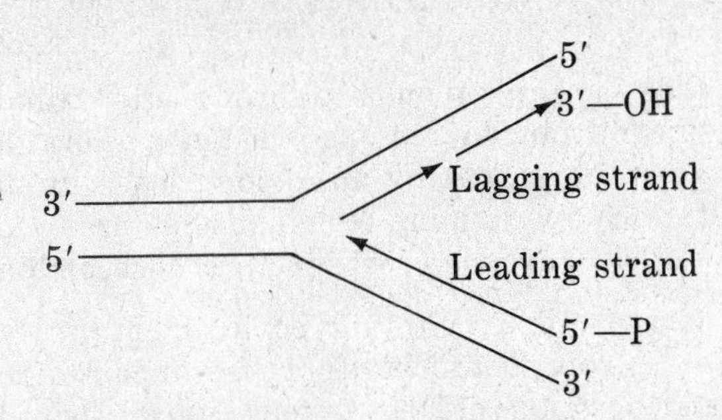


**Fig. 14-10.** Production of Okazaki fragments on the lagging strand during DNA replication.

direction from movement of the replication fork are often called *Okazaki fragments* (named after Reiji Okazaki) or *precursor fragments* (Fig. 14-10). Adjacent Okazaki fragments are rapidly joined by DNA ligase so that at any given time there is only a single incomplete fragment in the leading strand. The discontinuous replication of one strand of the DNA results in its seemingly paradoxical overall growth from 3′ to 5′.

DNA polymerases do not always distinguish uracils from thymines, so that some adenines in the template strand will be paired with U instead of T. Several enzymes aid in the removal of U from DNA and its replacement by T. These replacements produce segments called *uracil fragments*, and they become joined by the addition of T, thereby reforming continuous strands in both the leading and lagging strands.

### 2. DNA Repair Mechanisms.

One of the best understood repair mechanisms involves removal of pyrimidine dimers (usually covalently linked adjacent thymines in the same strand). Thymine dimers are easily induced in bacteria by ultraviolet (UV) light. These dimers are lethal if left unrepaired because they interfere with the normal replication of progeny DNA strands. There are at least three mechanisms known for repairing pyrimidine dimers.

(*a*) ***Photoreactivation.*** Some pyrimidine dimers can be removed by the action of an enzyme that becomes activated by absorption of blue light. This type of repair is more efficient if the bacteria are prevented from growing for a period of time after exposure to UV irradiation. Not much is known concerning the chemistry of this repair process.

(*b*) ***Dark Repair.*** This mechanism involves four steps. (1) A single-strand break is made on the 5′ side near the dimer by a specific endonuclease called *UV endonuclease*. (2) The 5′ to 3′ exonuclease activity of DNA polymerase I removes nucleotides near the cut, inlcuding the dimer. (3) One of the DNA polymerases (possibly I) synthesizes a correct replacement strand 5′ to 3′ using information from the intact complementary strand. (4) Polynucleotide ligase seals the break. Dark repair (excision repair) can begin as soon as a pyrimidine dimer is formed even if growth is not experimentally delayed.

(*c*) ***SOS Repair.*** This is a form of error-prone replication that repairs lesions in DNA without regard for restoring the original base sequence. This type of repair may be triggered by chemical mutagens that alter the hydrogen-bonding properties of bases or by radiation-induced mutations. Little is known concerning the nature of this emergency repair mechanism.

### 3. Recombination Enzymes.

There is no effective pairing force between homologous DNA double helices at specific points. Therefore, recombinations must involve the production of complementary single-stranded regions. This could be accomplished by an endonuclease nicking one strand of the double helix in each of two homologous DNA molecules. According to one theory (Fig. 14-11), DNA polymerase extends the broken ends to form single-stranded tails. Base pairing between complementary single-stranded regions creates a short double-stranded bridge. The other intact strands are then nicked by an endonuclease producing one recombinant molecule and two fragments with overlapping terminal sequences. DNA polymerase fills in the gaps and ligase seals the broken ends to create reciprocal recombinant genetic structures.

RNA molecules often have partially or totally complementary base sequences that tend to pair and thereby create "hairpin loops" in otherwise unpaired regions of the molecule. The tendency of single-stranded regions of DNA to form hairpin loops by the same mechanism is prevented by attachment of unwinding proteins. This allows extended single strands of complementary base sequences to form double helices.

According to a second theory (Fig. 14-12), the broken strands of parallel-aligned double helices are reciprocally joined by ligase, creating a cross-bridge. Equivalent bases on the two original molecules can exchange places, causing the cross connection to move along the complex in a zipperlike fashion ("branch migration"). This action commonly produces long regions of hybrid (*heteroduplex*) DNA containing some base sequences that may not be exactly complementary. Twisting the complex leads to steric rearrangement that converts

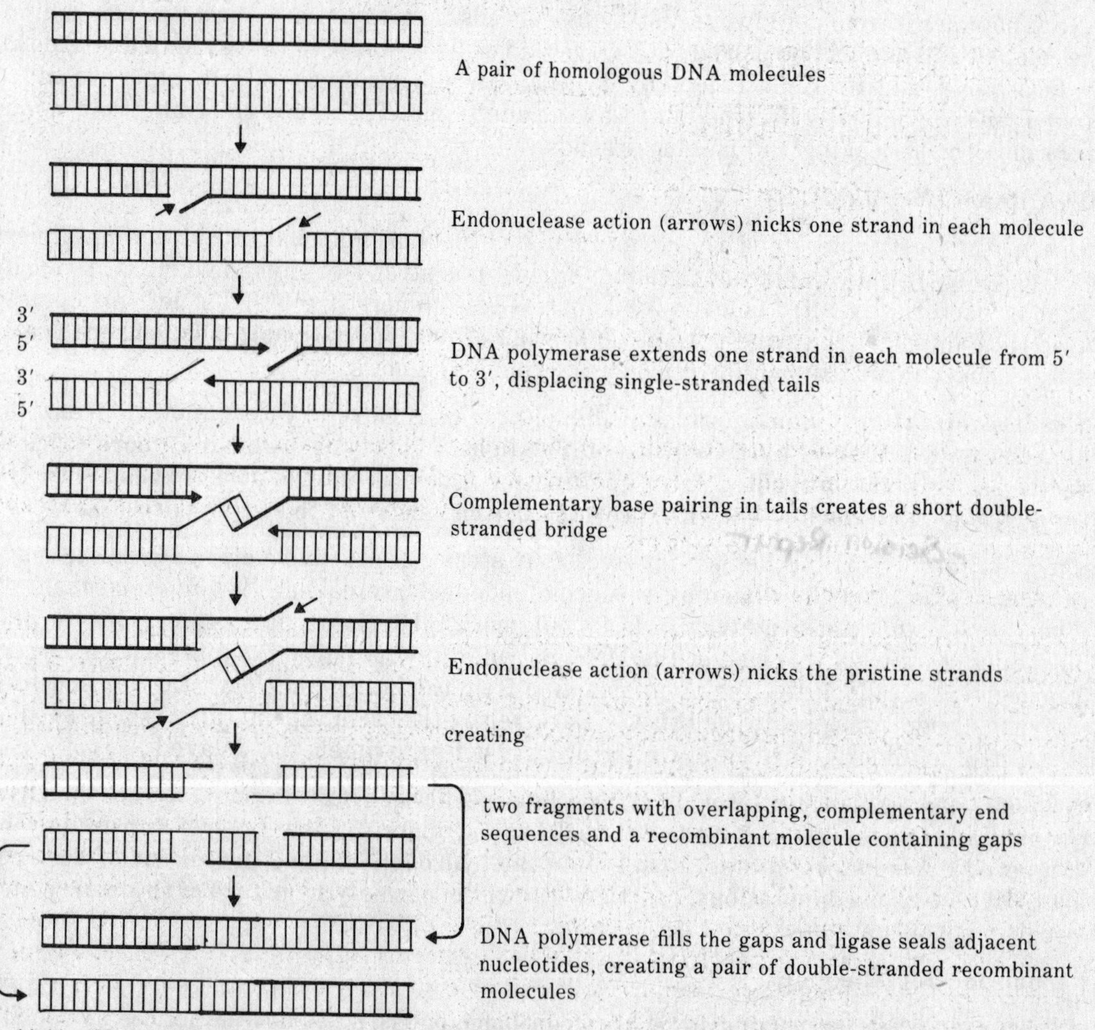


**Fig. 14-11.** Model of crossing over involving disruption of the individual DNA molecules. (Idea from J. D. Watson, *Molecular Biology of the Gene*, 3rd ed., p. 265, 1976, Benjamin-Cummings.)

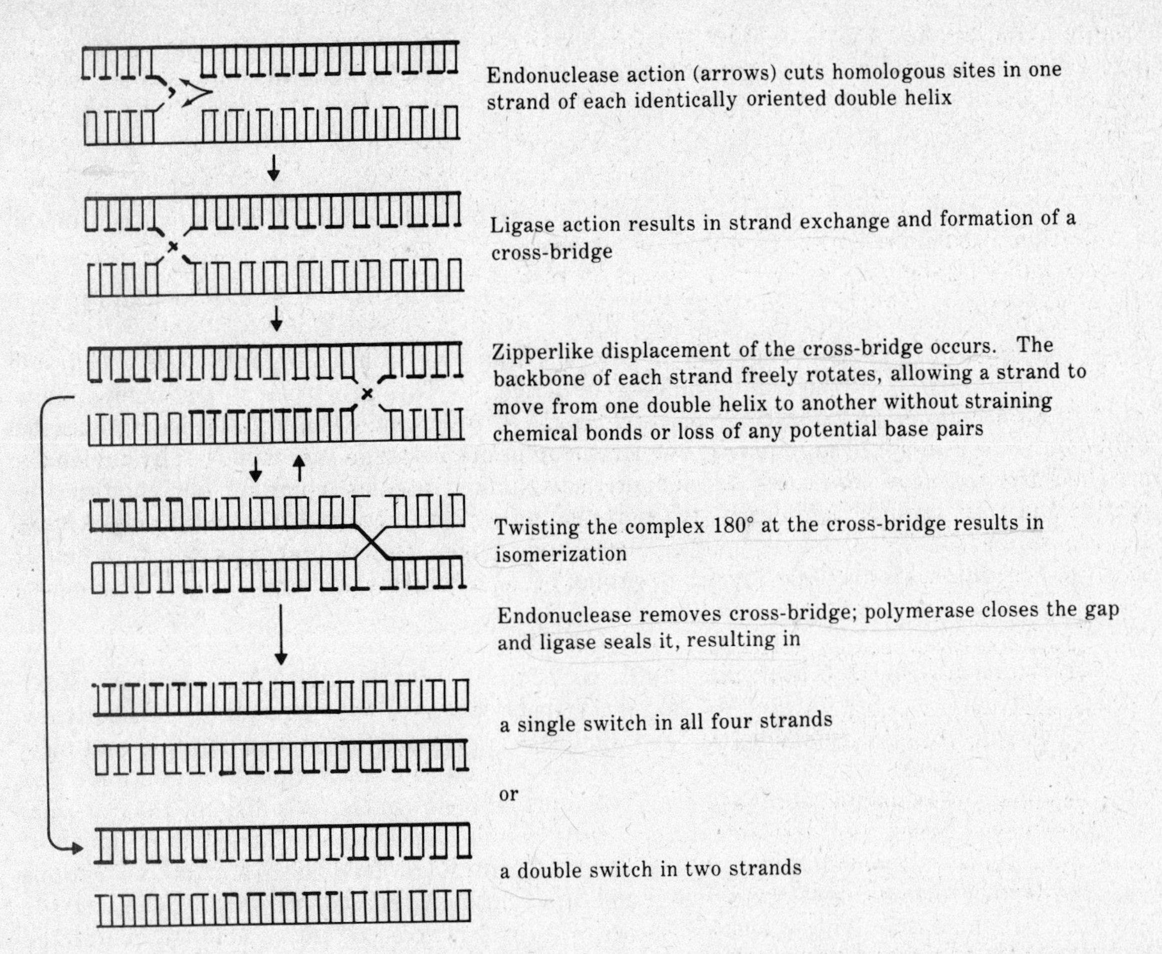


**Fig. 14-12.** Model of crossing over without disruption of the individual DNA molecules. (Idea from J. D. Watson, *Molecular Biology of the Gene*, 3rd ed., p. 267, Benjamin-Cummings.)

the bridging strands to outside strands and vice versa ("isomerization"). Cross-bridges are removed by nuclease cuts and the gaps are closed by DNA polymerase and polynucleotide ligase action. This type of crossing over can produce either a single switch in all four strands or a double switch in two of the strands.

**4. Enzymes of Transcription.**

RNA polymerase is the enzyme that makes RNA from DNA templates. All RNA molecules are synthesized in the 5′ to 3′ direction (the same direction in which mRNA's are translated into proteins). The portion of this enzyme that recognizes start signals on DNA is called *sigma* ($\sigma$) *factor*; the remainder consists of a complex of five different proteins called the *core enzyme*. The union of sigma factor with the core enzyme creates the *holoenzyme*. The sigma factor subunit of the holoenzyme binds to DNA regions called *promoters*. Each promoter consists of an *RNA-polymerase recognition site* where the holoenzyme initially binds and an *RNA-polymerase binding site* to which the holoenzyme subsequently moves and binds more firmly. During this movememt, the DNA double helix experiences local unwinding by poorly understood mechanisms. RNA synthesis begins (somewhat ambiguously) about six bases further from the RNA-polymerase binding site. Sigma factor is released after chain elongation commences; it can then complex with some other core enzyme and get it started in the same or a different promoter. Since different promoters may have different base sequences, they may vary in their affinities for RNA polymerase. An efficient promoter would favor synthesis of many RNA transcripts of the adjacent gene(s); an inefficient promoter might allow only a few RNA transcripts to be made during the same time. Thus, promoter efficiency is one way by which the cell can regulate the synthesis of various RNA's and proteins.

There are two kinds of termination signals: (1) a terminal sequence of poly-A read by RNA polymerase itself, and (2) a pallindromic sequence (in which an identical or nearly identical nucleotide sequence runs in opposite directions on the two DNA chains) to which a termination protein called *rho* (ρ) *factor* is bound.

In eukaryotic organisms (but not in prokaryotes), some genes consist of essential regions called *exons* and non-essential regions called *introns*, *spacer DNA* or *intervening sequences* (IVS). Both kinds of regions of a gene are transcribed into one continuous precursor RNA containing from one to about thirty gene lengths. This precursor is also known as a *primary transcript* or as *heterogeneous nuclear RNA* (hnRNA). The hnRNA is clipped into *monocystronic* (one-gene) segments. The RNA copies of the introns are removed by nucleases and the RNA copies of the exons are enzymatically joined to form a single chain that will be used as the translational template. After receiving a "cap" (Table 15.1) and a poly-A tail, the reassembled mRNA leaves the nucleus and moves to the cytoplasm. Functional polycystronic mRNA's are only known in prokaryotic organisms.

**5. Enzymes of Translation.**

Some "unusual" bases (other than A, U, G, C) exist in various tRNA molecules. Most of these unusual bases are derived by enzymatic methylation of the common bases after they have been joined by 3′–5′ phosphodiester linkages. A single RNA strand can fold and base pair over local regions wherever complementary segments exist in opposite directions that allow base pairing in antiparallel fashion. The unusual bases in tRNA's disrupt this internal base pairing tendency and are therefore usually found clustered in the unpaired loops of these molecules. Ribonucleotides in the loop regions have free keto and amino groups available for the formation of secondary bonds with mRNA, ribosomes or with the enzyme that attaches (loads, activates) the tRNA with its own species of amino acid. Each of the three loops of the cloverleaf model of tRNA (Fig. 14-4) are probably associated with one of these functions.

The 3′ end of each tRNA begins with an ACC sequence. A covalent bond is formed between the carbonyl group of an amino acid and the 3′ terminal adenosine by enzymes called *amino-acyl tRNA synthetases*. Each kind of tRNA becomes activated by a specific synthetase. The first step in this process involves the activation of the amino acid by the synthestase. The amino acid is coupled with adenosine triphosphate (ATP) to form an amino-acyl adenylate with the release of pyrophosphate (the two terminal phosphates of ATP). The same synthetase then transfers the adenylate to the 3′ A of the tRNA. The energy in the amino-acyl bond can later be used in the formation of the lower-energy peptide bond. One of the proteins of the 50S bacterial ribosome, an enzyme called *peptidyl transferase*, is responsible for cleaving the amino-acyl bond and simultaneously forming the peptide bond, thus freeing the tRNA to leave the ribosome and become reactivated.

All bacterial polypeptides are initiated by N-formyl-methioninyl tRNA. The formyl group is enzymatically added to methionine after the amino acid is attached to this kind of tRNA (a different met-tRNA is involved in elongation of polypeptides). An *amino peptidase* enzyme trims the terminal formylated methionine from some polypeptides, so that all functional bacterial proteins do not have a methionine at one end. In eukaryotes, a special initiator methionyl-tRNA recognizes the AUG start codon, but it is not formylated.

Every mRNA has a *ribosome binding site* consisting of several purine ribonucleotides at the 5′ end of the molecule (e.g. the Shine-DelGarno sequence AGGAGGU), some 8 to 13 nucleotides from the starting methionine codon (AUG). This untranslated segment of mRNA is referred to as a *leader sequence*. A complementary base sequence exists at the 3′ end of the 16S rRNA molecule of the 30S bacterial ribosomal subunit. There does not seem to be a comparable fixed nucleotide sequence in eukaryotic mRNA's for ribosomal binding. Several non-ribosomal proteins, collectively called *initiation factors*, are required for initiation of protein synthesis. The initiation complex consists of binding the 30S subunit, f-met-tRNA, and initiation factors to the 5′ end of mRNA. Finally the 50S subunit is bound, causing release of the initiation factors.

The elongation phase involves the successive movement of activated tRNA's from the "A" to the "P" sites on the 50S ribosomal subunit. This is accomplished by energy from the splitting of nucleotides GTP to GDP (guanosine tri- and diphosphate, respectively) by an *elongation factor* called *translocase*. The active movement of tRNA molecules from the "A" to the "P" sites (translocation) is responsible for the passive movement of the mRNA to which they are base paired.

The termination phase involves GTP-bound proteins called *release factors* that sit on the nonsense (stop) codons UGA, UAA or UAG. These release factors seem to modify the action of peptidyl transferase so that the carboxyl group of the terminal amino acid reacts with water. This results in the separation of the polypeptide chain from the tRNA and also releases the final tRNA from the ribosome. The 50S and 30S ribosomal subunits dissociate and can be recycled to a new initiation complex.

Ribosomal binding to mRNA somehow disrupts the double-stranded hairpin regions that commonly exist in mRNA molecules. This opens up single-stranded mRNA regions for the synthesis of proteins. There is evidence from certain RNA phages that the timing of expression of genes can be affected by the gene order. Ribosome-binding (initiation) sites within the viral genome may be inaccessible as long as the RNA is folded by internal base pairings. Genes distal to the 5′ end can only be read after the hairpin loops have been disrupted by ribosomes moving along the molecule from the 5′ end. *Ribonucleases* are enzymes that degrade RNA molecules. Linear segments of RNA molecules are especially subject to degradation by ribonucleases. When several ribosomes are simultaneously reading an mRNA molecule, the complex is called a *polysome*. The more ribosomes in a polysome, the less opportunity for ribonucleases to attack the mRNA. Longer-lived mRNA molecules do not have to be replaced as often as shorter-lived molecules, the difference often being primarily ascribed to differential affinities of the respective ribosomal-binding sites.

## MUTATIONS

Non-enzymatic or structural proteins constitute the bulk of organic matter in living systems. Most proteins are complex, high molecular weight molecules. The exact sequence of amino acids in proteins is known for only a few, including hemoglobin, insulin, bovine pancreatic ribonuclease, tobacco mosaic virus protein, lysozyme and tryptophan synthetase. The normal human hemoglobin (Hb A) has about 140 amino acid residues in each of its $\alpha$ and $\beta$ chains. The sequence in the $\beta^A$ chain has been determined to be

val - his - leu - thr - pro - glu - glu - lys - etc.
1 2 3 4 5 6 7 8

An abnormal hemoglobin (Hb S) is produced by individuals with a mutant allele, resulting in a deformity of the red blood cell called "sickling". In a heterozygous condition this allele produces a mild anemia; in a homozygous condition the severity of the anemia may be lethal. The difference between Hb A and Hb S is that the latter has valine substituted for glutamic acid in the sixth position of the $\beta^S$ chain. Another potentially lethal abnormal hemoglobin (Hb C) is known in which the glutamic acid of the sixth position is replaced by lysine. One of the codons for glutamic acid is GAA. If a mutation occurred which changed the first A to a U, then the codon GUA (a *missense* triplet) would be translated as valine. The substitution of A for G would produce the missense codon AAA which codes for lysine.

Thus a change in a single nucleotide in the hemoglobin gene can produce a substitution of one amino acid in a chain of about 140 residues with profound phenotypic consequences!

Fortunately most genes are relatively stable and mutation is a rare event. The great majority of genes have mutation rates of $1 \times 10^{-5}$ to $1 \times 10^{-6}$, i.e. one gamete in 100,000 to one gamete in a million would contain a mutation at a given locus. However in a higher organism containing 10,000 genes, one gamete in 10 to one gamete in 100 would be expected to contain at least one mutation. The rate at which a given gene mutates under specified environmental conditions is as much a characteristic of the gene as is its phenotypic expression. The mutation rate of each gene is probably dependent to some extent upon the residual genotype. The only effect which some genes exhibit is to increase the mutation rate of another locus. These kinds of genes are called "mutator genes".

**Example 14.2.** A dominant gene called "dotted" (*Dt*) on chromosome 9 in corn causes a recessive gene *a* governing colorless aleurone, on chromosome 3, to mutate quite frequently to its allele *A* for colored aleurone. Plants which are *aaDt-* often have kernels with dots of color in the aleurone produced by mutation of *a* to *A*. The size of the dot will be large or small depending upon how early or late respectively during development of the seed the mutational event occurred.

The vast majority of mutations are deleterious to the organism and are kept at low frequency in the population by the action of natural selection. Mutant types are generally unable to compete equally with wild type individuals. Even under optimal environmental conditions many mutants appear less frequently than expected. Mendel's laws of heredity assume equality in survival and/or reproductive capacity of different genotypes. Observed deviations from the expected Mendelian ratios would be proportional to the decrease in survival and/or reproductive capacity of the mutant type relative to wild type. The ability of a given mutant type to survive and reproduce in competition with other genotypes is an extremely important phenotypic characteristic from an evolutionary point of view.

**Example 14.3.** White-eyed flies may be only 60% as viable as flies with pigmented eyes. Therefore among 100 zygotes from the cross $w^+w$ ♀ (wild type) $\times$ $w$Y ♂ (white), the Mendelian zygotic expectation is 50 wild type : 50 white. If only 60% of white-eyed flies survive, then we would observe in the adult progeny

$$50 \times 0.6 = 30 \text{ white} : 50 \text{ wild type}$$

Ionizing radiations such as X-rays are known to increase the mutability of all genes in direct proportion to the radiation dosage. A linear relationship between dosage (in roentgen units) and the induction of sex-linked lethal mutations in *Drosophila* is shown graphically in Fig. 14-13. This indicates that there is no level of dosage which is safe from the genetic standpoint. If a given amount of radiation is received gradually in small amounts over a long period of time (chronic dose) the genetic damage is sometimes less than if the entire amount is received in a short time interval (acute dose). In most cases, dose rate effects are not demonstrable. Ionizing radiations produce their mutagenic effects most frequently by inducing small deletions in the chromosome.

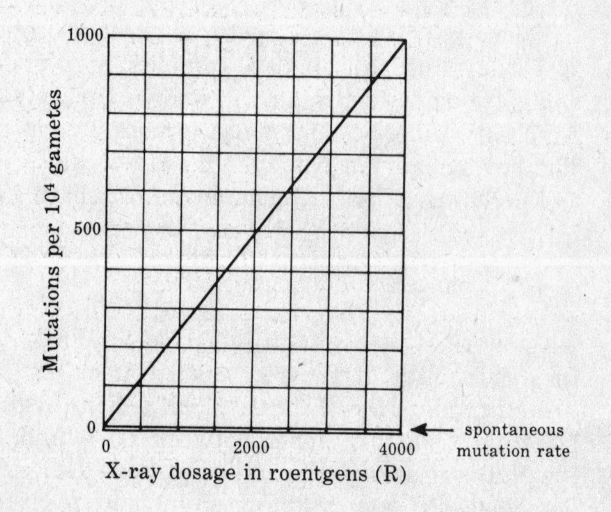


**Fig. 14-13.** X chromosome lethal mutations in *Drosophila* induced by X-rays.

The student's first encounter with the terminology involved in the study of mutations is sometimes a source of confusion. Mutations can be classified on the basis of several criteria. The following outline may be useful in showing the interrelationships of concepts and terms.

## A CLASSIFICATION OF MUTATIONS

**I. Size**

A. Point mutation – a change in a very small segment of DNA; usually considered to involve a single nucleotide or nucleotide pair.

1. Samesense (silent) mutation—change in a codon (usually at the third position) that fails to change the amino acid specificity from the unmutated state.
2. Nonsense mutation—a shortening of the protein product due to a chain-termination signal.
3. Missense mutation—a change in amino acid sequence with the wrong amino acid occupying a given position in the polypeptide chain.
4. Frameshift mutation—a shift of the reading frame, creating numerous missense or nonsense codons through the remainder of the cistron.

B. Gross mutations – changes involving more than one nucleotide pair; may involve the entire gene, the entire chromosome, or sets of chromosomes (polyploidy).

**II. Quality**

A. Structural mutations – changes in the nucleotide content of the gene.

1. Substitution mutations – substitution of one nucleotide for another.
   (a) *Transition* mutations substitute one purine for another or one pyrimidine for another.
   (b) *Transversion* mutations substitute a purine for a pyrimidine or vice versa.
2. Deletion mutants – loss of some portion of a gene.
3. Insertion mutants – addition of one or more extra nucleotides to a gene.

B. Rearrangement mutations – changing the location of a gene within the genome often leads to "position effects".

1. Within a gene – two mutations within the same functional gene can produce different effects, depending on whether they occur in the cis or trans position.
2. Number of genes per chromosome – different phenotypic effects can be produced if the numbers of gene replicas are non-equivalent on the homologous chromosomes.
3. Moving the gene locus may create new phenotypes, especially when the gene is relocated near heterochromatin.
   (a) Translocations – movement to a non-homologous chromosome.
   (b) Inversions – movement within the same chromosome.

**III. Origin**

A. Spontaneous mutation – origin is unknown; often called "background mutation".

B. Genetic control – the mutability of some genes is known to be influenced by other "mutator genes".

1. Specific mutators – effects limited to one locus.
2. Non-specific mutators – simultaneously affects many loci.

C. Induced mutations – through exposure to abnormal environments such as:

1. Ionizing radiations – changes in chemical valence through the ejection of electrons are produced by protons, neutrons, or by alpha, beta, gamma or X-rays.
2. Non-ionizing radiations – raise the energy levels of atoms (excitation), rendering them less stable (e.g. ultraviolet radiation, heat); UV often produces thymine dimers, i.e. bonding between thymines on the same strand.
3. Chemical mutagens – are chemical substances which increase the mutability of genes.
   (a) Copy errors – mutants arising during DNA replication (e.g. base analogue mutagens which are chemically similar to the nucleic acid bases may be incorporated by mistake; acridine causes single base additions or deletions possibly by intercalation between two sequential bases).
   (b) Direct gene change – produced in non-replicating DNA (e.g. nitrous acid by deamination directly converts adenine to hypoxanthine and cytosine to uracil).

**IV. Magnitude of Phenotypic Effect**

A. Change in mutation rate – some alleles can only be distinguished by the frequency with which they mutate.

B. Isoalleles – produce identical phenotypes in homozygous or heterozygous combinations with each other, but prove to be distinguishable when in combination with other alleles.

C. Mutants affecting viability:

1. Subvitals – relative viability is greater than 10% but less than 100% compared to wild type.
2. Semi-lethals – cause more than 90% but less than 100% mortality.
3. Lethals – kill all individuals before adult stage.

**V. Direction**

A. Forward mutation – creates a change from wild type to abnormal phenotype.

B. Reverse or back mutation – produces a change from abnormal phenotype to wild type.

1. Single site mutation – changes only one nucleotide in the gene (e.g. adenine $\xrightarrow{\text{forward}}$ guanine $\xrightarrow{\text{reverse}}$ adenine).
2. Mutation suppressor – a gene change which occurs at a different site from the primary mutation, yet reverses its effect.
   (*a*) Extragenic (intergenic) suppressor – occurs in a different gene from that of the mutant.
   (*b*) Intragenic suppressor – occurs at a different nucleotide within the same gene; shifts the reading frame back into register.

**VI. Cell Type**

A. Somatic mutation – occurs in non-reproductive cells of the body, often producing a mutant phenotype in only a sector of the organism (mosaic or chimera).

B. Gametic mutation – occurs in the sex cells, producing a heritable change.

## DEFINING THE GENE

Recall that the elementary concept of linked genes held them to be discrete elements analogous to beads on a string. The structure of DNA now indicates that a gene is one or a sequence of nucleotides. Formerly, a gene was assumed to be a genetic unit by three criteria: (1) the unit of physiological function leading to phenotypic expression, (2) the unit of structure which cannot be subdivided by crossing over, (3) the unit which can be changed by mutation. We now know that these criteria generally do not define the same amount of genetic material. Each of these criteria may be used to define a different genetic element.

A gene can be defined on the basis of its function. For example, the entire structure of an enzyme may be determined by a segment of DNA. The hypothesis that one gene produces an enzyme may hold true for some but not all enzymes. For example, the enzyme tryptophan synthetase has been found to consist of two structurally different protein chains (A and B), each chain produced by an adjacent segment of DNA. The "one gene – one enzyme" hypothesis has now been refined to the "one gene – one polypeptide chain" hypothesis. That portion of DNA specifying a single polypeptide chain is termed a *cistron* and is synonymous with the gene of function.

There are many positions or sites within a cistron where mutations can occur. As was shown previously in this chapter in the case of the hemoglobin molecule, a change in a single nucleotide can result in a mutant phenotype. A *muton* is the smallest unit of genetic material that when changed (mutated) produces a phenotypic effect. A muton may thus be delimited to a single nucleotide or some part of a nucleotide. Different forms of a mutationally defined gene are called *homoalleles.*

A mutation within a cistron may render its product defective. A fully active wild type cistron may be produced by recombination between two defective cistrons containing mutations at different sites. The smallest unit of recombination (the *recon*) is believed to involve two adjacent nucleotides. Recombinationally separable forms of a gene within a cistron are referred to as *heteroalleles.*

**Example 14.4.** An individual with a mutant phenotype has two mutant sites, $m_1$ and $m_2$, within homologous cistrons (diagrammed as boxes). These heteroalleles can recombine and be transmitted to the progeny as a functionally normal cistron.

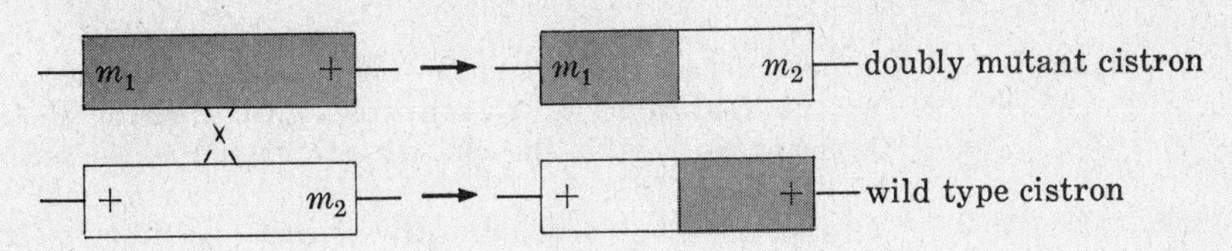


Two mutants can be assigned to the same or to different cistrons on the basis of the *complementation* or *cis-trans test*. Two mutants in the same cistron will not complement each other to produce wild type (or normal phenotype) when in the trans position. A normal phenotype (indicating complementation) can be produced by two mutants if they belong to different cistrons regardless of their linkage relationships.

**Example 14.5.** Two mutants ($m_1$ and $m_2$) in the same cistron are functionally allelic.

(*a*) Cis position (both mutant recons on one homologue)

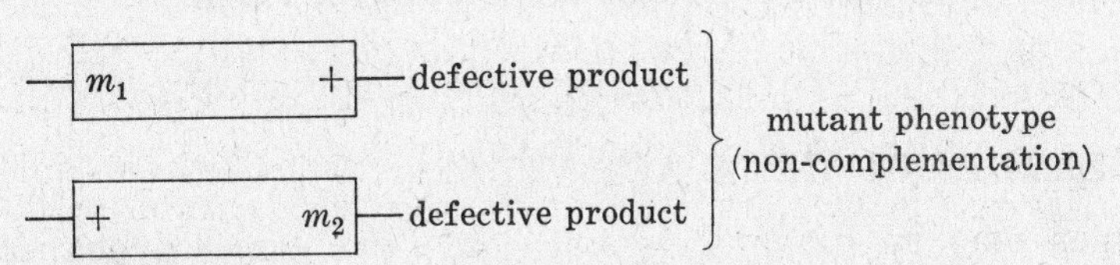


(*b*) Trans position (each homologue has a mutant and a non-mutant recon)

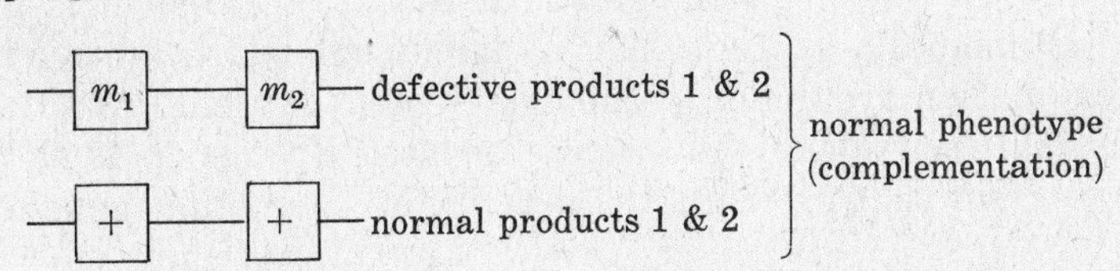


**Example 14.6.** Two mutants in different cistrons are functionally non-allelic.

(*a*) Coupling linkage

$m_1$ — $m_2$ — defective products 1 & 2

+ — + — normal products 1 & 2

normal phenotype (complementation)

(*b*) Repulsion linkage

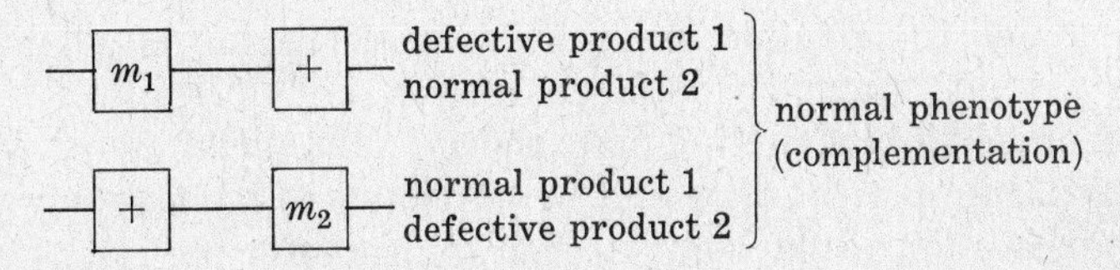


Certain genetic elements associated with duplicated segments of the *Drosophila* chromosome have proven to be separable by recombination (structurally non-allelic), but functionally allelic by the cis-trans test. These very closely linked loci with different, though related, phenotypic effects are called *pseudoalleles*. In some cases, pseudoalleles may actually be heteroalleles, i.e. different mutational sites within the same cistron. In other cases, pseudoallelism may be explained by functional interactions between mutations in different cistrons. Commonly, tens of thousands of offspring must be reared to find a rare crossover individual involving pseudoalleles which are of the magnitude 0.01 to 0.09 map units apart.

**Example 14.7.** A pair of pseudoalleles in *Drosophila*, called bithorax (*bx*) and post-bithorax (*pbx*), are approximately 0.02 map units apart. Each of these recessive genes when homozygous modify the club-shaped balancers (halteres) into a pair of wing-like structures. The cis form (+ +/*bx pbx*) has normal balancers; the trans form (+ *pbx*/*bx* +) has a slight post-bithorax "position effect". The homozygous genotype, *bx pbx*/*bx pbx*, has a fully developed second pair of wings. The bithorax series is composed of 5 separate pseudoallelic loci and is associated with two doublet bands in the salivary gland chromosomes.

**Example 14.8.** A dominant mutation called "Star", affecting eye morphology and mapping at locus 1.3 was found in the second chromosome of *Drosophila*. Later, a recessive mutant called "asteroid" which also affects eye morphology was found to map at the same position. When large numbers of progeny from Star × Asteroid parents were raised, about 1 in 5000 showed recombination. Trans heterozygotes ($S^+ \, ast/S \, ast^+$) have reduced eye size whereas cis heterozygotes ($S^+ \, ast^+/S \, ast$) have normal eye size.

All geneticists are not in agreement on the definition of a gene. Perhaps the unit which comes closest to the concept of the classical Mendelian gene is the cistron. The cistron is a relatively large segment of genetic material when compared with the muton or the recon. An attempt to provide for these various segments, defined on the basis of different criteria, is found in the following statement. "The gene of function is that sequence of nucleotides (providing numerous sites for intragenic mutation and/or recombination) which specifies the amino acid sequence of a specific polypeptide chain."

## REGULATION OF GENE ACTIVITY

Genes may be broadly classified into two major types: (1) structural genes and (2) regulatory genes. *Structural genes* encode the specifications for the synthesis of some type of chemical substance. The substance may be (*a*) an enzyme, (*b*) a non-enzymatic structural protein such as actin of muscle cells or collagen of connective tissue, (*c*) a transport protein such as oxygen-carrying hemoglobin of red blood cells, (*d*) a polypeptide hormone such as insulin, (*e*) immunity proteins such as immunoglobulins (antibodies), (*f*) a component of the blood coagulation system or of the complement system (involved in bacteriolysis by specific antimicrobial antibodies), (*g*) a non-translated RNA such as rRNA or tRNA or (*h*) a protein called *repressor* that prevents the activity of some other gene or gene complex.

*Regulatory genes* are not transcribed to RNA and hence have no chemical products. Rather they behave as "switching devices" to turn on or off one or more structural genes under their control. Included in the list of regulatory genes are (*a*) promoters, (*b*) terminators, (*c*) operators and (*d*) attenuators. A *promoter* region is where RNA polymerase initially binds prior to transcribing the adjacent segment of DNA into RNA. Not all promoters allow RNA polymerase to bind with equal facility. Bacterial genes whose products are required in relatively large quantities would be expected to have highly efficient promoters. A cell could compensate for a relatively inefficient promoter by having many copies of the requisite structural genes (each with their own promoters) in the genome. Almost all eukaryotic cells (plants, fungi and animals) have many *redundant DNA* sequences (sometimes as high as $10^6$-fold redundancy per genome), but redundant DNA sequences are rare or non-existent in prokaryotic cells (bacteria, blue-green algae). A *terminator* is a DNA sequence recognized by a protein termination factor called *rho* ($\rho$). RNA polymerase stops moving along the DNA when it encounters rho factor on the DNA. An *operator* is a region of DNA (between a promoter and a structural gene) to which repressor protein binds. When repressor protein is bound to the operator locus, RNA polymerase cannot pass and hence cannot transcribe the adjacent gene or genes. The promoter region, operator locus and adjacent structural gene(s) under control of the operator are collectively called an *operon*.

An *attenuator* is thought to be a ρ-factor-dependent region of an operon within which most RNA polymerase molecules prematurely stop elongation. However, transcription can proceed normally in the presence of an *antitermination factor* (either a specific protein such as the "N" gene product of bacteriophage λ or a specific deacylated or "unloaded" tRNA such as that for tryptophan). The attenuator region usually lies within the *leader sequence*, a segment of DNA between the operator locus and the starting codon for the translated protein product(s) of the operon. The leader sequence in the corresponding mRNA is not translated. Not all operons have attenuators. It is not clear how the terminating activity of rho factor can be nullified in the attenuator region but not at the termination sequences at the end of monogenic or polygenic structural gene sequences. However, in many cases RNA polymerase by itself can read the terminating signal, but presumably recognizes no stop signals within the attenuator region.

**1. Transcriptional Control Mechanisms.**

(*a*) ***Negative Control.***

(1) *Inducible Systems.* Some operons (e.g. genes making repressor proteins) may not have operator loci and hence make products *constitutively* (i.e. regardless of environmental conditions). Genes subject to regulation by repressor proteins are said to be under *negative control* because they only make products when a specific substance (repressor) is absent. There are two broad classes of operons, inducible operons and repressible operons—the functioning of which are influenced by substances in the environment.

An *inducible enzyme* is produced only when its substrate (inducer) is present in the environment. Most enzymes in this category are catabolic in their activity. The prototype of inducible operons is the "lactose system" of the bacterium *Escherichia coli*.

**Example 14.9.** Catabolism of the sugar lactose requires the cooperation of two enzymes: β-galactoside permease, which makes the cell permeable to the substrate (lactose), and β-galactosidase, an intracellular enzyme which splits the disaccharide into the simple sugars glucose and galactose. The operon consists of (from left to right) a promoter ($p$), the operator ($o$), the gene for galactosidase ($z$) and the gene for permease ($y$). A regulator gene ($i$) is located near this operon. Some of the alleles which have been found in this system and their functions are listed below.

**Operator Alleles.**

$o^+$ = this operator "turns on" the structural genes in its own operon, i.e. the $y^+$ and $z^+$ alleles in the same segment of DNA (cis position) can produce enzyme. This operator is sensitive to repressor substance, i.e. it will "turn-off" the synthetic activity of the structural genes in its operon if repressor is present.

$o^c$ = a constitutive operator which is insensitive to repressor substance and permanently "turns on" the structural genes in its own operon.

$o^o$ = a "defective" operator which permanently "turns off" the structural genes in its own operon (now known to be a nonsense mutation in the closely linked galactosidase cistron).

**Galactosidase Alleles.**

$z^+$ = makes β-galactosidase if its operon is "turned on".

$z^-$ = makes a modified enzymatically inactive substance called Cz protein.

**Permease Alleles.**

$y^+$ = makes β-galactoside permease if its operon is "turned on".

$y^-$ = no detectable product is formed regardless of the state of the operator; probably a nonsense mutation.

**Regulator Alleles.**

$i^+$ = makes a diffusible repressor substance which inhibits synthetic activity in any $o^+$ operon in the absence of lactose; in the presence of lactose, repressor substance is inactivated.

$i^-$ = a defective regulator unable to produce active repressor substance, due to a nonsense or missense mutation.

$i^s$ = superrepressor which is insensitive to the presence of lactose and inactivates any $o^+$ operon.

**Example 14.10.** Bacteria of genotype $i^+o^+z^+y^+$ grown on media devoid of lactose will produce neither galactosidase nor permease because $i^+$ makes repressor substance which inactivates the $o^+$ operator and "turns off" the synthetic activity of structural genes $y^+$ and $z^+$ in its own operon.

**Example 14.11.** Partial diploids can be produced in bacteria for this region of the chromosome. Cells of the genotype $\frac{i^-o^+z^+y^-}{i^+o^cz^-y^+}$ will produce Cz protein and permease constitutively (i.e. either with or without the presence of lactose inducer) because the allele $o^c$ permanently "turns on" the genes in its operon (i.e. those in cis position with $o^c$). Galactosidase will be produced only inductively because in the presence of lactose (inducer) the diffusible repressor substance from $i^+$ will be inactivated and allow the structural gene $z^+$ in cis position with the $o^+$ operator to produce enzyme.

(2) *Repressible Systems.* A *repressible* enzyme is normally present in the cell, but ceases to be synthesized when high concentrations of its end product are present. The end product is called *corepressor*. A regulator gene produces a substance called *aporepressor* which unites with corepressor to form a functional *repressor* molecule. The repressor inhibits synthesis of mRNA by all genes specifying enzymes in the synthetic pathway. Most repressible enzymes are found in anabolic pathways. Hormones may exert their phenotypic effects by depressing genes previously repressed.

**Example 14.12.** High concentrations of histidine in *Salmonella typhimurium* cause that amino acid to act as corepressor, thereby terminating synthesis of all 10 enzymes in the pathway to histidine (coordinate repression).

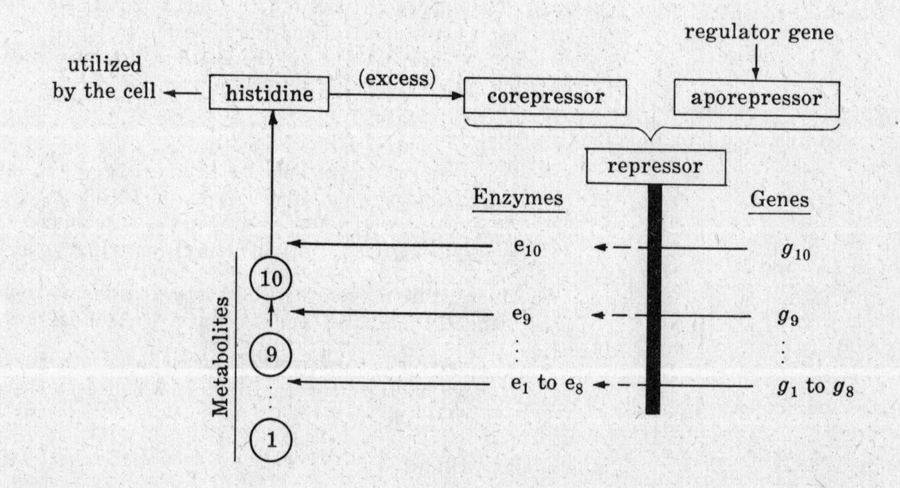


### (*b*) *Positive Control.*

Many bacterial genes are also under *positive control*. This mode of gene regulation is attributed to the presence of factors that enhance the attachment of RNA polymerase to promoters and the initiation of mRNA synthesis.

**Example 14.13.** The inducible lactose operon of *E. coli* consists of structural genes that make (1) galactoside permease (an enzyme that renders the cell more permeable to the entry of lactose) and

(2) β-galactosidase (the enzyme that splits lactose into glucose and galactose). The lactose operon is "turned off" (no transcription) when glucose is present in the growth medium regardless of the presence or absence of lactose. It is hypothesized that one of the products of glucose metabolism either (1) inhibits adenyl cyclase (the enzyme responsible for converting ATP to its cyclic derivative, cAMP) or (2) stimulates the hydrolysis of cAMP by the enzyme phosphodiesterase to yield noncyclic AMP. When glucose is absent from the growth medium, a constitutively produced protein called *catabolite activator protein* (CAP) binds to cAMP; this complex can then bind to the left end of the *lac* promoter region. RNA polymerase cannot bind to the right end of the promoter region unless the cAMP-CAP complex is bound at the left end of that same region. If no repressor is bound to the *lac* operator, RNA polymerase can begin transcribing the structural genes for β-galactosidase and permease.

**Example 14.14.** The operon for tryptophan synthesis in *E. coli* consists of a promoter, an operator and five contiguous structural genes (in that order, left to right). Like the histidine operon, the tryptophan operon is a repressible enzyme system. Attenuator control plays a secondary role to that of repressor-operator control in the regulation of the tryptophan operon. The attenuator site lies between the operator and the first structural gene in a non-translated region called the "leader sequence". When tryptophan is in ample supply within the cell, rho factor sits on the attenuator and prevents the movement of RNA polymerase into the adjacent structural genes. When tryptophan is in short supply, deacylated tryptophan-tRNA ($tRNA_{trp}$ uncomplexed with tryptophan; "unloaded") interferes with rho binding so that RNA polymerase can read through the attenuator region and on into the adjacent structural genes. The heart of this positive control mechanism is an *antitermination effect*. The histidine operon, by contrast, appears to be primarily regulated by attenuator control and secondarily by repressor-operator control.

### (c) *DNA Rearrangements.*

Some genetic regions in both prokaryotes and eukaryotes are highly mobile ("jumping genes"). Each transposable element is called a *transposon* and carries with it the capability to catalyze its own movement. Transposons can replicate enroute and integrate at numerous sites throughout the genome. This is thought to be accomplished by an enzyme (transposase) that facilitates genetic recombination of the ends of the transposon with other non-homologous DNA sequences. Movement of blocks of genetic information can produce the phenomena of gene instability, insertions, inversions and deletions. One or more of these phenomena may be essential to normal developmental processes in eukaryotes or to survival strategies in bacteria.

**Example 14.15.** An immunoglobulin molecule ("antibody") is composed of two identical *H* (heavy) polypeptide chains and two smaller identical *L* (light) chains.

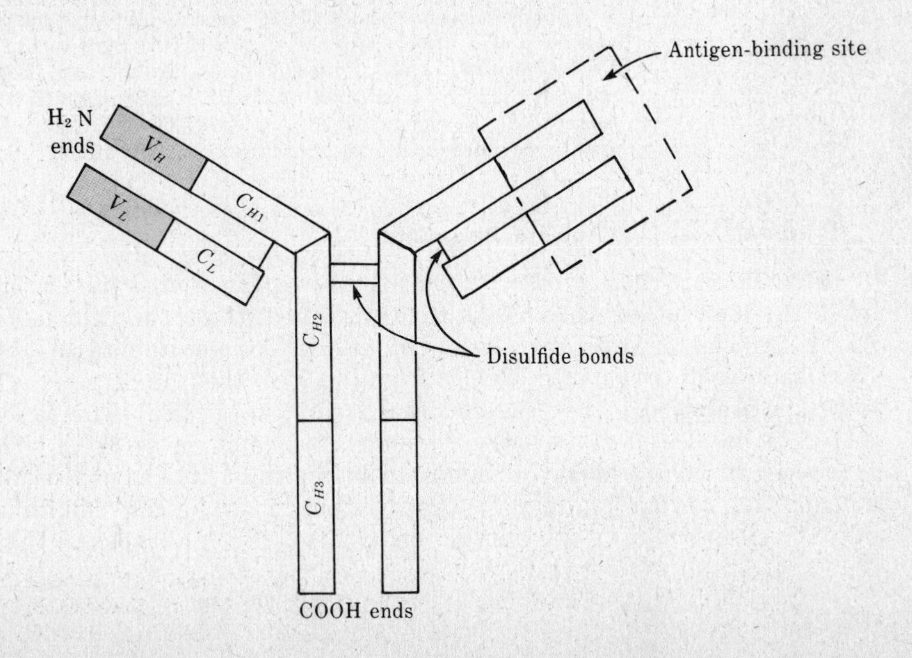

Heavy chains possess antigenic determinants in the "tail" (carboxyl) segments by which they can be classified as IgG, IgM, IgA, IgD or IgE. Light chains can likewise be typed as kappa or lambda. Within an $H$ chain class or $L$ chain type, these segments exhibit very little variation in primary structure from one individual to another and are called *constant regions* ($C$). The amino ends, however, are extremely diverse in primary structure, even within a class and are called variable ($V$) regions. The $V_H$ and $V_L$ regions together form an antibody-combining site for specific interaction with a homologous antigen molecule. The $C_H$ region consists of three or four similar segments, presumably derived evolutionarily by duplication of an ancestral gene and subsequent modification by mutations; the similar segments are called *domains* and are labeled $C_{H1}$, $C_{H2}$, $C_{H3}$, etc. A mature lymphocyte (plasma cell) produces antibodies with a single class of $H$ chain and a single type of $L$ chain, hence also a single antigen-binding specificity. The first antibodies produced by a developing plasma cell are usually of class IgM. Later the same antigen-binding specificity may be associated with $H$ chains of class IgG (or some other class).

There are three immunoglobulin *gene families* (kappa, lambda and heavy chain families), each on a different human chromosome. In embryonic cells, the DNA that specifies the $V$ and $C$ regions are widely separated (loosely linked) within a gene family. As the plasma cell matures, the $V$ and $C$ regions within a family become nearly adjacent by some kind of structural rearrangement (e.g. looping out of the intervening segment as in an interstitial deletion). The primary transcript from this rearranged DNA contains the information for one $V$ and one $C$ region of an immunoglobulin polypeptide chain. There are thought to be hundreds of different $V$ genes within each immunoglobulin gene family. There are four different $C_H$ genes for human IgG, two each for IgM and IgA, and one each for IgD and IgE; there are four $C_L$ genes in the human lambda family, one in the kappa family. Different $C_H$ classes endow the immunoglobulin molecule with special effector functions such as complement binding (IgG and IgM), placental passage (IgG), secretion into body fluids (IgA), binding to mast cells (IgE), etc. Thus, the union of one kind of $V$ region with one kind of $C$ region in both $L$ and $H$ chains creates an antibody-combining site that specifically binds antigen and also allows that union to become biologically active. The initial translocation event usually couples a $V_H$ gene to a $C_H$ gene of the IgM group. Later in plasma cell differentiation, other translocations may connect that same $V_H$ gene to a $C_H$ gene in one of the other Ig groups (e.g. $IgG_3$ or $IgA_1$, etc.). At any given time, however, a plasma cell is thought to synthesize primary mRNA transcripts of a single kind.

**Example 14.16.** Two antigenically different kinds of flagellar proteins (H1 and H2) can be made by *Salmonella* bacteria. Only one of these flagellin genes is expressed at a given time, and switching from one gene to the other results in a phenomenon called "phase variation". The controlling element in phase variation is an inversion of about a 1000 base pair (bp) DNA sequence adjacent to the *H2* gene. The invertible segment (a transposon) contains a promoter region for the initiation of transcription of the *H2* gene. When the *H2* gene is in the "on" position, both the *H2* gene and an adjacent gene (*rh1*) are coordinately expressed. The product of the *rh1* locus is a repressor of the *H1* gene. When the transposon is in the *H2* "off" configuration, neither the H2 flagellin nor the *rh1* repressor are produced and transcription can occur in the opposite direction so that the H1 flagellin is produced. This kind of "flip-flop" switch allows bacteria to alter their flagellar antigens and thereby escape (at least temporarily) immune attack by host antibodies.

## 2. Translational Control Mechanisms.

In addition to the regulatory mechanisms that function at the transcription level (genetic level) there are other mechanisms that function at the translation or posttranslation levels. At the translation level, the lifetime of an mRNA nolecule may be genetically determined. Enzymatic degradation of mRNA is from the 5′ to the 3′ end; i.e. the end of the RNA that is first synthesized is also the end that is first degraded. The average lifetime of many mRNA molecules of *E. coli* is only about two minutes at 37°C. The specific nucleotide sequences at the 5′ end may influence its susceptibility to enzymatic digestion (and thus may influence its average lifetime). More than one ribosome can simultaneously move along an mRNA molecule during the translation process. Such a ribosome-covered mRNA is called a *polyribosome* or *polysome*. Degradative enzymes are denied access to the mRNA when ribosome-coated at their 5′ ends. Hence, the lifetime of mRNA's may also be correlated with the number of free ribosomes available at any given moment to translate mRNA molecules. Once a bacterial ribosome has begun reading any mRNA molecule, it translates proteins at a relatively constant rate of about 15 amino acids per second at 37°C. Bacteria vary their rates of protein synthesis by varying their ribosomal content rather than by varying the translation rate. Multigenic (multicystronic) mRNA molecules (carrying information for more than one structural gene) are common in prokaryotes, but are rare or absent in eukaryotes. The efficiency of ribosome-binding sites within a multigenic mRNA may not be equal. Perhaps the different proteins of a polygenic messenger are translated as a function of ribosome detachment following the reading of each chain-terminating signal (nonsense codons to which chain-terminating proteins or "release factors" are bound).

**Example 14.17.** In the *E. coli* lactose system, there are three structural genes under control of a common operator locus governing production of (1) β-galactosidase, (2) galactoside permease and (3) galactoside acetylase. These three proteins are produced in the respective ratios $1 : \frac{1}{2} : \frac{1}{5}$, reflecting their respective locations relative to the 5′ (operator) end of the polycistronic mRNA in which they are coded. Thus, there is a *polarity gradient* within the polycistronic mRNA that reduces the probability of cistron translation as a function of its distance from the 5′ end. It is hypothesized that ribosomes attach to different starting points (ribosome-binding sites) along the polygenic mRNA at different rates as reflected by the relative amounts of the three proteins synthesized.

Different RNA molecules have unique secondary structures as a consequence of folding (e.g. hairpin loops) and internal base pairing of complementary segments. The polycistronic mRNA's of bacteria probably have little chance of forming secondary structures because ribosomes can begin reading at their 5′ ends before the entire mRNA is synthesized (there is no membrane separation between nucleus and cytoplasm in prokaryotic organisms). However, the RNA of an infective virus may have a complicated secondary structure that blocks certain ribosome-binding sites (perhaps even the site at the 5′ end). Translation begins at an open site(s) that is not base-paired and then other ribosome-binding sites become exposed as polyribosomes complex with the RNA and progressively destroy more of the secondary RNA structure. This mechanism provides one way for controlling the time at which different proteins need to appear and perhaps also accounts for some of the quantitative differences between these proteins.

**Example 14.18.** Viruses that rupture host cells (lytic viruses) do so by producing an enzyme (lysozyme). It would be disadvantageous for this enzyme to be produced early in the life cycle of a virus before it had replicated many times and made its "coat" proteins. Relatively few lysozyme molecules are required in comparison with those of coat proteins. Various timing devices are employed to ensure that virally coded proteins are made in the quantities needed (avoiding energy wastage) and in the sequence as needed to complete the life cycle.

### 3. Posttranslational Control.

Finally, the expression of genes can be regulated after proteins have been synthesized (posttranslational). *Feedback inhibition* (or end-product inhibition) is a regulatory mechanism which does not affect enzyme synthesis, but rather inhibits enzyme activity. The end product of a synthetic pathway may combine loosely (if in high concentration) with the first enzyme in the pathway. This union does not occur at the catalytic site, but it does modify the tertiary structure of the enzyme and hence inactivates the catalytic site. This *allosteric transition* of protein blocks its enzymatic activity and prevents overproduction of end products and their intermediate metabolites.

**Example 14.19.** The end product isoleucine in *E. coli*, when present in high concentration, unites with the first enzyme in its synthetic pathway and thus inhibits the entire pathway until isoleucine returns to normal levels through cellular consumption.

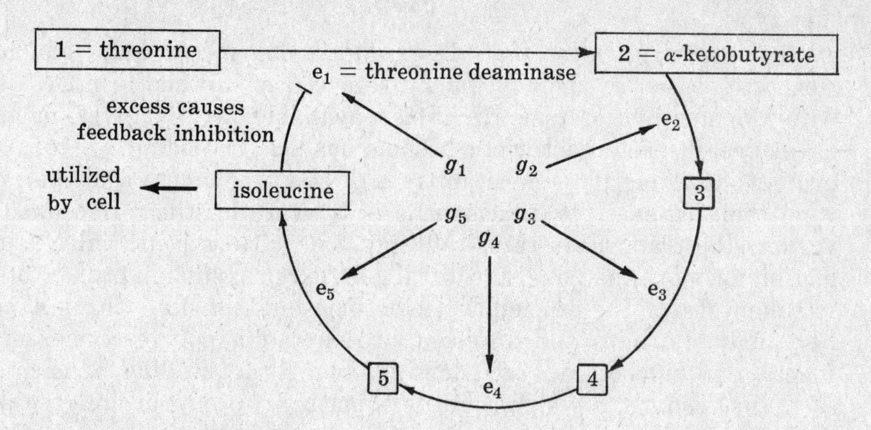


## 4. Posttranscriptional and Posttranslational Processing.

Products of transcription (RNA's) and translation (proteins) may be subject to enzymatic modifications before they become capable of performing their normal biological functions.

**Example 14.20.** There are two major kinds of rRNA molecules in bacterial ribosomes: (1) 16S rRNA associated with the smaller 30S ribosomal subunit, and (2) 23S rRNA associated with the larger 50S ribosomal subunit. A 30S rRNA precursor is transcribed from DNA. The precursor is enzymatically degraded stepwise through 17S and 25S intermediates and then further degraded to 16S and 23S chains respectively. All bacterial tRNA's are initially transcribed as larger precursors. Some pre-tRNA molecules contain more than one tRNA species, a mechanism that helps ensure equal amounts of the respective tRNA digestion products.

**Example 14.21.** Immunoglobulins ("antibody" molecules) are tetramers consisting of two identical light ($L$) chains and two longer identical heavy ($H$) chains. The amino end portion of these chains is designated $V$ and the carboxyl end portion is designated $C$ (Example 14.15). A $V$-region gene must be moved close to a $C$-region gene during differentiation of a plasma cell. The spliced gene that specifies a complete $L$ chain codes for a long pre-mRNA (primary transcript) with the information for the $V$ segment at the 5′ mRNA end and the information for the $C$ segment at the 3′ end. Between these two regions is a segment of mRNA called an *intervening sequence* (IVS) or an *intron* that is not translated into protein. The segments that are translated into proteins are called *exons*. The intron is enzymatically excised from the pre-mRNA and then the $V$ and $C$ exons are spliced together to form the template for making a complete $L$ chain. The same phenomenon occurs in the synthesis of the $H$ chain. Introns are also known in other proteins such as mouse β-globin and chicken ovalbumin, the latter having seven introns that must be cut from the pre-mRNA before becoming functional mRNA. In addition, the pre-mRNA molecules of ovalbumin are "capped" by a methylated guanine at the 5′ end and a string of adenine nucleotides is added at the 3′ end (poly-A "tail"). Primary transcripts of most genes seem to be modified in a similar manner prior to excision of introns. The biological role of introns is not yet known, but it has been suggested that they may be important in preventing errors in the inheritance of closely related genes and/or as enhancers of evolution. The various nearly identical members of the mouse globin gene family all have introns in exactly the same places, but these sequences are so different in base composition that they are barely able to "recognize" each other in *in vitro* molecular hybridization experiments. A single change in a nucleotide could alter the size of the intron excised and thereby generate greater changes in proteins, providing evolution with a broader spectrum of polypeptides from which to select adaptive forms.

**Example 14.22.** All bacterial polypeptide chains are initiated by a formylated methionine (formylmethionine), but most *E. coli* functional proteins do not have formylmethionine at their amino terminus. Enzymes usually remove one or more amino acids from that end, perhaps even before the polypeptide chain is completely translated.

The hormone insulin is synthesized as a single-chain precursor (proinsulin) with little or no hormonal activity. Two internal cuts remove 31 amino acids from proinsulin, producing the two polypeptide chains of the functional dimer. Likewise, human growth hormone that circulates in blood is a "clipped" version of the pituitary form of that hormone.

# Solved Problems

**14.1.** How many triplet codons can be made from the four ribonucleotides A, U, G and C containing (*a*) no uracils, (*b*) one or more uracils?

**Solution:**

(*a*) Since uracil represents 1 among 4 nucleotides, the probability that uracil will be the first letter of the codon is 1/4; and the probability that U will not be the first letter is 3/4. The same reasoning holds true for the second and third letters of the codon. The probability that none of the three letters of the codon are uracils is $(3/4)^3 = 27/64$.

(*b*) The number of codons containing at least one uracil is $1 - 27/64 = 37/64$.

**14.2.** A synthetic polyribonucleotide is produced from a mixture containing U and C in the relative frequencies of 5 : 1 respectively. Assuming that the ribotides form in a random linear array, predict the relative frequencies in which the various triplets are expected to be formed.

**Solution:**

The frequencies with which the different triplets are expected to be formed by chance associations can be predicted by combining independent probabilities through multiplication as follows:

UUU should occur with a frequency of $(5/6)(5/6)(5/6) = 125/216$.

Codons with 2U and 1C $= (5/6)^2(1/6) = 25/216$ each (UUC, UCU, CUU).

Codons with 1U and 2C $= (5/6)(1/6)^2 = 5/216$ each (UCC, CUC, CCU).

CCC $= (1/6)^3 = 1/216$.

**14.3.** Suppose that synthetic RNA's were made from a solution which contained 80% adenine and 20% uracil. The proteins produced in a bacterial cell-free system under direction of these mRNA's were found to contain amino acids in the following relative proportions: 4 times as many isoleucine as tyrosine residues, 16 times as many isoleucine as phenylalanine residues, 16 times as many lysine as tyrosine residues. What triplet codons were probably specifying each of the above amino acids?

**Solution:**

| | | Codon | | | Ratio | Amino Acid |
|---|---|---|---|---|---|---|
| AAA | = | $(0.8)^3$ | = | 0.512 | 64 | lysine |
| Some permutation of 2A and 1U | = | $(0.8)^2(0.2)$ | = | 0.128 | 16 | isoleucine |
| Some permutation of 1A and 2U | = | $(0.8)(0.2)^2$ | = | 0.032 | 4 | tyrosine |
| UUU | = | $(0.2)^3$ | = | 0.008 | 1 | phenylalanine |

phe—4x→tyr—4x→ileu—4x→lys

(phe —16x→ ileu; tyr —16x→ lys)

**14.4.** Phage MS2 is a single-stranded RNA virus of *E. coli*. When infected into a bacterium, the phage RNA (the "plus" strand) is made into a double-stranded replicative intermediate form ("plus–minus") from which "plus" RNA is synthesized. The "minus" strands when isolated are not infective. Phage X174 is a single-stranded DNA virus of *E. coli*. When infected into a bacterium, the same events as described for MS2 occur, but the "minus" strands when isolated are infective. Devise a reasonable hypothesis to account for these observations.

**Solution:**

The DNA "minus" strand can serve as a template and can utilize the bacterial enzyme DNA-polymerase for replication. The "minus" strand of RNA does not code for the enzyme which replicates RNA (RNA synthetase), and this enzyme is absent in uninfected bacteria. The "plus" strand of RNA carries the coded instructions for this enzyme and acts first as mRNA for enzyme synthesis. In the presence of the enzyme, single-stranded RNA can form a complementary strand and becomes a double helix replicative form.

**14.5.** H. J. Muller developed a system (*ClB* technique) for detecting recessive sex-linked lethal mutations induced by X-ray treatment in *Drosophila.* He used a heterozygous stock which had the dominant sex-linked gene "bar eye" ($B$) linked to a known recessive lethal gene ($l$). The segment of chromosome containing $B$ and $l$ was within an inversion (represented by the symbol $C$) which effectively prevented crossing over between the $B$ and $l$ loci. He began by mating these dihybrid bar eye females to wild type males which had been exposed to X-rays. He selected the $F_1$ bar eye females and crossed them to unirradiated wild type males in small vials, one female per vial. The incidence of sex-linked recessive lethals induced by the X-ray treatment was determined from inspection of the $F_2$. What criterion did he use for this determination?

**Solution:**

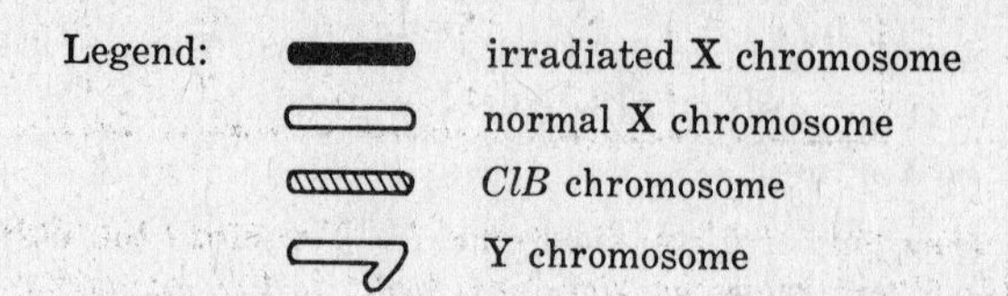


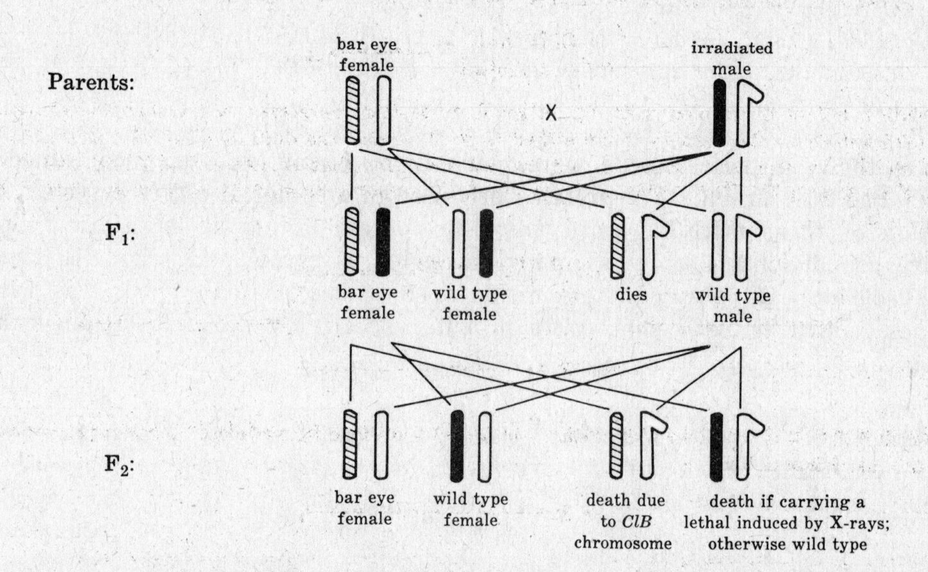


Any vials of the $F_2$ generation which are devoid of males can be scored as an X-ray induced sex-linked lethal as long as spontaneous mutations are considered negligible. The same procedure could be followed without irradiation treatment of the parental males to obtain an estimate of the rate of spontaneous sex-linked lethals. The reason that the $F_1$ bar eye females can survive even though carrying two recessive lethals is that the X-ray induced lethal is, in all probability, at some locus other than the one in the *ClB* chromosome.

**14.6.** If 54 mutations are detected among 723 progeny of males which received 2500 R (roentgens) and 78 mutations among 649 progeny of males which received 4000 R, how many mutants would be expected to appear among 1000 progeny of males which received 6000 R?

**Solution:**

The number of mutations induced by ionizing radiation is directly proportional to the dosage.

$$78/649 = 12.02\% \text{ mutations at } 4000\text{ R}$$

$$54/723 = 7.47\% \text{ mutations at } 2500\text{ R}$$

$$\text{Difference} = 4.55\% \text{ mutations for } 1500\text{ R}$$

Among 1000 progeny at 6000 R we expect $1000(6000/1500)(0.0455) = 182$ mutants.

**14.7.** Four single mutant strains of *Neurospora* are unable to grow on minimal medium unless supplemented by one or more of the substances A through F. In the following table, growth is indicated by + and no growth by 0. Both strains 2 and 4 grow if E and F or C and F are added to minimal medium. Diagram a biochemical pathway consistent with the data involving all six metabolites, indicating where the mutant block occurs in each of the four strains.

| Strain | A | B | C | D | E | F |
|---|---|---|---|---|---|---|
| 1 | + | 0 | 0 | 0 | 0 | 0 |
| 2 | + | 0 | 0 | + | 0 | 0 |
| 3 | + | 0 | + | 0 | 0 | 0 |
| 4 | + | 0 | 0 | 0 | 0 | 0 |

**Solution:**

Strain 1 will grow only if given substance A. Therefore the defective enzyme produced by the mutant gene in this strain must act sometime prior to the formation of substance A and after the formation of substances B, C, D, E and F. In other words, this mutation is probably causing a metabolic block in the last step of the biochemical sequence in the synthesis of substance A.

(B, C, D, E, F order unknown) —1→ A

Strain 2 grows if supplemented by either A or D and therefore its metabolic block must act somewhere earlier in the pathway than either A or D. Since the dual addition of substances E and F or C and F can also cause strain 2 to grow, we can infer that D may be split into two fractions E and F or C and F. The mutation in strain 2 is probably unable to convert substance B into substance D.

B —2→ D → (order unknown E, C) and → F; both —1→ A

Strain 3 grows if supplemented with A or C and therefore its metabolic block must precede the formation of C but not E.

B —2→ D → E —3→ C and D → F; both —1→ A

Strain 4 can grow if given dual supplementation of E and F or C and F but not if given D alone. The mutation in strain 4 apparently cannot split D into E and F.

B —2→ D —4→ (E —3→ C) and F; both —1→ A

**14.8.** The two hypothetical biosynthetic pathways shown below are available to an organism. At any one time, only product C or Z is being made. When the intermediate substance B reaches a threshold concentration, it inhibits enzyme 3 ($E_3$) by allosteric transition. Similarly substance Y inhibits $E_1$. (*a*) Propose some factors which might determine how one or the other reaction sequence could be "turned on" initially. (*b*) How is this system different from the operon models?

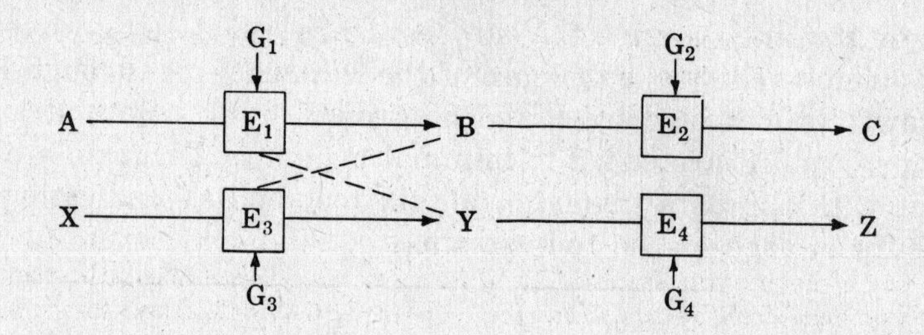


**Solution:**

(*a*) During the normal developmental process, substance A might become available to the organism before substance X, in which case only product C will be made. On the other hand, product B might later become a metabolite in some other reaction sequence thereby lowering its concentration below the threshold level necessary to inhibit $E_3$. In this case product Z would be made and synthesis of B and C would terminate. If the organism was making product Z normally and then came into an environment where substance B was present in relatively high concentration, synthesis of Y and Z would terminate and that of B and C would commence. Perhaps the enzymes of one sequence would be more sensitive to variations in temperature than those of the other sequence. Perhaps at high temperatures $E_1$ would become inactivated but $E_3$ would still be functional. The phenotype could thus be modified through manipulation of the environment, either by controlling the temperature or by the concentration of intermediates in one of the reaction sequences.

(*b*) Operon circuits are presumed to function at the "chromosomal" level and involve the sensitivity of an operator locus to the product of a regulator gene (repressor substance). In the model proposed in this problem, all of the interactions are removed from the gene; an intermediate metabolite in one reaction sequence directly inhibits the enzymatic activity in another sequence.

# Supplementary Problems

## DNA AND PROTEIN SYNTHESIS

**14.9.** Given a single strand of DNA ...$^{3'}$TACCGAGTAC$^{5'}$..., construct (*a*) the complementary DNA chain, (*b*) the mRNA chain which would be made from this strand.

**14.10.** If the ratio (A + G)/(T + C) in one strand of DNA is 0.7, what is the same ratio in the complementary strand?

**14.11.** How many different mRNA's could specify the amino acid sequence met-phe-ser-pro?

**14.12.** If the DNA of a species has the mole fraction of G + C = 0.36, calculate the mole fraction of A.

**14.13.** The size of a hemoglobin gene in humans is estimated to consist of approximately 450 nucleotide pairs. The protein product of the gene is estimated to consist of about 150 amino acid residues. Estimate the size of the codon.

**14.14.** Using the information in Table 14.1, page 292, convert the following mRNA segments into their polypeptide equivalents: (*a*) ...$^{5'}$ GAAAUGGCAGUUUAC $^{3'}$..., (*b*) ...$^{3'}$ UUUUCGAGAUGUCAA $^{5'}$..., (*c*) ...$^{5'}$ AAAACCUAGAACCCA $^{3'}$....

**14.15.** Given the hypothetical enzyme at the right with regions A, B, C and D (• = disulfide bond; shaded area = active site), explain the effect of each of the following mutations in terms of the biological activity of the mutant enzyme: (*a*) nonsense in DNA coding for region A, (*b*) samesense in region D, (*c*) deletion of one complete codon in region C, (*d*) missense in region B, (*e*) nucleotide addition in region C.

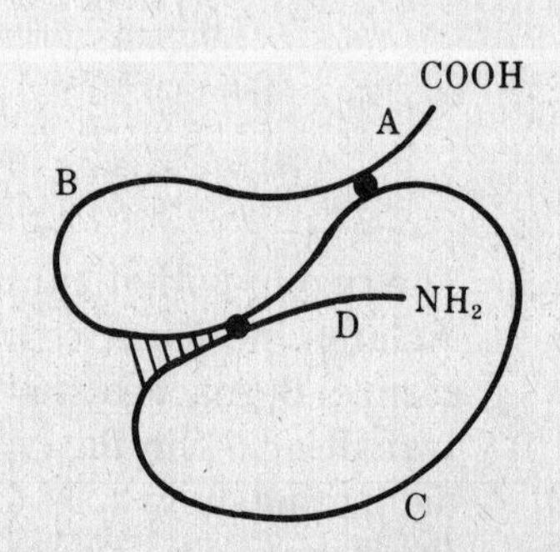


**14.16.** A large dose of ultraviolet irradiation can kill a wild type cell even if the DNA repair system is unsaturated. Under what circumstances would this lethality likely occur?

**14.17.** In a system containing 70% U and 30% C and an enzyme that links ribotides at random into a synthetic mRNA, determine the relative frequencies with which all possible triplet codons would be expected to be formed.

**14.18.** Acridine dyes can apparently cause a mutation in the bacteriophage T4 by the addition or subtraction (deletion) of a base in the DNA chain. A number of such mutants have been found in the $r_{II}$ region of T4 to be single base addition type (+) or deletion type (−) mutants. A normal or wild type $r_{II}$ region produces a normal lytic period (small plaque size) in the host bacteriuim *E. coli* strain B. Phage T4 mutant in the $r_{II}$ region rapidly lyses strain B, producing a larger plaque. Several multiple mutant strains of T4 have been developed. Suppose the mutant sites are close together and also in a region of the chromosome which is not essential to the normal functioning of the protein specified by this gene. Determine the lytic phenotypes (large plaque or small plaque) produced by the following rII single base mutations in *E. coli* B assuming a triplet codon: (*a*) (+), (*b*) (+)(−), (*c*) (+)(+), (*d*) (−)(−), (*e*) (+)(−)(+), (*f*) (+)(+)(+), (*g*) (−)(−)(+), (*h*) (−)(−)(−), (*i*) (+)(+)(+)(+), (*j*) (+)(−)(+)(+)(+).

**14.19.** A single base addition and a single base deletion approximately 15 bases apart in the mRNA specifying the protein lysozyme from the bacterial virus T4 caused a change in the protein from its normal composition ...lys-ser-pro-ser-leu-asn-ala-ala-lys... to the abnormal form ...lys-val-his-his-leu-met-ala-ala-lys.... (*a*) From the mRNA codons listed in Table 14.1, decipher the segment of mRNA for both the original protein and the double mutant. (*b*) Which base was added? Which was deleted?

**14.20.** In polynucleotides containing uracil and cytosine produced from a mixture of ribotides where uracil is in excess, more serine is incorporated into polypeptides than proline. However, when cytosine is in excess more proline is incorporated than serine. (*a*) Without reference to Table 14.1, list the codons containing uracil and cytosine which could possibly be coding for serine and for proline. (*b*) Using Table 14.1, page 292, determine the percentage of the various amino acids expected to be incorporated into polypeptides by a synthetic polynucleotide in which uracil constitutes 60% and cytosine 40% of the U-C mixture. (*c*) How much more phenylalanine is expected to be incorporated into protein than proline?

**14.21.** If the DNA of an *E. coli* has $4.2 \times 10^6$ nucleotide pairs in its DNA, and if an average cistron contains 1500 nucleotide pairs, how many cistrons does it possess?

**14.22.** The following experiment was performed: a short pulse of radioactive isotopes of the 20 amino acids is given to rabbit reticulocytes as they are synthesizing hemoglobin. The introduction of radioactive label occurs when some hemoglobin molecules are partly completed but unlabeled. Shortly after the pulse, completely finished hemoglobin molecules are isolated and analyzed for the location of the radioactive label. Where would you expect to find the label, and why?

**14.23.** The DNA of phage *lambda* has $1.2 \times 10^5$ nucleotides. How many proteins of molecular weight 40,000 could be coded by this DNA? Assume a molecular weight of 100 for the average amino acid.

**14.24.** In phage T4, deletion 1589 lacks part of the A and B cistrons of the $r_{II}$ region and shows no A but partial B activity in *E. coli* strain B. When a single defect (deletion or addition) is induced in the A cistron by acridine, the B cistron activity is suppressed. What hypothesis is supported by these observations?

## MUTATIONS

**14.25.** The "dotted" gene in maize (*Dt*) is a "mutator" gene influencing the rate at which the gene for colorless aleurone (*a*) mutates to its dominant allele (*A*) for colored aleurone. An average of 7.2 colored dots (mutations) per kernel was observed when the seed parent was *dt*/*dt*, *a*/*a* and the pollen parent was *Dt*/*Dt*, *a*/*a*. An average of 22.2 dots per kernel was observed in the reciprocal cross. How can these results be explained?

**14.26.** Assuming no intensity effect is operative, which individual would carry fewer mutations; an individual who receives 25 roentgens in 5 hours or an individual who receives only 0.5 roentgen per year for his or her normal lifetime (60 years)? In terms of percentage, how many more mutations would be expected in the individual with the higher total dosage?

**14.27.** If the mutation rate of a certain gene is directly proportional to the radiation dosage and the mutation rate of *Drosophila* is observed to increase from 3% at 1000 R to 6% at 2000 R, what percentage of mutations would be expected at 3500 R?

**14.28.** The frequency of spontaneous mutation at the *R* (plant color) locus in maize is very high (492 per $10^6$ gametes). The gene for red aleurone (*Pr*) is estimated to mutate in 11 out of $10^6$ gametes. How many plants must be investigated on the basis of probability to find one with mutations at both loci?

**14.29.** A strain of *Drosophila* called Muller-5 contains the dominant sex-linked mutation for bar eye ($B$) and the apricot allele of the sex-linked white eye locus ($w^{ap}$) together with an inversion to prevent crossing over between them. Homozygous bar-apricot females are crossed to X-irradiated wild type males. Each $F_1$ female is placed in a separate vial together with one or two unirradiated wild type males. Upon inspection of the $F_2$, what criterion can be used for scoring induced sex-linked lethals?

**14.30.** Single wild type female *Drosophila* possessing attached-X chromosomes ($\overset{\wedge}{XX}Y$) is mated to an irradiated wild type male. (*a*) In which generation can a sex-linked mutation be detected? (*b*) What criterion is used to detect a sex-linked viable mutation? (*c*) What criterion is used to detect a sex-linked lethal mutation? (*d*) What percentage death loss is anticipated among the $F_1$ zygotes due to causes other than lethal mutations?

**14.31.** A method is known for rendering chromosome 2 of *Drosophila* homozygous so that all recessive mutants induced in that chromosome by irradiation can be detected. This method uses a balanced lethal system involving three genes on the second chromosome: curly wings ($Cy$), plum eye color ($Pm$) and lobe eye shape ($L$). A ($Cy + L/+ Pm +$) female fly is mated to an X-rayed wild type male. A curly, lobed $F_1$ male is then backcrossed to a curly, plum, lobed female. From the backcross progeny, select and intercross curly-lobe flies. Predict the phenotypic ratios among the offspring when (*a*) no mutant was induced, (*b*) a viable recessive mutant was induced, (*c*) a semilethal or subvital was induced, (*d*) a lethal was induced.

**14.32.** The number of sex-linked lethals in *Drosophila* as detected by the *ClB* and Muller-5 techniques (Problems 14.5 and 14.29) increases in direct proportion to the amount of radiation at low dosage levels. However, at high dosage levels the amount of detectable lethal mutations falls below the linear expectations. How can this phenomenon be explained?

**14.33.** If the production of chromosomal rearrangements such as inversions or reciprocal translocations requires two hits on chromosomes within the same nucleus, then one might expect the number of such rearrangements to increase more nearly as the square of the radiation dosage rather than in direct proportion to the dosage. If 100 R produces 2.5% single-hit mutations, what percentage of double-hits would be expected at 150 R (neglecting background mutations)?

**14.34.** Assume that the rate of induced mutations is directly proportional to the radiation dosage. Further suppose that 372 individuals out of 6000 incur a mutation at 2000 R, and that 610 out of 5000 individuals incur a mutation at 4000 R. Estimate the spontaneous mutation rate.

**14.35.** Concerning a single nucleotide pair, list all possible (*a*) transitions, (*b*) transversions. (*c*) If purines and pyrimidines become replaced at random during evolution, what ratio of transversions to transitions is expected? (*d*) A comparison of homologous residues in the polypeptide chains of hemoglobins and myoglobins from various species indicates that 293 transitions and 548 transversions have probably occurred during evolution. Are these figures consistent with the hypothesis that transitions and transversions occur in a 1 : 2 ratio respectively? Test the hypothesis by chi-square.

**14.36.** A number of nutritional mutant strains was isolated from wild type *Neurospora* which responded to the addition of certain supplements in the culture medium by growth (+) or no growth (0). Given the following responses for single gene mutants, diagram a metabolic pathway which could exist in the wild type strain consistent with the data, indicating where the chain is blocked in each mutant strain.

(*a*)

| Mutant Strain | Supplements added to minimal culture medium | | | | |
|---|---|---|---|---|---|
| | Citrulline | Glutamic semi-aldehyde | Arginine | Ornithine | Glutamic acid |
| 1 | + | 0 | + | 0 | 0 |
| 2 | + | + | + | + | 0 |
| 3 | + | 0 | + | + | 0 |
| 4 | 0 | 0 | + | 0 | 0 |

(b)

| Strain | Growth Factors | | | |
|---|---|---|---|---|
| | A | B | C | D |
| 1 | 0 | 0 | + | + |
| 2 | 0 | 0 | 0 | + |
| 3 | + | 0 | + | + |
| 4 | 0 | + | + | + |

(c)

| Strain | Nutrients | | | |
|---|---|---|---|---|
| | E | F | G | H |
| 1 | + | 0 | 0 | + |
| 2 | 0 | 0 | + | + |
| 3 | 0 | 0 | 0 | + |
| 4 | 0 | + | 0 | + |

**14.37.** Point mutations correlated with amino acids in the active site in widely separated regions of a cistron can render its enzymatic or antibody product inactive. What inference can be made concerning the structure of the active sites in such proteins?

**14.38.** A nonsense point mutation in one cistron can sometimes be at least partially suppressed in its phenotypic manifestation by a point mutation in a different gene. Offer an explanation for this phenomenon of intergenic suppression.

**14.39.** In addition to the kind of mechanism accounting for intergenic suppression of nonsense mutations (see previous problem), give two other possible mechanisms for intergenic suppression of missense mutations.

**14.40.** Intracistronic (interallelic) *in vitro* complementation has been observed in alkaline phosphatase enzymes and other proteins. How can a diploid heterozygote or a heterocaryon bearing two point mutations within homologous cistrons result in normal or nearly normal phenotypes (complementation)?

**14.41.** Why are most mutations in structural genes recessive to their wild type alleles?

**14.42.** Forward mutation rates are usually at least an order of magnitude higher than back mutation rates for a given cistron. How can this be explained?

**14.43.** Bacterial cells that are sensitive to the antibiotic streptomycin ($Str^s$) can mutate to a resistant state ($Str^r$). Such "change of function" mutations, however, occur much less frequently than auxotrophic or sugar-fermenting "loss of function" mutations (e.g. $His^+ \rightarrow His^-$; $Lac^+ \rightarrow Lac^-$). Formulate a hypothesis that explains these observations.

## REGULATION OF GENE ACTIVITY

**14.44.** In addition to the $i^+$ allele, producing repressor for the lactose system in *E. coli* and the constitutive $i^-$ allele, a third allele $i^s$ has been found, the product of which is unable to combine with the inducer (lactose). Hence the repressor ("superrepressor") made by $i^s$ remains unbound and free to influence the operator locus. In addition to the repressor-sensitive $o^+$ operator and its constitutive $o^c$ allele, an inactive or "null" allele $o^o$ is known which permanently turns off or inactivates the operon under both induced and non-induced conditions. (*a*) Order the three alleles of the *i*-locus in descending order of dominance according to their ability to influence the lactose operator. (*b*) Order the three alleles of the *o*-locus in descending order of dominance according to their ability to "turn on" their operon. (*c*) Using + for production and 0 for non-production of the enzymes permease (P) and $\beta$-galactosidase ($\beta$-gal), complete the following table.

| Genotype | Inducer Absent | | Inducer Present | |
|---|---|---|---|---|
| | P | $\beta$-gal | P | $\beta$-gal |
| (1) $i^+o^+y^+z^+$ | | | | |
| (2) $i^-o^+y^+z^+$ | | | | |
| (3) $i^so^+y^+z^+$ | | | | |
| (4) $i^+o^cy^+z^+$ | | | | |
| (5) $i^+o^oy^+z^+$ | | | | |
| (6) $i^-o^cy^+z^+$ | | | | |
| (7) $i^-o^oy^+z^+$ | | | | |
| (8) $i^so^cy^+z^+$ | | | | |
| (9) $i^so^oy^+z^+$ | | | | |

**14.45.** For each of the following partial diploids, determine if enzyme formation is constitutive or inductive: (*a*) $i^+/i^+, o^+/o^+$, (*b*) $i^+/i^+, o^+/o^c$, (*c*) $i^+/i^+, o^c/o^c$, (*d*) $i^+/i^-, o^+/o^+$, (*e*) $i^-/i^-, o^+/o^+$.

**14.46.** In the lactose system of *E. coli*, $y^+$ makes permease, an enzyme essential for the rapid transportation of galactosides from the medium to the interior of the cell. Its allele $y^-$ makes no permease. The galactoside lactose must enter the cell in order to induce the $z^+$ gene to produce the enzyme $\beta$-galactosidase. The allele $z^-$ makes a related but enzymatically inactive protein called Cz. Predict the production or non-production of each of these products with a normal operator $o^+$ by placing a + or 0 respectively in the table below.

| Genotype | Inducer Absent | | | Inducer Present | | |
|---|---|---|---|---|---|---|
| | P | $\beta$-gal | Cz | P | $\beta$-gal | Cz |
| (*a*) $i^+y^+z^-$ | | | | | | |
| (*b*) $i^+y^+z^+$ | | | | | | |
| (*c*) $i^-y^+z^+$ | | | | | | |
| (*d*) $i^+y^-z^+$ | | | | | | |
| (*e*) $i^-y^-z^-$ | | | | | | |
| (*f*) $i^-y^-z^+$ | | | | | | |

**14.47.** The $i^-$ allele allows constitutive enzyme production by the $y^+$ and $z^+$ factors in an operon with a normal operator gene $o^+$. The action of $i^-$ might be explained by one of two alternatives: (1) $i^-$ produces an internal *inducer*, thus eliminating the need for lactose in the medium to induce enzyme synthesis; $i^+$ produces no inducer or (2) $i^+$ produces a *repressor* substance which, in the absence of lactose inducer, blocks enzyme formation, but in the presence of lactose inducer becomes inactivated and allows enzyme synthesis; $i^-$ produces no repressor. (*a*) Assuming dominance of the $i^-$ allele under the first alternative in an *E. coli* partial heterozygote of the constitution $i^+/i^-$, would internal inducer be produced? (*b*) Would enzymes be produced constitutively or inductively? (*c*) Assuming dominance of the $i^+$ allele under the second alternative in a partial heterozygote of the constitution $i^+/i^-$, would repressor be produced? (*d*) Would enzymes be produced constitutively or inductively? (*e*) From the pattern of reactions exhibited by different heterozygotes in the following table, determine which of the two assumptions is correct.

| Genotypes | Lactose Absent | | Lactose Present | |
|---|---|---|---|---|
| | P | $\beta$-gal | P | $\beta$-gal |
| (1) $o^+y^-z^-i^-/o^+y^+z^+i^-$ | + | + | + | + |
| (2) $o^+y^+z^+i^-/o^+y^-z^-i^-$ | + | + | + | + |
| (3) $o^+y^-z^-i^-/o^+y^+z^+i^+$ | 0 | 0 | + | + |
| (4) $o^+y^-z^-i^+/o^+y^+z^+i^-$ | 0 | 0 | + | + |

(*f*) Is the repressor substance diffusible, or can it only act on loci in cis position with the i-locus? How can this be determined from the information in the table of part (*e*)?

**14.48.** The repressor for an inducible operon has two binding sites. (*a*) What are the specificities of these two sites? (*b*) List four kinds of single mutations that change the functions of such a repressor.

**14.49.** List three kinds of single mutations that can change the function of an operator.

**14.50.** The entry of lactose into a bacterial cell is mediated by a permease enzyme. In cells that have not previously been exposed to lactose, how can lactose enter an uninduced $i^+z^+y^+$ cell to affect induction of β-galactosidase synthesis?

**14.51.** A bacterial mutation causes a cell to be incapable of fermenting many sugars (e.g. lactose, sorbitol, xylose) simultaneously. The operons of genes specifying the respective catabolic enzymes are wild type (unmutated). Offer an explanation for this phenomenon.

**14.52.** The enzymes necessary for glucose catabolism are made constitutively by bacterial cells. When both glucose and lactose are added to the growth medium, glucose enters the cell by its own permease molecules embedded in the cell membrane. The operons for catabolizing lactose and other sugars fail to be activated even though a few respective permease molecules for these other sugars are normally present in the cell membrane. Explain. *Hint:* See Example 14.13.

**14.53.** Shown below is a hypothetical biosynthetic pathway subject to feedback inhibition; letters represent metabolites; numbers represent enzymes. Identify the enzymes that are most likely to be subject to feedback inhibition and their inhibitor(s). *Note:* The inhibitor may consist of more than one metabolite.

$$A \xrightarrow{1} B \xrightarrow{2} C; \quad C \xrightarrow{3} D \xrightarrow{4} E; \quad C \xrightarrow{5} F \xrightarrow{6} G \xrightarrow{7} H \xrightarrow{8} I; \quad G \xrightarrow{9} J$$

**14.54.** An antibiotic is a microbial product of low molecular weight that specifically interferes with the growth of microorganisms when it is present in exceedingly small amounts. Specify some of the physiological activities that might be interrupted by an appropriate antibiotic and the reason why human cells are not harmed.

**14.55.** Give four reasons why knowledge of the genetic code cannot by itself allow construction of the eukaryotic gene from the amino acid sequence of its protein product.

# Answers to Supplementary Problems

**14.9.** (*a*) $^{5'}$ATGGCTCATG$^{3'}$ (*b*) $^{5'}$AUGGCUCAUG$^{3'}$

**14.10.** 1.43

**14.11.** $1 \times 2 \times 6 \times 4 = 48$

**14.12.** 0.32

**14.13.** Approximately 3 nucleotides code for each amino acid

**14.14.** (*a*) - glu - met - ala - val - tyr - (*b*) - phe - ala - arg - cys - asn - (*c*) - lys - thr - (nonsense), chain terminates prematurely

**14.15.** (*a*) The protein would be slightly shorter than normal. Since region A does not seem to interact with other portions of the polypeptide chain, the mutant enzyme should still function normally (barring unpredicted interaction of the side chain of the mutant amino acid with other parts of the molecule). If a nonsense mutation had occurred in region D, however, a very small chain would have been produced that would be devoid of a catalytic site, because proteins are synthesized beginning at the $NH_2$ end. (*b*) Samesense mutants produce no change in their polypeptide products from normal. (*c*) The polypeptide would be one amino acid shorter than normal. Since region C does not seem to be critical to the tertiary shape of the molecule, the mutant enzyme would probably function normally. (*d*) An incorrect amino acid would be present in region B. As long as its side chain did not alter the tertiary shape of the molecule, the mutant enzyme would be expected to functional normally. (*e*) A frameshift mutant in region C is bound to create many missense codons (or perhaps a nonsense codon) from that point on through the carboxyl terminus, including the enzymatic site. Such a protein would be catalytically active.

**14.16.** Nonfunctional DNA fragments might be produced if DNA replication occurs before all of the critical repairs have been made.

**14.17.** $3U = 0.343$, $2U + 1C = 0.441$, $1U + 2C = 0.189$, $3C = 0.027$

**14.18.** (*a*), (*c*), (*d*), (*e*), (*g*), (*i*) = large; (*b*), (*f*), (*h*), (*j*) = small

**14.19.** (*a*)

| | lys | ser | pro | ser | leu | asn | ala |
|---|---|---|---|---|---|---|---|
| normal mRNA | AA? | A GU | CCA | UCA | CUU | AAU | GC? |
| mutant mRNA | AA? | GUC | CAU | CAC | UUA | AUG | G C? |
| | lys | val | his | his | leu | met | ala |

(*b*) G was added, A was deleted (at shaded positions)

**14.20.** (*a*) serine: UUC, UCU, CUU; proline: CCU, CUC, UCC. (*b*) phenylalanine = 0.36, leucine = 0.24, proline = 0.16, serine = 0.24. (*c*) $2\frac{1}{4}$

**14.21.** 2800 cistrons

**14.22.** Most molecules would be labeled more in the COOH end because synthesis is unidirectional starting at the $NH_2$ end.

**14.23.** 50 proteins, assuming only one chain of the DNA is transcribed into mRNA.

**14.24.** Codons of the $r_{II}$ region are read unidirectionally beginning at the A end.

**14.25.** Seed parent contributes two sets of chromosomes to triploid endosperm; one *Dt* gene gives 7.2 mutations/kernel, two *Dt* genes increase mutations to 22.2/kernel

**14.26.** 20% more mutations in the individual receiving 0.5 R/yr.

**14.27.** $10\frac{1}{2}\%$

**14.28.** $1.848 \times 10^8$ (approx.)

**14.29.** Any vial which does not contain wild type male progeny is the result of a sex-linked lethal induced by X-ray treatment.

**14.30.** (*a*) $F_1$ (*b*) The appearance of the mutant trait in all $F_1$ males in a vial. (*c*) The absence of $F_1$ males in any vial. (*d*) 50%; superfemales $(\widehat{XXX})$ are usually inviable; all zygotes without an X chromosome die.

**14.31.** (*a*) 2/3 curly, lobe : 1/3 wild type (*b*) 2/3 curly, lobe : 1/3 recessive mutant (*c*) greater than 2/3 curly, lobe : less than 1/3 wild type (*d*) all curly, lobe

**14.32.** A single ionization ("hit") of a genetic element may destroy the functioning of a vital gene and result in death. Multiple hits at high dosage levels score as a single lethal event.

**14.33.** 0.140625%

**14.34.** 0.2%

**14.35.** (*a*)

| Original DNA | A : T | T : A | G : C | C : G |
|---|---|---|---|---|
| Transition DNA | G : C | C : G | A : T | T : A |

(*b*)

| Original DNA | A : T | A : T | T : A | T : A | G : C | G : C | C : G | C : G |
|---|---|---|---|---|---|---|---|---|
| Transversion DNA | T : A | C : G | A : T | G : C | T : A | C : G | A : T | G : C |

(*c*) 2 transversions : 1 transition

(*d*) Yes. $\chi^2 = 0.86$, $p = 0.3$–$0.5$

**14.36.** (*a*) GA ⟶ GSA ⟶ O ⟶ C ⟶ A (blocked by 2, 3, 1, 4 respectively)

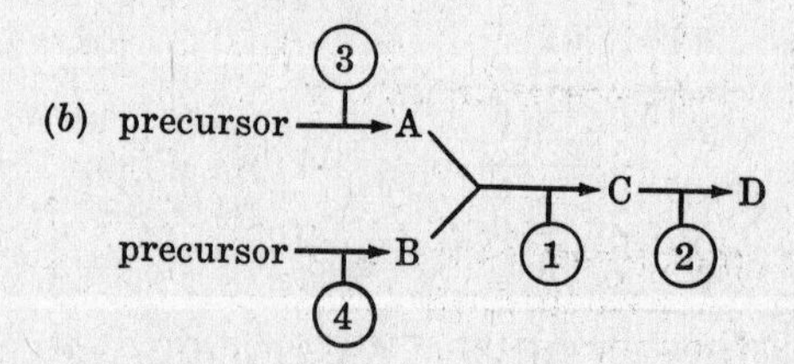


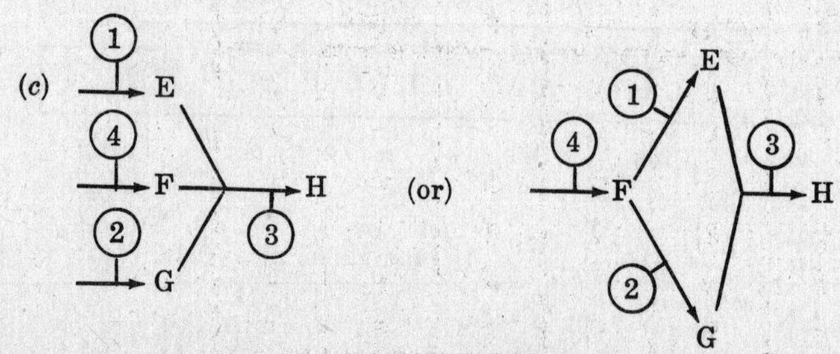

**14.37.** The polypeptide chain folds into a configuration such that noncontiguous regions form portions of the catalytic or antibody-combining sites.

**14.38.** The suppressing mutation could be in that portion of a gene specifying the anticodon region of a tRNA molecule. For example, a tyrosine suppressor gene changes the anticodon of $tRNA_{tyr}$ from $^{3'}AUG^{5'}$ to $^{3'}AUC^{5'}$ thereby allowing it to recognize UAG mRNA nonsense codons. If the genes for $tRNA_{tyr}$ exists in multiple copies and only one of the $tRNA_{tyr}$ genes was mutated to a suppressor form, there would still be other normal (non-suppressor) $tRNA_{tyr}$ genes to make some normal proteins. The efficiency of suppression must be low to be compatible with survival of the organism.

**14.39.** (1) A change in one of the ribosomal proteins in the 30S subunit could cause misreading of the codon-anticodon alignment, resulting in substitution of an "acceptable" (although perhaps not the normal) amino acid in a manner analogous to the misreading induced by the antibiotic streptomycin. In a cell-free system with synthetic poly-U mRNA, streptomycin causes isoleucine tRNA to be substituted for that of phenylalanine tRNA. (2) A mutation in a gene coding for an amino-acid-activating enzyme (amino-acyl synthetase) causes a different amino acid to occasionally be attached to a given species of tRNA. For example, if AUU (isoleucine mRNA codon) is mutated to UUU (phenylalanine mRNA codon), its effect may be suppressed by the occasional misattachment of isoleucine to $tRNA_{phe}$ by a mutant amino-acyl synthetase that is less than 100% specific in its normal action of attaching phenylalanine to $tRNA_{phe}$.

**14.40.** Such proteins are normally homopolymers (quaternary complexes consisting of two or more identical polypeptide chains). If two mutant polypeptide chains contain compensating amino acid substitutions, they may aggregate into a heterodimer that exhibits at least partial enzymatic activity.

**14.41.** Wild type alleles usually code for complete, functional enzymes or other proteins. One active wild type allele can often cause enough enzyme to be produced so that normal or nearly normal phenotypes result (dominance). Mutations of normally functioning genes are more likely to destroy the biological activities of proteins. Only in the complete absence of the wild type gene product would the mutant phenotype be expressed (recessiveness).

**14.42.** A cistron contains numerous mutons, many of which if altered would destroy the biological activity of the gene product. Once a point mutation has occurred it usually requires a very specific back mutation of that same nucleotide to restore normal or wild type gene activity.

**14.43.** Loss of function can potentially occur by point mutations at a number of sites within a cistron coding for a given fermentation enzyme or in any gene coding for one of the multiple enzymes in a common biosynthetic pathway such as those in histidine synthesis. The loss of such an indispensable function is lethal. Streptomycin distorts ribosomes, causing misreading of the genetic code. Only a limited number of changes in the ribosomal proteins or rRNA could render the ribosome immune from interference by streptomycin and still preserve the way these components normally interact with mRNA, tRNA, initiation factors, etc. during protein synthesis.

**14.44.** (*a*) $i^s, i^+, i^-$ (*b*) $o^c, o^+, o^o$
(*c*)

| | Inducer Absent | | Inducer Present | |
|---|---|---|---|---|
| | P | β-gal | P | β-gal |
| (1) | 0 | 0 | + | + |
| (2, 4, 6, 8) | + | + | + | + |
| (3, 5, 7, 9) | 0 | 0 | 0 | 0 |

**14.45.** (*a*), (*d*) = inductive; (*b*), (*c*), (*e*) = constitutive

**14.46.**

| | Inducer Absent | | | Inducer Present | | |
|---|---|---|---|---|---|---|
| | P | β-gal | Cz | P | β-gal | Cz |
| (*a*) | 0 | 0 | 0 | + | 0 | + |
| (*b*) | 0 | 0 | 0 | + | + | 0 |
| (*c*) | + | + | 0 | + | + | 0 |
| (*d*) | 0 | 0 | 0 | 0 | + | 0 |
| (*e*) | 0 | 0 | + | 0 | 0 | + |
| (*f*) | 0 | + | 0 | 0 | + | 0 |

**14.47.** (*a*) Yes (*b*) Constitutively (*c*) Yes (*d*) Inductively (*e*) Note that in (3) and (4) the $i^-$ allele fails to produce enzymes in the absence of the external inducer (lactose). It fails to exhibit dominance. Therefore the first assumption is incorrect. Under the second assumption, $i^+$ is dominant and produces repressor which, in the absence of lactose inducer, blocks enzyme synthesis as seen in (3) and (4). (*f*) Heterozygote (4) has $y^+$ and $z^+$ on one DNA segment and the $i^+$ allele on the other. Yet in the absence of lactose inducer, the repressor made by $i^+$ still prevents the production of enzymes by $y^+$ and $z^+$. Therefore repressor must be able to act at a distance (diffusible).

**14.48.** (*a*) One site is for the operator locus; the other site is for the inducer. (*b*) (1) Inactivation of the operator binding site causes the repressor to be incapable of binding to the operator. (2) Increasing the operator binding affinity results in permanent repressor binding even in the presence of inducer. (3) Inactivation of the inducer binding site makes derepession impossible. (4) Increasing the inducer binding affinity results in permanent allosteric change, and once the inducer is bound the repressor can never bind to an operator.

**14.49.** (1) A change that prevents repressor binding. (2) Modifications that increase repressor binding so that operons cannot be derepressed even when inducer is bound. (3) A conformational change (allosteric transformation) occurs in a normal repressor when it binds to the mutant operator so the bound repressor cannot recognize the inducer.

**14.50.** Occasionally a repressor molecule will momentarily become dissociated from the operator and RNA polymerase will attach and begin transcription of the β-galactosidase and permease genes before the repressor reattaches. This so-called "sneak synthesis" endows the cell with enough enzymes to transport a few lactose molecules through the plasma membrane; these will be catabolized to the true inducer (allo-lactose) so that derepression can occur.

**14.51.** The mutation could be in the gene for adenyl cyclase or in the gene for catabolite activator protein (CAP).

**14.52.** Glucose enters the cell via its own permease and is metabolized. One or more glucose metabolites somehow decrease the intracellular level of cAMP. In the absence of cAMP, CAP cannot bind to the left end of the lactose (or other sugar) promoter(s) and hence the binding of RNA polymerase to the right end of the promoter(s) is not fostered. Some lactose gains entry into the cell via its permease and inactivates the lactose represssor so that the lactose operator locus is open. However, little or no mRNA's are made from such operons if RNA polymerase binding is inefficient at the respective promoter site(s).

**14.53.** I inhibits 7; J inhibits 9; G alone or (I, J) together inhibits 5; E inhibits 3; enzyme 1 could be inhibited by (I, J, E), (G, E), (I, J, C), or (C, G).

**14.54.** Cell wall formation is interfered with by penicillins and cephalosporins. DNA replication is prevented by the bleomycins and anthracyclines. Rifamycins interrupt the transcription of DNA into RNA. Translation is disrupted by erythromycin, the tetracyclines, chloramphenicol and streptomycin. Antibiotics are toxic for microorganisms but safe for humans because all of these metabolic processes are subtly different in bacteria and humans.

**14.55.** (1) The code is degenerate; most amino acids can be specified by two or more codons. (2) Introns are enzymatically removed from the primary mRNA transcript before being translated. (3) Some segments of a protein may be enzymatically removed to convert it into the biologically active form from which its primary structure was determined. (4) The promoter region and other control sites of the gene that are essential for its functioning in the cell are not translated into protein structure.

# Chapter 15

## Genetics of Bacteria and Viruses

### BACTERIA

#### 1. Characteristics of Prokaryotes.

There are two fundamentally different kinds of cellular organisms in the biological world. The cells of protozoans (single-celled animals), fungi (non-photosynthesizing plants), plants and animals are called *eukaryotic* ("truly nucleated") cells. Bacteria and blue-green algae (Cyanophyta: "cyanobacteria") are more primitive cell types described as *prokaryotic* ("before nuclei"). There are more fundamental differences between a bacterial cell and the simplest eukaryotic cell than between a protozoan and a human cell (Table 15.1).

#### 2. Trophic (Nutritional) Categories of Bacteria.

Most bacteria obtain energy by absorbing and degrading organic nutrients from the living or dead tissues of other organisms or from their secretions or wastes. A few kinds of photosynthetic bacteria contain bacteriochlorophyll and can utilize light energy for producing their own organic molecules from inorganic sources. Chemosynthetic bacteria can utilize the energy released from oxidizing inorganic chemicals (e.g. converting nitrate to nitrite, sulfide to sulfate, or ferrous iron to ferric iron) for producing their own organic molecules from inorganic or organic sources. Blue-green algae contain the same kind of chlorophyll found in green eukaryotic plants and photosynthesize in the same way, releasing oxygen as a byproduct (no oxygen is evolved by bacterial photosynthesis).

Organisms that can subsist on inorganic substances by either photosynthesis or chemosynthesis are said to be *autotrophic* ("self-nourished"). All other organisms that require organic nutrients are *heterotrophic* ("other nourished"). Most bacteria require some organic compound (such as sugar) for their carbon and energy sources. Wild type (*prototrophic*) strains are able to grow on a very simple diet. A *minimal medium* for many heterotrophic bacteria includes water, a few simple inorganic salts and an organic compound as a source of carbon and chemical energy. Mutant strains of bacteria that require supplementation of the minimal medium for growth are said to be *auxotrophic*.

The generation time for the colon bacillus *Escherichia coli* growing on minimal medium is about 60 minutes at 37°C. If the medium is supplemented with all of the amino acids and nucletoides normally made by the bacteria (*complete medium*), the generation time is reduced to 20 minutes. When nutritional conditions are good, each bacterial cell typically contains between two and four genetically identical chromosomes. When nutritional conditions are poor, chromosome duplication lags behind cell division so that viable cells contain only one chromosome.

#### 3. Phenotypic Variations.

Bacteria are barely visible in the most powerful light microscopes; they are about the size of mitochondria in eukaryotic cells. Their individual gross morphologies are classified as either spherical (cocci), rod-shaped (bacilli), or helical (spirilla). Internal structures of bacteria can only be revealed by the electron microscope. All bacteria have a cell membrane

**Table 15.1. Some Distinctive Differences between Eukaryotes and Prokaryotes**

| Eukaryotes | Prokaryotes |
|---|---|
| (1) Nucleus, mitochondria, chloroplasts, lysosomes, etc. | (1) Membrane-bound organelles absent |
| (2) Elaborate endoplasmic reticulum (not associated with DNA replication) | (2) Few internal membranes (e.g. mesosome associated with DNA replication) |
| (3) Five kinds of histones are tightly bound to DNA | (3) Few proteins (mostly enzymes) are loosely bound to DNA; no histones |
| (4) Even the smallest cell is 5-10 times larger than *Escherichia coli* | (4) Appoximately the size of eukaryotic mitochondria |
| (5) Actin and myosin proteins | (5) Actin or myosinlike proteins absent |
| (6) Microtubules (e.g. spindle fibers subject to degradation by colchicine) | (6) No microtubules |
| (7) Mitosis and meiosis used for asexual and sexual reproduction | (7) Nucleoprotein chromosomes absent; amitotic division (binary fission) |
| (8) Usually more than one linear DNA molecule (chromosome) in genome | (8) Single circular DNA molecule ("chromosome") sufficient for life; no centromeres or kinetochores |
| (9) 80S ribosomes (40S and 60S subunits) | (9) 70S ribosomes (30S and 50S subunits) |
| (10) Monocystronic mRNA's only | (10) Polycystronic mRNA's common |
| (11) Three major forms of RNA polymerase (one each for rRNA, mRNA and tRNA) | (11) Only one kind of RNA polymerase |
| (12) One or more nucleoli (tandomly linked DNA copies of genes coding for rRNA) | (12) No nucleolus counterpart |
| (13) Repetitive genes common (sometimes in thousands of copies) | (13) Each gene present only once |
| (14) Introns and exons common | (14) No intervening sequences |
| (15) DNA replication initiated at many sites on a chromosome (e.g. Okazaki fragments) | (15) One initiation site for DNA replication |
| (16) Unformylated methionine initiates all polypeptides | (16) Formylated methionine initiates all polypeptides |
| (17) Most mRNA's contain long poly-A sequences at 3′ ends (function unknown); not coded by nuclear DNA; added by poly-A polymerase | (17) Poly-A at 3′ ends of mRNA's rare |
| (18) A different protein is bound to each end of mRNA's (function unknown); virtually no naked mRNA's exist in the cell | (18) No end-bound proteins |
| (19) cap = 7′-methyl-guanosine-5′ppp5′ purine nucleoside (usually A) at 5′ end of mRNA's; not to be confused with CAP (catabolite gene activator protein) | (19) No cap |
| (20) Hogness box (TATA box) is the RNA polymerase II promoter analogous to the Pribnow box in prokaryotes | (20) Pribnow box (TATAATG) exists near the RNA start point of promoter regions |

and most bacteria possess an outer cell wall of lipids, proteins and polysaccharides. The cell walls of eukaryotic plants and prokaryotic blue-green algae contain cellulose, but cellulose is not found in bacterial cell walls. The DNA of bacteria is not complexed with histone proteins and therefore is not strictly comparable to a eukaryotic chromosome. Nonetheless, the "naked", circular DNA of a bacterium is called a "chromosome". There are no membrane-bound organelles (e.g. nuclei or mitochondria) in prokaryotes. However, the localized region of the bacterial DNA is referred to as the "nuclear region"; the area outside the bacterial nucleus is called the "cytoplasm".

In contrast to the morphologies of individual bacterial cells, the morphologies of bacterial colonies growing on a nutrient agar plate can easily be observed by the naked eye (macroscopically). Bacterial colonies may exhibit variations in size, shape or growth habit, texture, color, etc. Genetically different colonies or broth cultures may respond differentially to nutrients, dyes, drugs, antibodies or viral pathogens in the culture medium.

**4. Reproduction.**

Prokaryotes reproduce vegetatively by an asexual process called "binary fission" ("splitting in two"). All of the asexually derived descendants of a parent bacterium constitute a *clone* of genetically identical cells. At least two models have been proposed for replication of the circular bacterial chromosome. In the "rolling circle" model, a linear DNA molecule is produced by replication from a circle for genetic transfer from donor to recipient cells during bacterial mating (conjugation). A site-specific break occurs in one strand of the DNA circle, creating a free 3′-OH end and a free 5′-P end. The enzyme DNA polymerase adds nucleotides to the 3′ end by base pairing with the other DNA strand and thereby displaces a single-stranded tail ending in 5′-P. A complementary strand is then synthesized using the "tail" as a template. The circle may revolve several times, creating *concatemers* containing repetitive genome sequences. A nuclease enzyme cuts genome-sized segments containing "sticky ends" (single-stranded complementary regions) from the concatemer. The linear genomes circularize by base pairing of the sticky ends. DNA ligase seals each gap, resulting in a double-stranded DNA circle.

According to a second model, bacterial, mitochondrial and viral DNA circles may replicate without forming linear concatemers. Replication begins at one specific position, and two replication forks move in opposite directions around the circle. Some device must exist to allow free rotation for one of the strands; this device is referred to as a *molecular swivel.* The enzyme *DNA gyrase* is known to have nicking and sealing activities and may precede the replication forks as the swivel enzyme. Partway through the cycle, the two replicas have an intermediate structure resembling θ (Greek theta). This replication mechanism eventually creates supercoils of double-stranded circular DNA because the single-stranded nicks that allow untwisting are soon resealed. Chain growth proceeds until sterically hindered by supercoils. *Swivelase* ("relaxing protein") then makes a nick that relaxes the molecule and also repairs the cut so that replication can again resume.

The bacterial chromosome is attached to an invagination of the cell membrane, called the *mesosome*, at the two nearly adjacent replicating Y-forks. During replication of the chromosome, the cell elongates by growth of the sector between the two attachment points, causing the two chromosomal replicas to move apart. A septum of new cell wall material is then synthesized in this elongation region, creating two daughter cells. Even before the septum is completed, each chromosome replica may acquire a second Y-fork on the mesosome near the original point of attachment, and a second round of chromosome replication may have already begun. This asexual process repeats over and over again to produce the millions of cells that constitute a clone or colony visible to the naked eye on a nutrient agar plate.

**5. Bacterial Mutations.**

Bacteria are very useful for studying genetic mutations because of their short generation times and the ease with which very large populations ($10^6$ to $10^9$) of cells can be produced within a test tube or on an agar plate. For example, $10^9$ bacteria can be grown in liquid media (broth culture) and then distributed evenly over an agar plate so that a solid mass of cells, called a "lawn", covers the surface of the agar. If one mutant cell in a million is present in this lawn it can be detected by the use of selective media. For example, if a mutation to streptomycin resistance has occurred in a sensitive strain, we could plate the bacteria on agar containing streptomycin. All cells without the mutation would be sensitive to this antibiotic and unable to grow; only the streptomycin resistant mutant would form a colony. Furthermore, the spontaneous mutation to streptomycin resistance would be produced regardless of the presence or absence of streptomycin in the environment. Some kinds of spontaneous mutations are *preadaptive*, i.e. they would be advantageous to the organism in a different environment. This can be demonstrated by a technique called *replica plating*. A sterile piece of velvet is impressed onto a lawn of bacteria grown on agar without streptomycin (master plate) and transferred to any number of agar plates containing streptomycin (replica plates). The nap of the velvet picks up representatives of each clone from the master plate and inoculates them onto each replica plate. The position of the clones which grow on each replicate plate is always the same, indicating that the mutation to streptomycin resistance occurred in the master plate prior to any contact with streptomycin, i.e. the mutation is preadapted to surviving in a different environment, namely in the presence of streptomycin.

**Example 15.1.**

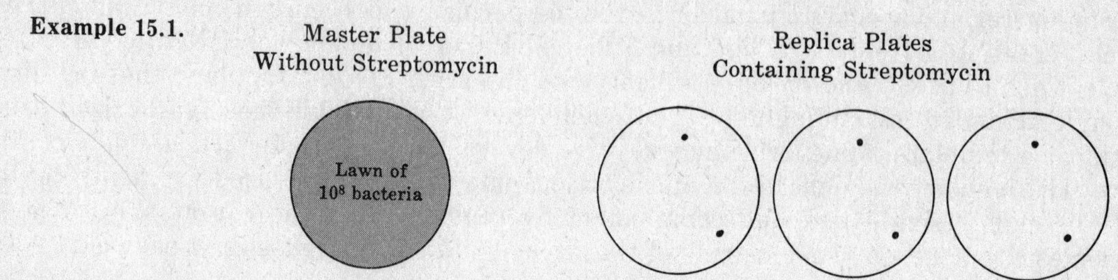


## 6. Genetic Recombination.

### (a) *Transformation.*

When pneumococci are surrounded by polysaccharide capsules, they are *virulent* (capable of producing disease) and form colonies with a smooth border on solid medium. Unencapsulated pneumococci are *avirulent* and plate as rough shaped colonies. When a rough strain is exposed to a purified DNA extract from a smooth strain, some rough cells are *transformed* into smooth cells. This is one line of evidence in the proof that DNA is genetic material. Transformation may occur in some species with frequencies as high as 25% but more commonly occurs at very low frequencies. Aging cultures of some bacteria spontaneously incur cell rupture (autolysis) which releases DNA into the environment with other cells, and thereby offers the opportunity for transformation to occur. A cell must be in a physiologically receptive state, called *competence*, which occurs only during a fraction of its growth cycle. During competence, DNA is transiently bound to the cell surface. Each cell has a number of sites at which DNA can attach. These receptor sites can be saturated with nontransforming DNA (e.g. DNA from an unrelated organism), thus preventing the appearance of transformants. Only relatively large DNA fragments can penetrate the competent cell, but once inside, the DNA is permanently bound. One explanation of the integration process assumes that DNA penetrates only at two sites corresponding to the two replication sites located internally on the cell membrane. As transforming DNA rotates past the stationary replication fork, the synthesis of host DNA is interrupted at the homologous region and transforming DNA becomes integrated into the host chromosome. Usually only a small segment of transforming DNA becomes integrated. Transforming DNA is never found free in the cytoplasm, and is capable of producing a phenotypic effect only after integration. The frequency of transformation increases with the concentration of DNA up to the saturation level. If two markers are closely linked on the same transforming piece of DNA, they may both attain phenotypic expression. Such *double transformation* events are rare and their frequency is a function of the degree of linkage.

### (b) *Conjugation.*

The genetic material of bacteria and viruses is organized into a single circular structure rather than into several linear structures as in higher organisms. *Conjuga-*

*tion* is a one way transfer of DNA from a donor cell (male) to a recipient cell (female). In *Escherichia coli* strain K-12, the male possesses many replicas of the sex factor F and is symbolized $F^+$ when the sex factor is in its extrachromosomal phase. Females do not have the sex factor and are symbolized $F^-$. When $F^+$ cells conjugate with $F^-$ cells, one or more sex factors are transmitted across the "cytoplasmic bridge" (via a tubular appendage on male cells called a *pilus*) to infect the $F^-$ recipients and convert them into $F^+$ (male) cells. No other genetic material is transferred in $F^+ \times F^-$ conjugations.

A few of the $F^+$ cells in a culture may integrate one of the sex particles into its bacterial "chromosome". These cells are then called Hfr (high frequency of recombination) cells with potential for genetic transfer. Extrachromosomal F particles soon disintegrate and disappear in newly formed Hfr cells. When an F particle takes up residence in the chromosome, it breaks the ring at that point. The end of the chromosome distal to the F factor is the first to be transferred during conjugation. The F factor at the terminal end of the Hfr chromosome is seldom transferred, since the cytoplasmic bridge is maintained for only a short period of time. Rupture of the bridge stops the transfer of donor genetic material into the recipient cell. Thus unless the entire chromosome is transferred (with F particle at its terminal end) the exconjugant (recipient cell after conjugation is completed) of Hfr $\times$ $F^-$ is always $F^-$. Each Hfr strain transfers its markers in a particular order.

For a few generations after conjugation the exconjugant will be a partial diploid (*merozygote*) for that portion of the chromosome it received from the Hfr. Unless the extra piece (*exogenote* or *merogenote*) becomes integrated into the host chromosome, it is eventually lost and the merozygote again becomes haploid.

**Example 15.2.**

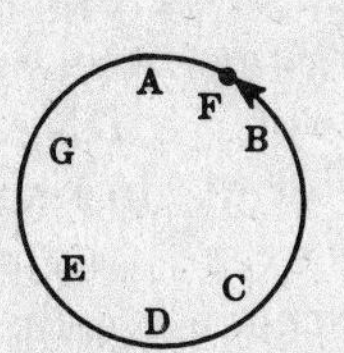


This Hfr strain transfers its markers in the order BCDEGAF.

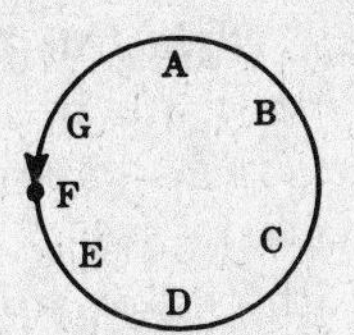


This Hfr strain transfers its markers in the order GABCDEF.

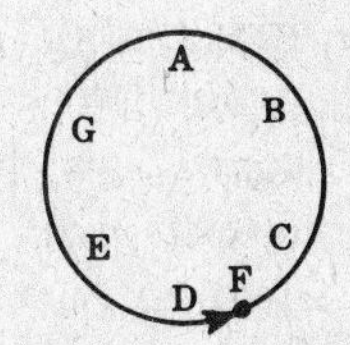


This Hfr strain transfers its markers in the order DEGABCF.

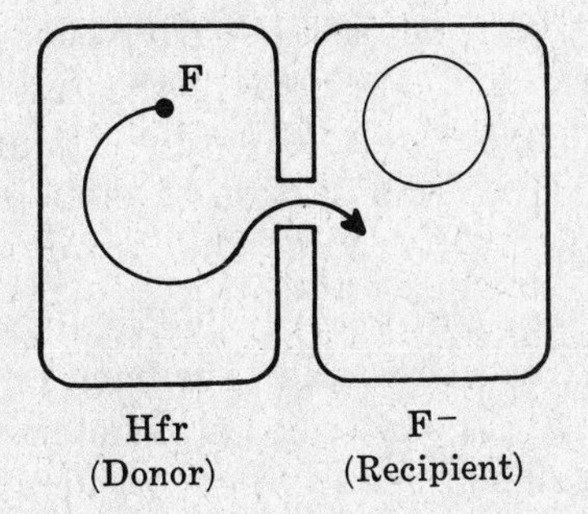


Hfr (Donor) $F^-$ (Recipient)

(*c*) ***Transduction.***

In order to understand transduction, we must first become familiar with the life cycle of bacteriophage. The best known of the transducing systems is the lambda ($\lambda$) phage in its host *E. coli* and phage P22 in its host *Salmonella typhimurium*. The lytic cycle of the virulent T-even phages (2, 4, 6) which infect *E. coli* is represented in Fig. 15-1.

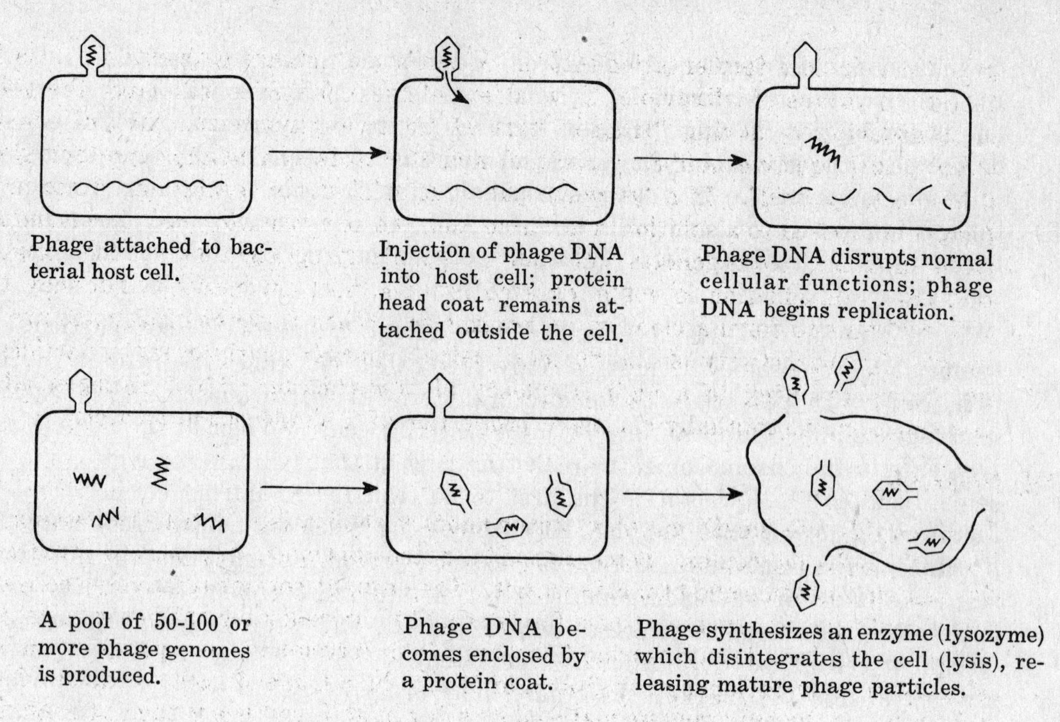


**Fig. 15-1.** Life cycle of T-even phages.

When a virus infects a cell, it has two mutually exclusive choices: (1) it can act as a *virulent* phage and enter the vegetative (lytic) cycle or (2) it can act as a nonvirulent (temperate) phage, becoming integrated into the host chromosome as a *prophage* and replicating in synchrony with the bacterial chromosome. A cell which harbors a prophage is said to be *lysogenic* because the prophage DNA may, under appropriate conditions, be released or *induced* from the bacterial chromosome (*deintegration*) and returned to the lytic phase. Lysogenic bacteria usually cannot be distinguished from normal bacterial cells with the exception that lysogenic cells are immune to infection by other phage of the same species. Some lysogenic bacteria can be induced to release their prophage from the host chromosome by treatment with ultraviolet light (UV induction) or by the process of conjugation (zygotic induction).

When a prophage deintegrates from the bacterial chromosome, it may take with it a small adjacent segment of host DNA, losing some of its own genome in the process. Such a phage particle can mediate the transfer of a bacterial gene or a portion of a bacterial gene from one cell to another in a process called *transduction*. All transducing phages are defective for a portion of their own genome. There are two major types of transducing phages: (1) generalized transducers and (2) restricted transducers.

*Generalized transducers* do not have a specific site of attachment to the chromosome; they may attach anywhere, and thus have an equal probability of transducing any gene. If the exogenote of a generalized transducing phage is incorporated into the genome of the recipient, a clone of recombinant haploid cells is produced. This is termed *complete transduction*. If the exogenote fails to be integrated, but is maintained as a nonreplicating particle, it will exist as a merozygote in only one cell of a clone (*abortive transduction*).

*Restricted transduction* occurs only in cells which possess a prophage integrated into the chromosome at a specific locus called the "prophage site". Only the locus immediately adjacent to the prophage site can be transduced. The best example is the phage λ (lambda) in *E. coli* strain K-12. The prophage site of λ is adjacent to the bacterial gene for galactose fermentation ($gal^+$). When a $gal^+$ strain lysogenic for $\lambda$ is induced to lyse by means of ultraviolet light a few of the $\lambda$ particles will carry the $gal^+$ gene. A transducing $\lambda$, having "traded" part of its own genome for the $gal^+$ marker, is called $\lambda$dg (defective lambda galactose) and therefore can neither lyse nor lysogenize a host cell by itself. If a lysogenic $gal^-$ bacterial strain (containing a normal $\lambda$ prophage) is exposed to a solution containing $\lambda$dg, the few resulting $gal^+$ transductants become unstable heterogenotes $gal^+/gal^-$ ("double lysogenization"). The exogenote in restricted transduction is not integrated into the host chromosome, but may replicate with the host to form a clone of heterogenotes. When these heterogenotes are induced to lyse (by the activity of the normal "helper" phage) nearly every $\lambda$ particle carries *gal* genes, resulting in a high frequency of transduction (Hft). Lysogenization and transduction are mutually exclusive properties of a single phage particle.

(*d*) ***Plasmids.***

*Plasmids* are small, circular, autonomously replicating, extrachromosomal ("cytoplasmic") DNA molecules. Some plasmids, called *episomes*, may become integrated into the host chromosome and replicate with it. They usually contain relatively "nonessential" genetic information such as genes for "sexuality" or genes for antibiotic resistance. A *vector* used in genetic engineering (Chapter 16) for introducing new genes into a bacterial cell may be a bacteriophage such as $\lambda$ or it may be a plasmid such as the sex factor (F) of *E. coli*. If the only genetic material existing in an F particle is the sex gene, it would probably possess little homology with the bacterial chromosome and hence would have little opportunity to become integrated into the host chromosome (converting it into an Hfr cell). When an F particle in an Hfr cell deintegrates, it may include a portion of adjacent host chromosome. Such an F particle is said to possess *chromosomal memory* and would be expected, upon infection, to become integrated into the $F^-$ chromosome with relatively high frequency. Occasionally the memory piece of an F particle is long enough to allow the sex factor to autonomously replicate as an infectious particle and it may also be capable of producing a phenotypic effect if it carries a complete cistron. This process is called *sexduction* and may produce phenotypic effects in merozygotes by complementation or by becoming integrated into the host chromosome.

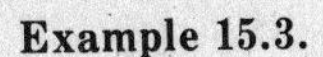

**Example 15.3.**

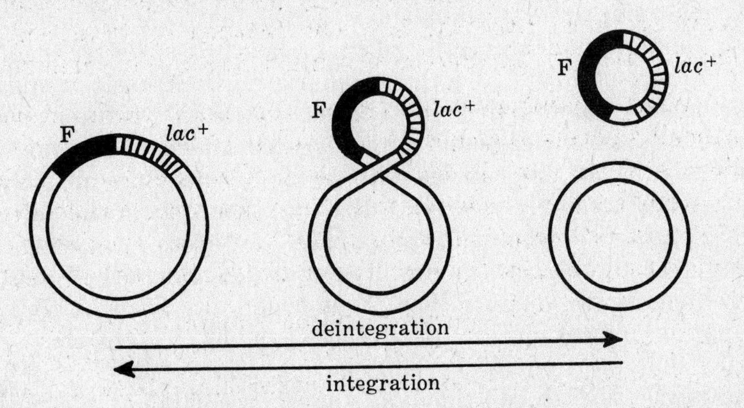


If the lactose locus is adjacent to the F locus, an F-$lac^+$ particle can be formed by deintegration (breakage and reunion).

## 7. Mapping the Bacterial Chromosome.

(*a*) ***Interrupted Conjugation.***

When Hfr and $F^-$ cultures are mixed, conjugation can be stopped at any desired time by subjecting the mixture to the shearing forces of a Waring blender which artificially disrupts the conjugation bridge. The sample is diluted immediately and plated on selective media, incubated, and then scored for recombinants. In addition to the selected marker, and Hfr s train must also carry a distal auxotrophic or sensitivity market which prevents the growth of Hfr cells on the selective medium and thereby allows only recombinant cells to appear. This technique is called *contraselection*. Because of the polarity with which the Hfr chromosome is transferred, the time at which various genetic markers appear in the recipient indicates their linear organization in the donor chromosome. At a given temperature, the transfer of the first half of the Hfr chromosome proceeds at a relatively uniform rate. Therefore the time of entry of different markers into a recipient ($F^-$) cell is a function of the physical distance between them. Because of errors introduced by experimental manipulations, this method is best suited for markers which are more than two minutes apart.

**Example 15.4.** An Hfr strain carrying the prototrophic markers $a^+$, $b^+$, $c^+$ is mixed with a $F^-$ strain carrying the auxotrophic alleles *a, b, c*. Conjugation was interrupted at 5 minute intervals and plated on media which revealed the presence of recombinants.

| Time (minutes) | Recombinants |
|---|---|
| 5 | $ab^+c$ |
| 10 | $ab^+c^+$ |
| 15 | $a^+b^+c^+$ |

The order of the genes in the Hfr donor strain is $b^+$-$c^+$-$a^+$; *b* is less than 5 time units from the origin; *c* is less than 10 time units from *b*; *a* is less than 10 time units from *c*.

(*b*) ***Uninterrupted Conjugation.***

When conjugation is allowed to proceed without artificial interruption, the time of rupture of the cytoplasic bridge is apparently randomized among the mating pairs. The nearer a marker is to the origin (leading end of donor chromosome) the greater its chances of appearing as a recombinant in a recipient cell. Donor and recipient cells are mixed for about an hour in broth and then placed on selective media which only allows growth of $F^-$ recombinants for a specific marker. Contraselection against Hfr must also be part of the experimental design. The contraselective marker should be located as distally as possible from the selected marker so that unselected recombinants will not be lost by its inclusion. The frequencies with which unselected markers appear in selected recombinants are inversely related to their distances from the selected marker, provided they lie distal to it. Obviously, any unselected marker between the selected marker and the origin of the chromosome will always be transferred ahead of the selected marker. Proximal markers more than three time units apart exhibit approximately 50% recombination, indicating that the average number of exchanges between them is greater than one. Just at the point where gross mapping by conjugation becomes ineffective, i.e. for markers less than two time units apart, recombination mapping becomes very effective, permitting estimation of distances between closely linked genes or between mutant sites within the same gene. Distances between genes can be expressed in three types of units: (1) time units, (2) recombination units, or (3) chemical units.

**Example 15.5.** If 1 minute of conjugation is equivalent to 20 recombination units in *E. coli*, and the entire chromosome is transferred in 100 minutes, then the total map length is 2000 recombination units. If $10^7$ nucleotide pairs exist in the chromosome, then 1 recombination unit represents $10^7/2000 = 5000$ nucleotide pairs.

### (c) *Recombination Mapping.*

Virtually all of the opportunities for recombination in bacteria involve only a partial transfer of genetic material (*meromyxis*) and not the entire chromosome. One or more genes have an opportunity to become integrated into the host chromosome by conjugation, depending upon the length of the donor piece received. Exogenotes usually must become integrated if they are to be replicated and distributed to all of the cells in a clone. Only a small segment of DNA is usually integrated during transformation or transduction. Thus if a cell becomes transformed for two genetic markers by the same transforming piece of DNA (double transformation) the two loci must be closely linked. Similarly, if a cell is simultaneously transduced for two genes by a single transducing phage DNA (*cotransduction*) the two markers must be closely linked. The degree of linkage between different functional genes (intercistronic) or between mutations within the same functional gene (intracistronic) may then be estimated from the results of specific crosses.

In merozygotic systems where the genetic contribution of the donor parent is incomplete, an even number of crossovers is required to integrate the exogenote into the host chromosome (*endogenote*).

**Example 15.6.**

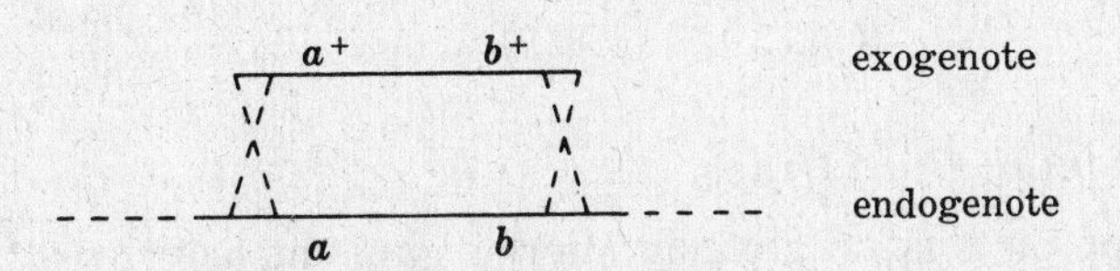


Prototrophic recombinants must integrate the exogenote from somewhere left of the $a$ locus to right of the $b$ locus. Two crossovers (an even number) are required for this integration.

**Example 15.7.**

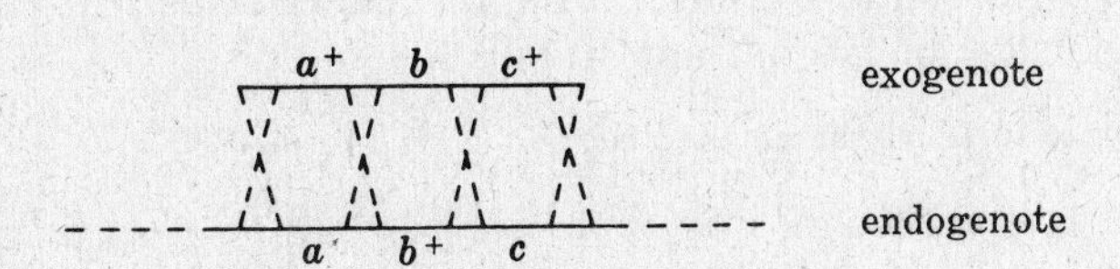


A prototrophic recombinant in this example requires a quadruple (even number) crossover for integration of all wild type genes.

The total number of progeny is unknown in merozygotic systems so that recombination frequency cannot be expressed relative to this base. Therefore recombination frequencies must be made relative to some standard which is common to all crosses. For example, the number of prototrophic recombinants produced by crossing two mutant strains can be compared to the number emerging from crossing wild type by mutant type. However, many sources of error are unavoidable when comparing the results of different crosses. This problem can be circumvented by comparing the number of prototrophic recombinants to some other class of recombinants arising from the same cross.

**Example 15.8.** Ratio test for different functional genes. Suppose we have two mutant strains, $a$ and $b$, where the donor strain ($a^+b$) can grow on minimal medium supplemented with substance B, but the recipient strain ($ab^+$) cannot do so.

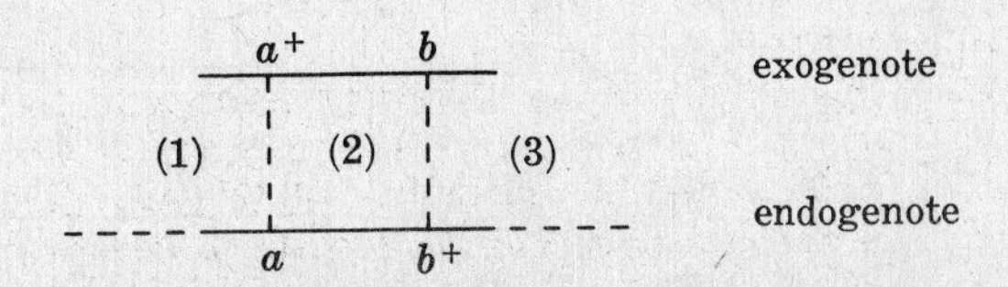

Crossing over in regions (1) and (2) produces prototrophic recombinants ($a^+b^+$) able to grow on unsupplemented medium. If the medium is supplemented with substance B, then $a^+b$ recombinants arising by crossing over in regions (1) and (3) can grow in addition to the prototrophs.

$$\text{Standardized recombination ratio} = \frac{\text{number of prototrophs}}{\text{number of recombinants}}$$

**Example 15.9.** Intracistronic ratio test. Consider two intracistronic mutations, $b_1$ and $b_2$, unable to grow in medium without substance B. The recipient strain contains a mutation in another functionally different gene ($a$), either linked or unlinked to $b$, which cannot grow unless supplemented by substances A and B.

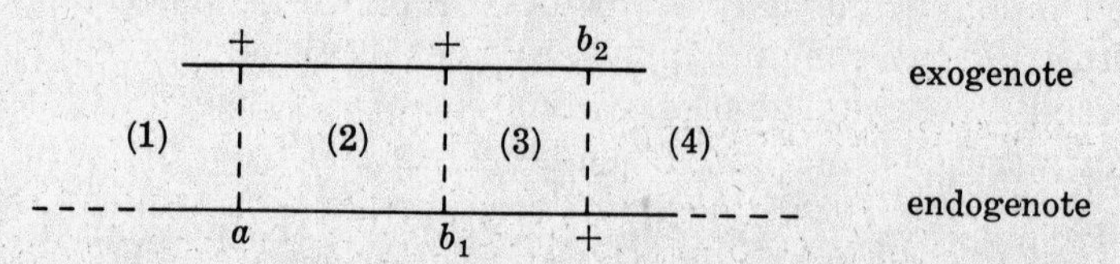


On unsupplemented medium, only prototrophs arising through crossovers in regions (1) and (3) appear. On medium supplemented only by substance B, recombinants involving region (1) and any of the other three regions can survive.

$$\text{Standardized recombination ratio} = \frac{\text{number of colonies on unsupplemented medium}}{\text{number of colonies on B-supplemented medium}}$$

### (*d*) *Establishing Gene Order.*

Mapping small regions in microorganisms has revealed that multiple crossovers often occur with much greater than random frequency, a phenomenon called "localized negative interference." The only unambiguous method for determining the order of very closely linked sites is by means of three factor reciprocal crosses. Suppose that the location of gene $a$ is known to be to the left of gene $b$ but that the order of two mutants within the adjacent $b$ cistron is unknown. Reciprocal crosses will yield different results, depending upon the order of the mutant sites.

**Example 15.10.** Assume the order of sites is $a$-$b_1$-$b_2$.

Original Cross:

+ + $b_2$ exogenote
(1) (2) (3) (4)
$a$ $b_1$ + endogenote

Reciprocal Cross:

$a$ $b_1$ + exogenote
(1) (2) (3) (4)
+ + $b_2$ endogenote

In the original cross, prototrophs (+ + +) can be produced by crossovers in regions (1) and (3). In the reciprocal cross, prototrophs arise by crossovers in regions (3) and (4). The numbers of prototrophs should be approximately equivalent in the two crosses.

**Example 15.11.** Assume the order of sites is $a$-$b_2$-$b_1$.

Original Cross:

+ $b_2$ + exogenote
(1) (2) (3) (4)
$a$ + $b_1$ endogenote

Reciprocal Cross:

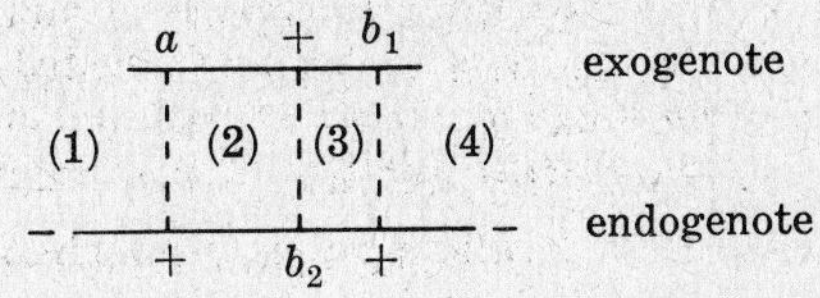

```
          a      +   b1
          ——————————————        exogenote
   (1)    ¦  (2) ¦(3)¦   (4)
   – ——————————————————— –      endogenote
          +      b2  +
```

In the original cross, the production of prototrophs requires four crossovers, one in each of the regions (1), (2), (3) and (4). Only two crossovers (in regions (2) and (3)) in the reciprocal cross are needed to produce prototrophs. Therefore many more prototrophic recombinants are expected from the reciprocal cross than from the original cross.

### (*e*) *Complementation Mapping.*

An F particle which carries another bacterial gene other than the sex factor produces a relatively stable $F^+$ merozygote. These partial diploids can be used for complementation tests of mutants affecting the same trait.

**Example 15.12.** An Hfr strain of *E. coli* is unable to ferment lactose ($z_1^-$) and can transfer the $z_1^-$ gene through sexduction to a mutant ($z_2^-$) recipient, forming the heterogenote $z_2^-/(F\text{-}z_1^-)$. If $z_1$ and $z_2$ belong to the same cistron (functional alleles), then complementation does not occur and only mutant phenotypes are produced. If $z_1$ and $z_2$ are mutants in different cistrons, complementation could produce wild types able to ferment lactose.

*Intracistronic complementation* may sometimes be possible when the enzyme product is composed of two or more identical polypeptide chains. Experimental evidence has shown that an *in vitro* mixture of inactive enzymes from some complementing mutants can "hybridize" to produce an enzyme with up to 25% normal activity. Mutants that fail to complement with some but not all other mutants are assumed to overlap in function. A complementation map can be constructed from the experimental results of testing all possible pairs of mutants for complementary action in bacterial merozygotes or in fungal heterokaryons. A complementation map cannot be equated in any way with a crossover map, since the gene is defined by different criteria. A complementation map tells us nothing of the structure or location of the mutations involved. Complementation maps are deduced from merozygotes or heterokaryons; crossover maps from recombination experiments.

**Example 15.13.** Three mutants map by complementation as follows:

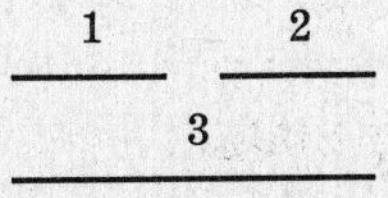

```
   1        2
 ——————   ——————
        3
 ———————————————
```

This indicates that mutants 1 and 2 are complementary and do not overlap in function. Hence 1 and 2 are nonallelic mutations by this criterion. Mutant 3 fails to complement with either 1 or 2 and hence must overlap (to some degree) with both 1 and 2. Hence 3 is functionally allelic with both 1 and 2.

### (*f*) *Mapping by Deletion Mutants.*

A deletion in some segment of a functional gene cannot recombine with point mutations in that same region even though two point mutations at different sites within this region may recombine to produce wild type. Another distinctive property of deletion mutants is their stability, being unable to mutate back to wild type. The use of overlapping deletions can considerably reduce the work in fine structure analysis of a gene.

**Example 15.14.** Determining the limits of a deletion. Suppose that a series of single mutants (1, 2, 3, 4) has already been mapped as shown below:

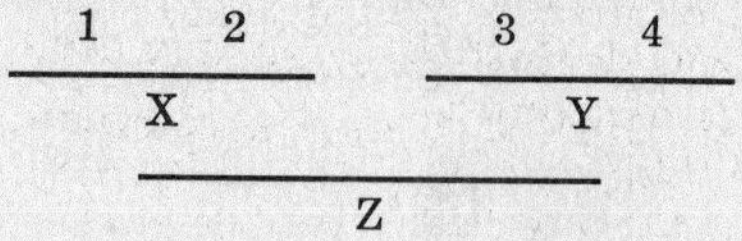

```
  1      2          3      4
 ____________    ____________
      X                Y
      ________________
              Z
```

A deletion that fails to recombine with point mutants 1 and 2 but does produce wild type with 3 and 4 extends over region X. A deletion that yields no recombinants with 3 and 4 has the boundaries diagrammed as Y. A deletion mutant which produces wild type only with point mutants 1 or 4 has the limits of Z.

**Example 15.15.** Assigning point mutations to deletion regions.

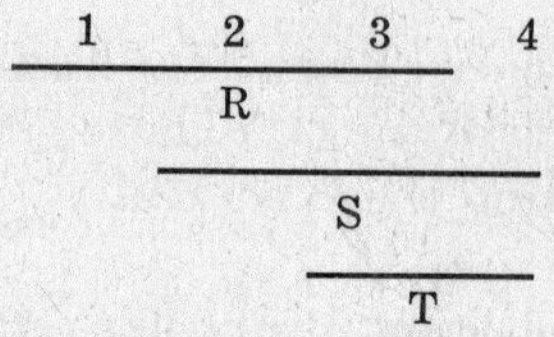

```
  1     2     3     4
 _______________
        R
        _______________
              S
              _________
                  T
```

Given the deletions R, S and T as shown above, the point mutation which recombines to give wild type with deletions S and T, but not with R, is 1. Number 3 is the only one of the four mutants which fails to recombine with one of the three deletions.

## VIRUSES

### 1. Characteristics of Viruses.

Viruses are noncellular, obligate intracellular parasites. All infective viruses (virions) except viroids* consist of a genetic component (either DNA or RNA, but not both) complexed with (often surrounded by) "coat proteins" collectively called the *capsid*. The capsid usually consists of repeating identical subunits called *capsomeres*. The combination of viral nucleic acid and its coat proteins is called a *nucleocapsid*. Viral components (nucleic acid and proteins) assemble themselves spontaneously into complete virions inside host cells without the aid of special enzymes. Viruses that infect bacteria are called *bacteriophages* ("bacterial consumer") or simply *phages*. The nucleocapsid constitutes the infective stage of all phages. Most phages escape from their host cells by causing their dissolution (*lysis*). Some animal and plant viruses do not rupture or lyse their host cells, but rather are "secreted" from the host cell by a "budding" process. The nucleocapsid thereby acquires an outer layer of host cell membrane (containing host proteins and lipids) called the *peplos*. Proteins produced by viral instructions may be exposed on the outer layer of the peplos; these viral proteins are called *peplomeres* and often serve as specific attachment sites for infecting other host cells. The T-series of phages attach to their bacterial host cells by their tails or tail fibers and inject their nucleic acid into the interior of the cell; the capsid and tail proteins remain outside as "ghosts". Other kinds of viruses seem to enter cells by a reversal of the budding process (a form of endocytosis or phagocytosis). Inside the host cell, the peplos (if present) and the capsid must be removed before the viral nucleic acid can either replicate ("vegetative growth" or "lytic cycle") or lysogenize the cell (become integrated into the host chromosome). The number of phages in a culture can be assayed by plating a suitable dilution of the culture on a lawn of host cells. Each lytic phage particle will eat a hole (*plaque*) in such a lawn.

---

* Viroids are small (about 225 nucleotides), infectious segments of naked single-stranded RNA. How they replicate is a mystery. Viroids were first discovered as the cause of spindle tuber disease of potatoes and tomatoes.

Most virions range in size from about 10 nanometers (nm) (e.g. the bacteriophages F2, R17 and MS2) to 300 nm (e.g. the animal pox viruses) and will pass through filters (0.2 to 0.45 micrometer ($\mu$m)) that prohibit most bacteria (other than *Chlamydia, Rickettsia* and *Mycoplasma*) from passing through. The limit of resolution of the light microscope is about 200 nm, so that most viruses cannot be visualized without the aid of the electron microscope.

Viruses differ from cellular organisms in the following major respects:

(1) Viruses have only one kind of nucleic acid (either DNA or RNA), whereas cells have both kinds.

(2) Viruses do not have the capacities to synthesize proteins (e.g. they do not contain ribosomes), to develop their own source of chemical energy (no metabolism), or to generate their own ATP (no cytochrome system).

(3) Viruses have no lipoprotein outer membranes or membrane-bound subcompartments (organelles) other than the host-derived peplos.

(4) Viruses are not affected by levels of antibiotics that are nontoxic to cells.

(5) Viruses are sensitive to certain host-produced proteins of higher vertebrates called *interferons* that interfere with viral replication, but not viral attachment or entry into host cells.

(6) Viruses have no means of motility other than diffusion.

(7) Viruses do not "grow" in the classical sense of increasing in mass; i.e. once the virion is formed, it does not increase in size.

Viral phenotypes that are useful for classification include nucleic acid type (RNA or DNA), nucleic acid form (single-stranded or double-stranded), molecular weight, molecular base composition (G + C content), symmetry (cubic, helical or complex), possession of a peplos vs. naked, size (determined by filtration, electron microscopy and ultracentrifugation), presence or absence of the enzyme reverse transcriptase (RNA-dependent DNA polymerase) in certain RNA viruses, antigenic composition (serological activity), hemagglutination properties and host range (species infected).

## 2. Genetic Structure and Replication of Viruses.

Whereas the genetic material of all cellular organisms is linear double-stranded DNA, viruses may have either DNA or RNA in either single-stranded (SS) or double-stranded (DS) forms. Each of these forms requires a different mode of replication. There are six major patterns of replication and transcription known in animal viruses, the details of which can be found in any complete virology textbook. RNA viruses do not need to synthesize DNA in order to replicate, but DNA viruses must make RNA to produce new virions. Single-stranded oncogenic RNA viruses (capable of inducing tumors or cancers) must make a DNA copy and the copy DNA (cDNA) must become integrated into the chromosome of the host cell as a provirus before it is able to *transform* a normal cell into a cancerous state. A special enzyme called RNA-dependent DNA polymerase (reverse transcriptase) is required to make DNA from RNA. Uninfected cells do not contain this enzyme; it must be produced by translation from instructions in the viral RNA (behaving as mRNA). These oncogenic viruses contain reverse transcriptase as well as their own DNA-dependent DNA polymerase in infective virions (synthesized in the host from which the virions were least liberated). Such viruses belong to a group called "retroviruses" (Retroviridae) because they appear to reverse the normal cellular flow of genetic information from DNA to RNA. Reverse transcriptase can be utilized *in vitro* to synthesize a cellular gene from its purified mRNA transcript.

There are five major steps in productive infection (release of virions) of bacteriophage (Fig. 15-1).

(1) Adsorption of the phage by specific viral proteins to host cell receptors.
(2) Penetration of viral nucleic acid (usually by injection) into the host cell.
(3) Replication of the viral "genome" and viral protein synthesis.
(4) Spontaneous assembly of new nucleocapsids.
(5) Release of mature virions from the host cell.

Essentially the same steps are followed by animal viruses with the additional step of uncoating the virion (nucleocapsid) after it has pentrated the host cell. Soon after phage infection, synthesis of cellular mRNA ceases; phage nucleases destroy the host DNA. Certain phages (e.g. T4) possess the enzyme hydroxymethylase that modifies the phage cytosines to hydroxymethylcytosines. Such modified bases are resistant to degradation by nucleases of either the host or the phage. Some phages possess or make repressor proteins that turn off host DNA functions. Others make or possess specific nucleic acid polymerases for replicating or trancribing only viral genomes. Some viruses possess or contain the information for synthesizing new tRNA's that preferentially translate only viral proteins.

**Example 15.16.** The *Rous sarcoma virus* (RSV) that causes tumors in bone and connective tissues of chickens consists of two identical single-stranded RNA molecules. Four to eight tRNA molecules are bound to each of the 35S viral genomes and help to "glue" the two chains together into a 70S complex. Like all DNA polymerases, reverse transcriptase cannot initiate DNA chains, but can only add deoxyribonucleotides to nascent DNA chains or to short RNA "primers" that initiate DNA synthesis. The primers for DNA synthesis from RSV genomes seem to be some of the tRNA molecules bound to the 70S RNA complex. The dual genome of RSV is a rather unique characteristic not found in many other viruses.

### 3. Mapping the Viral Genome.

During replication of phage DNA, a pool of perhaps 100 or more phage genomes may be present within the host cell. There is an opportunity for each DNA molecule to encounter several recombinational events, called "rounds of matings". This aspect of its life cycle renders the interpretation of recombination data considerably more difficult than in most higher organisms. Compensation can be made for some of these difficulties by appropriate experimental design and the acceptance of their concurrent assumptions. By these techniques, analysis of the phage genome has approached the nucleotide level of resolution.

Perhaps the most definitive fine structure analysis ever performed on a gene was Benzer's study of the $r_{II}$ region in phage T4. When wild type ($r^+$) T4 infects its host *E. coli,* it produces relatively small plaques with fuzzy edges. Many mutations in the $r$ region have been found, all of which produce more rapid lysis of their host than wild type, yielding larger plaques with relatively smooth edges. These $r$ mutants may be classified into three phenotypic groups, depending on their ability to lyse three strains of *E. coli.*

| Strain of Phage | Plaque type on *E. coli* strains | | |
|---|---|---|---|
| | B | S | K |
| $r^+$ | wild | wild | wild |
| $r_I$ | large | large | large |
| $r_{II}$ | large | wild | no plaques |
| $r_{III}$ | large | wild | wild |

The sites of $r_I$, $r_{II}$ and $r_{III}$ map at distinctly different noncontiguous locations. The $r_{II}$ region is about 8 recombination units long, representing approximately 1% of the phage chromosome. By means of complementation tests, the $r_{II}$ region is shown to consist of two adjacent and contiguous cistrons, A and B.

When two different $r_{II}$ strains (containing mutations in different cistrons) are added to strain K-12 in sufficiently large numbers to ensure that each cell is infected with at least one representative of each mutant, all of the cells will rapidly lyse, releasing a normal number of progeny phage (most of which are defective like the parental strains). If the two phage strains contain mutations in the same $r_{II}$ cistron, only the rare wild type recombinants will be capable of causing lysis and the number of cells so affected will be very small. Thus, the results of complementation are easily distinguished from those of recombination.

Overlapping deletions have been used to delimit a number of mutant sites to a small segment of a cistron. Benzer did not attempt to order the mutant sites within each of these small segments, but he did subject them to recombination tests to ascertain identity or nonidentity. This was accomplished by doubly infecting strain B with a pair of $r_{II}$ mutants ($r_{IIa}$ and $r_{IIb}$) in broth and allowing them to lyse the culture. The total number of progeny phage is estimated by plating dilutions of this lysate on *coli* B and counting the resulting plaques. Wild type recombinants are scored by plating the lysate on *coli* K. For every wild type ($r^+_{IIa}r^+_{IIb}$) plaque counted on K, we assume that an undetected double mutant ($r_{IIa}r_{IIb}$) reciprocal recombinant was also formed.

$$\text{Recombination Percentage} = \frac{200\ (\text{number of plaques on K})}{\text{number of plaques on B}}$$

The smallest reproducible recombination frequency observed between two sites in the $r_{II}$ region is about 0.02%, corresponding to approximately 1/400 of a gene whose total length is only 8 recombination units!

## 4. Lambda (λ) Bacteriophage.

In order to transform a normal cell into a cancer cell, an oncogenic virus must become integrated into the host chromosome in a manner analogous to the lysogenization of a bacterial cell by a template phage. Much more is known at present about the control mechanisms for lysogenization by phage λ than for any other virus. Comparatively little is known about the analogous process of virogeny in animal cells. It is hoped that some of the knowledge from the study of lysogenic phages such as lambda can eventually be used to help understand the mechanism(s) whereby oncogenic viruses transform a normal cell into a cancerous one or how the fertilized egg differentiates into the various cell types of a multicellular organism.

Figure 15-2 is a genetic map of the lambda phage genome and should be referred to during all of the following discussion. A repressor protein produced by the *c*I gene prevents the lytic cycle and allows lysogenization to be established or maintained. Repressor also makes the lysogenic cell immune to superinfection by exogenous lambda phages. Two operators are sensitive to the repressor, $o_R$ (right operator) and $o_L$ (left operator). Repressor blockage of $o_L$ prevents expression of gene *N*; blockage of $o_R$ prevents expression of genes *O* and *P*. The protein products of genes *N*, *O*, and *P* are required for expression of almost all other structural lambda genes. Each operator has its own promoter site, $p_L$ and $p_R$.

$p_{RE}$

$p_{RM}$

*ABCDEFGHIJ int xis red* cIII *N* $p_Lo_L$ *rex* cI $p_Ro_R$ *cro* cII *OP* *Q* $p_{LATE}SR$

(continued at A)

**Fig. 15-2.** Map of the lambda genome. Solid arrows indicate the direction of cotranscribed genetic regions subject to termination by rho factor. Dashed arrows indicate extensions of cotranscription when antitermination factors are present to nullify rho factor activity; *N* gene product functions to nullify termination at (•) and *Q* gene product nullifies at (:). (After Gunther S. Stent and Richard Calendar, *Molecular Genetics: An Introductory Narrative*, W. H. Freeman and Company, San Francisco, 1978.)

Each operator consists of three partly symmetric tandem nucelotide sequences of 16 bases. Thus, three molecules of repressor can be bound at each operator locus.

Lambda repressor controls its own synthesis; it induces its own synthesis when its concentration is low and it represses its own synthesis when its concentration is high. The *c*I gene that produces repressor is transcribed from right to left beginning at the promoter site labeled $p_{RM}$ (promoter of repressor, maintenance). The leftmost of the three tandem repressor-binding sites of the $p_R o_R$ region overlaps with $p_{RM}$ so that when repressor is bound there, the *c*I gene is turned off. Repressor is bound most readily at the rightmost of the three tandem sites of the $p_R o_R$ region and it has least affinity for the leftmost site. Thus, when repressor is in short supply, only the rightmost site is complexed with repressor and this situation actually stimulates repressor transcription (analogous to the positive control exerted by the CAP-cAMP complex when bound to the CAP-binding site within the *lac*P region). If excess repressor is present, the leftmost binding site will be covered by repressor and synthesis of repressor will be inhibited by negative control. Therefore, the concentration of repressor can be maintained within a relatively narrow range.

A much higher concentration of repressor is needed to establish the lysogenic state than for its maintenance. A 50- to 100-fold higher rate of repressor synthesis can be achieved when transcription is initiated at the site labeled $p_{RE}$ (promoter of repressor, establishment) that is several hundred nucleotide base pairs (nbp) to the right of the $p_{RM}$ promoter. The products of the *c*II and *c*III genes exert positive control essential for initiation of *c*I mRNA synthesis at the $p_{RE}$ site, but not essential for initiation at the $p_{RM}$ site. The longer mRNA molecules initiated at the $p_{RE}$ site possess an efficient ribosome-binding sequence for translating the *c*I gene. The shorter mRNA molecules initiated at the $p_{RM}$ site probably do not contain all of the ribosome-binding nbp and hence are very inefficiently translated into repressor proteins. When transcription begins at the $p_{RE}$ site, RNA polymerase is unrestricted in its movement across the $o_R$ locus into the *c*I gene even if all three tandem sequences of $o_R$ have bound repressor. This positive control of repressor synthesis occurs both at the *transcriptional level* (by *c*II and *c*III products stimulating mRNA initiation at the $p_{RE}$ site) and at the *translational level* (by the high efficiency of ribosome binding to the mRNA product of the $p_{RE}$ site).

If a nonlysogenic cell is infected by one or more lambda phages, a productive lytic response may be evoked. The first gene synthesized by the right operon is *cro* ("control of repressor and other things"); its product is a repressor of the lytic immunity repressor (*c*I gene product). The *cro* gene product is thought to repress initiation of transcription of the *c*I gene at the $p_{RM}$ promoter site. If immunity repressor is present in the cell, it will turn off the right operon and no "repressor of repressor" will be made; this is the normal condition during the prophage state. However, if no repressor is present in the cell, the repressor from *cro* prevents the immunity repressor from being synthesized, thus favoring the lytic response.

Attenuator sites (discussed in Chapter 14 in regard to the tryptophan operon of *E. coli*) exist between the genes *N* and *c*III, between *cro* and *c*II, and between *P* and *Q*. The product of gene *N* is an *antitermination factor* that prevents rho factor from binding to the termination signals. Thus, the presence of *N* gene product allows transcription of distal genes otherwise subject to termination in the attenuator regions. However, the antitermination factor of gene *N* apparently does not allow transcription of "late genes" from *S* to *J* into a single polygenic mRNA molecule initiated at a promoter designated $p_{LATE}$. Most of these gene products are protein structural components of the infective lambda phage (e.g. head and tail proteins). The antitermination factor produced by gene *Q* is required for expression of genes under control of the $p_{LATE}$ promoter.

The linear lambda genome of infective phage becomes circularized prior to replication so that genes *R* and *A* are actually adjacent. This is accomplished by the "early enzyme" products of genes *int, xis, red* and *rex* that are concerned with genetic recombination and excision or integration of the phage genome. Genes *O* and *P* appear to determine components of the enzymatic complexes that replicate the lambda genome.

When a nonlysogenic cell is infected by a lambda phage, either the cell becomes lysed or becomes lysogenic. Which of these mutually exclusive states is attained may depend upon a race between synthesis of the immunity repressor and synthesis of one or more early proteins of vegetative phage development such as the product of the *cro* gene. Alternatively, the decision to follow the lytic or lysogenic responses may be primarily determined by the level of the *c*III gene product reached before further transcription of the *c*III gene is repressed by immunity repressor at the $p_L o_L$ locus.

# Solved Problems

## BACTERIA

**15.1.** The discipline of bacterial genetics began in 1943 when S. E. Luria and M. Delbrück published a paper entitled "Mutations of bacteria from virus sensitivity to virus resistance". Before this time, it was not known if the heredity of bacteria adaptively changed in specific ways as a consequence of exposure to specific environments, or whether specific mutants existed in the population prior to an environmental challenge, the latter acting as a selective agent to increase the numbers of the adaptive mutants. The former idea was Lamarckian, the latter was neodarwinian. Luria and Delbrück found that there was great variation from one trial to another in the number of *E. coli* that were resistant to lysis by phage T1. In order to determine which of the two hypotheses was correct, they devised the following "fluctuation test". Twenty 0.2-ml "individual cultures" and one 10-ml "bulk culture" of nutrient medium were incubated with about $10^3$ *E. coli* cells per milliliter. The cultures were incubated until they contained about $10^8$ cells/ml. The entire 0.2 ml of each individual culture was spread on a nutient agar plate heavily seeded with T1 phages. Ten 0.2-ml samples from the bulk culture were also treated in similar fashion. After overnight incubation, the total number of T1-resistant (Ton$^r$) bacterial cells was counted; the results are presented in the following table. What inferences can be drawn from this "fluctuation test"?

| Individual Cultures | | Samples from Bulk Culture | |
|---|---|---|---|
| Culture Number | Number of Resistant Colonies | Sample Number | Number of Resistant Colonies |
| 1 | 1 | 1 | 14 |
| 2 | 0 | 2 | 15 |
| 3 | 3 | 3 | 13 |
| 4 | 0 | 4 | 21 |
| 5 | 0 | 5 | 15 |
| 6 | 5 | 6 | 14 |
| 7 | 0 | 7 | 26 |
| 8 | 5 | 8 | 16 |
| 9 | 0 | 9 | 20 |
| 10 | 6 | 10 | 13 |
| 11 | 107 | | |
| 12 | 0 | | |
| 13 | 0 | | |
| 14 | 0 | | |
| 15 | 1 | | |
| 16 | 0 | | |
| 17 | 0 | | |
| 18 | 64 | | |
| 19 | 0 | | |
| 20 | 35 | | |

**Solution:**

Variances for each experiment can be calculated from the square of formula (*11.2*); the individual cultures have a variance of 714.5, whereas the variance of samples from the bulk culture is 16.4. In a Poisson distribution, the mean and the variance are essentially identical; hence, the variance/mean ratio should be unity (1.0). The variance/mean ratio for the bulk culture samples is 16.4/16.7 = 0.98 or nearly 1.0, as expected from a random distribution of rare events. The samples from the bulk culture collectively serve as a control for the individual cultures. The same ratio for the individual cultures, however, is 714.5/11.3 = 63.23, indicating that there are extremely wide fluctuations of the numbers of $Ton^r$ cells in each culture around the mean. If resistance to the T1 phages occurs with a given probability only after contact with the phages, then each culture from both the individual and batch experiments should contain approximately the same average number of resistant cells. On the other hand, if $Ton^r$ mutants occurred prior to contact with the phages, great variation around the mean is expected from one individual culture to another because some will incur a mutation early and others late (or not at all) during the incubation period. This experiment argues in favor of the mutation hypothesis and against the induced resistance hypothesis.

Certain mutations, such as that to phage resistance, are *preadaptive* in that their selective advantage only becomes manifest when phages are in the environment as a selective agent; in this case, T1-sensitive bacteria ($Ton^s$) are killed by T1 phages, allowing only the few $Ton^r$ cells to survive and multiply. Phage resistance depends upon altering the structure of the bacterial receptor sites to which T1 phages normally attach. Immunity to superinfection by a specific phage is based upon production of a repressor of phage replication by a lysogenic cell.

**15.2.** Two triple auxotrophic strains of *E. coli* are mixed in liquid medium and plated on complete medium which then serves as a master for replica plating onto 6 kinds of media. From the position of the clones on the plates and the ingredients in the media, determine the genotype for each of the 6 clones. The gene order is as shown.

$$T^-L^-B_1^-\,B^+Pa^+C^+ \times T^+L^+B_1^+\,B^-Pa^-C^-$$

Nutritional supplements are abbreviated as follows:

T = threonine B = biotin

L = leucine Pa = phenylalanine

$B_1$ = thiamin C = cystine

4 1
5 3 2
6

Master Plate
(complete medium)

Replica plates: Each dish contains minimal medium plus the supplements listed below each dish.

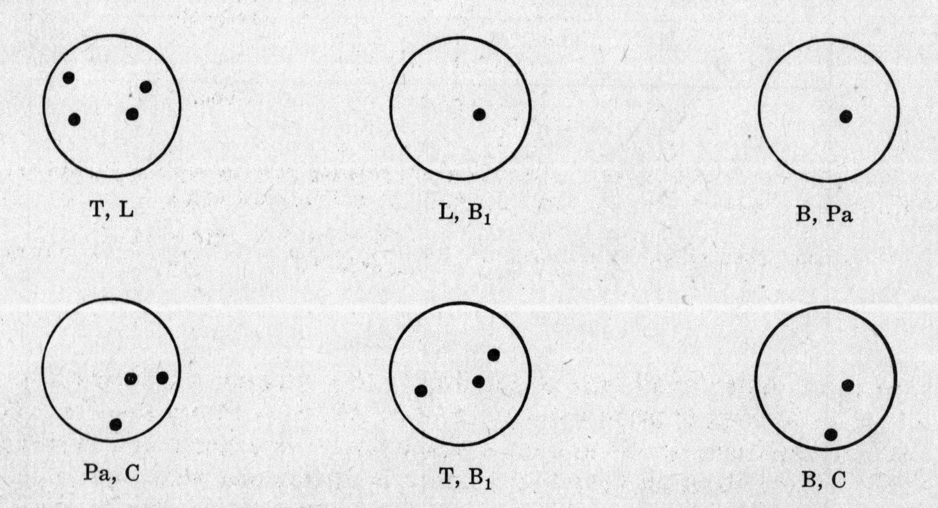


**Solution:**

#1 grows when supplemented with T and L or T and $B_1$, but not with L and $B_1$. Therefore this is auxotrophic for T alone ($T^-L^+B_1^+\,B^+Pa^+C^+$).

#2 appears only on the plate supplemented with Pa and C. This is a double auxotrophic clone of genotype $T^+L^+B_1^+\ B^+Pa^-C^-$.

#3 appears on all replica plates and therefore must be prototrophic ($T^+L^+B_1^+\ B^+Pa^+C^+$).

#4 grows only when supplemented with T and L and must be a double auxotroph of genotype $T^-L^-B_1^+\ B^+Pa^+C^+$.

#5 and #1 always appear together on the replica plates and therefore have the same genotype.

#6 can grow in the presence of Pa and C or B and C. The common factor is C for which this strain is singly auxotrophic ($T^+L^+B_1^+\ B^+Pa^+C^-$).

**15.3.** Under optimal conditions, some bacteria can divide every 20 minutes. Suppose each cell has a mass of $2 \times 10^{-9}$ milligrams. The mass of the earth is approximately $5.97 \times 10^{27}$ grams. Determine the time (in hours) required for the progeny of a single cell dividing without restriction at the above rate to equal the weight of the earth.

**Solution:**

At time zero we have 1 cell; 20 minutes later we have 2 cells; at 40 minutes there are 4 cells; at 60 minutes there are 8 cells; etc. The number of cells at any hour, $t$, is obviously $2^{3t}$. The number of cells equivalent to the weight of the earth is

$$(5.97 \times 10^{27})/(2 \times 10^{-12}) = 2.98 \times 10^{39} = 2^{3t}$$

from which $3t \log 2 = \log 2.98 + \log 10^{39}$, $\quad t = \dfrac{\log 2.98 + 39}{3 \log 2} = \dfrac{39.475}{3(0.301)} = 43.7$ hours.

**15.4.** A strain of *E. coli* unable to ferment the carbohydrate arabinose ($ara^-$) and unable to synthesize the amino acids leucine ($leu^-$) and threonine ($thr^-$) is transduced by a wild type strain ($ara^+\ leu^+\ thr^+$). Recombinants for leucine are detected by plating on minimal medium supplemented with threonine. Colonies from the transduction plates were replicated or streaked onto plates containing arabinose. Out of 270 colonies which grew on the threonine supplemented plates, 148 could also ferment arabinose. Calculate the amount of recombination between *leu* and *ara*.

**Solution:**

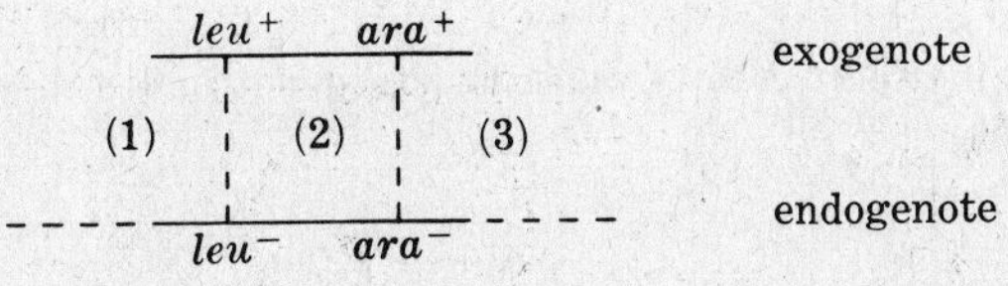

In order for a transductant to be $leu^+\ ara^+$, crossing over in regions (1) and (3) must occur; for $leu^+\ ara^-$ to arise, crossing over in regions (1) and (2) must occur.

$$\text{Standardized recombination ratio} = \frac{\text{no. of } leu^+\ ara^-}{\text{no. of } leu^+} = \frac{(270-148)}{270} = 0.45 \text{ or } 45\%$$

**15.5.** Several $z^-$ mutants, all lacking the ability to synthesize $\beta$-galactosidase, have been isolated. A cross is made between Hfr ($z_1^-\ Ad^+\ S^s$) $\times$ F$^-$ ($z_2^-\ Ad^-\ S^r$) where $Ad^-$ = adenine requirement, $S^s$ and $S^r$ = streptomycin sensitivity and resistance respectively. After about an hour, the mixture is diluted and plated on minimal medium containing streptomycin. Many of the $Ad^+$ clones were able to ferment lactose, indicating $\beta$-galactosidase activity. Only a few $Ad^+$ clones from the reciprocal cross Hfr ($z_2^-\ Ad^+\ S^s$) $\times$ F$^-$ ($z_1^-\ Ad^-\ S^r$) were able to ferment lactose. What is the order of the markers relative to the $Ad$ locus?

**Solution:**

Assume that the order is $z_2$-$z_1$-$Ad$. In the first mating, four crossovers are required to produce a streptomycin resistant prototroph able to ferment lactose ($z_2^+ \, z_1^+ \, Ad^+ \, S^r$).

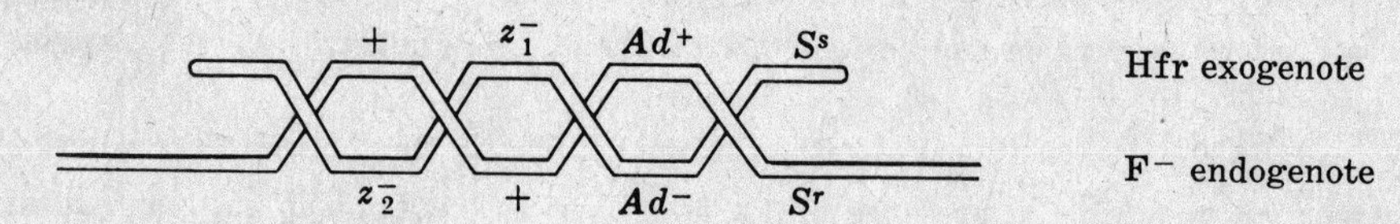


The reciprocal mating requires only two crossovers to produce a prototroph able to ferment lactose.

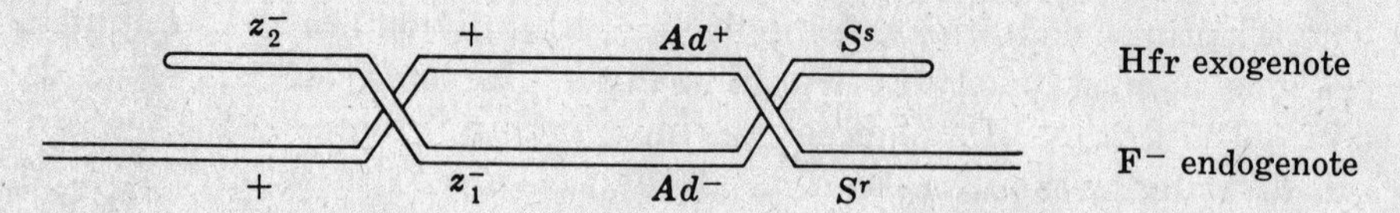


Double crossovers are expected to be much more frequent than quadruple crossovers. The above scheme does not fit the data because the first mating was more frequent than the reciprocal mating. Our assumption must be wrong.

Let us assume that the order is $z_1$-$z_2$-$Ad$. The first cross now requires a double crossover.

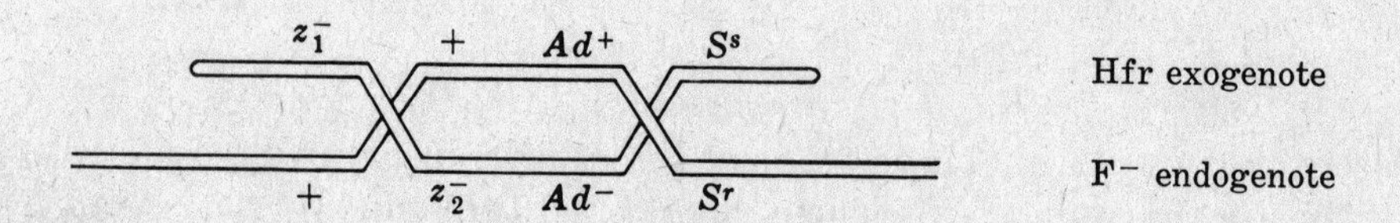


The reciprocal cross requires four crossover events.

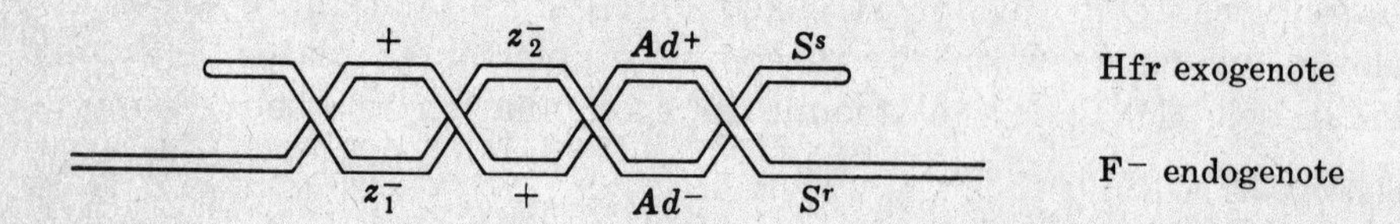


The reciprocal cross is expected to be much less frequent under this assumption and is in agreement with the observations.

**15.6.** Six mutations are known to belong to three cistrons. From the results of the complementation tests, determine which mutants are in the same cistron.

| | 1 | 2 | 3 | 4 | 5 | 6 |
|---|---|---|---|---|---|---|
| 1 | 0 | + | + | | | |
| 2 | | 0 | | + | + | |
| 3 | | | 0 | + | 0 | |
| 4 | | | | 0 | | + |
| 5 | | | | | 0 | + |
| 6 | | | | | | 0 |

+ = complementation
0 = non-complementation
blank = not tested

**Solution:**

Obviously mutations 3 and 5 are in the same cistron, since they fail to complement each other. Mutations 1 and 3 are in different cistrons, since they do complement each other. We will arbitrarily assign these to cistrons A and B.

| (3, 5) | 1 | |
|---|---|---|
| Cistron A | Cistron B | Cistron C |

1 and 2 are in different cistrons, but we do not know whether 2 is in A or C. However, 5 and 2 complement and therefore 2 cannot be either in cistron A or B and thus must be in C.

| (3, 5) | 1 | 2 |
|---|---|---|
| Cistron A | Cistron B | Cistron C |

3 and 4 complement; thus 4 must be in either B or C. But 2 and 4 also complement; thus 4 cannot be in C and must reside in B.

| (3, 5) | (1, 4) | 2 |
|---|---|---|
| Cistron A | Cistron B | Cistron C |

6 cannot be in A since it complements with 5. Thus 6 is either in B or C. Since 6 and 4 complement, they are in different cistrons. If 6 cannot be in A or B, it must be in C. The mutants are grouped into cistrons as shown below.

| (3, 5) | (1, 4) | (2, 6) |
|---|---|---|
| Cistron A | Cistron B | Cistron C |

## VIRUSES

**15.7.** In an attempt to determine the amount of recombination between two mutations in the $r_{II}$ region of phage T4, strain B of *E. coli* is doubly infected with both kinds of mutants. A dilution of $1:10^9$ is made of the lysate and plated on strain B. A dilution of $1:10^7$ is also plated on strain K. Two plaques are found on K, 20 plaques on B. Calculate the amount of recombination.

**Solution:**

In order to compare the numbers of plaques on B and K, the data must be corrected for the dilution factor. If 20 plaques are produced by a $10^9$ dilution, the lesser dilution ($10^7$) would be expected to produce 100 times as many plaques.

$$\text{Recombination percentage} = \frac{200(\text{no. of plaques on K})}{\text{no. of plaques on B}} = \frac{200(2)}{20(100)} = 0.2\%$$

**15.8.** Seven deletion mutants within the A cistron of the $r_{II}$ region of phage T4 were tested in all pairwise combinations for wild type recombinants. In the adjacent table of results, + = recombination, 0 = no recombination. Construct a topological map for these deletions.

| | 1 | 2 | 3 | 4 | 5 | 6 | 7 |
|---|---|---|---|---|---|---|---|
| 1 | 0 | + | 0 | 0 | + | 0 | 0 |
| 2 | | 0 | 0 | 0 | + | + | 0 |
| 3 | | | 0 | 0 | + | + | 0 |
| 4 | | | | 0 | + | 0 | 0 |
| 5 | | | | | 0 | 0 | 0 |
| 6 | | | | | | 0 | 0 |
| 7 | | | | | | | 0 |

**Solution:**

If two deletions overlap to any extent, no wild type recombinants can be formed.

(1) Deletion #1 overlaps with 3, 4, 6 and 7 but not with 2 or 5.

1 2, 5

3, 4, 6, 7

(2) Deletion #2 overlaps with 3, 4 and 7 but not with 1, 5 or 6.

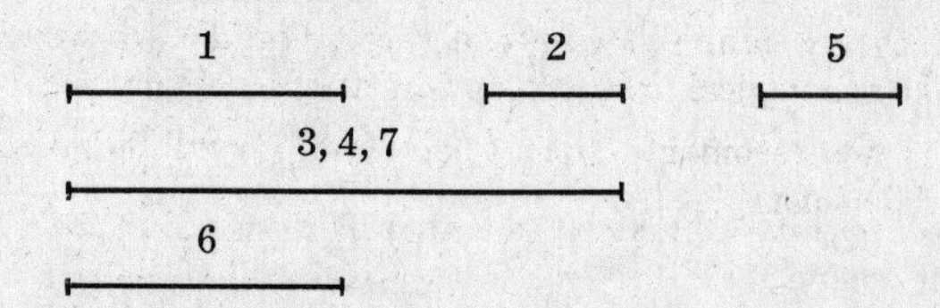


(3) Deletion #3 overlaps with 1, 2, 4 and 7 but not with 5 or 6.

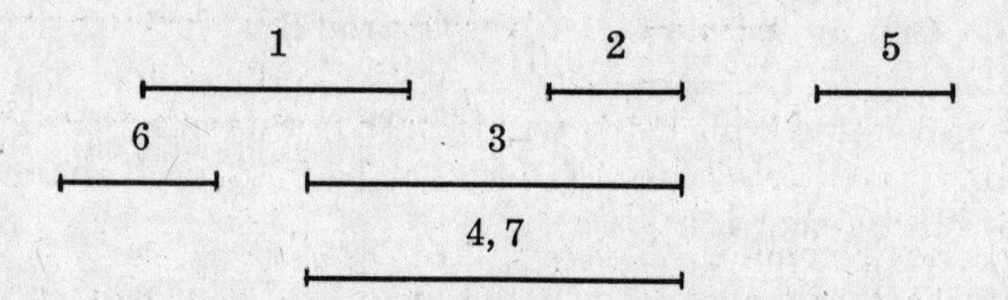


(4) Deletion #4 overlaps with 1, 2, 3, 6 and 7 but not with 5.

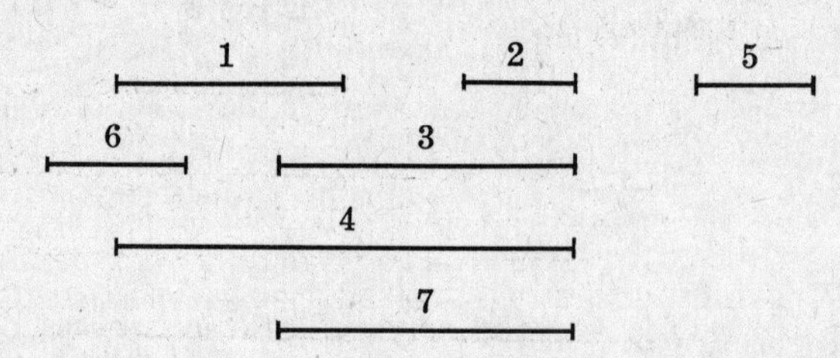


(5) Deletion #5 overlaps with 6 and 7 but not with 1, 2, 3 or 4. To satisfy these conditions we will shift 5 to the left of 1 and extend 7 into part of region 5 so that the overlaps in step (4) have not changed. Now segment 5 can overlap 6 and 7 without overlapping 1, 2, 3 or 4.

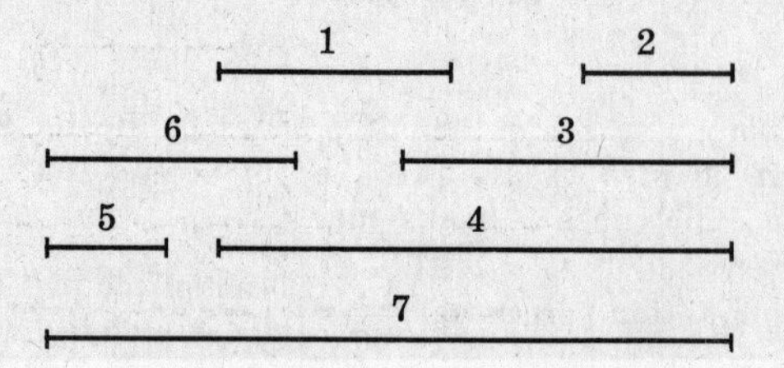


(6) Deletion #6 overlaps 1, 4, 5 and 7 but not 2 or 3. No change is required.

(7) Deletion #7 overlaps all other regions, just as it was temporarily diagrammed in step (5). This completes the topological map. Although the overlaps satisfy the conditions in the table, we have no information on the actual lengths of the individual deletions.

**15.9.** Five point mutations (*a* through *e*) were tested for wild type recombinants with each of the seven deletion mutants in the above problem. Determine the order of the point mutations and modify the topological map accordingly.

| | 1 | 2 | 3 | 4 | 5 | 6 | 7 |
|---|---|---|---|---|---|---|---|
| *a* | 0 | + | 0 | 0 | + | + | 0 |
| *b* | + | + | + | 0 | + | 0 | 0 |
| *c* | + | + | + | + | 0 | 0 | 0 |
| *d* | 0 | + | + | 0 | + | 0 | 0 |
| *e* | + | 0 | 0 | 0 | + | + | 0 |

**Solution:**

A deletion mutation cannot recombine to give wild type with a point mutant which lies within its boundaries. The topological map was developed in Problem 15.8.

(1) Mutant *a* does not recombine with 1, 3, 4 and 7 and therefore must be in a region common to all of these deletions.

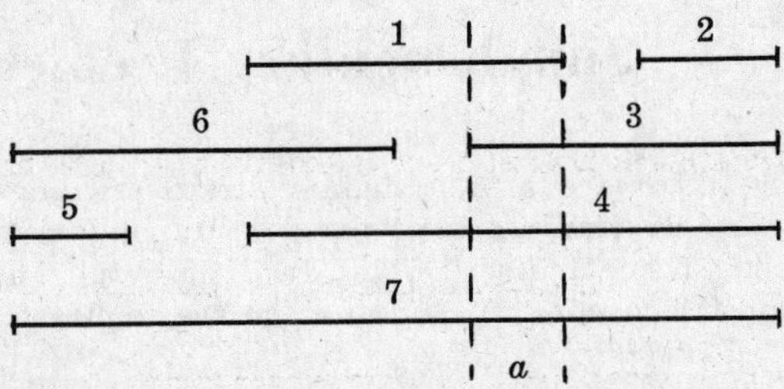


(2) Mutant *b* does not recombine with 4, 6 and 7 and thus lies in a region common to these three deletions. As the topological map stands in step (1), a point mutant could not be in regions 4, 6 and 7 without also being in region 1. Therefore this information allows us to modify the topological map by shortening deletion 1, but still overlapping deletion 6. This now gives us a region in which *b* can exist.

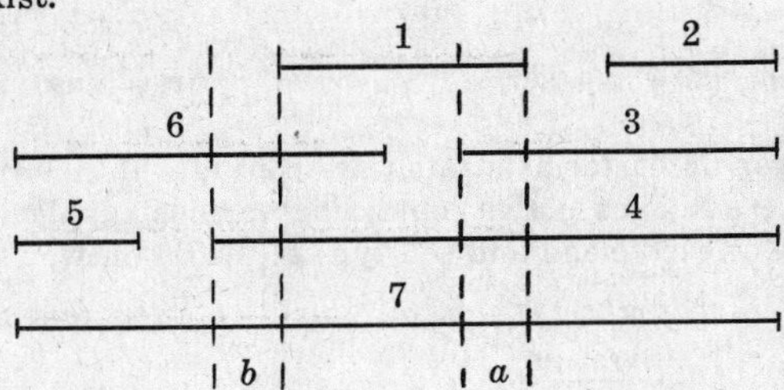


(3) Mutant *c* lies in a region common to deletions 5, 6 and 7. Mutant *d* lies in a region common to deletions 1, 4, 6 and 7. Mutant *e* lies in a region common to deletions 2, 3, 4 and 7.

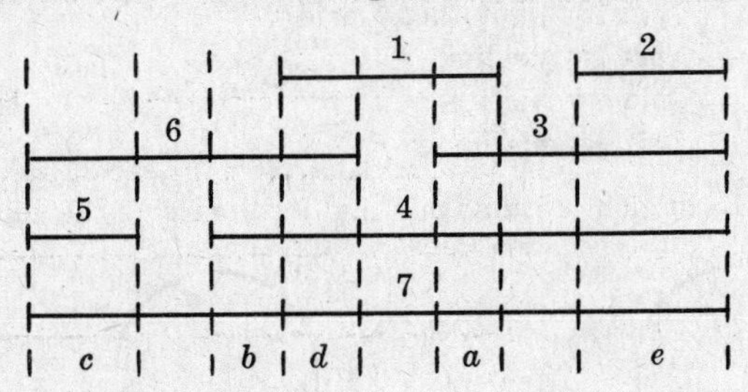


Thus the order of these point mutations is *c-b-d-a-e* and the topological map is modified as shown in step 2.

**15.10.** Propose a procedure for establishing the location of a lambda prophage with respect to other bacterial genes.

**Solution:**

Cross a donor Hfr cell (lysogenic for lambda) with a nonlysogenic $F^-$ recipient. Once the lambda genome has been transferred via conjugation to the $F^-$ recipient cell, it often will die by lysis (a phenomenon known as *zygotic induction*). Using the interrupted mating technique (blender treatment), it should be possible to determine the point at which the frequencies of origin-proximal recombinants decrease with time. At this point, all of the lambda genome has been donated and it can therefore be related on the temporal map to the location of other bacterial genes.

Since recipient cells contain no repressor (and no unbound repressor is likely to be transferred during conjugation), the exogenous prophage has a good chance of entering the lytic cycle because relatively high levels of repressor are required to establish the lysogenic state within the $F^-$ recipient cell. A lysogenic cell usually contains sufficient unbound ("cytoplasmic") repressor to inhibit superinfection by one or a few lambda phages, but a nonlysogenic cell has no immunity to infection by exogenous lambda phages or lambda prophages. Hence, in the latter case, zygotic induction of exogenous lambda prophages has a good chance of occurring. When lambda enters a nonlysogenic cell, a race occurs between the production of immunity repressor and "repressor of immunity repressor". The outcome of this race determines whether the cell becomes lysogenic or lytic respectively.

An alternative procedure for mapping the location of prophage lambda is by mating a nonlysogenic Hfr strain with an $F^-$ strain lysogenic for lambda and studying the loss of immunity to superinfection by lambda through recombination. However, this would be much more laborious than the previous procedure.

# Supplementary Problems

## BACTERIA

**15.11.** Approximately $10^8$ *E. coli* cells of a mutant strain are plated on complete medium forming a bacterial lawn. Replica plates are prepared containing minimal medium supplemented by the amino acids arginine, lysine and serine. (*a*) From the results, determine the genotype of the mutant strain. (*b*) Explain the colonies which appear on the replica plates.

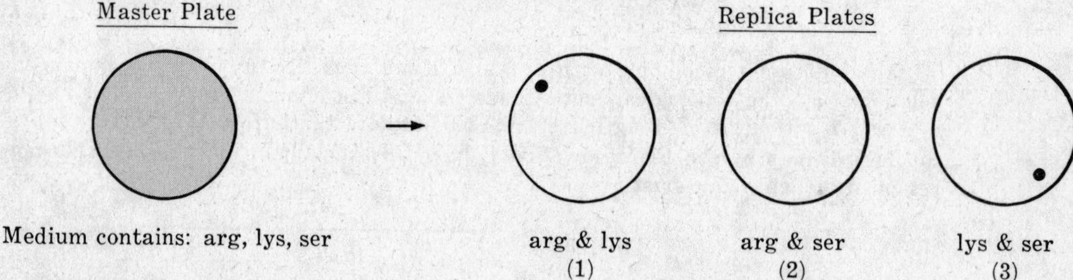


**15.12.** Two triple auxotrophic bacterial strains are conjugated in broth, diluted and plated onto complete agar (master plate). Replica plates containing various supplements are then made from the master. From the position of each clone and the type of media on which it is found, determine its genotype.

$$met^- thr^- pan^- pro^+ bio^+ his^+ \times met^+ thr^+ pan^+ pro^- bio^- his^-$$

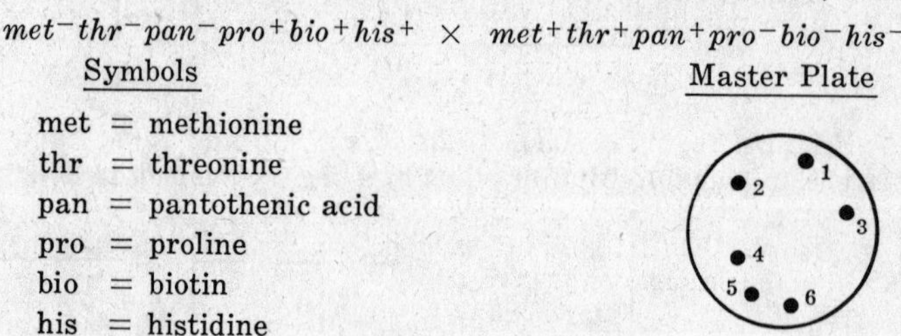


Replica plates: Each dish contains minimal medium plus the supplements shown at the bottom.

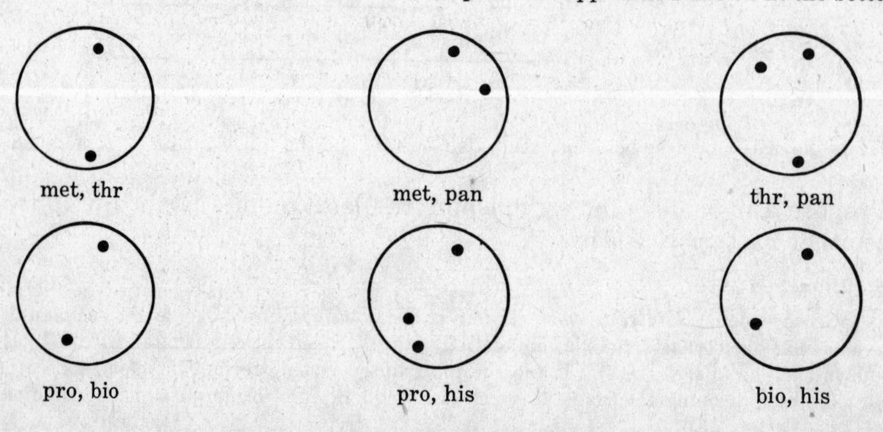

**15.13.** A bacterial strain unable to synthesize methionine ($met^-$) is transduced by a strain unable to synthesize isoleucine ($ileu^-$). The broth culture is diluted and plated on minimal medium supplemented with isoleucine. An equivalent amount of diluted broth culture is plated on minimal medium. Eighteen clones appeared on the minimal plates and 360 on the isoleucine plates. Calculate the standardized recombination ratio.

**15.14.** Calculate the standardized recombination ratio between two mutants of the arginine locus, using the following information. A bacterial strain which is doubly auxotrophic for thymine and arginine ($thy^-\ arg^-_{E1}$) is transformed by using a high concentration of DNA from a single auxotrophic strain ($thy^+\ arg^-_{E2}$). Identical dilutions are plated on minimal medium and on minimal plus arginine. For every colony which appears on the unsupplemented plates there are about 120 colonies on the serine-supplemented plates.

**15.15.** Two mutants at the tryptophan locus, $try^-_A$ and $try^-_B$, are known to be to the right of a cysteine locus (*cys*). A bacterial strain of genotype $cys^+\ try^-_A$ is transduced by phage from a strain which is $cys^-\ try^-_B$. The reciprocal cross is also made wherein the strain $cys^-\ try^-_B$ is transduced by phage from a strain which is $cys^+\ try^-_A$. In both cases the number of prototrophic recombinants are equivalent. Determine the order of the tryptophan mutants relative to the cysteine marker.

**15.16.** A cross is made between the streptomycin resistant ($S^r$) $F^-$ strain marked with the characters $Gal^-T^-Az^rLac^-T1^rMal^-Xyl^-L^-$ and the prototrophic Hfr strain having opposite characters. After 60 minutes of contact, samples are transferred to plates with minimal medium plus streptomycin. The original mixture is in the ratio of $2 \times 10^7$ Hfr to $4 \times 10^8$ $F^-$. The frequencies (percentages) of the Hfr characters recovered in the $T^+L^+S^r$ recombinants are as follows: 72% $T1^s$, 0% $Mal^+$, 27% $Gal^+$, 91% $Az^s$, 0% $Xyl^+$, 48% $Lac^+$. (*a*) How many $F^-$ cells exist in the original mixture for every Hfr cell? (*b*) What is the counterselective agent which prevents the Hfr individuals from obscuring the detection of recombinants? (*c*) In what order are these genes probably being transmitted by the Hfr?

**15.17.** Four Hfr strains of *E. coli* are known to transfer their genetic material during conjugation in different sequences. Given the time of entry of the markers into the $F^-$ recipient, construct a genetic map which includes all of these markers and label the time distance between adjacent gene pairs.

| | | | | | | | |
|---|---|---|---|---|---|---|---|
| Strain 1. | Markers: | *arg* - | *thy* - | *met* - | *thr* | | |
| | Time in minutes: | 15 | 21 | 32 | 48 | | |
| Strain 2. | Markers: | *mal* - | *met* - | *thi* - | *thr* - | *try* | |
| | Time in minutes: | 10 | 17 | 22 | 33 | 57 | |
| Strain 3. | Markers: | *phe* - | *his* - | *bio* - | *azi* - | *thr* - | *thi* |
| | Time in minutes: | 6 | 11 | 33 | 48 | 49 | 60 |
| Strain 4. | Markers: | *his* - | *phe* - | *arg* - | *mal* | | |
| | Time in minutes: | 18 | 23 | 35 | 45 | | |

**15.18.** Abortive transductants are relatively stable merozygotes which can be used for complementation tests. Six mutants were tested in all pairwise combinations, yielding the results shown in the table (+ = complementation, 0 = non-complementation). Construct a complementation map consistent with the data.

| 1 | 2 | 3 | 4 | 5 | 6 | |
|---|---|---|---|---|---|---|
| 0 | + | 0 | + | + | + | 1 |
| | 0 | 0 | + | + | + | 2 |
| | | 0 | + | + | + | 3 |
| | | | 0 | 0 | + | 4 |
| | | | | 0 | 0 | 5 |
| | | | | | 0 | 6 |

**15.19.** Six point mutants are known to reside in three cistrons. Complete the following table where + = complementation and 0 = non-complementation.

| | 1 | 2 | 3 | 4 | 5 | 6 |
|---|---|---|---|---|---|---|
| 1 | 0 | + | | + | + | + |
| 2 | | 0 | + | | | + |
| 3 | | | 0 | 0 | | |
| 4 | | | | 0 | | |
| 5 | | | | | 0 | + |
| 6 | | | | | | 0 |

**15.20.** Given the topological map of 6 deletion mutants shown below, predict the results of recombination experiments involving the 5 point mutants ($a$ thru $e$) with each of the 6 deletions (1 thru 6). Complete the accompanying table of results by using + for recombination and 0 for no recombination.

Deletions

| | 1 | 2 | 3 | 4 | 5 | 6 |
|---|---|---|---|---|---|---|
| *a* | | | | | | |
| *b* | | | | | | |
| *c* | | | | | | |
| *d* | | | | | | |
| *e* | | | | | | |

**15.21.** Five point mutations ($a$ thru $e$) were tested for wild type recombinants with each of the five deletions shown in the topological map below. The results are listed in the table below (+ = recombination, 0 = no recombination). Determine the order of the point mutations.

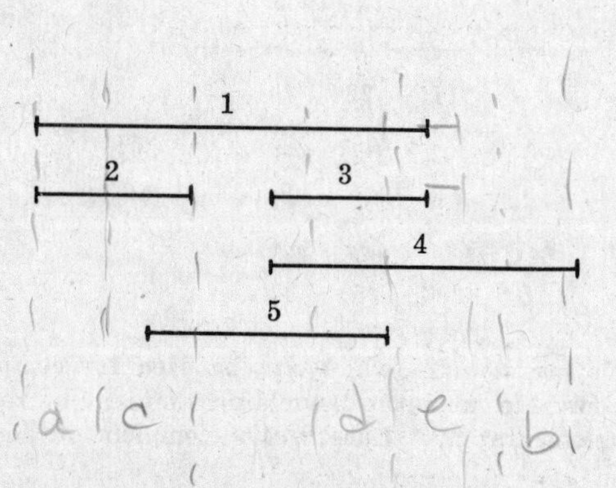


Deletions

| | 1 | 2 | 3 | 4 | 5 |
|---|---|---|---|---|---|
| *a* | 0 | 0 | + | + | + |
| *b* | + | + | + | 0 | + |
| *c* | 0 | 0 | + | + | 0 |
| *d* | 0 | + | 0 | 0 | 0 |
| *e* | 0 | + | 0 | 0 | + |

**15.22.** Several lines of evidence suggest that the circular chromosome of *E. coli* has only one or two replicating Y-forks. The length of one whole unreplicated nucleus is 1300 μm (about 500 times longer than the *E. coli* cell). There are ten base pairs per one complete turn of the DNA double helix, equivalent to 34 Å or $3.4 \times 10^{-3}$ μm. (*a*) How many nucleotide base pairs are in the *E. coli* DNA complement or genome? (*b*) If the *E. coli* genome is replicated in 40 minutes at 37°C by two replicating forks, how many revolutions per minute (rpm) must the parental double helix make to allow separation of its complementary nucleotide strands during replication?

**15.23.** DNA damage (mutation) is an essential initiation event for a cell to transform into a cancerous state, but it is not the only event causing cancer. Therefore, DNA-damaging agents (mutagens) are only *potential* carcinogens (agents causing cancer). Most chemical carcinogens are not biologically active in their original form; they must first be metabolized to carcinogenic metabolites. Bruce Ames devised a test for screening chemicals for their potenital carcinogenic properties. The *Ames test* is currently the standard test for a quantitative estimate of the mutagenic potency of a chemical. This test employs an auxotrophic strain of *Salmonella typhimurium* that cannot make the amino acid histidine (*his*$^-$). To increase the sensitivity of the tester strain, (1) it carries a mutation that makes the cell envelope more permeable to allow penetration of the test chemicals, (2) its capacity for excision repair is eliminated so that most of the primary lesions remain unhealed, and (3) a genetic element that makes DNA replication more error prone is introduced via a plasmid. Rat liver extract is added to a minimal medium culture plate coated with a thin layer of these bacteria. The chemical to be tested is impregnated in a disc of filter paper; the paper is placed in the center of the plate. After two days of incubation, the number of colonies are counted. (*a*) What events are being scored by the colony counts? (*b*) Why was mammalian liver extract added to the test? (*c*) Diagram the expected distribution of colonies on a plate containing a known carcinogen. Explain why this distribution develops. (*d*) Suppose that the test chemical (e.g. nitrosoguanidine) is mixed with the bacteria prior to plating at two dosages (low and high). A control is run simultaneously with these two doses. Diagram the expected distribution of colonies on these three plates.

**15.24.** When bacterial DNA is damaged by a mutagenic agent, excision repair normally operates to repair the lesion. This process is less than 100% efficient, however, so that some residual lesions remain unrepaired. If these lesions delay replication of DNA, an error prone "SOS repair" system becomes operative involving activation and increased production of a multifunctional protein called RecA protein (for "recombination"). RecA protein interferes with cell partition, resulting in elongation of cells into filaments. RecA protein also cleaves lambda repressor; this repressor must remain intact for the virus to remain dormant as a prophage. *E. coli* strain B is lysogenic for lambda; strain A is not lysogenic for lambda. This knowledge led Moreau, Bailone and Devoret to devise a "prophage induction test" or "inductest" for potential carcinogens. Lysogenic strain B of *E. coli* is made defective in its excision repair system and genetically modified to make the cell envelopes permeable to a wide variety of test chemicals. This special strain is mixed with indicator strain A and rat liver extract; the mixture is then plated; the medium is covered with a thin layer of indicator bacteria interspersed with a few lysogenic bacteria. The test chemical is applied to a filter paper disc and placed in the center of the plate for a "spot test". (*a*) After incubation, how is DNA damage assayed? (*b*) Why is strain A required as an indicator? (*c*) What advantage does an inductest have over an Ames test? (*d*) Explain the selective advantage of lysogenic induction. (*e*) Genetic engineers have spliced the gene for galactokinase into a bacterial chromosome, thereby creating an organism for assaying mutagens by an enzymatic activity test. Where was this gene inserted into the chromosome and how does the system work?

## VIRUSES

**15.25.** Six deletion mutants within the A cistron of the *r*II region of phage T4 were tested in all pairwise combinations for wild type recombinants. In the following table, + = recombination, 0 = no recombination. Construct a topological map for these deletions.

| | 1 | 2 | 3 | 4 | 5 | 6 |
|---|---|---|---|---|---|---|
| 1 | 0 | 0 | 0 | 0 | 0 | 0 |
| 2 | | 0 | 0 | 0 | 0 | + |
| 3 | | | 0 | + | 0 | 0 |
| 4 | | | | 0 | + | + |
| 5 | | | | | 0 | + |
| 6 | | | | | | 0 |

**15.26.** Five deletion mutants within the B cistron of the *r*II region of phage T4 were tested in all pairwise combinations for wild type recombinants. In the following table of results, + = recombination, 0 = no recombination. Construct a topological map for these deletions.

| | 1 | 2 | 3 | 4 | 5 |
|---|---|---|---|---|---|
| 1 | 0 | + | + | 0 | 0 |
| 2 | | 0 | + | 0 | + |
| 3 | | | 0 | + | 0 |
| 4 | | | | 0 | 0 |
| 5 | | | | | 0 |

**15.27.** The DNA of bacteriophage T4 contains approximately 200,000 nucleotide pairs. The $r_{\text{II}}$ region of the T4 genome occupies about 1% of its total genetic length. Benzer has found about 300 sites are separable by recombination within the $r_{\text{II}}$ region. Determine the average number of nucleotides in each recon.

**15.28.** The molecular weight of DNA in phage T4 is estimated to be $160 \times 10^6$. The average molecular weight of each nucleotide is approximately 400. The total genetic map of T4 is calculated to be approximately 2500 recombination units long. With what frequency are $r^+$ recombinants expected to be formed when two different $r$ mutants (with mutations at adjacent nucleotides) are crossed?

**15.29.** A number of mutations were found in the $r_{II}$ region of phage T4. From the recombination data shown in the table on the right, determine whether each mutant is a point defect or a deletion (+ = recombination, 0 = no recombination). Two of the four mutants have been known to undergo back mutation; the other two have never been observed to back mutate. Draw a topological map to represent your interpretation.

| | 1 | 2 | 3 | 4 |
|---|---|---|---|---|
| 1 | 0 | 0 | 0 | + |
| 2 | | 0 | + | 0 |
| 3 | | | 0 | + |
| 4 | | | | 0 |

**15.30.** *E. coli* strain B is doubly infected with two $r_{II}$ mutants of phage T4. A $6 \times 10^7$ dilution of the lysate is plated on *coli* B. A $2 \times 10^5$ dilution is plated on *coli* K. Twelve plaques appeared on strain K, 16 on strain B. Calculate the amount of recombination between these two mutants.

**15.31.** Propose a simple mechanism by which ultraviolet light could induce lambda prophages into a lytic state.

**15.32.** A nonlytic response usually is observed in lysogenic (λ) *E. coli* cells when conjugated with nonlysogenic Hfr donors or in crosses between Hfr (λ) × $F^-$(λ). The donated prophage is almost never inherited by the recombinants. Lysis is very anomalous in crosses of Hfr (λ) × $F^-$. Explain these observations.

**15.33.** Temperate phages such as lambda produce turbid plaques; virulent phages always produce clear plaques. (*a*) Offer an explanation for the turbid plaques. (*b*) Some lambda mutants produce clear plaques. What genetic locus is most likely mutant in these cases?

**15.34.** What is expected to happen to an *E. coli* cell in which: (*a*) the level of lambda *c*III gene product is sufficiently high to permit initiation of transcription of the *c*I gene at the $p_{RE}$ promoter? (*b*) the level of *c*III gene product is not sufficient to permit initiation of transcription at the $p_{RE}$ promoter?

## Answers to Supplementary Problems

**15.11.** (*a*) $arg^- lys^- ser^-$ (triple auxotroph) (*b*) plate (1) contains a mutation to $ser^+$, plate (3) contains a mutation to $arg^+$.

**15.12.**

| Clone | Genotype of Clone | | | | | |
|---|---|---|---|---|---|---|
| | *met* | *thr* | *pan* | *pro* | *bio* | *his* |
| 1 | + | + | + | + | + | + |
| 2 | + | − | − | + | + | + |
| 3 | − | + | − | + | + | + |
| 4 | + | + | + | + | + | − |
| 5 | + | + | + | − | + | + |
| 6 | + | − | + | + | + | + |

**15.13.** 20% recombination

**15.14.** 0.83% recombination

**15.15.** *cys* - $try_B$ - $try_A$

**15.16.** (*a*) 20 (*b*) streptomycin; Hfr is streptomycin sensitive ($S^s$) (*c*) origin - ($T^+L^+$) - $Az^s$ - $T1^s$ - $Lac^+$ - $Gal^+$ - $S^s$ - ($Mal^+Xyl^+$). *Note*: The order of markers within parentheses cannot be determined from the data.

**15.17.** *arg* - *thy* - *mal* - *met* - *thi* - *thr* - *azi* - *bio* - *try* - *his* - *phe* - *arg*

6 4 7 5 11 1 15 8 14 5 12

**15.18.**

1 ——— 2 ——— 4 ——— 6 ———

3 ——————— 5 ———————

**15.19.**

| 2, 5 | 1 | 3, 4, 6 |
|---|---|---|
| Cistron A | Cistron B | Cistron C |

| | 1 | 2 | 3 | 4 | 5 | 6 |
|---|---|---|---|---|---|---|
| 1 | 0 | + | + | + | + | + |
| 2 | | 0 | + | + | 0 | + |
| 3 | | | 0 | 0 | + | 0 |
| 4 | | | | 0 | + | 0 |
| 5 | | | | | 0 | + |
| 6 | | | | | | 0 |

*Note*: Name of cistron is arbitrary.

**15.20.**

| Point Mutations | Deletions | | | | | |
|---|---|---|---|---|---|---|
| | 1 | 2 | 3 | 4 | 5 | 6 |
| *a* | 0 | + | + | 0 | 0 | + |
| *b* | + | + | 0 | + | + | 0 |
| *c* | + | 0 | 0 | + | + | + |
| *d* | 0 | 0 | + | + | 0 | + |
| *e* | + | 0 | 0 | + | + | 0 |

**15.21.** *a* - *c* - *d* - *e* - *b*

**15.22.** (*a*) $\dfrac{1300\ \mu\text{m} \times 10\ \text{bp/turn}}{3.4 \times 10^{-3}\ \mu\text{m/turn}} = 3.9 \times 10^6$ or 3900 kilobase (kb) pairs

(*b*) Rate of chain growth $= \dfrac{3900\ \text{kb}}{2(2400\ \text{sec})} = 0.8\ \text{kb/sec}$

$$\frac{800\ \text{bp/sec}}{10\ \text{bp/rev}} \times 60\ \text{sec/min} = 80\ \text{rev} \times 60\ \text{min}^{-1} = 4800\ \text{rev/min}$$

or about as fast as a laboratory centrifuge.

**15.23.** (*a*) Back mutations (reverse mutations) from *his*$^-$ to *his*$^+$. (*b*) It supplies the mammalian metabolic functions that are usually required to convert a chemical into its carcinogenic metabolites. (*c*) After two days, most of the *his*$^-$ bacteria have died for lack of histidine. Back mutation rates are expected to be proportional to concentration of the chemical that forms a radially diminishing concentration gradient around the paper disc. Close to the disc there is a zone in which no cells grow because of toxic levels of the chemical. Beyond this zone there may be so many *his*$^+$ revertants that the cells almost form a continuous lawn. At the periphery are a few larger clones (because they are isolated) representing spontaneous *his*$^+$ mutants that have not been exposed to the chemical.

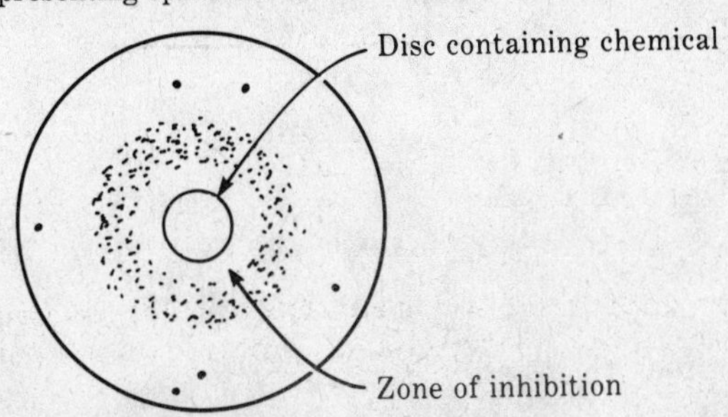

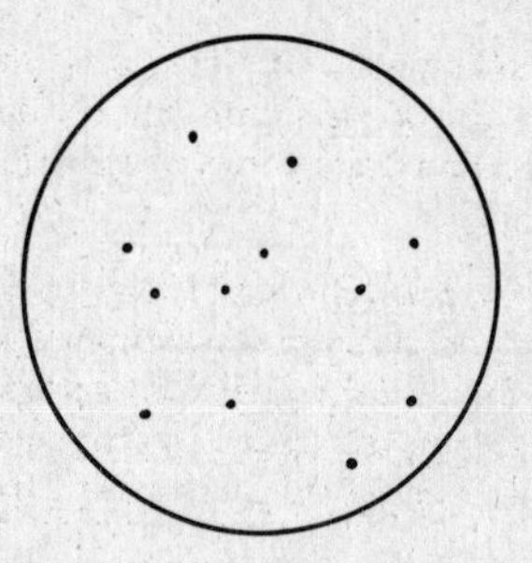
Control: no nitrosoguanidine; spontaneous $his^+$ revertants

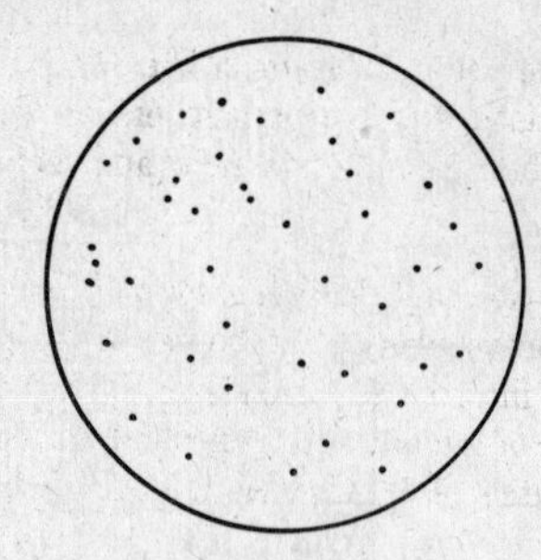
Low dosage of nitrosoguanidine

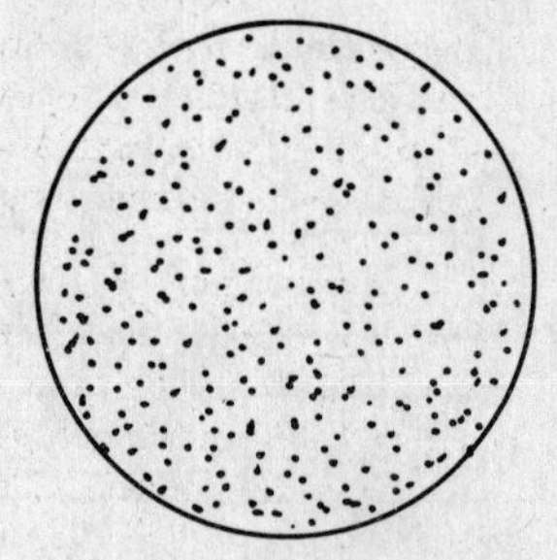
High dosage of nitrosoguanidine

**15.24.** (*a*) DNA damage activates RecA protein that then cleaves lambda repressor and opens up the viral genome for replication (induction). The cell bursts and releases viruses that infect and lyse the indicator strain A, causing plaques (holes) to appear in the bacterial "lawn" surrounding the paper disc. (*b*) If a cell of strain B is induced to lysis, the viruses cannot multiply in other cells of the same strain because active lambda repressor is present in these cells as a product of their prophages. Therefore, a nonlysogenic strain (A) is required to indicate how many viruses have been induced by the chemical treatment. (*c*) The inductest can assay a potential carcinogen at doses that would kill the tester bacteria in an Ames test (giving a false negative reaction). The Ames test only detects the rare back mutations of $his^-$ to $his^+$, whereas DNA damage at any site can initiate lysogenic induction (a mass effect, independent of cell survival by toxic chemicals). (*d*) If DNA of the host cell cannot replicate, the cell is likely to die. Under these conditions it would be advantageous for a prophage to enter the lytic cycle and thereby possibly infect a "healthier" cell (like a "rat leaving a sinking ship"). (*e*) The gene for galactokinase was inserted adjacent to (and under the control of) the lambda repressor. When the mutagen damages DNA, RecA protein is activated and cleaves the repressor; this opens the operon to RNA polymerase and allows synthesis of the enzyme galactokinase, the activity of which can be quantitated spectrophotometrically when supplied with its substrate.

**15.25.**

1

2

3

4 5 6

**15.26.**

3 1 2

5

4

**15.27.** Approximately 7 nucleotides per recon

**15.28.** 0.00625% of all progeny are expected to be $r^+$ recombinants.

**15.29.**

$m_1$

$m_2$

$m_3$

$m_4$

$m_1$ and $m_2$ are deletions, $m_3$ and $m_4$ are point mutations.

**15.30.** 0.5% recombination

**15.31.** UV light destroys the immunity repressor.

**15.32.** Repressor is already present in the cytoplasm of the $F^-$ (λ) recipient cell. It binds to $o_R$ and $o_L$ operators and prevents vegetative reproduction of the prophage either in the $F^-$ (λ) or in the Hfr (λ) donated chromosome segment. Early enzymes are therefore not produced and recombination (leading to the "inheritance" of the donated lambda genes) cannot occur.

Nonlysogenic $F^-$ cells do not contain repressor. There are so few repressor molecules in a lysogenized cell that it is unlikely that free repressor would be bound to the newly synthesized donor fragment that moves almost immediately through the pilus into the $F^-$ recipient. When the prophage from Hfr (λ) enters the $F^-$ cell, a race occurs between the production of lambda repressor and an early protein of vegetative phage development (such as the *cro* gene product). The outcome of this race is not predictable; hence lysis is anomalous in such crosses.

**15.33.** (*a*) Turbid plaques are due to secondary growth of lysogenized bacteria of the lambda-sensitive indicator strain. (*b*) Mutations of the *c*I locus are likely to produce a defective repressor; hence these mutants cannot lysogenize the indicator strain.

**15.34.** (*a*) Immunity repressor synthesis is removed from the negative control by the *cro* gene product at the $p_{RM}$ promoter. High enough concentrations of repressor can be attained to allow establishment of lysogenization. (*b*) Synthesis of repressor continues to depend on transcription of the *c*I gene initiated at the $p_{RM}$ promoter, and hence is subject to repression by the *cro* gene product. This condition favors the lytic response.

# Chapter 16

## Molecular Genetics

### HISTORY

Prior to discovery of the chemical structure of the genetic material, the "gene" was an abstract, indivisible unit of heredity (comparable to the old concept of the indivisible atom). This period in history is referred to as *classical* or *formal genetics*. The word "formal" pertains to the extrinsic aspect of something as distinguished from its substance or material. Classical genetics has been extremely successful in elucidating many basic biological principles without understanding the nature of the gene. The era of *molecular genetics* followed the discovery of DNA structure when the fundamental unit of heredity was determined to be the DNA nucleotide and the "gene" was found to consist of an aggregate of nucleotides.

The histories of most scientific disciplines are generally characterized by relatively long periods of stagnation punctuated by bursts of rapid progress. Most of these flurries of research are initiated by new technical developments. This is certainly true of biochemistry and molecular biology. At least three major areas of technology have been influential in this respect: (1) instrumentation and techniques, (2) radioactive tracers and (3) nucleic acid enzymology.

#### 1. Instrumentation and Techniques.

The analytical ultracentrifuge was developed in the 1920's by Theodor Svedberg. The sedimentation rate of a substance during ultracentrifugation is mainly a function of its size and secondarily of its shape. The unit of sedimentation (S, in honor of Svedberg) is an expression of these parameters. This instrument has been modified for isolating organelles such as nuclei, ribosomes, mitochondria and chloroplasts. It can be used for determining the minimum number of kinds of macromolecules in a biological specimen and for estimating the molecular weights of macromolecules.

The electron microscope was invented in the 1930s, and eventually enabled the direct visualization not only of cellular substructures but also of viruses and macromolecules. Circular genetic maps of microorganisms have been shown by electron microscopy to have a corresponding circular physical structure. Multiple ribosomes attached to an mRNA molecule (polysomes) have also been visualized by this instrument.

Electrophoresis is a technique that separates molecules according to their net charge in an electric field, usually on solid or semisolid support media such as paper or agar. Linus Pauling used this technique to differentiate sickle cell hemoglobin from normal hemoglobin and determined (by protein sequence analysis) that the difference in electrophoretic mobilities of these proteins was due to a single amino acid difference in the β-chains. Walter Gilbert and Allan Maxam developed an electrophoretic technique for rapid nucleotide sequencing of DNA fragments up to about 100 base pairs in length. Electrophoresis has been extensively used to differentiate *isozymes*, i.e. proteins possessing the same enzymatic properties but differing in primary structure.

X-ray diffraction data from crystalline materials have been analyzed by electronic computers to help elucidate the three-dimensional shapes of nucleic acids (e.g. DNA, tRNA's) and proteins (e.g. myoglobins, viral capsomeres, enzymes).

During the mid 1940s and early 1950s, various forms of chromatography were perfected, enabling molecules to be separated by differences in solubilities in organic solvents, electrical charge, molecular weight, and specific binding properties for the support medium, or combinations of these factors. Erwin Chargaff used paper chromatography to determine the base compositions of DNA's from various sources. He found that the molecular ratio of adenine was equivalent to that of thymine and the ratio of guanine equals that of cytosine. This was a vital clue in the search for DNA structure utilized by Watson and Crick.

Several techniques have been developed to separate, rejoin or break nucleic acid molecules. Separation of the complementary chains of a DNA molecule is known as *denaturation*. DNA is denatured if placed in distilled water or when boiled. The latter process is referred to as *melting*. Separation of DNA strands can be detected by spectrophotometric instruments; *optical density* (OD) or *absorbance* at 260 nm increases during the melting process. The temperature at which the increase in $Od_{260}$ is 50% of that attained when strand separation is complete is known as the *melting temperature* ($T_m$). Because G and C base pair by three hydrogen bonds, whereas A and T base pair by two hydrogen bonds, the higher the G-C content in DNA the higher the melting temperature. Melting is enhanced where there are clusters of A's and T's, and also when the purines (A, G) are on one strand and all the pyrimidines (T, C) are on the other strand.

If DNA is boiled and then quickly cooled, the strands will remain single; if cooled slowly, complementary strands will base pair and reform double-helical DNA molecules. This process is called *renaturation* or *annealing*. Hybrid DNA-RNA molecules can be produced by analogous processes from single strands. RNA can be totally hydrolyzed to nucleotides by exposure to high pH (alkali). This property can be used to purify DNA from a mixture of DNA and RNA. Single-stranded DNA will bind to membranous filters made of nitrocellulose; RNA will pass through such filters. However, if single-stranded RNA is complementary to nitrocellulose-bound single strands of DNA, it will form DNA-RNA hybrid molecules and be retained by such a filter.

There are two main methods for breaking long DNA molecules into fragments of suitable size for base sequencing or for recombinant engineering: (1) shear degradation and (2) restriction endonuclease treatment. If a solution of DNA is subjected to the stirring forces of a Waring blender or forced through a narrow tube or orifice, the ends of long DNA strands will usually move at different speeds; this stretches the DNA and tends to break it near the middle. This phenomenon is called *shear degradation*. The higher the stirring speed or velocity of flow through an orifice, the greater the shearing force. The effectiveness of any shearing force increases with molecular size of the DNA, but decreases with concentration (because entanglement of DNA molecules reduces the effective stretching).

The use of restriction endonucleases to fragment DNA at specific sites is discussed elsewhere in this chapter under the heading of Nucleic Acid Enzymology.

## 2. Radioactive Tracers.

Radioactive elements can be used as highly sensitive labels for detecting minute amounts of specific macromolecules.

**Example 16.1.** A. D. Hershey and M. Chase differentially labeled the nucleic acid and the protein components of T2 phages. They used radioactive $^{32}P$ in place of normal $^{31}P$ to label DNA; radioactive $^{35}S$ was used in place of normal $^{32}S$ to label protein (cysteine and methionine are two amino acids that contain sulfur). Since there is no phosphorous in proteins and no sulfur in nucleic acids, the fate of both viral components could be followed during the viral life cycle. After allowing the phages to become attached to sensitive *Escherichia coli* host cells, the mixture was subjected to the shearing forces of a Waring blender. The mixture was centrifuged to sediment the cells and then activity characteristic of each radionuclide was assayed in the pellet and in the supernatant fluid. All of the $^{32}P$ activity was found in the bacterial pellet and virtually all of the $^{35}S$ was found in the supernate. $^{32}P$ was found in some progeny phages, but no $^{35}S$ was found. The inference is that phages inject their DNA into host cells. Blender treatment shears the phage tail fibers from receptor sites on host cells; the empty phage protein capsids (ghosts) are therefore left free in the supernate. Semiconservative replication from the infecting $^{32}P$-labeled DNA caused some progeny phages to be released with one of the original radioactively labeled infecting strands. This experiment was the first to demonstrate that DNA and not protein is genetic material in phages.

DNA labeled with radioactive nuclides can reveal its own presence in a photographic technique called *autoradiography*. A preparation of the DNA on a plate is covered with a thin layer of photographic emulsion. Electrons released by disintegration of the radioactive elements cause darkening of the silver grains in the emulsion. After several weeks in the dark, the preparation is developed. It was by autoradiography that John Cairns found the theta-intermediate of circular DNA replication.

Radioactively labeled thymine can be used to differentiate DNA from RNA molecules because uracil usually replaces thymine in RNA. Tritium ($^{3}H$) is a radioactive isotope of hydrogen commonly used to label thymidine. The product is called *tritiated thymidine*. By allowing thymine deficient ($Thy^-$) *E. coli* to grow in the presence of tritiated thymidine, its DNA becomes radioactively labeled.

Any organic compound can be labeled with radioactive $^{14}C$. This isotope of carbon-12 has a relatively long *half-life* (its radioactivity decreases by one-half each 5730 years). All living things have incorporated a predictable amount of $^{14}C$ while alive. After death, $^{14}C$ disintegrates or *decays* to $^{14}N$ at the predictable rate of its half-life. This knowledge allows $^{14}C$-dating of organic remains up to about 25,000 years old. $^{14}C$ has a relatively low *specific activity* (few disintegrations per minute) and may therefore require a longer assay period in a scintillation counter to attain the same statistical confidence limits as radionuclides with higher specific activities.

Radioactive iodine ($^{125}I$) is extensively used to label proteins of medical interest (e.g. hormones, antigens, viral proteins, etc.). This isotope is easily coupled to the amino acid tyrosine. Quantitation of small amounts (nanograms or picograms/milliliter) of these proteins is accomplished by sophisticated techniques such as competitive protein binding assays or radioimmunoassays.

### 3. Nucleic Acid Enzymology.

*Nucleases* are enzymes that hydrolyze nucleic acids. Those that detach terminal nucleotides are *exonucleases*; those that break the sugar-phosphate backbone at nonterminal sites are *endonucleases*. Deoxyribonucleases attack DNA molecules; ribonucleases degrade RNA's, especially their single-stranded regions where they are not internally base paired. Some of the endonucleases act nonspecifically, cleaving the phosphodiester bonds of many different nucleotide sequences. Others, such as the bacterial *restriction endonucleases*, act very specifically, breaking bonds only within DNA nucleotide sequences that are symmetrical around a given point (*pallindromes*) and form inverted base sequences.

**Example 16.2.** A restriction endonuclease called EcoRI (derived from the bacterium *E. coli*) cuts bonds within the pallindromic sequence at the arrows shown below.

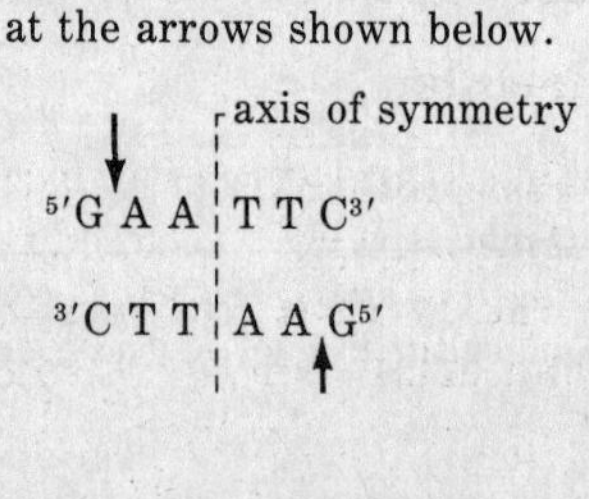

axis of symmetry

$$
\begin{array}{l}
\quad\downarrow \\
{}^{5'}\mathrm{G\,A\,A} \,\vdots\, \mathrm{T\,T\,C}^{3'} \\
{}^{3'}\mathrm{C\,T\,T} \,\vdots\, \mathrm{A\,A\,G}^{5'} \\
\qquad\qquad\quad\uparrow
\end{array}
$$

Notice that the 5′ to 3′ nucleotide sequences within the pallindrome is the same on both strands of the DNA. Another restriction endonuclease (Hae III), derived from the bacterium *Haemophilus aegypticus*, snips DNA in the axis of symmetry as shown below.

$$\begin{array}{c} ^{5'}\text{G G}\downarrow\vdots\ \text{C C}^{3'} \\ ^{3'}\text{C C}\ \vdots\uparrow\text{G G}^{5'} \end{array}$$

The cuts made by many restriction endonucleases create *cohesive* ("sticky") *ends*, i.e. complementary single-stranded tails projecting from otherwise double-helical DNA molecules that are terminally redundant. DNA molecules with sticky ends can be covalently rejoined by action of the enzyme *polynucleotide ligase*. Many other enzymes are known to be involved in the replication, recombination, repair, modification, transcription and translation of nucleic acids, but those already discussed are the main ones utilized by recombinant DNA technology.

## GENETIC ENGINEERING

During the late 1970s, the science of genetics entered a new era dominated by the use of *recombinant DNA technology* or *genetic engineering* to produce novel life forms not found in nature. Through this technology, it has been possible to transfer genes from mammals into bacteria, causing the microbes to become tiny factories for making (in relatively large quantities) proteins of great economic importance such as hormones (insulin, growth hormone) and interferons (lymphocyte proteins that prevent replication of a wide variety of viruses). These proteins are produced in such small quantities in humans that the cost of their extraction and purification from tissues has been very expensive, thus restricting their medical use in prophylaxis (prevention) and therapeutics (treatment) of disease. By genetic engineering, it should be possible to produce various blood clotting factors, complement proteins (part of the immune system), and other substances for the amelioration of genetic deficiency diseases (*euphenics*). In 1980, the Supreme Court decreed that new life forms created by genetic engineering could be patented. This will undoubtedly contribute to the investment of large sums of money by private corporations into the development of many useful genetic recombinants. One company has already developed a new microbe that can degrade oil and thereby might be used in the biological cleanup of oil spills that otherwise would harm the environment. It is hoped that recombinant DNA technology can someday infuse the genes from nitrogen-fixing bacteria into cereal crops. This would allow these plants to become "self-fertilized" from the boundless source of nitrogen in the atmosphere. These few examples suffice to demonstrate the possibilities of this new technology and why there is such great excitement in both the scientific and lay communities concerning the accomplishments to be attained in the very near future.

To make recombinant DNA molecules function *in vivo*, it is necessary to cut the desired nucleotide sequence from the donor source and splice it into the recipient bacterial chromosome. The vehicle for transferring the donor gene into the recipient cell is a circular plasmid such as an F factor or a phage. The plasmid is extracted from the bacterial cells and purified. It is then cut at one position by a restriction endonuclease. The donor DNA may be derived by either of the following two methodologies. In "shotgun" experiments, the donor DNA is unselectively broken into many pieces and attempts are made to isolate the desired protein from all of the potential recombinants. In more selective experiments, only one or a few isolated donor genes are exposed to recipient vehicles.

### 1. "Shotgun" Experiments.

Donor DNA is isolated and cut into many pieces by the same restriction endonuclease used to cleave the plasmid. The two kinds of fragments are mixed *in vitro* and allowed to randomly rejoin their sticky ends to form circles. DNA ligase is added to seal the broken ends. Plasmid-free bacterial cells are treated with a dilute solution of calcium chloride, causing them to become more permeable to the uptake of the plasmids. Clones of cells that contain the desired recombinant plasmid and manufacture the donor protein can be identified by various techniques (enzymatic, serological). It may be desirable to couple the cloned protein to some bacterial protein that is normally excreted from the cell so that both can be recovered from the culture medium. The hybrid protein would usually need to be enzymatically treated to separate the desired protein from the carrier protein. The final steps involve isolation and purification, followed by testing the desired protein for biological activity.

**2. Selective Experiments.**

If the desired protein is very small (15 to 20 amino acids) and its primary structure is known, it should be possible (using genetic coding information) to chemically synthesize a corresponding DNA molecule. This has actually been done for the hormone somatostatin (14 amino acids; 42 bases in the sense strand). Most proteins, however, are too long to allow chemical synthesis of the corresponding DNA. In this case, the homologous mRNA may be isolated from tissues in which the protein is made. For example, insulin is produced only by the pancreas, despite the fact that the insulin gene is present in all nucleated body cells. A purified preparation of insulin mRNA is isolated and treated with reverse transcriptase to make a single-stranded DNA copy (*cDNA* = complementary DNA). The mRNA template is then destroyed and the remaining cDNA serves as a template for extension synthesis of a complementary DNA strand. Another enzyme (*S1 nuclease*) breaks the covalent linkage between the two DNA strands at one end of the molecule. A short segment of identical bases (e.g. all cytosines) is added to the 3′ ends by the enzyme *terminal transferase*. The recombinant vehicle (e.g. plasmid) is isolated, opened at one position by a suitable restriction endonuclease, and then treated with terminal transferase and deoxyguanidine triphosphates. This renders the poly-G plasmid ends complementary ("sticky") to the poly-C ends of cDNA. Repair enzymes can be used to seal the connection between the plasmid DNA and the inserted DNA. The recipient bacterial cells are then made permeable to the plasmids by treatment with dilute calcium chloride. The plasmid should contain one or more genes that allows counterselection against cells that have not taken in the plasmid.

**Example 16.3.** If a plasmid contains the gene for tetracycline resistance, all bacterial cells will die in the presence of tetracycline except those that have received the plasmid. The endonuclease that opens the plasmid DNA for recombination with cDNA should not cut within the gene for tetracycline resistance because this would destroy the counterselective marker. Perhaps less than 1% of treated cells will have taken in the recombinant plasmid. Plating the culture on nutrient agar containing the counterselective agent (tetracycline in this example) produces clones of cells that have received the plasmids. The clones can then be tested for the presence of the desired protein. Once such a clone is found, pure cultures of the recombinant bacteria can be propagated in liter quantities for commercial production of the product.

One of the most difficult problems in genetic engineering involves the isolation of a single structural gene and connecting it with an appropriate promoter so that it can be constitutively transcribed into mRNA. The promoter must be highly efficient (i.e. readily bind RNA polymerase) in order to make many copies of the desired mRNA. Likewise, the 5′ end of the mRNA must also contain a highly efficient ribosome-binding site for maximal reading (translation) of the mRNA's. If the promoter is part of an inducible operon, the foreign gene must be connected to the operon in such a way that it comes under the most efficient control of the operator.

**Example 16.4.** In the lactose operon of *E. coli*, the gene order is P (promoter), O (operator), Z (β-galactosidase), Y (permease), A (acetylase). There is about ten times less Y protein made than Z protein, and about ten times less A protein made than Y protein (one hundred times less than Z protein). Thus, for maximal efficiency of production in such an operon, the foreign gene should be spliced as close to the operator as possible.

## NUCLEOTIDE SEQUENCING

There are many nontranslated regions within nucleic acids, i.e. segments that do not code for polypeptide products. It is crucial for efficient production of foreign structural genes that they possess or become connected to efficient promoters, operators, leader sequences and the like. Several techniques have been developed for analyzing the nucleotide sequences of such regions.

### 1. RNA Sequence Analysis.

Many RNA molecules are too long to be sequenced in one piece. Limited treatment with certain ribonucleases can break long RNA chains at a few specific sites into fragments suitable for sequencing. Various ribonucleases can digest these major fragments differentially to yield overlapping sets of smaller fragments. The original sequence can be inferred by matching the overlaps in the different digestion groups.

**Example 16.5.** "There are two major classes of nucleic acids: DNA's and RNA's." If the previous sentence is cut by one pair of scissors following each verb or noun, the following fragments are produced.

A1 "There are
A2 two major classes
A3 of nucleic acids:
A4 DNA's
A5 and RNA's."

If another pair of scissors cuts the sentence following adjectives and conjunctions, the following fragments are produced.

B1 "There are two
B2 major
B3 classes of nucleic
B4 acids: DNA's and
B5 RNA's."

The entire sentence can be reconstructed by matching overlapping sequences from the two sets of fragments. The left end of the sentence must begin with a capital letter, just as the left (5′) end of an RNA chain must begin with a free phosphate group. The right end of the sentence must end with a period, just as the right (3′) end of an RNA chain must end with a free hydroxyl group.

| | | | | | | | | | |
|---|---|---|---|---|---|---|---|---|---|
| A1 | "There are | | | | | | | | |
| B1 | "There are | two | | | | | | | |
| A2 | | two | major | classes | | | | | |
| B2 | | | major | | | | | | |
| B3 | | | | classes | of nucleic | | | | |
| A3 | | | | | of nucleic | acids: | | | |
| B4 | | | | | | acids: | DNA's | and | |
| A4 | | | | | | | DNA's | | |
| A5 | | | | | | | | and | RNA's." |
| B5 | | | | | | | | | RNA's." |

The first RNA molecule to be sequenced was that of alanine tRNA from yeast. The transfer RNA's are among the shortest of natural RNA molecules (approximately 75 nucleotides). In addition to the standard bases (A, G, C, U), seven other unusual bases are found in tRNA's: inosinic acid (I), 1-methylinosinic acid ($I^{\mathrm{m}}$), 1-methylguanilic acid ($G^{\mathrm{m}}$), $N^2$-dimethylguanilic acid ($G^{m}$), ribothymidylic acid (T), dihydrouradylic acid ($U^{h}$) and pseudouridylic acid ($\psi$). These unusual bases are sparsely distributed along the RNA chain and serve as distinctive markers for aiding in sequence analysis.

Pancreatic ribonuclease cleaves RNA chains to the right (at the 3′ end) of all pyrimidine nucleotides (C, U, T, $\psi$, $U^{h}$). Takadiastase ribonuclease T1 cleaves RNA chains to the right of certain purine nucleotides (G, $G^{\mathrm{m}}$, $G^{m}$, I), but not others (A, $I^{\mathrm{m}}$).

Digestion fragments are separated according to size by passing the mixture through a long glass column packed with diethylaminoethyl (DEAE) cellulose. Mononucleotides pass through the column ahead of dinucleotides, dinucleotides ahead of trinucleotides, etc. Nucleotides within each di-, tri-, and oligo- (short) polynucleotide fragment are released by hydrolyzing with alkali. The individual nucleotides are then identified by paper chromatography, paper electrophoresis and spectrographic analysis. The sequence of dinucleotides is easily established by the nature of the cleaving enzyme.

**Example 16.6.** A dinucleotide consisting of G and C produced by takadiastase ribonuclease T1 must be CG and not GC because this enzyme cleaves to the right of G, not C.

Special techniques are used to establish the nucleotide sequence in larger fragments (5 to 8 residues). A phosphodiesterase enzyme from snake venom removes nucleotides sequentially from one end, leaving a mixture of smaller fragments of all possible intermediate lengths. The mixture can then be separated into fractions of homogeneous length on a DEAE cellulose column. The nucleotide at the 3′ end of each fraction of homogeneous length is identified, allowing reconstruction of the entire fragment. This method will not resolve segments of much greater length.

**Example 16.7.** An octanucleotide (8 bases) when completely degraded by snake venom phosphodiesterase produces 5 G's, 2 A's and 1 U.

| Fragment Size | Nucleotide at right (3′) end |
|---|---|
| 8 | U |
| 7 | G |
| 6 | A |
| 5 | G |
| 4 | A |
| 3 | G |
| 2 | G |
| 1 | ? |

Since 4 of the 5 G's are accounted for in fragments of two or more nucleotides, the 1 remaining G must be at the left (5′) end of the fragment.

$^{5'}GGGAGAGU^{3'}$

## 2. DNA Sequence Analysis.

One of the most popular methods for rapid sequencing of relatively short DNA segments of about 100 nucleotides is that of Maxam and Gilbert (Fig. 16-1). The DNA is cut with one or more restriction endonucleases and the desired restriction fragment is isolated by gel electrophoresis. Its 5′ ends are enzymatically labeled with radioactive phosphorus ($^{32}P$). The two strands of the DNA molecule are separated and purified by electrophoresis. One of the two purified strands is then subjected to chemicals that break the chain at specific bases. For example, one treatment cleaves G bases alone; another treatment cuts off both G's and A's; a third method destroys T's and C's; a fourth treatment removes only C's. These chemical reactions are controlled so that on average only about one base in fifty is subject to removal in a random fashion. Thus, the restriction fragments are sliced into segments of all possible sizes by chemical treatment. The chemical fragments are then separated according to size by gel electrophoresis. The smaller the fragment, the further it moves along the gel. The gel is covered with a photographic emulsion and stored in the dark. Radioactive disintegrations of $^{32}P$ are detected by X-ray film (autoradiography). When the four sets of reactions are run side by side, the order of nucleotides can be read sequentially from the autoradiograph. The sequence in the complementary strand can also be determined by the same technique for checking the accuracy of the sequence in the other chain (by rules of base pairing) or for extending the size of the fragment capable of being analyzed up to about 100 nucleotides (50 in each chain from the 5′ end). RNA molecules can also be sequenced by a similar technique using a variety of riboendonucleases and riboexonucleases to partially digest at each of the four bases to produce a nested set of fragments.

1. Radioactive labeling at 5′ ends of each strand of a restriction fragment.

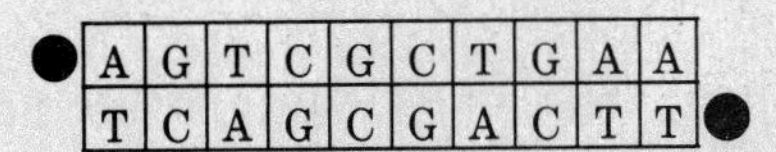


2. Separation and isolation of the two strands. Sequence analysis can be performed separately on each strand up to about 50 nucleotides.

3. Chemical treatments cleave the strand at specific bases. For example, a strand with three G's would be cut randomly at one of three places (arrows), producing radioactive fragments of three lengths.

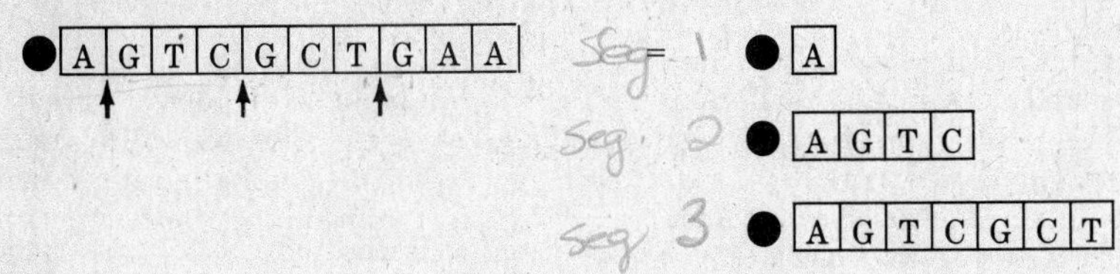


4. The shorter fragments move faster during electrophoresis. Only the fragments bearing the radioactive tag will be revealed by autoradiography.

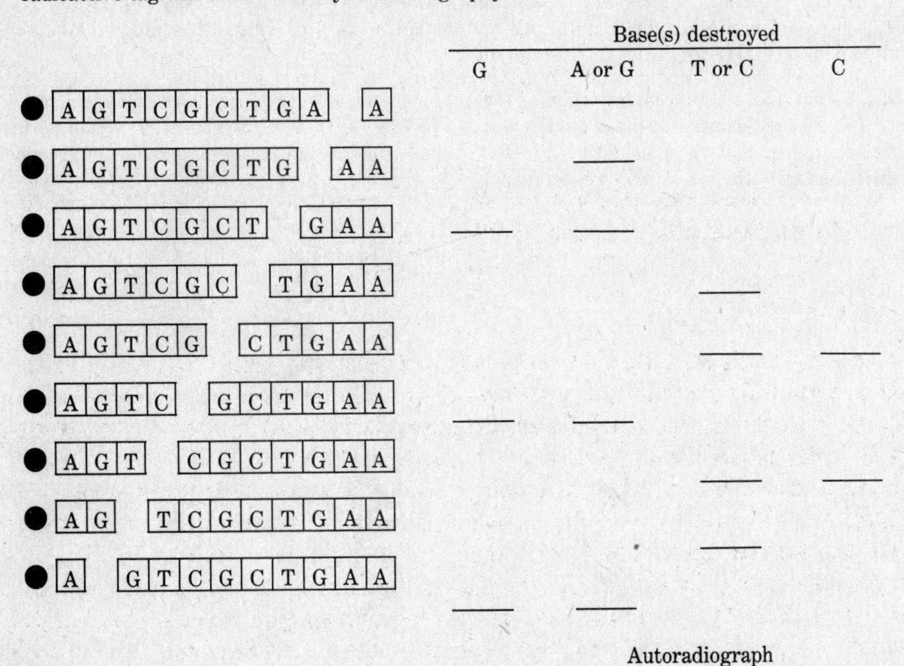


**Fig. 16-1.** Major steps in DNA sequence analysis of Maxam and Gilbert.

## Solved Problems

**16.1.** In 1953, Watson and Crick proposed that DNA replicates *semiconservatively*; i.e. both strands of the double helix become templates against which new complementary strands are made so that a replicated molecule would contain one original strand and one newly synthesized strand. A different hypothesis proposes that DNA replicates *conserva-*

*tively*; i.e. the original double helix remains intact so that a replicated molecule would contain two newly synthesized strands. Bacterial DNA can be "labeled" with a heavy isotope of nitrogen ($^{15}N$) by growing cells for several generations in a medium that has $^{15}NH_4Cl$ as its only nitrogen source. The common "light" form of nitrogen is $^{14}N$. Light and heavy DNA molecules can be separated by high-speed centrifugation (50,000 rpm = $10^5$ × gravity) in a $6M$ (molar) CsSl (cesium chloride) solution, the density of which is 1.7 g/cm$^3$ (very close to that of DNA). After several hours of spinning, the CsSl forms a density gradient, being heavier at the bottom and lighter at the top. In 1957, Matthew Meselson and Franklin W. Stahl performed a density-gradient experiment to clarify which of the two replication hypotheses was correct. How could this be done, and what results are expected after the first, second and third generations of bacterial replication according to each of these hypotheses?

**Solution:**

Bacteria from $^{15}N$-labeled culture are transferred into medium containing $^{14}N$ as the only source of nitrogen. A sample is immediately taken and its DNA is extracted and subjected to density-gradient equilibrium centrifugation. The DNA forms a single band relatively low in the tube where its density matches that of the CsCl in that region of the gradient. After each generation of growth and replication of all DNA molecules, DNA is again extracted and measured for its density. According to the semiconservative theory, the first generation of DNA progeny molecules should all be "hybrid" (one strand containing only $^{14}N$ and the other strand only $^{15}N$). Hybrid molecules would form a band at a density intermediate between fully heavy and fully light molecules. The second generation of DNA molecules should be 50% hybrids and 50% totally light, the latter forming a band relatively high in the tube where the density is lighter. After three generations, the ratio of light:hybrid molecules should be 3:1 respectively. The amount of hybrid molecules should be decreased by 50% in each subsequent generation.

According to the conservative replication scheme, the first generation of DNA molecules should be 50% heavy : 50% light. The second generation should be 25% heavy : 75% light. The third generation should be $12\frac{1}{2}$% heavy : $87\frac{1}{2}$% light. No hybrid molecules should be detected. The results of the Meselson and Stahl experiment supported the semiconservative theory of DNA replication.

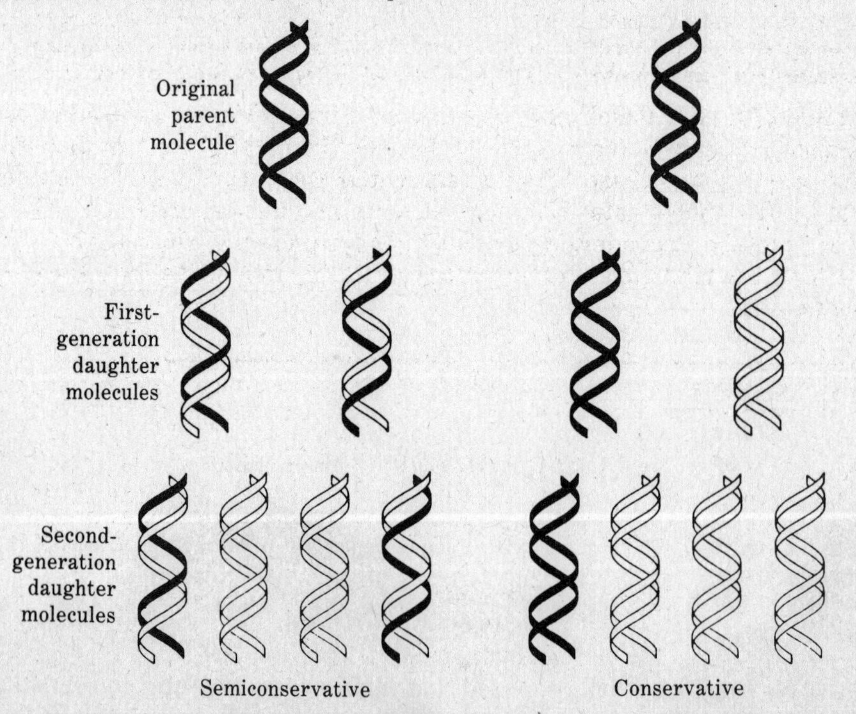


**16.2.** The Meselson-Stahl density-gradient experiment demonstrated that some elementary DNA unit replicates conservatively, but it could be argued that it failed to prove conclusively that this unit applied to the entire DNA molecule. For example, at least two other models of DNA replication could have produced the same first-generation results as those observed by Meselson-Stahl: (1) dispersive and (2) end-to-end conservative. In both of these models, about half of each strand is newly synthesized and half is old parental material.

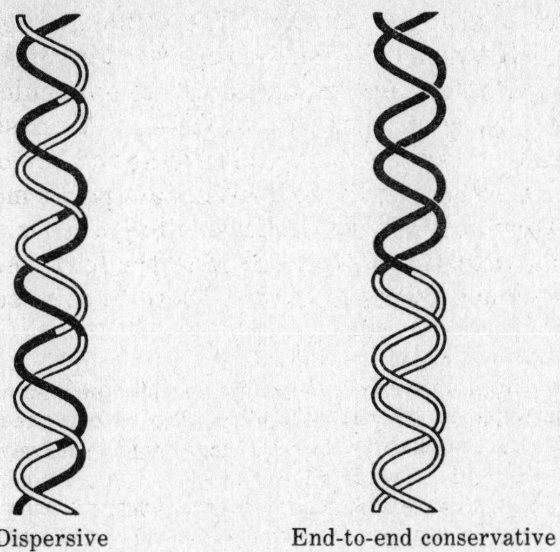

Dispersive  End-to-end conservative

Using the first-generation hybrid double helices, devise a method for confirming that semiconservative replication applies to the entire DNA molecule, not just to certain segments of the molecule.

**Solution:**

Denature $^{14}N$-$^{15}N$ hybrid DNA (strand separation) and subject the isolated strands to density-gradient equilibrium analysis. If semiconservative replication applies to the entire DNA molecule, half of the strands should find equilibrium at a density identical to that of isolated $^{15}N$ strands and the other half should form a band at the same position as isolated $^{14}N$ strands. If the DNA replicates according to either the dispersive or end-to-end conservative models, all single strands would seek an intermediate density equilibrium between those for totally light and totally heavy single strands. The actual results of such an experiment confirmed that the theory of semiconservative replication applies to the entire DNA molecule.

**16.3.** A portion of a tRNA molecule, containing 22 nucleotides, was broken into two major fragments, A and B (each with 11 bases), by limited treatment with the enzyme takadiastase ribonuclease (T1). Fragment A was later found to contain a nucleotide bearing a free phosphate; fragment B contained inosinic acid (I). The entire region was completely degraded by both T1 and pancreatic ribonuclease (PR) enzymes. Determine the ribonucleotide sequence in this region of the tRNA molecule from the following data:

| T1 Fragments | | PR Fragments |
|---|---|---|
| 1 pG, 2 G, 3 G | mononucleotides | 7 C, 8 C, 9 C, 10 C, 11 U |
| 4 AI | dinucleotides | 12 pGC, 13 AU, 14 AT, 15 AG |
| | trinucleotides | 16 AAC |
| | hexanucleotides | 17 GCAIGC |
| 5 CATCCAG | heptanucleotides | |
| 6 CCCCAUUAACG | decanucleotides | |

p = nucleotide bearing a free 5′ phosphate

**Solution:**

(1) The free phosphate group in fragments 1 and 12 establish the left (5′) end of the tRNA molecule.

```
T1:    1
       pG
PR:    pGC
       12
```

(2) One of the two T1 fragments (5 or 6) beginning with C must follow fragment 1. If 5 followed 1, the following fragments overlap.

```
        1           5
T1:    pG C A T C C A G
PR:    pG C A T C C ? ?
         12   14 7 8
```

PR does not cleave RNA molecules to the right of G. The reason why fragment 15 ends in G is because this segment of the tRNA was initially cleaved from the molecule by T1 and then subjected to complete digestion by both enzymes. Thus, fragment 15 establishes the right terminus of this segment. Since there is no PR fragment beignning with AG other than 15, our assumption must be wrong. Let 6 follow 1 to produce the following overlapping fragments.

```
        1                  6 2   4 3             5
T1:    pG C C C A U U A A C G G A I G C A T C C A G
PR:    pG C C C A U U A A C G G A I G C A T C C A G
         12 7 8   13 11     15        17  14 9 10  15
       \_______ fragment A _______/\_____ fragment B _____/
```

(3) Fragment A contains 11 nucleotides and a free phosphate, so it must be at the left (5′) end of the tRNA molecule followed by fragment B (containing I).

(4) The reason why there is no free 3′ hydroxyl (OH) group at the right end of this segment is because it was cleaved from the left end of a tRNA molecule. If the entire tRNA molecule had been sequenced, there should have been found in both T1 and PR digests a nucleotide bearing a free 3′ hydroxyl group, thus establishing the right terminal nucleotide.

**16.4.** A DNA restriction endonuclease fragment is treated by the method of Maxam and Gilbert. From the autoradiograph shown below, determine the double-stranded DNA sequence of this fragment including polarity of the strands.

Bases Destroyed

| G | A or G | T or C | C |
|---|---|---|---|
| | | — | |
| | — | | |
| | — | | |
| — | — | | |
| — | — | | |
| | | — | — |
| | | — | — |
| | | — | — |
| | — | | |
| — | — | | |
| — | — | | |
| | — | | |
| — | — | | |
| — | — | | |
| | | — | |

Direction of Migration ⟶

**Solution:**

Since the distance of movement of fragments from the origin on the gel increases with decreasing fragment size, the fragment at the bottom of the gel is the smallest one containing the radioactive label at its 5′ end. Any band appearing only in the T or C column indicates that the corresponding fragment must have been derived by cleaving at T. Any fragment in the C column was cleaved at C. Similarly, a fragment that appears only in the A or G column must have been cut at A. Any fragment in the G column must have been cut at G. Therefore, the nucleotides in this strand of DNA can be read sequentially from the 5′ end starting at the bottom of the gel.

$$^{5'}\text{T G G A G G A C C C G G A A T}^{3'}$$

The complementary strand runs in an antiparallel direction and its base sequence is determined by the conventional base pairing rules (A with T; G with C).

$$^{5'}\text{T G G A G G A C C C G G A A T}^{3'}$$
$$^{3'}\text{A C C T C C T G G G C C T T A}^{5'}$$

**16.5.** The relative position of recognition sites for various restriction endonucleases can be determined by a procedure known as *restriction-enzyme mapping*. The 3′ ends of a DNA molecule are labeled with radioactive $^{32}$P. The DNA is then completely digested in separate experiments with two restriction endonucleases (X and Y), the resulting fragments are separated by polyacrylamide gel electrophoresis (PAGE), and the labeled end fragments are identified by autoradiography. The mobilities of nucleic acid fragments in PAGE are inversely proportional to the logarithms of their lengths. Treatment with enzyme X produced fragments A*, B and C* (* = radioactively labeled); treatment with enzyme Y produced fragments D*, E and F*. Fragments A-C were then digested by enzyme Y into subfragments 1-5; fragments D-F were digested by enzyme X into subfragments, some of which overlap with those of Y. These subfragments can be homologized between the two enzyme digests because they occupy similar positions after PAGE. Fragment A contains a single subfragment 1; fragment F contains only subfragment 5. B was digested into subfragments 2 and 3; likewise C into 4 and 5, D into 1 and 2, and E into 3 and 4. Reconstruct the order of subfragments in this DNA molecule and show where the recognition sites for enzymes X and Y reside.

**Solution:**

Since three fragments were generated by each enzyme, there must be two recognition sites for each enzyme (cut a string twice and three pieces are produced). Fragment B was unlabeled, indicating it must be between labeled end fragments A and C; similarly E must be between end fragments D and F. Fragment B contains a site for enzyme Y because treatment with Y produced subfragments 2 and 3. Likewise C also contains a site for Y. But A does not contain a site for Y because Y could not digest A (contains only subfragment 1). Similarly D and E each have a site for enzyme X, but F does not. Because subfragment 4 is produced by digestion of C and E, they must overlap. By the same token, B and E must overlap (both contain subfragment 3), C and F must overlap (both contain 5), B and D must overlap (both contain 2), and A and D must overlap (both contain 1). These facts can now be used to reconstruct the original DNA molecule.

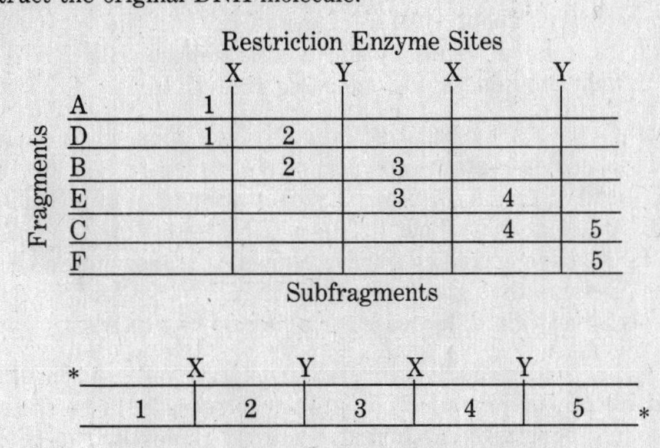


## Supplementary Problems

**16.6.** The buoyant density (ρ) of DNA molecules in 6*M* CsCl solution increases with the molar content of G + C nucleotides according to the following formula:

$$\rho = 1.660 + 0.00098\,(G + C)$$

Find the molar percentage of (G + C) in DNA from the following sources: (*a*) *Escherichia coli:* $\rho = 1.710$, (*b*) *Streptococcus pneumoniae:* $\rho = 1.700$, (*c*) *Mycobacterium phlei:* $\rho = 1.732$.

**16.7.** Given two DNA molecules, the overall composition of which is represented by the segments shown below, determine which molecule would have the highest melting temperature. Explain.

(*a*) T T C A G A G A A C T T
A A G T C T C T T G A A

(*b*) C C T G A G A G G T C C
G G A C T C T C C A G G

**16.8.** Two DNA molecules having identical (G + C)/(A + T) ratios are shown below. If these molecules are melted and subsequently annealed, which one would require a lower temperature for renaturation of double helices? Explain. *Hint:* Consider the effects of intrastrand interactions.

(*a*) A A T A G C C C C A T G G G G C T A
T T A T C G G G G T A C C C C G A T

(*b*) C T G C A T C T G A T G C A G C T C
G A C G T A G A C T A C G T C G A G

**16.9.** The primary mRNA transcript for chicken ovalbumin contains seven introns (light, A-G) and eight exons (dark) as shown below.

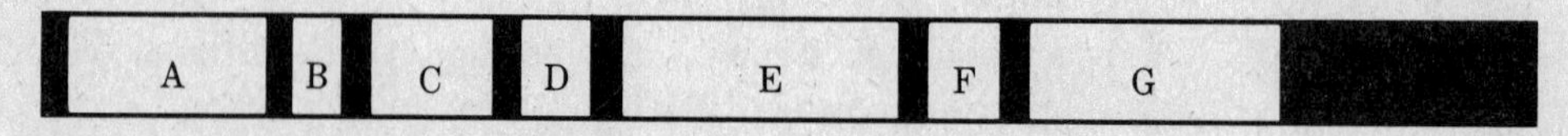


If the DNA for ovalbumin is isolated, denatured to single strands and hybridized with cytoplasmic mRNA for ovalbumin, how would the hybrid structure generally be expected to appear in an electron micrograph? *Note:* Double-stranded regions appear thicker than single-stranded regions.

**16.10.** RNA can be translated into protein only when it is single-stranded; if DNA hybridizes with RNA, no translation occurs. This fact suggests a way by which one can identify those recombinant bacterial clones that are synthesizing rat insulin. Bacterial DNA is first isolated from the clones being tested and then denatured to single strands. Unpurified RNA (the same source used to make insulin cDNA) is added to the single-stranded DNA under conditions that promote annealing between the RNA's and any homologous DNA. A "translation system" containing radioactive amino acids, ribosomes, tRNA's, enzymes, energy sources, etc. is added to the mixture. Small plastic beads (coated with antibodies specifically reactive with rat insulin) are later introduced into the system. The tube is centrifuged and the supernatant fluid is discarded leaving the antibody-coated beads in the tube. The tubes are then assayed for radioactivity. Rat insulin was detected in some of the tests; it was not detected in other tests. How does one know which clones contained the insulin gene?

**16.11.** The *E. coli* genome contains about 4000 kilobase pairs (kb; kilo = 1000); there are about 1.5 kb in 16S rRNA. If 0.14% of the genome forms hybrid double helices with RNA complementary to one strand of DNA, estimate the number of genetic loci encoding 16S rRNA.

**16.12.** About half the weight of RNA synthesized at any given time within a bacterial cell is rRNA. The 30S subunit of bacterial ribosomes contains one 16S rRNA molecule (1.5 kb); the 50S subunit contains one 23S rRNA (3 kb) and one small 5S rRNA (0.1 kb). Hybridization tests of 16S and 23S rRNA's with complementary single strands of DNA reveal that about 0.14% of DNA is coding for 16S rRNA and about 0.18% for 23S rRNA. Estimate the relative activity of rRNA genes as transcription templates compared to the average gene of the bacterial genome that gives rise to mRNA. *Note:* Assume that the amount of DNA allocated to 5S rRNA synthesis is negligible; likewise for all kinds of tRNA's.

**16.13.** Some bacterial proteins are normally secreted from the cell. If insulin could be attached by genetic engineering to such a bacterial protein, it too might be secreted from the cell. Suppose that you are given an agar plate containing several recombinant bacterial clones known to contain the gene for rat insulin. Propose an autoradiographic method for identifying those clones that are secreting rat insulin. *Hint:* Antibodies can be attached to certain kinds of plastic in a way that leaves their antigen-combining sites free to react.

**16.14.** Restriction nuclease EcoRI makes staggered cuts in a six-nucleotide DNA pallindrome; Hae III nuclease cleaves at one point in the middle of a four-nucleotide DNA pallindrome. If different aliquots of a purified DNA preparation are treated with these enzymes, which one would be expected to contain more restriction fragments?

**16.15.** Only about 200 molecules of lambda repressor are made by lysogenic bacteria when lambda is normally integrated at its specific attachment site between *E. coli* genes *gal* and *trp*. Some bacterial genes such as *lac* can be induced to produce more than 20,000 molecules of an enzyme per cell. If you could cut and splice structural and regulatory genes at your discretion, how would you design a bacterial cell for maximum synthesis of lambda repressor protein?

**16.16.** A fragment of serine tRNA was cleaved by enzyme T1 and then this fragment was completely degraded by both T1 and PR enzymes to give the overlapping sets of fragments shown below. Reconstruct this segment of serine tRNA.

T1 digest: AG, G, UCCUG, G, TCCCG, pG, G
PR digest: U, pGGGGU, C, C, C, C, C, G, GAGT

**16.17.** Digesting a purified species of RNA by two different enzymes (T1 and PR) gave the overlapping sets of fragments shown below. Reconstruct this RNA molecule.

T1 digest: G, pCAUUG, AUCUCG-OH, AAUCCAG
PR digest: pC, C, C, C, C, U, U, G-OH, AU, AGAU, GGAAU

**16.18.** A fragment cleaved from an RNA molecule by brief exposure to T1 enzyme was then thoroughly digested by both T1 and PR enzymes yielding the following sets of overlapping fragments. Reconstruct the original sequence.

PR fragments: AG$^m$GGAAU, AAAGU, AAG, GAT, IGC, C, C, C, C, U, U
T1 fragments: ATUCCAG$^m$, CAAAG, CCAAG, AAUI, UUG, G, G, G

**16.19.** Sanger and Coulson developed a technique for DNA sequencing known as a "primed synthesis method" or the "plus and minus method". Bacterial DNA polymerase I can only extend polydeoxyribonucleotide chains; it cannot initiate DNA synthesis without either a DNA or RNA primer. A primer fragment of the DNA to be sequenced can be obtained by digestion with restriction endonucleases. After purification, the primer fragment can be annealed (by complementary base pairing) to a single-stranded DNA template. The primer undergoes limited extension by DNA polymerase I when given a restricted supply of all four deoxyribonucleoside triphosphates (one of which is radioactively labeled with $^{32}$P). Different primed complexes are extended in a random manner so that ideally every chain length over the region to be sequenced should be present. The unincorporated labeled nucleosides and DNA polymerase I are then removed.

(*a*) In the "plus technique", one aliquot of the reaction mixture is then exposed to only one of the four deoxyribonucleoside triphosphates (unlabeled). T4 DNA polymerase is added. This enzyme has 3′ to 5′ exonuclease activity on double-stranded DNA, removing one nucleotide at a time. If adenine was the added base (+A), the enzyme would chew away the 3′ end until it removed an A residue. At that point, the 5′ to 3′ polymerase activity of this enzyme being much greater than its nuclease activity, A would be immediately replaced. Thus, each length of primed chain would terminate at the 3′ end with an A. The same process is repeated in three other systems, adding only one of the other three nucleoside triphosphates. The extension products are separated from the template by heating in formaldehyde. All four systems are then electrophoresed simultaneously on acrylamide gel and autoradiographed. Determine the base sequence from the following autoradiograph and specify its polarity.

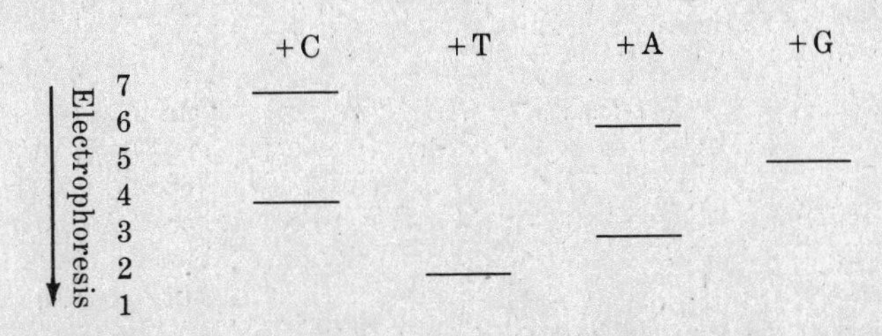


(*b*) In the "minus technique", T4 DNA polymerase and three of the four deoxyribonucleoside triphosphates are added to the complementary DNA (cDNA). The enzyme extends cDNA until the missing base is required, at which point extension stops. If adenine is missing (−A), the random-length cDNA will thus have their 3′ ends terminating immediately before an A residue. Four separate minus reactions are performed, each with a different missing deoxyribonucleoside triphosphate. The chains are denatured (separated) and all four reaction products are electrophoresed simultaneously on acrylamide gel and autoradiographed. In four columns adjacent to those in part (*a*), diagram the minus bands on an autoradiograph from the DNA segment $^{5'}$TACGAC$^{3'}$.

(*c*) When more than one identical nucleotide is adjacent in a DNA sequence (e.g. AA or CCC), one or more bands will be missing from the plus and minus columns. Only the first (5′) nucleotide of the run is present in the minus system, and only the last (3′) nucleotide is represented in the plus system. The total number of residues in a sequence is revealed by running an aliquot of the initial extension alongside the plus and minus reactions on the same gel. Diagram the bands expected from the DNA segment $^{5'}$TAACGGGATCCCC$^{3'}$.

**16.20.** DNA segments can be sequenced by using chain-terminating inhibitors. This method is similar in principle to the plus and minus methods described in the previous problem, but the 3′ end of each cDNA fragment is specifically terminated with one of four analogues of DNA bases. A suitable primer is annealed to the single-stranded template DNA. In each of four different reactions, the primer is extended by DNA polymerase I in the presence of all four deoxyribonucleoside triphosphates (at least one of which is $^{32}P$ labeled) and one of the four specific chain-terminating base analogues. For example, 2′3′-dideoxyribonucleoside triphosphate has no 3′ hydroxyl group. Therefore, this analogue would not allow further extension of the chain by Pol I. The ratio of an analogue (e.g. ddGTP) to its DNA counterpart (dGTP) is adjusted so that only partial incorporation of the terminator occurs. Thus, extended chains of varying lengths ending with ddG at their 3′ ends are produced. The three other extension reactions use one of the analogues ddCTP, ddTTP or ddATP. The extended chains are separated from the template and all four systems are electrophoresed side by side on the same acrylamide gel and autoradiographed. (*a*) From the autoradiograph shown below, determine the sequence of this DNA fragment.

G A T C

Electrophoresis

7
6
5
4
3
2
1

(*b*) Diagram all possible extension chains in each of the four reaction systems.

**16.21.** A purified segment of DNA is labeled at its 5′ ends with $^{32}P$ and partially digested with the restriction enzyme Alu (from *Arthrobacter luteus*). The concentration is adjusted so that only about one of every fifty sites is recognized by the enzyme and cleaved. Therefore, each DNA molecule, if broken, is only broken once. This creates five fragments of various sizes. The fragments were electrophoresed, autoradiographed and the longest one (slowest migrating band) is selected for partial digestion (in separate experiments) with Alu and with another restriction endonuclease Hae III (from *Haemophilus aegyptius*). These new fragments are separated by electrophoresis and located on the gel by autoradiography. The bands appear as shown below.

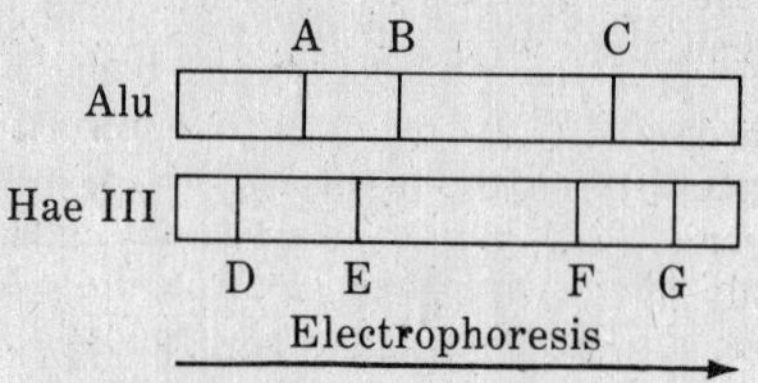


(*a*) Determine the sequence of restriction sites in this DNA segment. (*b*) How can the relative distances between cleavage sites be determined?

## Answers to Supplementary Problems

**16.6.** (*a*) 51.02 (*b*) 40.82 (*c*) 73.47

**16.7.** (*b*), because it has a higher (G + C)/(A + T) ratio.

**16.8.** (*b*) would renature at a lower temperature because the long stretches of G-C pairs in (*a*) will tend to form *intrastrand* hydrogen bonds during cooling and therefore will require a higher temperature to disrupt these intrastrand interactions.

**16.9.**

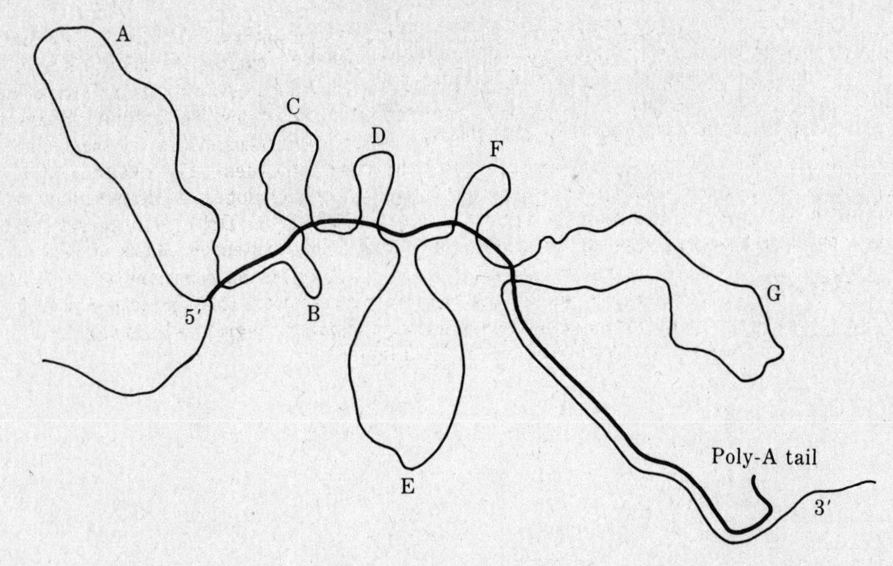


**16.10.** The fact that insulin was detected in some tests indicates that the RNA source contained the necessary information for *in vitro* insulin synthesis. Those clones that did not possess the insulin gene allowed single-stranded insulin mRNA's to make insulin; insulin was then specifically bound to the antibody-coated beads. Tubes that contain insulin are radioactive. However, if a clone contains rat insulin cDNA homologous with insulin mRNA's in the unpurified mixture, hybrid DNA-RNA molecules form, preventing such RNA's from making insulin. If no insulin is made, radioactive amino acids will not be bound by the antibody-coated beads. Little or no radioactivity should be detected in the tube after decantation of the supernate.

**16.11.** $(0.0014)(4000)/1.5 = 4$

**16.12.** Of all the RNA's that hybridize with DNA, the 16S and 23S rRNA's account for only $0.0014 + 0.0018 = 0.0032$ or 0.32%. Since these rRNA's and mRNA's are about equally represented in a cell, the ratio $1/0.0032 = 312.5$ expresses how much more active in transcription are the genes for the rRNA's.

**16.13.** Attach anti-rat insulin antibodies to a plastic disc about the size of an agar plate. Impress the disc onto the plate and allow any secreted insulin to be specifically bound by the antibodies. Remove the plastic disc and expose it to radioactive anti-insulin antibodies, forming an "immunological sandwich" with the antigen (insulin) between two antibody molecules. Wash away any unattached radioactive antibodies and then make an autoradiograph of the disc. Images on the film can be used to identify the locations of insulin-secreting clones on the agar plate.

**16.14.** More fragments are expected from Hae III because the probability of a specific four-base sequence is greater than the probability of a specific six-base sequence if the nucleotides are distributed along a chain in essentially a random order.

**16.15.** Insert the gene for lambda repressor protein immediately adjacent to the *lac* promoter in a plasmid. With no operator locus between these two genes, thousands of repressor molecules should be made per cell constitutively.

**16.16.** $^{5'}$pG G G G U C C U G A G T C C C G$^{3'}$

**16.17.** $^{5'}$pC A U U G G A A U C C A G A U C U C G-OH$^{3'}$

**16.18.** C A A A G U U G A T U C C A G$^{m}$ G G A A U I G C C A A G

**16.19.** (*a*) $^{5'}$TACGAC$^{3'}$

(*b*)

| | +C | +T | +A | +G | −C | −T | −A | −G |
|---|---|---|---|---|---|---|---|---|
| 7 | — | | | | | | | |
| 6 | | | — | | — | | | |
| 5 | | | | — | | | — | |
| 4 | — | | | | | | | — |
| 3 | | | — | | — | | | |
| 2 | | — | | | | | — | |
| 1 | | | | | | — | | |

Notice that the "negative bands" are shifted one step lower than the "positive bands". For example, bands at positions 3 and 6 in the −C column indicate that these fragments were terminated just prior to C, hence bands should appear at positions 4 and 7 in the +C columns. DNA base sequencing can be done by either the plus or minus system, but they are usually analyzed together on the same gel to provide mutual confirmation of the sequence.

(*c*)

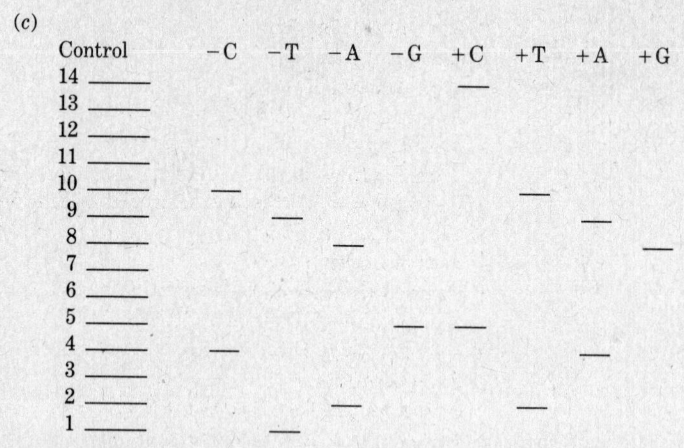


**16.20.** (*a*) $^{5'}$CGACGGT$^{3'}$

(*b*) ddGTP-terminated fragments (*d*)

$^{5'}$CGACGGT
CGACG*d*
CGAC*d*
C*d*

ddATP-terminated fragments (*d*)

$^{5'}$CGACGGT
CG*d*

ddTTP-terminated fragments (*d*)

$^{5'}$CGACGGT
CGACGG*d*

ddCTP-terminated fragments (*d*)

$^{5'}$CGACGGT
CGA*d*

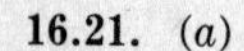

**16.21.** (*a*)

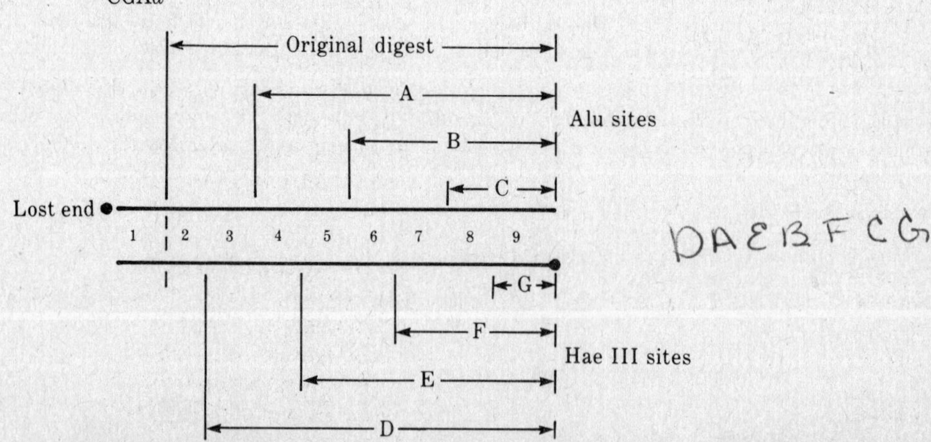


(*b*) Since the distances the various fragments travel in the gel is related (by a log function) to their sizes, the relative distances between cleavage sites can be determined.

# INDEX

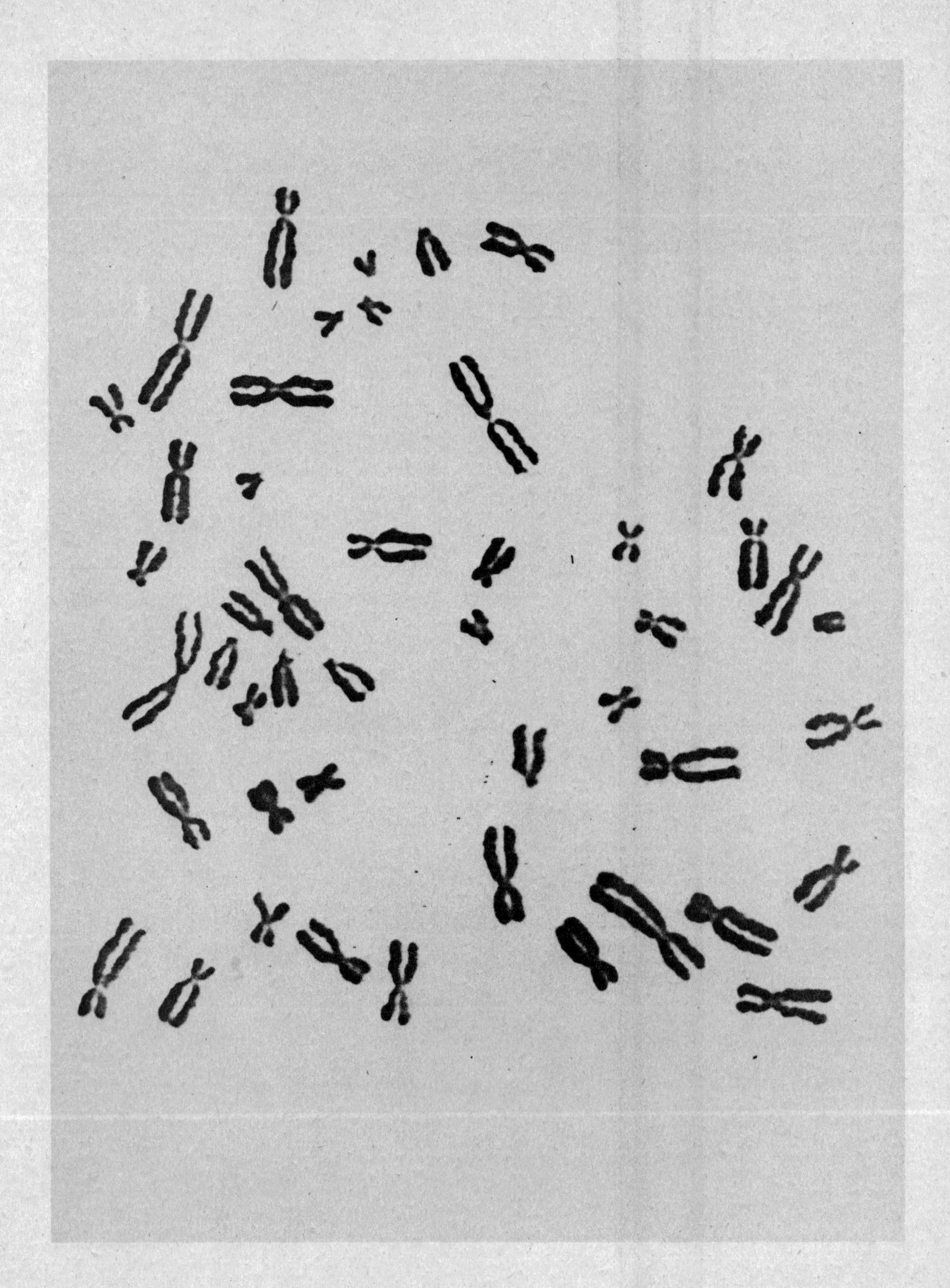

Photograph accompanying Problem 9.45. See p. 194.